Ecological Studies, Vol. 208

Analysis and Synthesis

Ecological Studies

Further volumes can be found at springer.com

Volume 190
Wetlands and Natural Resource Management (2006)
J.T.A. Verhoeven, B. Beltman, R. Bobbink, and D.F. Whigham (Eds.)

Volume 191
Wetlands: Functioning, Biodiversity Conservation, and Restoration (2006)
R. Bobbink, B. Beltman, J.T.A. Verhoeven, and D.F. Whigham (Eds.)

Volume 192
Geological Approaches to Coral Reef Ecology (2007)
R.B. Aronson (Ed.)

Volume 193
Biological Invasions (2007)
W. Nentwig (Ed.)

Volume 194
***Clusia:* A Woody Neotropical Genus of Remarkable Plasticity and Diversity** (2007)
U. Lüttge (Ed.)

Volume 195
The Ecology of Browsing and Grazing (2008)
I.J. Gordon and H.H.T. Prins (Eds.)

Volume 196
Western North American Juniperus Communites: A Dynamic Vegetation Type (2008)
O. Van Auken (Ed.)

Volume 197
Ecology of Baltic Coastal Waters (2008)
U. Schiewer (Ed.)

Volume 198
Gradients in a Tropical Mountain Ecosystem of Ecuador (2008)
E. Beck, J. Bendix, I. Kottke, F. Makeschin, R. Mosandl (Eds.)

Volume 199
Hydrological and Biological Responses to Forest Practices: The Alsea Watershed Study (2008)
J.D. Stednick (Ed.)

Volume 200
Arid Dune Ecosystems: The Nizzana Sands in the Negev Desert (2008)
S.-W. Breckle, A. Yair, and M. Veste (Eds.)

Volume 201
The Everglades Experiments: Lessons for Ecosystem Restoration (2008)
C. Richardson (Ed.)

Volume 202
Ecosystem Organization of a Complex Landscape: Long-Term Research in the Bornhöved Lake District, Germany (2008)
O. Fränzle, L. Kappen, H.-P. Blume, and K. Dierssen (Eds.)

Volume 203
The Continental-Scale Greenhouse Gas Balance of Europe (2008)
H. Dolman, R. Valentini, and A. Freibauer (Eds.)

Volume 204
Biological Invasions in Marine Ecosystems: Ecological, Management, and Geographic Perspectives (2009)
G. Rilov and J.A. Crooks (Eds.)

Volume 205
Coral Bleaching: Patterns, Processes, Causes and Consequences (2009)
M.J.H van Oppen and J.M. Lough (Eds.)

Volume 206
Marine Hard Bottom Communities: Patterns, Dynamics, Diversity, and Change (2009)
M. Wahl (Ed.)

Volume 207
Old-Growth Forests: Function, Fate and Value (2009)
C. Wirth, G. Gleixner, and M. Heimann (Eds.)

Volume 208
Functioning and Management of European Beech Ecosystems (2009)
R. Brumme and P.K. Khanna (Eds.)

Rainer Brumme · Partap K. Khanna
Editors

Functioning and Management of European Beech Ecosystems

Springer

Editors
Dr. Rainer Brumme
Department of Soil Science
of Tropical Ecosystems
Georg-August- University Göttingen
Büsgenweg 2
37077 Göttingen, Germany
rbrumme@gwdg.de

Dr. Partap K. Khanna
Department of Soil Science of Temperate
and Boreal Ecosystems
Georg- August-University Göttingen
Büsgenweg 2
37077 Göttingen, Germany
partap.khanna@waldbau.uni-freiburg.de

ISBN 978-3-642-00339-4 e-ISBN 978-3-642-00340-0
DOI: 10.1007/978-3-642-00340-0

Ecological Studies ISSN 0070-8356

Library of Congress Control Number: 2009926190

Cover design: WMXDesign GmbH, Heidelberg, Germany

Design of the cover illustration and the figures in Chapter 22 by W. Tambour.
Concept of cover illustration by Prof. Dr. F. Beese.

Printed on acid-free paper

Springer-Verlag Berlin Heidelberg (www.springer.com)

This book is dedicated to Prof. Dr. Dr. h.c. mult. Bernhard Ulrich on his 83rd birthday for his lifelong contributions to the understanding of forest ecosystem functioning especially in relation to nutrient cycling processes. His nutrient-balance approach has described many interactions among inter- and intra-ecosystems components in relation to the effects of atmospheric emissions on soils, plants and water quality, leading to emission reductions in the industrial countries through appropriate policy measures. He was involved in establishing many long term ecosystem studies, some of which are the subject of this book. He has been an inspiration to most of the contributory authors of this book.

Preface

Temperate forests cover large areas of Europe and perform a number of important functions such as the regulation of energy and matter, production of wood and other resources, and conservation of biodiversity and habitats; they also have special significance in social and cultural contexts. Initiated in 1960s, the first International Biological Program (IBP) focused on "the biological basis of productivity and human welfare." As the German contribution to the IBP, ecosystem research has been carried out since 1966 in the Solling area (Ellenberg H., Ecological Studies 2, 1971), an upland region in Northwest Germany. This study provided clear evidence that the stability of forest ecosystems was threatened by the high inputs of atmospheric pollutants. This promoted many interdisciplinary research programs which were coordinated by Prof. Dr. Bernhard Ulrich and the Forest Ecosystems Research Center of the University of Göttingen. This involved, in addition to the Solling site, the establishment of two other sites for long-term monitoring of ecosystem processes. The two contrasting sites were established in 1980 at Göttinger Wald on base-rich calcareous soil and in 1989 at Zierenberg on volcanic soil.

These projects were funded initially by the Federal Ministry of Research and Technology (BMBF) as interdisciplinary projects under the titles: "Conditions of Stability of Forest Ecosystems" (1989–1993), and "Dynamics of Forest Ecosystems" (1993–1998). The primary goal of these studies was to quantify the ecological condition of forests in a changing environment and element fluxes.

The BMBF-funded interdisciplinary projects consisted of over 50 individual research projects to study element cycling and budgets in mature forest ecosystems and after experimental manipulations, to assess the atmospheric depositions and their effects, to evaluate trace gas exchanges between forests and atmosphere, to develop models of primary production, to study the biological element transformations in soils, and to describe the effects of soil biota and root turnover processes. These studies provided the basis of developing an integrated theory of forest ecosystem by Prof. Ulrich (Ulrich B., Forest ecosystem theory based on material balance. Ecological Modelling 63, 163–183, 1992). This ecosystem theory was used in this volume to interpret the long-term observations at Solling, Göttinger Wald, and Zierenberg beech sites together with the data from the national forest soil condition inventory in Germany, and with those of input–output budgets of an additional 53 forest sites in Germany.

Another program on "Survey of Soil Conditions under Forests" (Bodenzustandserhebung im Walde, BZE) was funded by the federal States of Germany and the Federal Ministry of Food, Agriculture, and Consumer Protection (BMELV). It was coordinated by the Johann Heinrich von Thünen Institute (Eberswalde, formerly Federal Research Centre for Forestry and Forest Products, BFH) and was conducted by various Forest Research Institutes of different federal States of Germany. It was aimed at assessing the current conditions of forest soils in Germany. The soil profiles were sampled during the same period as the process studies were carried out (1987–1993) and they were analyzed using the same sampling protocols and inter-calibrated instruments. Because of the integrated nature of this study, more confidence can be given to any relationship between C- and N-concentrations, climate and soil conditions as they may reflect real conditions rather than artefacts of different methodologies or analytical procedures.

With research support from various agencies including the European Union, monitoring of forest state variables and ecosystem processes at Solling, Göttinger Wald, and Zierenberg beech sites has continued under the umbrella of the Intensive Forest Monitoring Programme (Level II) of the International Co-operative Programme on Assessment and Monitoring of Air Pollution Effects on Forests (ICP Forests) operating under the UNECE Convention on Long-range Transboundary Air Pollution. The monitoring is aimed at developing strategies to protect forests against air pollution by increasing our knowledge about their functioning, maintenance of biodiversity, and C sequestration under changing climate conditions. This program started in 1994 on more than 800 sites representing the diversity of forests in Europe.

Our special thanks go to the above-mentioned forest research institutes for their continuing efforts to monitor ecosystem processes, to various organizations involved in data management, to many funding agencies and finally to Prof. Dr. F. Beese for his encouragement and help at various levels including his support as the Director of Ecosystem Research Centre at the University of Göttingen.

Without these combined efforts we would not have been able to compile the data presented in this volume on the three beech forests at Solling, Göttinger Wald, and Zierenberg. We had the access to the countrywide input–output data to study the current N status of forest ecosystems and to the data of forest soil condition inventory for describing the issues on functioning and management of temperate forest ecosystems in this volume.

April 2009

Rainer Brumme
Partap K. Khanna

Contents

Contributors

T.-H. Anderson
Johann Heinrich von Thünen-Institut, Federal Research Institute for Rural Areas, Forestry and Fisheries, Institute of Biodiversity, Bundesallee 50, 38116 Braunschweig, Germany, heidi.anderson@vti.bund.de

C.T. Aydin
State Agency for Agriculture, Environment and Rural Areas of the Federal State Schleswig-Holstein, Department Geology and Soil, Soil Conservation, Hamburger Chaussee 25, 24220 Flintbek, Germany, cihan-tarih.aydin@LLur.Landsh.de

N. Bartsch
Department of Silviculture and Forest Ecology of the Temperate Zones, Georg-August-University Göttingen, Büsgenweg 1, 37077 Göttingen, Germany
n.bartsch@forst.uni-goettingen.de

J. Block
Research Institute for Forest Ecology and Forestry Rhineland-Palatinate, Hauptstrasse 16, 67705 Trippstadt, Germany, joachim.block@wald-rlp.de

A. Bolte
Johann Heinrich von Thünen-Institute (vTI), Federal Research Institute for Rural Areas, Forestry and Fisheries, Institute for Forest Ecology and Forest Inventory, A.-Möller-Straße 1, 16225 Eberswalde, Germany, andreas.bolte@vti.bund.de

W. Borken
Department of Soil Ecology, University of Bayreuth, Dr.-Hans-Frisch-Str. 1-3, 95448 Bayreuth, Germany, werner.borken@uni-bayreuth.de

M. Bredemeier
Forest Ecosystems Research Center, Georg-August University Göttingen, Büsgenweg 2, 37077 Göttingen, Germany, mbredem@gwdg.de

R. Brumme
Department of Soil Science of Tropical Ecosystems, Georg-August-University Göttingen, Büsgenweg 2, 37077 Göttingen, Germany, rbrumme@gwdg.de

M. Egenolf
Department of Soil Science of Temperate and Boreal Ecosystems, Georg-August-University Göttingen, Büsgenweg 2, 37077 Göttingen, Germany
megenolf@online.de

J. Eichhorn
Northwest German Forest Research Station, Grätzelstr. 2, 37079 Göttingen, Germany, johannes.eichhorn@nw-fva.de

H. Fortmann
Northwest German Forest Research Station, Grätzelstr. 2, 37079 Göttingen, Germany, heike.fortmann@nw-fva.de

G. Gravenhorst
Department of Bioclimatology, Georg-August University Göttingen, Büsgenweg 2, 37077, Göttingen, Germany, ggraven@gwdg.de

M. Guericke
University of Applied Science Eberswalde, Alfred-Möller-Str. 1, 16225 Eberswalde, Germany, martin.guericke@fh-eberswalde.de

A. Horn
Department of Silviculture and Forest Ecology of the Temperate Zones, Georg-August University Göttingen, Faculty of Forest Sciences and Forest Ecology, Büsgenweg 1, 37077 Göttingen, Germany

C. Jacobsen
Fachbüro für Waldhydrologie, Goblerstr. 44, 37075 Göttingen, Germany
Carsten-Jacobsen@t-online.de

W. Ke-Hong
Department of Silviculture and Forest Ecology of the Temperate Zones, Georg-August University Göttingen, Faculty of Forest Sciences and Forest Ecology, Büsgenweg 1, 37077 Göttingen, Germany

P.K. Khanna
Department of Soil Science of Temperate and Boreal Ecosystems, Georg-August-University Göttingen, Büsgenweg 2, 37077 Göttingen, Germany
partap.khanna@waldbau.uni-freiburg.de

H. Kreilein
Department of Bioclimatology, Georg-August University Göttingen, Büsgenweg 2, 37077, Göttingen, Germany hkreile@gwdg.de

H. Meesenburg
Northwest German Forest Research Station, Grätzelstr. 2, 37079 Göttingen, Germany, henning.meesenburg@nw-fva.de

K.J. Meiwes
Northwest German Forest Research Station, Grätzelstr. 2, 37079 Göttingen, Germany, karl-josef.meiwes@nw-fva.de

H. Meyer
Department of Soil Science of Temperate and Boreal Ecosystems, Georg-August-University Göttingen, Büsgenweg 2, 37077 Göttingen, Germany, hmeyer@ngi.de

S. Migge-Kleian
Ecology Group, Johann Friedrich Blumenbach Institute of Zoology and Anthropology, University of Göttingen, Berliner Str. 28, 37073 Göttingen, Germany smigge@gwdg.de

D. Murach
Faculty of Forest and Environment, Fachhochschule Eberswalde, Alfred-Möller-Strasse 1, 16225 Eberswalde, Germany, dmurach@fh-eberswalde.de

O. Panferov
Department of Bioclimatology, Büsgen-Institute, Georg-August University Göttingen, Büsgenweg 2, 37077, Göttingen, Germany, opanfy@gwdg.de

J. Prenzel
Department of Soil Science of Temperate and Boreal Ecosystems, Georg-August-University Göttingen, Büsgenweg 2, 37077 Göttingen, Germany jprenze@gwdg.de

J.A. Priess
Helmholtz Centre for Environmental Research – UFZ, Department for Computational Landscape Ecology, Permoserstr. 15, 04318 Leipzig, Germany joerg.priess@ufz.de

P. Rademacher
Department of Wood Biology and Wood Products, Georg-August-University Göttingen, Büsgenweg 4, 37077 Göttingen, Germany, pradema@gwdg.de

C. Rapp
Department of Silviculture and Forest Ecology of the Temperate Zones,

Georg-August-University Göttingen, Büsgenweg 1, 37077 Göttingen, Germany
chr.rapp@t-online.de

M. Raubuch
Department of Soil Biology, Faculty of Ecological Agricultural Sciences, University of Kassel, Nordbahnhofstraße 1a, 37213 Witzenhausen, Germany
raubuch@uni-kassel.de

E. Röhrig
Department of Silviculture and Forest Ecology of the Temperate Zones, Georg-August-University Göttingen, Büsgenweg 1, 37077 Göttingen, Germany
waldbau@gwdg.de

M. Schaefer
Ecology Group, Johann Friedrich Blumenbach Institute of Zoology and Anthropology, Georg-August University Göttingen, Berliner Str. 28, 37073 Göttingen, Germany, mschaef@gwdg.de

J. Schauermann
Ecology Group, Johann Friedrich Blumenbach Institute of Zoology and Anthropology, Georg-August University Göttingen, Berliner Str. 28, 37073 Göttingen, Germany, jschaue@rauschenwasser.de

S. Scheu
Ecology Group, Johann Friedrich Blumenbach Institute of Zoology and Anthropology, Georg-August University Göttingen, Berliner Str. 28, 37073 Göttingen, Germany, sscheu@gwdg.de

W. Schmidt
Department of Silviculture and Forest Ecology of the Temperate Zones, Georg-August University Göttingen, Faculty of Forest Sciences and Forest Ecology, Büsgenweg 1, 37077 Göttingen, Germany, wschmid1@gwdg.de

E. Schönfelder
Northwest German Forest Research Station, Grätzelstr. 2, 37079 Göttingen, Germany, egbert.schoenfelder@nw-fva.de

I.A. Schulze
Northwest German Forest Research Station, Grätzelstr. 2, 37079 Göttingen, Germany

G. Steinmetz
Department of Soil Science of Temperate and Boreal Ecosystems, Georg-August-University Göttingen, Büsgenweg 2, 37077 Göttingen, Germany

C.P. Wang
Taiwan Forest Research Institute, Division of Silviculture, Nan-Hai Rd #53, 100, Taipei, Taiwan, cpwang@tfri.gov.tw

K.V. Wilpert
Forest Research Station Baden-Württemberg, Department of Soil and Environment, Wonnhaldestraße 4, 79100 Freiburg, Klaus.Wilpert@forst.bwl.de

Introduction

R. Brumme and P.K. Khanna

Development of management strategies to achieve sustainable forest ecosystems has remained an important issue in forestry. An important component of sustainability is the maintenance of structure and functions of ecosystems. Beese (1996) outlined four primary functions of forest use which would need due attention to achieve a long term sustainable system: (a) production and use of resources (wood and other forest products, recreation, water conservation and other functions), (2) maintenance of ecosystem-internal cycles of matter and energy to minimise any effect on the neighboring systems, (3) habitat conservation (biodiversity – structure and functioning), and (4) maintenance of cultural and social role (landscape and heritage conservation).

European beech is a dominating tree species in central Europe covering a wide ecological spectrum of site conditions (climate, soil type, soil pH) (Ellenberg 1988, 1996; Leuschner et al. 2006; Bartsch and Röhrig, Chap. 22, this volume). Figure 1 shows some distribution pattern of beech forests in relation to chemical soil properties. Without human influence, forests would have covered most of Germany, while at present only about 30% of land area are covered by forests of which only 15% is under beech and another 25% under other deciduous species (BMVEL 2004). There is a broad consensus among forest managers that in future the area of beech forests in central Europe should increase, mainly as mixed broadleaf stands and as admixtures with conifers. The argument in favour of growing beech, especially in areas which are presently covered by conifers, is based on the fact that conifers had been planted on beech sites in the past, on considerations of forest health, and preservation and enhancement of biodiversity in a changing climate to provide long-term sustainability of forests and the productivity of such stands. The phenotypic plasticity and evolutionary adaptability of European beech to drought and its competitiveness appear to be underestimated (Bolte et al. 2007).

Ecosystems of central Europe, especially of Germany, have been strongly influenced by human activities in historical and in recent times. They have undergone significant changes with respect to their distribution, structure and functioning (for a detailed description of the topic refer to Jahn 1991). About 700–800 years ago in the medieval ages, the beech and other broadleaf deciduous forests dominated the landscape of central Europe. Most of these forests were harvested and the fertile and easily accessible land was converted into arable use. The remaining forests, mainly on the marginal soils, were intensively used as sources of energy and raw materials

R. Brumme and P.K. Khanna (eds.), *Functioning and Management of European Beech Ecosystems*, Ecological Studies 208,
DOI: 10.1007/978-3-642-00340-0, © Springer-Verlag Berlin Heidelberg 2009

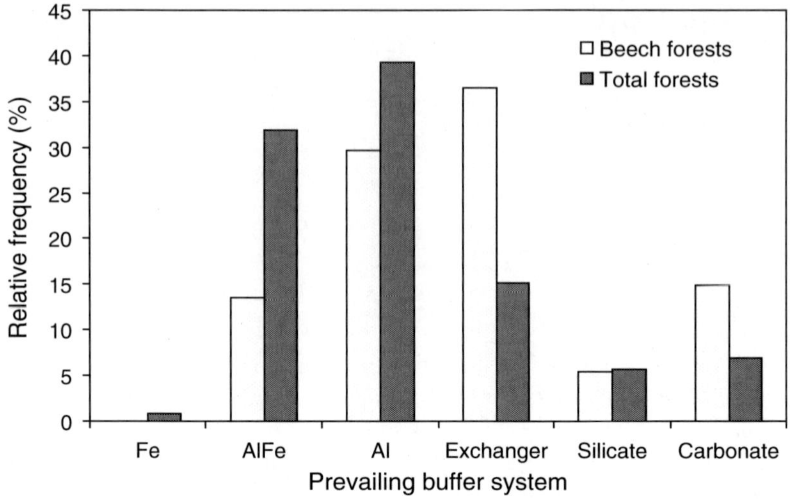

Fig. 1 Relative frequency of occurrence of German forests (*n* =1,848) and German beech forests (*n* =74) on soils of different buffer ranges (Fe: Iron buffer range, pH <3.2; AlFe: Aluminium/iron buffer range, pH 3.2–3.8; Al: Aluminium buffer range, pH 3.8–4.2; Exchanger buffer range, pH 4.2–5; Silicate buffer range, pH 5–6.2; Carbonate buffer range, pH >6.2) (Data from W. Riek on the Soil Survey of German Forests)

for living and for the production of goods, as animal fodder and as sources of nutrients for agriculture (leaf raking). Export of biomass from forests lasted for many centuries and resulted in the continual loss of soil nutrients, lowering their soil fertility and increasing the acidity levels in soils (Puhe and Ulrich 2001).

During the last few decades, industrial development has accelerated the processes of soil degradation in forests by causing additional changes through high atmospheric inputs of nutrients, acidity and heavy metals. High inputs of N and S and the subsequent losses of base cations from soils in association with percolating nitrates and sulphates increased the soil acidity. Use of fossil fuels and conversion of terrestrial ecosystems have increased greenhouse gases causing changes in temperature and precipitation conditions (IPCC 2007). These factors of high atmospheric depositions and climate change have to be considered in the development of future management strategies for forest stands. Moreover, the future management practices of forest stands would include past land-use history for sustainable and site-specific options. Some of the essential management components were described by Puhe and Ulrich (2001).

A number of studies have been undertaken in the last few years to analyse the structure, and to understand the functioning of various forest ecosystems in the temperate region, especially under changing conditions of atmospheric depositions and climate change (Ellenberg et al. 1986; Adriano and Johnson 1989; Schulze et al. 1989; Röhring and Ulrich 1991; Tamm 1991; Johnson and Lindberg 1992; Godbold and Hüttermann 1994; Ellenberg 1996; Schulze 2000; Mickler et al. 2000; Puhe and Ulrich 2001; Dohrenbusch and Bartsch 2002; Veblen et al. 2003; Hanson and Wullschleger 2003; Valentini 2003; Matzner 2004). Of recent interest in this context is the role of soil organic matter as a sink for CO_2 and nitrogen and a source for nutrients for tree growth. This pool has gained importance for forest growth since the mineral soil pools as sources for nutrients have been reduced by acid depositions. Importance of soil organic matter is related to its soil chemical state which influences the soil microbial and faunal activity (bioturbation) and its decomposition in the short-term, and the amount, quality, and distribution of soil organic matter in the soil profile in the long-term. Atmospheric inputs during the last few decades have led to long-lasting changes in the soil chemical and biological states but the resulting effects on soil organic matter remain unclear. Issues of concern are the description of long-term changes using short-term measurements, which will depend upon finding suitable indicators to predict the changes. One-time observations or observations covering very short-term periods may not provide a true picture of the long-term developments which occur in the processes and functioning of ecosystems. Prediction of those long-term changes is only possible when the long-term observations are made to cover climatic variations and life cycles of various organisms involved, as well as various states of soil development. These are some of the essential elements towards an understanding of ecosystem functioning. This volume provides some of the long-term monitoring data in three beech forests with following aims:

(1) Relating the differences in site factors and input conditions to soil processes and plant growth on three beech stands which cover a wide range of chemical soil conditions. These sites have been monitored for decades to assess differences in the growth rates of aboveground and belowground components of over-storey and herbaceous plants, decomposition processes, nutrient uptake and turnover, fluxes between soil, hydrosphere and atmosphere, methane oxidation and general health and vitality of beech stands. Selection of beech as a potential tree species for future is based on its versatility to grow well on a wide spectrum of suitable stands (soil pH, rainfall and temperature conditions).
(2) Studying those processes at the three beech stands which are primarily related to the carbon cycling and nutrient uptake, litter production, decomposition and mineralisation, so as to provide background information about the quality and quantity of the biotic components (trees, herbs, soil fauna, microbial biomass) and the environmental conditions (soil organic matter content, soil properties, climate). This would provide the level of matter exchange with the neighbouring systems (matter input, matter output by leaching, soil and root respiration, N_2O emissions, methane uptake).
(3) Studying the long-term perspective of C- and N-mobilisation or immobilisation, using regional-scale inventories of microbial biomass-C and organic carbon and nitrogen concentrations and stocks of forest soils in Germany, which may provide relationships to the soil chemical state and to any past or expected future changes by soil acidification. In relation to C and N, management options for beech forests would be developed.
(4) Synthesising the stand-specific information on C- and N-balances of the three beech stands, the regional-scale inventories of soil carbon and nitrogen, and the

input–output balance of nitrogen for a wide range of stands to describe the present N-status of forests and to present future perspectives for forest development and forest management.

References

Adriano DC, Johnson AH (1989) Acidic precipitation, vol 2. Biological and ecological effects. Springer, Berlin

Beese F (1996) Indikatoren für eine multifunktionelle Waldnutzung. Forstw Cbl 115:65–79

Bolte A, Czajkowski T, Kompa T (2007) The north-eastern distribution range of European beech – a review. Forestry 80:413–429

BMVEL (2004) Die zweite Bundeswaldinventur. Bundesministerium für Verbraucherschutz, Ernährung und Landwirtschaft, Bonn

Dohrenbusch A, Bartsch N (2002) Forest development, succession, environmental stress and forest management. Springer, Berlin

Ellenberg H (1988) Vegetation ecology of Central Europe, 4th edn. Cambridge University Press, Cambridge

Ellenberg H (1996) Vegetation Mitteleuropas mit den Alpen. Ulmer, Stuttgart, Germany

Ellenberg H, Mayer R, Schauermann J (1986) Ökosystemforschung, Ergebnisse des Solling-Projekts. Ulmer, Stuttgart

Godbold DL, Hüttermann A (1994) Effects of acid rain on forest processes. Wiley, New York

Hanson PJ, Wullschleger SD (2003) North American temperate deciduous forest responses to changing precipitation regimes, Ecological Studies 166. Springer, Berlin

IPCC (2007) In: Solomon S, Qin D, Manning M, Marquis M, Averyt K, Tignor MMB, Miller Jr LeR, Chen Z (eds) Climate Change 2007, Fourth Assessment Report (AR4), Working Group I Report "The physical science basis". Cambridge University Press, Cambridge

Jahn G (1991) Temperate deciduous forests of Europe. In: Röhring E, Ulrich B (eds) Temperate deciduous forests, Ecosystems of the world 7. Elsevier, Amsterdam, pp 377–502

Johnson DW, Lindberg SE (1992) Atmospheric deposition and forest nutrient cycling. Springer, Berlin

Leuschner C, Meier IC, Hertel D (2006) On the niche breadth of *Fagus sylvatica*: soil nutrient status in 50 Central European beech stands on a broad range of bedrock types. Ann For Sci 63:355–368

Matzner E (2004) Biogeochemistry of forested catchments in a changing environment, A German case study, Ecological Studies 172. Springer, Berlin

Mickler RA, Birdsey RA, Hom J (2000) Responses of Northern U.S. forests to environmental change, Ecological Studies 139. Springer, Berlin

Puhe J, Ulrich B (2001) Global climate change and human impacts on forest ecosystems: postglacial development, present situation, and future trends in central Europe, Ecological Studies 143. Springer, Berlin

Röhrig E, Ulrich B (1991) Temperate deciduous forests, Ecosystems of the world 7. Elsevier, Amsterdam

Schulze ED (2000) Carbon and nitrogen cycling in European forest ecosystems, Ecological Studies 142. Springer, Berlin

Schulze ED, Lange OL, Oren R (1989) Forest decline and air pollution, Ecological Studies 77. Springer, Germany

Tamm CO (1991) Nitrogen in terrestrial ecosystems, Ecological Studies 81. Springer, Germany

Valentini R (2003) Fluxes of carbon water and energy of European forests, Ecological Studies 163. Springer, Berlin

Veblen TT, Baker WL, Montenogro G, Swetnam TW (2003) Fire and climate change in temperate ecosystems of the Western Americas, Ecological Studies 160. Springer, Berlin

Part A
Description of Long-term Observation Sites

Chapter 1
General Description of Study Sites

H. Meesenburg and R. Brumme

The three case studies provided in this book describe mature and almost pure beech stands (*Fagus sylvatica*) of 130–160 years old. They are located in the hilly country around Göttingen (Fig. 1.1) which was formed by different geological formations of Triassic limestone (Göttinger Wald site), and Triassic sandstone (Solling site), with locally interspersed tertiary volcanic stones (Zierenberg site) which are covered by loess of variable thickness. These sites cover a wide amplitude of beech forests under similar climatic conditions (Chap. 2, this volume). The sites differ considerably with respect to the acid–base status of the soils which affects the composition of the ground vegetation, the soil biota, litter quality and the surface humus type. These site differences have considerable implications for the nutrient status of these forests, the element cycling processes, the carbon and nitrogen sequestration, and the management options. The three sites have been included as intensive monitoring plots of the Forest Ecosystem Research Center of the University of Göttingen and are part of the Intensive Forest Monitoring Program (Level II) of the International Co-operative Program on Assessment and Monitoring of Air Pollution Effects on Forests (ICP Forests) operating under the UNECE Convention on Long-Range Transboundary Air Pollution. Selected results from these sites have been published in several publications cited in the individual chapters and they contributed largely to the volumes of Ellenberg et al. (1986), Röhrig and Ulrich (1991), and Godbold and Hüttermann (1994). The main features of the sites are summarised in Table 1.1.

The beech forest at the Göttinger Wald site is situated close to the rim of the Triassic limestone plateau of the "Göttinger Wald" 8 km east of Göttingen (Fig. 1.1) at an elevation of about 420 m a.s.l. (Table 1.1). The almost pure beech (*F. sylvatica*) stand has a small admixture of ash (*Fraxinus excelsior*), maple (*Acer platanoides, Acer pseudoplatanus*), oak (*Quercus robur, Quercus petraea*), and elm (*Ulmus glabra*), and is about 130 years old (Chap. 5, this volume). A dense and diverse herb layer covers the soil surface in May each year which consists primarily of *Allium ursinum*, *Mercurialis perennis* and *Aconitum vulparia* (Chap. 10, this volume). These geophytes use the higher amount of radiation available on the organic layer during mid-March to May for photosynthesis and plant growth before the trees start developing their foliage. The soils at the site are shallow with a depth of between 20

R. Brumme and P.K. Khanna (eds.), *Functioning and Management of European Beech Ecosystems*, Ecological Studies 208,
DOI: 10.1007/978-3-642-00340-0_1, © Springer-Verlag Berlin Heidelberg 2009

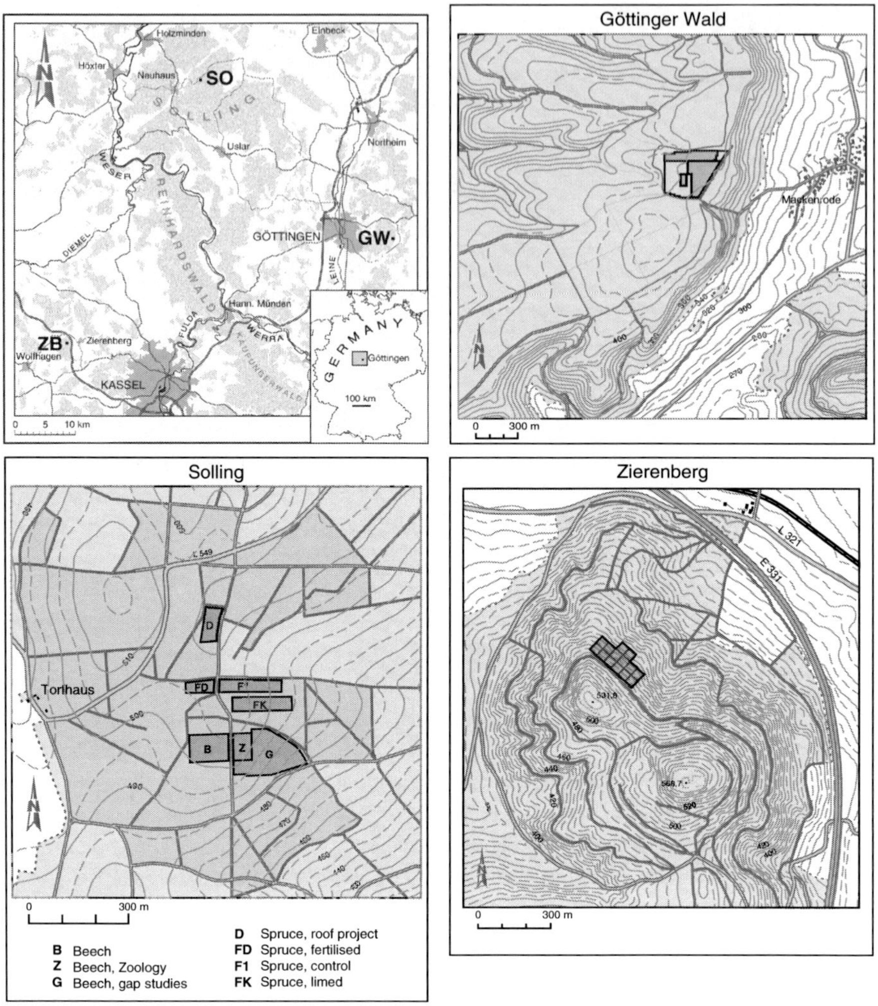

Fig. 1.1 Locations of the three beech sites in Germany (drawn by Tambour)

and 50 cm, have high soil pH(H_2O) of 5.5–7.4 (0–20 cm), and are rich in base cations and carbonate content (Chap. 3, this volume). The soil parent material is calcareous bedrock with a calcite content of about 95%. The biological activity in this soil is very high (Chaps. 6 and 7, this volume) and has caused the development of a mull type litter layer and a humus-rich surface mineral soil. This site was established in 1980 for an intensive monitoring programme and has been included as an intensive permanent soil monitoring plot in the soil protection programme of the state of Lower Saxony.

The beech forest site at Zierenberg is located on the north-eastern slope of the "Kleiner Gudenberg" 50 km south-west of Göttingen in the north of Hesse State

Table 1.1 Description of study sites (climate for 1990–2002)

	Unit	Göttinger Wald	Zierenberg	Solling
Latitude	(°)	51°32′N	51°22′N	51°46′N
Longitude	(°)	10°03′E	09°16′E	09°34′E
Elevation	(m) a.s.l.	420	450	504
Slope	(°)	0–3	15	0–3
Aspect		Plain	North-east	South
Stand age (2000)	(years)	132	156	153
Stand density (2000)	(trees ha^{-1})	245	132	199
Mean diameter (2000)	(cm)	46	60	47
Mean height (2000)	(m)	34	37	30
Timber volume (2000)	(m^3 ha^{-1})	686	595	519
Natural vegetation		Hordelymo-Fagetum	Hordelymo-Fagetum	Luzulo-Fagetum
Bedrock		Limestone	Basalt over limestone	Sandstone with loess cover
Soil type FAO		Rendzic Leptosol/ Calcaric Cambisol	Eutric Cambisol	Dystric Cambisol
Humus type		F-Mull	F-Mull	Typical Moder
Precipitation	(mm)	709	754	1,193
Precipitation (May–Oct.)	(mm)	410	406	553
Mean temperature	(°C)	7.4	6.9	6.9
Mean temperature (May–Oct.)	(°C)	12.6	11.9	12.0

(Fig. 1.1). The "Kleiner Gudenberg" is part of a hilly landscape with elevations of 200–550 m a.s.l. Small hills were created through volcanic activity during the Tertiary period. Basaltic material (Limburgite) at the hilltop of the "Kleiner Gudenberg" overlies middle Triassic limestone. Basaltic debris of about 90 cm depth covers the limestone at the mid region of the slope (the so-called 'Basalt site') (450 m a.s.l.) where the main research area of the study site is situated. For some of the investigations, two other neighbouring sites have been included which are located below the main research area. The sites are distinguished as a limestone-dominated site at the lower part of the hill (Limestone site), and a transitional site which is located between the Limestone and Basalt sites. The sites carry almost pure beech (*F. sylvatica*) stands of about 160 years old. A dense and diverse herb layer covers the soil. It is dominated by *M. perennis* or *Urtica dioica* which grow up to 1 m high during the summer. The soil at the main research area (Basalt site) has high pH(H_2O) between 5.2 and 5.7 (0–20 cm), is free of lime and rich in magnesium and phosphorous. The biological activity in the soil is very high. The surface organic layer is F-mull type. Surface mineral soil is rich in organic matter. The experimental monitoring was started in 1989 as an intensive study site of the Forest Ecosystems Research Center of the University of Göttingen and of the Forest Ecosystem Study Hesse (Waldökosystemstudie Hessen WÖSSH).

The large forested area of the Solling is enclosed by the basins of the river Weser to the west and the river Leine to the east covering heights of 250 to over 500 m a.s.l. (Fig. 1.1). The Solling massive consists of Triassic sandstone, locally known as "Solling-Folge", which is covered with 60 to 80 cm thick solifluction layers of loess material. The study site is located at the centre of the Solling plateau at 504 m a.s.l. about 33 km northwest of Göttingen. The pure beech (*F. sylvatica*) stand is about 150-years-old (Table 1.1). The soil is covered by a sparse herb layer, dominated by *Oxalis acetosella* and *Luzula luzuloides*. The soil is very acid with pH(H_2O) values of 3.4–4.0 (0–20 cm), and has a low base saturation of $< 7\%$. The soil biological activity is very low and has developed a thick moder-type surface organic layer. Monitoring on this site was initiated in 1966 as part of the international biological programme (Ellenberg 1971). There are two main sites: Beech stand (B1) and an adjacent spruce stand (F1). A number of additional plots were

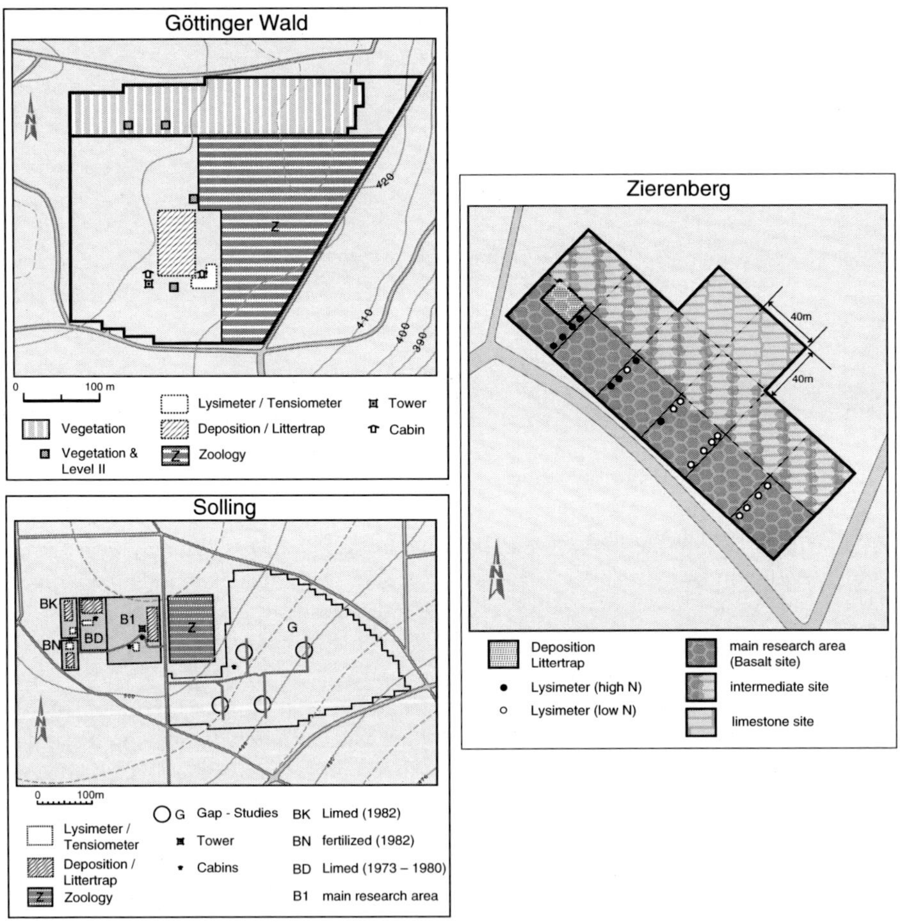

Fig. 1.2 Plot design of the three beech sites in Germany (drawn by Tambour)

established in the neighbourhood; of special interest are two which were established in 1982, the first was limed with 30 Mg per ha of finely ground dolomitic limestone (BK plot), and the second was fertilised annually with 140 kg N ha^{-1} with solid ammonium sulfate from 1982 until 1993 (BN plot) (Fig. 1.2). The two main sites, B1 and F1, have been included as intensive permanent soil monitoring plots in the soil protection programme of the state of Lower Saxony since 1992. A comprehensive summary of the results collected from 1966 to 1986 at the B1 and F1 sites and other stands of the Solling area was published by Ellenberg et al. (1986).

References

Ellenberg H (1971) Integrated experimental ecology – Methods and results of ecosystem research in the German Solling Project. Ecological Studies 2. Springer, Berlin

Ellenberg H, Mayer R, Schauermann J (1986) Ökosystemforschung: Ergebnisse des Solling–Projekts. Ulmer, Stuttgart

Godbold DL, Hüttermann A (1994) Effects of acid rain on forest processes. Wiley, New York

Röhrig E, Ulrich B (1991) Temperate deciduous forests. Ecosystems of the World 7. Elsevier, Amsterdam

Chapter 2
Climatic Condition at Three Beech Forest Sites in Central Germany

O. Panferov, H. Kreilein, H. Meesenburg, J. Eichhorn, and G. Gravenhorst

2.1 Climate of Germany

Climate is one of the major factors controlling the growth and functioning of plant ecosystems in general and of beech forests in particular. In the present Chapter climatic conditions at three experimental beech forest sites are described.

The territory of Germany extends for about 8° (55°–47°N) from the coastal line of the North and Baltic Seas to the Alps, within the range of maritime temperate climate sub-zones Cfb and Cwb according to Koeppen classification (e.g. McKnight and Hess 2000). It is described as a transition zone between the maritime climate of western Europe and the continental one of eastern Europe. Major agents of the general atmospheric circulation pattern contributing to the maritime character of the climate are the Icelandic low-pressure system, the Azores high-pressure system and often the Asian (or Siberian) high-pressure system. The first two of them furnish western Europe with moisture-loaded air masses propelled by dominant westerly winds. Other major factors are: (1) the maritime influence of the Atlantic Ocean adding to climate 'oceanicity'; (2) the mountain ranges hindering the southward advance of maritime air masses and, thus, adding to climate 'continentality', and (3) the altitude effects in highland regions. Therefore, in general, the oceanicity of climate increases to the north and the west whereas the continentality increases to the south and east. During winter time, the air temperature isotherms run from north to south with values increasing westward, and in summer the isotherms run generally parallel to latitudes (from west to east) with temperatures increasing southward. The mean annual temperature thus decreases generally from north-west to south-east and the annual precipitation from west to east (Fig. 2.1) Table 2.1 summarises the mean climatic conditions over the area of Germany based on data from the German Weather Service (Deutsche Wetterdienst – DWD) for the standard climatological period of 1961–1990. Mean annual air temperature for Germany is +8.4°C with July being the warmest month and January – the coldest. DWD data indicate a gentle increase of annual temperatures after 1988 when compared to the 30-year mean value. Mean annual precipitation for Germany is

R. Brumme and P.K. Khanna (eds.), *Functioning and Management of European Beech Ecosystems*, Ecological Studies 208,
DOI: 10.1007/978-3-642-00340-0_2, © Springer-Verlag Berlin Heidelberg 2009

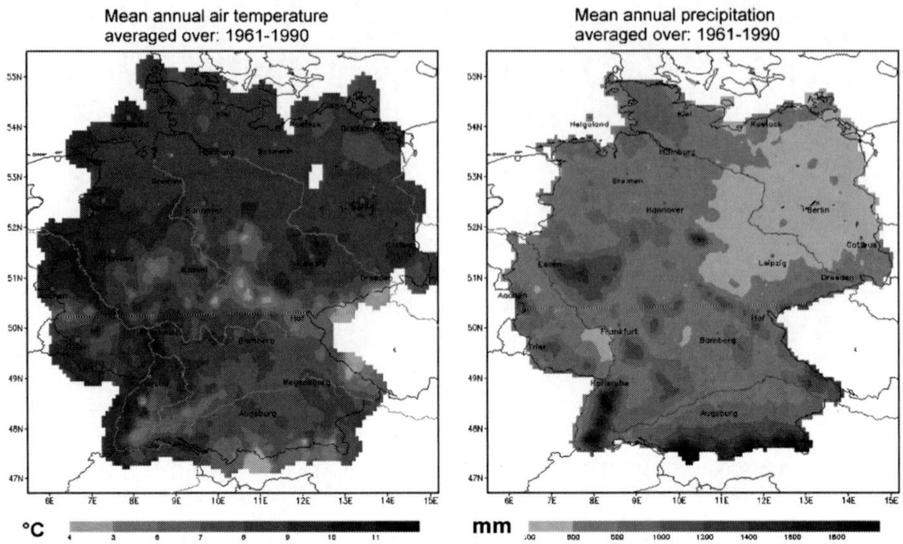

Fig. 2.1 Spatial distribution of mean annual T_a (°C, *left panel*) and mean annual precipitation amount (mm year^{-1}, *right panel*) in Germany. (Maps are derived from the Climate maps of Germany presented in www.klimadiagramme.de and are based on the data of the German Weather Service)

Table 2.1 Multiyear (1960–1990) monthly, seasonal and annual mean values and standard deviations (σ) of air temperature, T_a (°C) and precipitation rates, P (mm per period) averaged over the area of Germany (Mitchell et al. 2004). Season abbreviations: *MAM* March, April, May; *JJA* June, July, August; *SON* September, October, November; *DJF* December, January, February; *ANN* annual

	T_a	σ_T	P	σ_P
JAN	−0.5	2.9	50.8	19.5
FEB	0.4	2.7	40.5	19
MAR	3.7	2.1	47.9	17.7
APR	7.6	1.1	50.9	19.9
MAY	12.2	1.2	64.7	19.8
JUN	15.5	1	77.3	20.5
JUL	17.2	1.3	72	23.7
AUG	16.9	0.9	71.5	16.8
SEP	13.8	1.2	57.2	19.6
OCT	9.4	1.2	50.3	23.1
NOV	4.2	1.3	58.4	16.7
DEC	0.8	2.2	58.6	25.9
MAM	7.8	0.9	163.4	35
JJA	16.5	0.7	220.7	41.3
SON	9.1	0.7	165.9	32.2
DJF	0.2	2	149.9	69.7
ANN	8.4	0.7	699.9	92.8

700 mm with the pronounced seasonal peak in summer (June–August) with 31.4% of annual precipitation. Seasonal relative minimum value was observed for December–February – 21.5% of annual precipitation. Based on these climatic characteristics most of the land area in Germany could be considered as almost optimal for beech (*Fagus silvatica L.*) growth as beech prefers oceanic climates with at least 600 mm per year precipitation, with annual mean temperature of about +8° (though wide temperature variations are tolerated) and a growing season of at least 5 months (Mayer 1992). The climatic conditions in Germany fall around these mean values and are classified into four major climatic regions (Schüepp and Schirmer 1977), namely: (1) Northern Lowlands, (2) German Highlands, (3) The German Alpine Foreland and (4) The German Alps. Since the investigated sites are located within the German Highlands, a brief description of climatic conditions of this area is provided below.

The German Highlands belong to the Central European Uplands which start from the Massif Central in France and stretch to Poland and the Czech Republic. The Highlands are generally moderate in height with few peaks exceeding 1,100 m a.s.l. The area of German Highlands includes Saarland, Hessen and Thuringia, the north of Rhineland and the south of North Rhine Westphalia, the southernmost parts of Lower Saxony and Saxony Anhalt and western parts of Saxony.

In the Highlands, the climate continentality increases southward and eastward and well-pronounced orographic effects are observed. The influence of high-pressure systems gets stronger in this region especially during winter periods. The decrease of air temperature with altitude is well expressed in summer but is reduced in winter because of the stagnation of cold air in valleys enhanced by its radiative cooling. The moist Atlantic air masses are transported into this region by westerly winds so that western slopes receive more precipitation from orographic rain clouds than eastern slopes which are located in a rain shadow. Generally, the amount of precipitation at the same altitude decreases southward and eastward. Thus, two major factors that modify the climate at regional and local levels in the German Highlands are the height and the orientation of mountain ranges, and the variability of local topography (Schüepp and Schirmer 1977). The influence of topography on climatic conditions of the German Highlands, namely the increase of orographic precipitation and decrease of mean air temperature, is very well illustrated in the north by the Harz Mountains and Thuringian Forest. The mean air temperature at Bad Harzburg (300 m a.s.l) is 8.5°C and the mean annual precipitation is 813 mm, while at the highest point of the Harz Mountains, the Brocken (1,142 m a.s.l.), the mean annual air temperature is low (2.9°C) and the mean annual precipitation is high (1,594 mm) (www.klimadiagramme.de). The typical rain shadow area extends eastward of the Thuringian Forest with annual precipitation of less than 500 mm. The southern parts of the highlands, like the Bohemian Forest and Bavarian Forest belonging to the Bohemian Massif, receive the remaining Atlantic-originated moisture. At higher altitudes, an increase of total precipitation is accompanied by an increased share of snowfall.

2.2 Experimental Sites

The three investigation sites: Solling Mountains, Zierenberg and Göttinger Wald in Central Germany are quite similar in their topography and landscape features. They are situated in the south of Lower Saxony and in the north of Hessen, and being quite close together form a triangle with a side length of approximately 70 km. The sites are situated either on plateaus (Solling and Göttinger Wald) or on a slope (Zierenberg, see Chap. 1 "Site description") and are surrounded by forested rural areas. It should be emphasised that the weather data collected on these three sites significantly differ from standard synoptic and climate stations of the DWD as the meteorological stations of Solling Mountains, Zierenberg and Göttinger Wald are placed in the beech forest stands underneath tree canopies and not in open field surroundings as prescribed by the World Meteorological Organisation for its standardised meteorological stations.

2.2.1 Climatic Variables

To compare the microclimates of the three beech sites, three measured climate characteristics were chosen: (1) daily mean air temperature at 2 m above the surface organic layer (T_a); (2) precipitation amount at a nearby opening in the trees, and (3) daily mean forest soil temperature (T_s) at three different depths (5 cm, 10 cm and 20 cm – $T_{s,5}$, $T_{s,10}$ and $T_{s,20}$, respectively). Global radiation (sum of direct solar and diffuse sky radiation flux components in short wave spectral range from 300 to 3,000 nm) was measured at the Solling site only during the period from 1990 to 1999. For that reason, the 30-year mean values provided by DWD on sunshine duration and annual global radiation sums were used to characterise the energy input to beech ecosystems. The mean annual global radiation sums measured at Solling are used for partial validation of DWD data.

Air and soil temperatures were measured by means of a PT–100 resistance thermometer. The amount of precipitation was recorded using a Thies tipping bucket rain gauge (Adolf Thies, Göttingen, Germany) and a heated Hellmann rain gauge recorder. The global radiation was measured by means of a CM11 Kipp and Zonen pyranometer. The sensors were connected to Orion data loggers so that the measurements were performed automatically. The values were sampled with a 10-Hz frequency and averaged over 10 min at Göttinger Wald (e.g. Kreilein 1987) and over 15 min at Solling and Zierenberg.

Three time periods were chosen for calculating mean values: (1) annual; (2) the summer half-year period (May–October), which can be characterised as a period with foliage cover of deciduous forests and defined as the "growing season", and (3) the winter half-year period (November–April), which represents a leafless period or the "non-growing season". Thus, 'seasonal average' of a variable describes the mean values over 'growing' or 'non-growing' half-year periods. However, the limits of growth conditions of plants do not solely depend on monthly or seasonal average values of climatic variables. Extreme values of meteorological variables, their different combinations as well as the duration of stress conditions are also of great importance. The quantification of the damage on trees caused by weather extremes in terms of the tree vitality reduction still remains a challenging task and is a subject of numerous scientific discussions. The present article serves a descriptive purpose. It includes no detailed discussion on extreme values and their consequences but merely mentions the temperatures minima and maxima to give the reader some impression of the variability ranges at the study sites.

The measurements at the three beech sites were carried out from 1969 to 2002. However, not all parameters were measured continuously for the entire period. Air temperature was measured from 01.01.2001 to 31.12.2002 at the Zierenberg site, from 01.01.1990 to 31.12.1998 at the Göttinger Wald site, and from 01.01.1990 to 31.12.2002 at the Solling site. Any missing values for these stations were interpolated or extrapolated from the measured values at nearby DWD stations, e.g. from Kassel for Zierenberg, Silberborn or Holzminden for Solling, and Göttingen for Göttinger Wald, and thus are not independently derived.

2.3 Climatic Conditions at Beech Sites

Because of their location within the northern part of German Highlands, all three studied forest sites experience a smoothing maritime influence on air temperature variability through the mild and moisture-loaded Atlantic air masses which are transported by prevailing westerly and north-westerly winds. The additional effects of elevation and forest cover alter the microclimate of the measuring plots. The combination of various influencing factors results generally in mild air temperatures in summer and winter, high annual precipitation which exceeds potential and actual evapotranspiration rates, and the absence of a well-pronounced dry period. Fog occurs on about 70 days per year.

2.3.1 Solar Radiation

According to the climatological data of the 30 years 1961–1990, as provided by DWD (www.dwd.de), the sunshine duration in Germany ranges from 1,200 to 1,920 h year^{-1} with the mean value of 1,542 $\pm$ 127 h year^{-1}. The mean values of sunshine duration within the area covering the experimental beech sites are well below the mean value for Germany (Table 2.2). The Solling and Zierenberg sites experience less sunshine than Göttinger Wald, The annual sums of global radiation on the three sites, however, differ only slightly whereas the insolation in the Göttinger Wald and Zierenberg sites is higher than in Solling. The results of global radiation measurements for Solling show an even lower value of annual radiation

Table 2.2 Mean annual sunshine duration (hours), radiation sums, GJ per year, (values in kWh per year m^{-2} are given in parenthesis) and the corresponding irradiances, W m^{-2}, (period 1961–1990) at Göttinger Wald (GW), Zierenberg (ZB), and Solling (SO). The mean annual radiation sums for the period of (1990–1999) measured at Solling (SO_{meas}) and correspondent mean irradiance are also shown

Site	Sunshine duration (hours)	Radiation sums	Irradiance (W m^{-2})
GW	1,400–1,450	3.4–3.5 GJ m^{-2} a^{-1}(960–980 kWh m^{-2} a^{-1})	109.6–111.9
ZB	1,300–1,350	3.4–3.5 GJ m^{-2} a^{-1}(960–980 kWh m^{-2} a^{-1})	109.6–111.9
SO	1,300–1,350	3.3–3.4 GJ m^{-2} a^{-1}(940–960 kWh m^{-2} a^{-1})	105–107.3
SO_{meas}	–	3.29 ± 0.24 GJ m^{-2} a^{-1} (915 ± 65.7 kWh m^{-2} a^{-1})	104.5

sum than that given by DWD. However, the difference is within the range of expected variability.

2.3.2 Air Temperature

The highest value of average annual air temperature, T_a, (Table 2.3) was observed at the lowest station – Göttinger Wald (420 m a.s.l.). However, the values of T_a observed at both Zierenberg (450 m a.s.l.) and Solling (504 m a.s.l.) sites are equal despite the 50 m difference in altitude. The T_a gradient between Göttinger Wald and Solling (80 m altitude difference), both of them located on plateaus, is about 0.6°C/100 m which is in a good agreement with annual averages observed at meteorological stations of the DWD. At the same time, the T_a value at Zierenberg which is on a slope is equal to that at Solling located 50 m higher on a plateau, indicating that besides the altitude effect local environmental and/or topographic factors are important for T_a gradients.

The annual courses of T_a are quite similar at the three sites, though their absolute values differ from one another. August is the warmest month at all sites (Table 2.3). This is rather untypical for the more continental climatic region of the German Highlands, where the maximum monthly T_a value usually occurs in July. It would be more characteristic for the Northern Lowlands with their stronger oceanicity. One of the factors causing the T_a maxima to occur in August is that the investigated forests are situated in the northernmost region of the German Highlands which forms a transition zone to the Northern Lowlands, having the cooling ocean influence in summer. Another factor may be related to the dense canopy of beech forests where the stations are installed. During the growing season, the canopy blocks the direct radiative heating of the underlying surface in the daytime and the radiative cooling at night, smoothing the daily amplitudes (Ellenberg 1996). The seasonal increasing of foliage density also slows down the spring to summer increase of the mean monthly T_a. Without the protective influence of foliage during winter, the T_a values are minimal corresponding to the expected values of the

Table 2.3 Mean T_a (°C) and precipitation rates (mm per year) for the observation period 1990–2002 averaged over: entire year, the growing season (May–October), and the non-growing season (November–April) at Göttinger Wald (GW), Zierenberg (ZB), and Solling (SO) (min and max values for the observation period in are parentheses)

Site	Air temperature			Precipitation		
	Annual	May–October	November–April	Annual	May–October	November–April
GW	7.4 ± 0.8 (5.5, 8.3)	12.6 ± 0.63 (11.3, 13.4)	2.2 ± 1.18 (−0.4, 3.7)	709 ± 193 (537, 973)	410 ± 156 (233, 596)	299 ± 143 (170, 453)
ZB	6.9 ± 0.72 (5.0, 7.7)	11.9 ± 0.59 (10.6, 12.6)	1.8 ± 1.0 (−0.6, 3.2)	754 ± 201 (535, 981)	406 ± 112 (281, 594)	348 ± 97 (237, 438)
SO	6.9 ± 0.75 (5.0, 7.7)	12.0 ± 0.67 (10.7, 12.8)	1.7 ± 1.01 (−0.8, 3.0)	1,193 ± 215 (8,621,571)	553 ± 254 (370, 915)	641 ± 238 (391, 896)

Table 2.4 Mean monthly temperatures (°C) and monthly precipitation rates (mm per month) for the period 1990 – 2002 at the Göttinger Wald, Zierenberg and Solling sites

Month	Göttinger Wald		Zierenberg		Solling	
	Temperature	Precipitation	Temperature	Precipitation	Temperature	Precipitation
Jan	−0.3 ± 2.1	40 ± 28	−0.6 ± 1.9	53 ± 48	−0.8 ± 1.83	111 ± 102
Feb	0.8 ± 3.1	41 ± 158	0.5 ± 2.7	54 ± 50	0.3 ± 2.9	103 ± 88
Mar	3.5 ± 1.5	45 ± 19	3.0 ± 1.4	53 ± 31	2.7 ± 1.6	106 ± 194
Apr	6.7 ± 1.7	44 ± 31	6.1 ± 1.5	50 ± 70	6.0 ± 1.4	72 ± 78
May	11.3 ± 1.6	60 ± 69	10.6 ± 1.6	62 ± 51	10.7 ± 1.9	66 ± 70
Jun	13.5 ± 1.1	65 ± 26	12.7 ± 1.2	70 ± 67	12.7 ± 1.4	81 ± 40
Jul	15.6 ± 1.7	79 ± 68	14.9 ± 1.8	81 ± 44	15.2 ± 2.1	104 ± 87
Aug	16.1 ± 1.3	69 ± 56	15.2 ± 1.2	61 ± 85	15.5 ± 1.3	82 ± 50
Sep	11.6 ± 1.5	70 ± 45	10.9 ± 1.5	69 ± 54	11.0 ± 1.8	107 ± 101
Oct	7.6 ± 1.7	67 ± 53	7.0 ± 1.7	63 ± 41	7.1 ± 2.1	113 ± 63
Nov	2.8 ± 1.7	67 ± 66	2.5 ± 1.6	69 ± 65	2.6 ± 1.7	116 ± 104
Dec	−0.2 ± 2.2	61 ± 59	−0.4 ± 1.8	69 ± 71	−0.5 ± 1.8	133 ± 101

climatic region of the German Highlands. Average air temperatures in December and January are below 0°C with January as the multiyear average – the coldest month at all sites (Table 2.4). At all stations, the minimum temperatures in winter were about −11°C (±4°C) and the maximum temperatures in summer were on average around +22°C (±1.5°C). The absolute minimum for the observation period was as low as −20.6°C in Göttinger Wald and was recorded on 01.01.1997. At Solling and Zierenberg, the lowest air temperatures were recorded on 31.12.1978 as −18.6°C (Solling) and −17.5°C (Zierenberg). The warmest day at all stations was 09.08.1992 when the air temperature reached 27.8°C at Göttinger Wald, 27.5°C at Solling and 25.9°C at Zierenberg.

Multiyear average values for seasonal (growing/non-growing seasons) patterns of T_a at three studied sites (Table 2.3) show Göttinger Wald with the highest mean temperatures for all seasons. Mean values at Zierenberg and at Solling are equal to each other and lower than at Göttinger Wald. Similar differences are observed for seasonal maxima and minima values of T_a. At all stations, the amplitudes of inter-annual courses are higher for the non-growing season than for the growing one suggesting the highest variability of winter conditions from year to year. The T_a values at the Göttinger Wald site show the highest among the stations' amplitudes (4.1°C) of inter-annual course of values for the non-growing season. The temperatures of both the other stations had amplitudes of 3.8°C. The discussion on the inter-annual variability and its causes is outside the scope of this chapter.

2.3.3 *Precipitation*

The annual and seasonal mean values of precipitation for the three sites are given in Table 2.3. The annual mean rainfall is highest at Solling, which is about 1.5 times higher than at both Zierenberg and Göttinger Wald. The values presented in Table 2.3 indicate that Göttinger Wald is the warmest and driest of all three sites, which is probably related to its highest values for annual sunshine duration and radiation sums among the three stations. Highest precipitation at the Solling site is related to its location: the Solling ranges are the first massive obstacle for the dominant westerly winds from the North Atlantic and the North Sea. As described in Section 2.1, these westerlies usually bring moisture-laden air masses with high relative humidity and, therefore, with a big chance for cloud and precipitation formation when the air is orographically uplifted. The contribution of orographically-induced precipitation is significantly lower at the Zierenberg and Göttinger Wald sites because they are located leeward, i.e. in a rain shadow of the Solling Mountains. Being located on lower and narrower ranges than Solling, their own ability to block and lift the on-flowing air masses is weak.

The orography has also a significant effect on seasonal distribution of precipitation (Table 2.3). For example, the major part of annual precipitation at Göttinger Wald and Zierenberg (ca. 58% of annual values at Göttinger Wald and 54% at Zierenberg) occurs during the growing season, whereas at Solling most of the precipitation occurs during the non-growing season (53.5%). Mean annual courses of precipitation at the three sites follow generally the patterns of stations in the German Highlands (www.klimadiagramme.de) for the periods 1961–1990 and 1971–2000 (Table 2.4): namely, the two main maxima occur in midsummer and in midwinter and a local maximum in March. The midwinter maximum is rather weakly expressed at both Göttinger Wald and Zierenberg. The midsummer maximum at all three beech sites is shifted to July as against June for the DWD stations. Amount of precipitation at the Solling site during the non-growing season has a clearly expressed pattern of a monotonous increase from October to December and then a decrease until March. This pattern is typical for highland stations like Braunlage and Clausthal-Zellerfeld. Thus, at Solling, a pronounced precipitation peak occurs in December. About 15–20% of the annual precipitation at Solling site falls as snow. A continuous snow cover during winter periods seldom occurs, as the intermediate warm weather events often cause a thaw. The spring rainfall minimum in April, typical of highlands in general, occurs at all three forest sites, and extends to May at Solling. The secondary precipitation maximum in December–January is not clearly expressed at the Göttinger Wald and Zierenberg sites. This maximum is characteristic for all low-altitude stations of the Northern Lowlands and the northern part of the German Highlands and is strongly positively correlated to cyclonic activity of the North Atlantic Oscillation (NAO) (Wibig 1999). Thus, the precipitation courses result from the combined influence of orography and of atmospheric circulation patterns described in Section 2.1. During the summer months of June–August, convective precipitation prevails, with similar contributions at all three forest sites though the orographic effect causes higher absolute values at Solling. Summer monthly precipitation at Göttinger Wald and Zierenberg range between 60 and 80 mm $month^{-1}$ and at Solling between 80 and 100 mm $month^{-1}$.

2.3.4 Variations of Air Temperature and Precipitation During the Observation Period

Annual mean values of air temperature, T_a, and the precipitation rates are shown (Fig. 2.2) for the entire observation period at the Solling site, because continuous measurements were carried out over the entire 33-year period only at this site. Both T_a and precipitation increased slowly from 1969 until 2002. This increasing trend becomes more evident in the patterns of 5-year running means of T_a and precipitation (Fig. 2.2). The patterns of both air temperature and precipitation for the observation period are almost sinusoidal with two clearly marked periods. According to running mean values the air temperature increased by 1.4°C and precipitation increased by 230 mm from the beginning of the 1970s to the end of the 1990s, i.e. the climatic conditions have become warmer and more pluvious. Running mean curves indicate that oscillations of air temperature and precipitation were running in reversed phases. The oscillations were correlated with the NAO index and especially with the winter index (as defined by Hurrell and van Loon 1997), positively for T_a and negatively for precipitation. This result is not unexpected as the positive phases of the NAO have often been associated with positive temperature anomalies and below-average precipitation over southern and central Europe (e.g. Hurrell 1995). The influence of the NAO is more pronounced in wintertime (Mareş et al. 2002). The amplitude of T_a oscillations increased with time toward 2002 and the variations of precipitation (Fig. 2.2) decreased. The minimal variation, i.e. the period with more or less constant, though high (around 1,000 mm year^{-1}), value of annual precipitation was observed during 1981–1994.

Values of T_a and precipitation are averaged over three unequal sub-periods, namely 1969–1980, 1981–1989 and 1990–2002, and the results are presented in Table 2.5. The table also contains the mean values for the entire period of 33 years

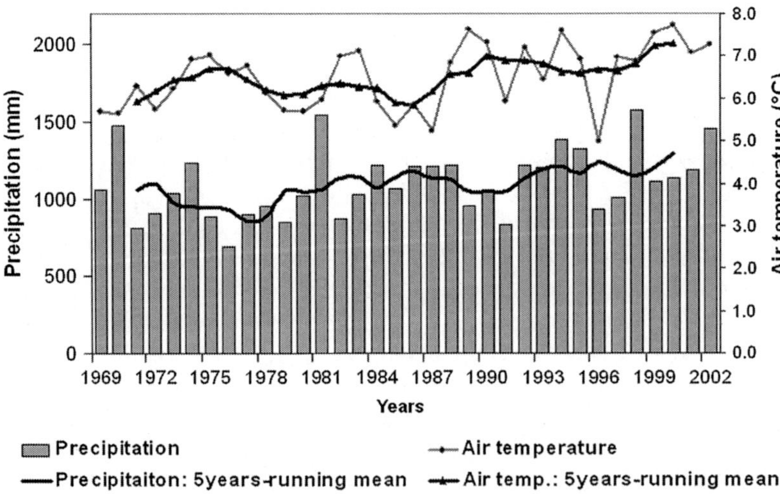

Fig. 2.2 The inter-annual variability of T_a and precipitation are shown for the Solling site. Running means of 5-year periods for air temperature and for precipitation are also shown

Table 2.5 Mean annual air temperatures (°C, 2 m above ground) and annual precipitation rates (mm $year^{-1}$) at the Solling site from 1969–2002 for the entire year, the growing (May–October), and the non-growing season (November–April) (min, max and amplitude values for corresponding periods in are parentheses)

Period	Air temperature			Precipitation		
	Annual	May–October	November–April	Annual	May–October	November–April
1969–2002	6.5 (5.0, 7.7, 2.7)	11.7 (10.2, 12.8, 2.6)	1.2 (−0.8, 3.4, 4.2)	1,095 (6, 72, 1,571, 899)	524 (276, 915, 639)	571 (391, 896, 505)
1969–1980	6.2 (5.7, 7.0, 2.7)	11.4 (10.2, 12.5, 2.3)	1.0 (−0.7, 3.4, 4.1)	951 (672, 1,440, 768)	487 (276, 681, 405)	464 (396, 759, 363)
1981–1989	6.3 (5.2, 7.6, 2.4)	11.7 (10.8, 12.8, 2.0)	0.8 (−0.7, 2.7, 3.4)	1,146 (878, 1,527, 649)	532 (379, 744, 365)	613 (450, 800, 350)
1990–2002	6.9 (5.0, 7.7, 2.7)	12.0 (10.7, 12.8, 2.1)	1.7 (−0.8, 3.0, 3.8)	1,193 (862, 1,571, 709)	553 (370, 915, 545)	641 (391, 896, 505)

from 1969 to 2002, and the seasonal means of growing and non-growing seasons to describe forest growth conditions. The results show that the annual mean air temperature for the first two decades remained quite constant, but increased considerably (by 0.6°C) during the third decade (1990 – 2002). The T_a for the growing season showed a steady increase throughout the entire observation period with a total increment of 0.6°C. However, the mean T_a for the non-growing season decreased slightly (0.2°C) from the 1970s to the 1980s and then dramatically rose by 0.9°C during the 1990s. The amplitudes in mean *annual* T_a (calculated as period's maximal minus period's minimal values of annual temperature) for the sub-periods showed an increase of variability during the observation period. However, the amplitudes for *seasonal* mean T_a showed a weak minimum in the 1980s for both growing and non-growing seasons, i.e. a decrease by 0.3°C since the 1970s and then a slight increase by 0.1°C to the 1990s. It should be noted, however, that during the period of 1981–1989 two major warm ENSO events of 1982 and 1988 considerably altered the pattern of general atmospheric circulation.

The annual precipitation during the observation period increased continuously from 1969–1980 to 1990–2002 with a total increment of 242 mm year^{-1} which was mainly due to an increase in precipitation during the non-growing season, especially during the 1980s (Table 2.3). November–April precipitation increased by 149 mm year^{-1} whereas the May–October values increased by only 45 mm year^{-1}. The increase of annual and seasonal precipitation from the 1980s to the 1990s was notably less: 28 mm year^{-1} for the non-growing season and 21 mm year^{-1} for the growing season.

This general increase in precipitation for the winter months could be explained by changes of the atmospheric circulation pattern over Europe during the period of 1960–1990. Zonal flow (westerlies) was generally strengthened, especially during winter periods. The rise of anti-cyclonic activity, however, has slowed down the precipitation increase during the last decade (Kyselý and Huth 2006).

The period of 1969–1980 was somewhat abnormal for the Solling forest site as the pattern of seasonal distribution of precipitation was reversed – more than half of the annual precipitation (51%) occurred in the growing season. A possible reason could be the decrease in frequency of the 'north' types of atmospheric circulation over Europe in winter (prevailing northern flows) and their increase in summer during the decade (Kyselý and Huth 2006). Table 2.5 also illustrates the fact that the decadal amplitudes (maximum minus minimum annual precipitation for a decade) during the 33-year period were always higher for growing seasons than for non-growing ones. The values point to a reduction of intra-decadal variability during the 1980s whereas the amplitudes increased further in the 1990s.

2.3.5 Soil Temperature

The annual mean values of soil temperature at three depths ($T_{s,5}$, $T_{s,10}$, $T_{s,20}$,) are presented in Table 2.6. Changes in soil temperatures at the investigated sites did not always follow the pattern of air temperatures. The Solling station is the 'coldest' site.

Table 2.6 Mean daily soil temperatures (°C) at 5, 10 and 20 cm depths at the Zirenberg, Göttinger Wald and Solling sites (1998–2002) (min, max and amplitude values in parentheses)

Depth	Soil temperature		
	Göttinger Wald	Zierenberg	Solling
5	7.2 ± 4.33 (−3.9, 17.7, 21.6)	8.4 ± 4.27 (1.3, 17.0, 15.7)	6.0 ± 4.02 (−2.5, 16.2, 18.7)
10	7.0 ± 4.18 (−3.4, 16.9, 20.3)	8.2 ± 4.19 (1.3, 16.4, 15.1)	6.0 ± 3.94 (−2.2, 15.8, 18.0)
20	7.1 ± 3.93 (−2.5, 16.1, 18.6)	8.2 ± 4.15 (1.4, 16.0, 14.6)	6.4 ± 3.75 (−0.5, 13.0, 13.5)

However, the warmest one is not Göttinger Wald but Zierenberg despite its higher altitude and the north-east exposure of the slope. The Göttinger Wald station almost exactly represents the transition state, but also shows the highest T_s variability during the observation period. Temperature amplitudes (differences between maximal and minimal values are given in parentheses in the table) in all three soil layers at the Göttinger Wald site exceeded 18.5°C reaching 21.6°C at 5 cm depth, while at the Zierenberg site all amplitudes were smaller than 16°C. Highest amplitude at the Solling site was 18.7°C in 5 cm depth.

Seasonal variations of temperatures at different soil layers ($T_{s,5}$, $T_{s,10}$, $T_{s,20}$) were not in phase with the variations of T_a. The largest variability and the highest amplitude of the temperature variation are expected at the surface of bare soil and this temporal variability should decrease with soil depth (e.g. Scharringa 1976). In a forest, the variability should be smoother even at the soil surface because the forest canopy reduces the short-wave solar radiation input to and the thermal radiation loss from the surface organic layer as compared to the open field. A general decrease of variability with depth was also illustrated by a decrease in the period of amplitudes presented in Table 2.4. In Fig. 2.3, the seasonal variations of the vertical profiles of the soil temperature at the three sites are shown. As expected, the highest temperatures ($T_{s,5}$, $T_{s,10}$, $T_{s,20}$) at all sites were observed during the summer months, before the cooling period followed in the autumn. The spring mean temperatures at all depths were considerably lower than the autumn ones and followed the same pattern as T_a values.

Generally, with increasing solar and thermal radiation input to the soil surface, the upper soil surface is warmed up first, then the temperatures of the deeper soil layers follow. Therefore, in summer when the soil surface receives an increasing amount of energy from the sun and from the atmosphere, its energy balance becomes positive and soil temperature decreases with depth. During the winter season, when the incoming short-wave solar radiation is low in the temperate climate zone, the soil surface energy balance is usually negative and thus the temperature increases with soil depth. During the transition periods of spring and autumn when the short-wave radiative heating and long-wave cooling are almost balanced, the soil conditions could be considered as isothermal (Fig. 2.3). This ideal temporal pattern of soil temperature profiles holds true at the Göttinger Wald site but not at the Zierenberg and Solling sites. At Zierenberg, the mean seasonal $T_{s,5}$

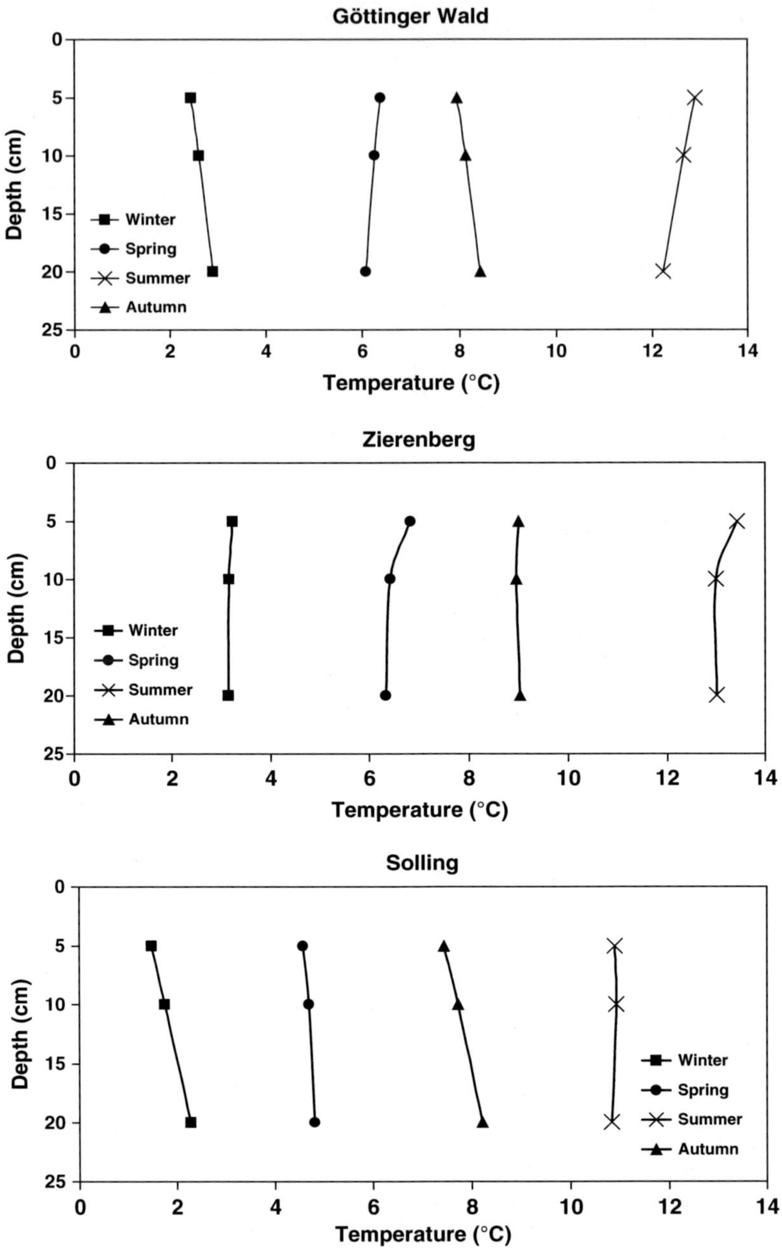

Fig. 2.3 Means for different seasons of vertical profiles of soil temperature at three different beech forest sites are given for the period 1990–2002: Göttinger Wald (*upper panel*), Zierenberg (*middle panel*) and Solling (*lower panel*)

Table 2.7 Vertical gradients of mean seasonal soil temperatures (°C m^{-1}) between the layers of 0.05 m and 0.20 m at sites Zirenberg (ZB), Göttinger Wald (GW) and Solling (SO) (1998–2002)

	Vertical gradients			
Site	Winter	Spring	Summer	Autumn
GW	3.1	2.0	−4.5	3.1
ZB	−0.6	−3.3	−2.9	0.26
SO	5.5	1.6	−0.04	5.2

were always slightly higher (from 0.4°C in summer to 0.04°C in autumn) than in the deeper layers ($T_{s,10}$ and $T_{s,20}$). At the Solling site, in contrast, the upper soil layers were colder than lower ones during all seasons except summer. The higher mean T_s values in the upper soil layers were also observed by Holst et al. (2000) in intensively thinned beech stands growing on north-east slopes, while the unmanaged control stand showed a temperature increase with depth or at least an isothermal vertical profile. This influence of forest structure was confirmed by the fact that the Zierenberg stand had a considerably lower stem density than the Solling and especially Göttinger Wald stands (see Chap. 1).

The absolute values of vertical gradients of soil temperature between $T_{s,5}$ and $T_{s,20}$ varied differently at all stations (Table 2.7). While at the Solling site the highest gradients of approximately the same magnitude were observed during autumn and winter, the maximal values at the Zierenberg site occurred in spring and summer and the lowest values in autumn and winter. At the Göttinger Wald site, the highest gradient was observed in summer, as also at the Zierenberg site, but the lowest in spring. The annual gradient was also the highest at the Solling site as well (2.96°C m^{-1}, directed upward) and the lowest at the Göttinger Wald site (−0.08°C m^{-1}).

2.4 Comparison with Other Climatic Regions

The three investigated beech sites, Göttinger Wald, Solling and Zierenberg, are located in the climatic region of the German Highlands (Central Germany). In order to position the climate conditions of these sites against the general climatic situation in Germany the beech stations are plotted within the Precipitation-T_a space (Fig. 2.4) based on the 30-year mean values of DWD. The figure also shows that the mean air temperature at all three studied beech stations was slightly lower than the average for Germany (8.1 ± 1.3°C). The same holds true for the annual precipitation rates at the Göttinger Wald and Zierenberg sites which were lower than average for Germany (700 ± 237 mm $year^{-1}$). The annual precipitation rate at the Solling site was among the highest in Germany exceeding the mean value by far.

To compare the growth conditions on the experimental sites with not only to the general climate in Germany but specifically to other beech sites growing in different German climatic regions, the air temperature (T_a), soil temperature ($T_{s,10}$) and precipitation values of five beech sites were compiled. These sites were located to the north, southeast, south and southwest of the studied ones. The northernmost station under the beech canopy at the Belauer forest about 30 km south of Kiel represents the Northern Lowlands. The southwest is represented by beech forests near Tuttlingen which is located in Schwabian Alb – a part of the Jura Mountains. To the southeast, the Bavarian Forest site Mitterfels is situated within the southern part of the German Highlands. A southern but not the southernmost site is the beech forest near Freising which is located at a slightly higher latitude than the Tuttlingen forests. All stands with the exception of the Belauer forest are highland forests, i.e. above 400 m a.s.l (Table 2.8). The observation periods varied from station to station which complicated a comparison, but it still provided sufficient information of a descriptive nature. To position these stations against the general climatic situation in Germany and against the three studied sites, the additional beech stations are also plotted within the Precipitation-T_a space at Fig. 2.4.

Annual mean T_a values in beech forests varied strongly in Germany with an amplitude of 2.9°C (Table 2.8). However the differences between forest stations were much lower than the absolute amplitude across Germany which was 8.8°C (Fig. 2.4) or even as high as 15.7°C if the Zugspitze station was included. It should also be noted that generally the mean T_a value averaged over the considered beech forests across Germany is below the 'geographical' mean value of T_a for Germany. Considering the dry potential temperature θ_d[1] to filter out the differences caused by different altitudes one can observe the expected latitudinal effects. θ_d increases from the northernmost station at Belauer to Central Germany and then to three southeast/southwest stations. The southern station at Freising shows an 'intermediate' θ_d value of 13°C though its latitude is within the range of the other southern stations. Thus, the elevation effects are not clearly expressed, although the highest station of Mitterfels has the lowest T_a value. Considering the absolute values of T_a at all beech forest stations (Table 2.8), the three studied beech forest sites are close to the average value, but are somewhat 'colder' than all other stations (even those at higher altitudes) except the one at Mitterfels.

The annual rainfall rates at the selected sites range between 700 mm and 1,600 mm with highest values at Solling and Mitterfels and the lowest at Göttinger Wald and generally exceed the actual evapotranspiration rates.

Pattern of soil temperature values ($T_{s,10}$) does not fully correspond to the pattern of air temperature. The values of $T_{s,10}$ depend upon many factors such as inclination and orientation of slopes, the density of tree canopy which controls the transmission of solar and thermal radiation to and from the soil surface, the temperature conductivity of the litter and soil layers above the temperature sensor, the litterfall rate changing the depth of the sensor with time, and the heat flow from the layers under the temperature sensor. The highest altitude site, Mitterfels, showed $T_{s,10}$

[1]The dry potential temperature of air parcel at pressure P is the temperature that the parcel would acquire if dry adiabatically brought to a standard reference pressure P_0 (e.g. at sea level). It is denoted as θ_d and is often roughly calculated assuming 1°C per 100 m altitude

Table 2.8 Comparison of air temperature (°C, 2 m above the ground, except at Tuttlingen at 1.5 m), dry potential temperature, soil temperature at 10 cm depth (°C), precipitation (mm $year^{-1}$), in beech forests at different sites across Germany (from north to south)

Site name	Coordinates	Altitude	Time period	Air temp.	Dry potential temp.	$T_{s,10}$	Precipitation
Belauer Forst	54°6′N, 10°16′E	50 m	1989–1998	7.3	7.7	7.1	769
Solling	51°46′N, 09°34′E	504 m	1990–2002	6.9	11.8	6.1	1,193
Göttinger Wald	51°32′N, 10°03′E	420 m	1990–2002	7.4	11.6	7.0	709
Zierenberg	51°22′N, 09°16′E	450 m	1990–2002	6.9	11.3	8.2	754
Freising	48°2′N, 11°4′E	508 m	1998–2005	8.1	13.1	8.8	840
Mitterfels (Bayerischer Wald)	48°6′N, 12°4′E	1,025 m	1998–2005	5.2	15.2	6.5	1,598
Tuttlingen (NE)	47°59′N, 8°45′E	800 m	2000–2004	7.5	15.4	7.6	783
Tuttlingen (SW)	47°59′N, 8°45′E	800 m	2000–2004	7.9	15.7	8.6	800
Mean of all stations				7.15 ± 0.89	12.7 ± 2.7	7.5 ± 0.98	931 ± 309.5

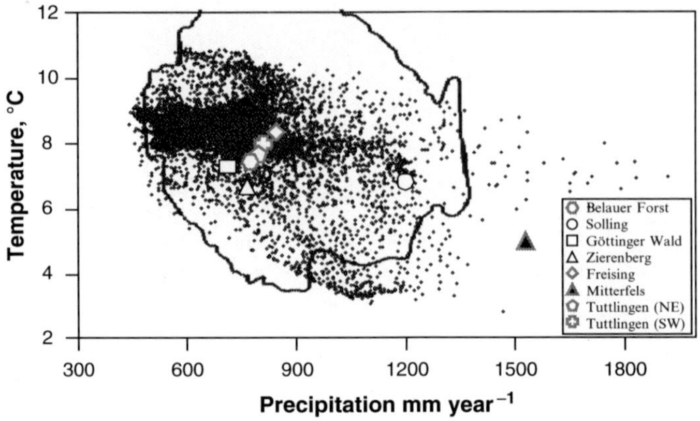

Fig. 2.4 Mean annual T_a (°C) versus mean annual precipitation sums (mm year^{-1}) (data of the German Weather Service, DWD) over Germany. The *geometrical symbols* denote the stations in beech forest as given in Table 2.1. The *solid black line* denotes the 'climate envelope' (Klimahülle) of beech according to Kölling and Zimmermann (2007)

values below the mean but not the lowest $T_{s,10}$ value of all sites. The lowest value was observed at Solling and the highest at Freising, while Tuttlingen (north-east) was very close to the mean value over all stations. The $T_{s,10}$ values for the Göttinger Wald and Zierenberg sites differed strongly from the mean, where the first one was considerably below and the latter one considerably above the mean $T_{s,10}$ value of all the stations.

Figure 2.4 shows that most beech stands in Germany lie within the 'climate envelope' of beech as defined by Kölling and Zimmermann (2007). The considered stations are not exactly but very close to the centre of this climatic envelope or optimum (Mayer 1992), where the competition strength of beech has to be the strongest. The exceptions are the beech sites in Germany, located at high altitudes, such as the highland beech stand at Mittelfels in the Bayerischer Wald (1,025 m), which fall outside the climate envelope of beech. Growth on such sites may be limited due to severe climatic conditions. Still, they are inside the critical limitations for *F. sylvatica* given by Manthey et al. (2007) with the lower limit: 471 mm year^{-1} and the upper limit: 2,000 mm year^{-1}. The Solling, which has the almost optimal air temperature, but lowest mean soil temperature of all the considered beech forests, is still well within the envelope, though the amount of precipitation puts it close to the limits of the envelope. The site of Freising appears to be the most optimal one with both $T_{s,10}$ and T_a above +8°C and with annual precipitation above 840 mm year^{-1}.

2.5 Conclusions

The climatic conditions of three beech sites from the German Highlands, Solling, Zierenberg and Göttinger Wald, are described. It is shown that Göttinger Wald is distinctly the warmest site of the three with a mean annual air temperature of 7.4°C and with the highest absolute amplitude for the period of observation. Zierenberg, however, is the warmest in terms of soil temperature (8.2°C). Solling has the highest annual precipitation (1,193 mm), whereas other forests are drier (around 700 mm), being in the rain shadow of the Solling mountains.

All the three studied stands are to some extent untypical for the German Highlands because the warmest month there is not July but August, which is due to oceanic influence and the effect of vegetation cover. January is the coldest month at these sites.

There is a clear evidence of an increase of air temperature and annual precipitation during the 33-year period of observations at the Solling site. However, the increase of total annual precipitation was mainly due to the increase of precipitation during the non-growing ('winter') season.

Changes in climate will have significant effects on forest ecosystems, as discussed by Gravenhorst (1993). However, the present values of climatic variables at the eight beech forests representing all main climate regions of Germany and including the three study sites in Central Germany are quite close to the optimum for the growth of beech. However, the Solling Forest is close to the limits of the beech climate envelope in terms of precipitation values, and the projected increase of winter precipitation under climate change may move it outside the envelope. The climatic conditions of Göttinger Wald and Zierenberg are sufficiently close to the average values for Germany, and therefore may be considered as representative of present conditions of beech forest growth in Germany.

Acknowledgements The data for Belauer Forest were provided by Dr. G. Hörmann, Ökologie-Zentrum Kiel; for Tuttlingen by Dr. Th. Holst, Institut für Meteorologie, Universität Freiburg; for Mittelfels (Bayerischer Wald) and Freising by Dr. S. Raspe, Bayerische Landesanstalt für Wald und Holzwirtschaft, Freising. We thank our colleagues for their generous and rapid cooperation. We thank Dr. B. Mühr, Uni Karlsruhe, for permission to use and modify the maps of mean temperature and precipitation distributions in Germany. Special thanks to Dr. R. Brumme for his valuable comments on the structure and content of this chapter.

References

Ellenberg H (1996) Vegetation Mitteleuropas mit den Alpen in ökologischer Sicht. Ulmer, Stuttgart, Germany

Gravenhorst G (1993) Klimaänderungen und Waldökosysteme. In: Schellnhuber HJ, Sterr H (eds) Klimaänderung und Küste: Einblick ins Treibhaus. Springer, Berlin, Heidelberg, New York, pp 276–298

Holst T, Rost J, Schindler D, Matzarakis A, Mayer H (2000) Mikroklimatische Untersuchungen in südwestdeutschen Buchenbeständen. Ber Meteor Inst Univ Freiburg 5:123–135

Hurrell JW (1995) Decadal trends in the North Atlantic oscillation regional temperatures and precipitation. Science 269:676–679

Hurrell JW, van Loon H (1997) Decadal variations associated with the North Atlantic oscillation. Clim Change 36:301–326
Kölling C, Zimmermann L (2007) Die Anfälligkeit der Wälder Deutschlands gegenüber Klimawandel. In: Gefahrstoffe – Reinhaltung der Luft 67: 6/2007 pp 259–268
Kreilein H (1987) Energie- und Impulsaustausch in der atmosphährischen Grenzschicht über einem Waldbestand. Diploma thesis, Faculty of Physics, University of Goettingen
Kyselý J, Huth R (2006) Changes in atmospheric circulation over Europe detected by objective and subjective methods. Theor Appl Climatol 85:19–36
Manthey M, Leuschner Ch, Härdtle W (2007) Beech forests and climate change. Natur und Landschaft 82 (9/10):441–445
Mareş C, Mareş I, Mihailescu M (2002) Testing of NAO and ENSO signals in the precipitation field in Europe. In: Beniston M (ed) Climatic change: implications for the hydrological cycle and for water management. Advances in Global Change Research, 10. Kluwer, Dordrecht, pp 113–121
Mayer H (1992) Waldbau auf soziologisch-ökologischer Grundlage. Fischer, Stuttgart, Germany
McKnight TL, Hess D (2000) Climate zones and types: marine West Coast climate (Cfb, Cfc), Physical geography: a landscape appreciation. Prentice Hall, Upper Saddle River, NJ, pp 226–229
Mitchell T., Carter TR, Jones P, Hulme M (2004) A comprehensive set of high-resolution grids of monthly climate for Europe and the globe: the observed record (1901–2000) and 16 scenarios (2001–2100): Tyndall Centre Working Paper 55, pp30
Scharringa M (1976) On the representativeness of soil temperature measurements. Agric Meteorol 16:263–276
Schüepp M, Schirmer H (1977) Climates of Central Europe. In: Landsberg HE (ed) World survey of climatology 6, Climates of Central and Southern Europe. Elsevier, Amsterdam, pp 3–73
Wibig J (1999) Precipitaiton in Europe in relation to circulation patterns at the 500 hPa level. Int J Climatol 19:253–269

Chapter 3
Soil Properties

H. Meesenburg, R. Brumme, C. Jacobsen, K.J. Meiwes, and J. Eichhorn

3.1 Introduction

The soils at the three beech forest sites, Göttinger Wald, Zierenberg, and Solling, are derived from different bedrocks and cover the whole acidity range (pH(H_2O) 3–7) of German soils (Wolff and Riek 1997). The bedrock of the Göttinger Wald soil is limestone. Its carbonate buffer is responsible for the high soil pH(H_2O) between 5.7 and 6.9 in the surface 10 cm of the mineral soil. According to Wolff and Riek (1997), only 9% of the German forest soils have soil pH(H_2O) >6.2 and are buffered by carbonate. Of German forest soils, 59% have a pH(H_2O) <4.2, and 32% a pH(H_2O) between 4.2 and 6.2. A representative soil of the lower pH range is the Solling soil, derived from loess, with a pH(H_2O) of 3.4–3.8 and a very high content of aluminum and protons. Representative of the third group is the soil at the Zierenberg site on basaltic rock which is free of carbonate, has a slightly lower pH(H_2O) than Göttinger Wald soil (5.2–5.7) but is very rich in Mg and P, two of the most important plant nutrients in forest ecosystems. The soil properties of the three sites will be described in detail regarding the parent material, the mineral content of the soils, soil texture, the chemical status, the prevailing systems for buffering of produced and deposited acidity, the nutrient status and the organic layer types of the soils.

3.2 Parent Material, Mineral Composition, Soil Texture

The study sites were not covered by glaciers during the Pleistocene, but were influenced by peri-glacial processes such as cryoturbation, solifluction, erosion and loess deposition. The soils have developed during the last 10,000 years (Holocene) and are thus relatively young.

The minerals of the soil forming Triassic limestone at *Göttinger Wald* are calcite ($CaCO_3$, 940 g kg^{-1}), dolomite (Ca, $MgCO_3$, 2 g kg^{-1}), and siderite ($FeCO_3$, 4 g kg^{-1}) (Thöle and Meyer 1979). The horizontally bedded rock consists of

R. Brumme and P.K. Khanna (eds.), *Functioning and Management of European Beech Ecosystems*, Ecological Studies 208,
DOI: 10.1007/978-3-642-00340-0_3, © Springer-Verlag Berlin Heidelberg 2009

Table 3.1 Soil texture and gravel content of the soils at Göttinger Wald (Brumme 1986), Zierenberg, and Solling (Beese 1986) sites

	Depth	Soil texture[a]			Gravel
	cm	Clay %	Silt %	Sand %	%
Göttinger Wald	0–5	36	61	3	21
	5–10	41	57	2	12
	10–20	39	59	2	23
	20–30	54	45	1	39
	30–60	30	53	17	40
	60–90	23	61	16	57
Zierenberg	0–5	13	79	8	5
	5–10	12	78	10	15
	10–20	14	74	12	20
	20–40	14	76	10	25
	40–80	9	77	14	30
Solling	0–10	17	57	26	15
	10–20	18	55	27	9
	20–30	17	55	28	9
	30–40	19	51	30	10
	40–50	16	53	31	10

[a]As fraction of fine earth, clay (<2 μm), silt (2–60 μm), sand (60 μm–2 mm), gravel (>2 mm) (British Standards Institution)

solid plates of a few centimetres thickness, which are separated by thin silty layers of a few millimetres thickness, and are vertically broken by endogenic processes below 100 cm depth. In the upper 100 cm, the nearly horizontal stratification of the limestone is completely disturbed by cryoturbation. Dissolved limestone leached out as $Ca(HCO_3)_2$. Only 5% of the limestone remained as non-calcareous residue and formed the soil at Göttinger Wald. The predominant particle size of the non-calcareous residue (analyzed after dissolving the rock with HCl) consists of clay (58%), silt (41%), and of quartz (310 g kg^{-1}), illite (200 g kg^{-1}), montmorillonite (150 g kg^{-1}), mica (140 g kg^{-1}), feldspar (100 g kg^{-1}), chlorite (50 g kg^{-1}), and vermiculite (40 g kg^{-1}) (Thöle and Meyer 1979).

The skeleton content decreases from the subsoil upward, the clay content of the soil reaches its maximum between 20 and 30 cm (Table 3.1). The decrease of gravel content in the surface soil (<20 cm) indicates an admixture of loess which is reflected also in the higher content of quartz in the fine soil (Table 3.2). Within the upper 20 cm, the calcite content is negligible, but calcite is the most abundant mineral below 20 cm depth.

The parent material at *Zierenberg* is tertiary basaltic debris overlying Triassic limestone with a thickness of the basalt decreasing downslope from 100 cm to around 10 cm (Jochheim 1992). The main research area is located on the upper midslope. Its soil is moderately gravelly and dominated by silt (70–80%) which is partly derived from basalt weathering but also from loess deposits. This is indicated by the high quartz content but of the illite/muscovite components (Table 3.3). Of basaltic origin are the Mg-bearing minerals pyroxene, hornblende, olivine and

Table 3.2 Mineral composition of the soils (g kg^{-1}) (fine earth <2 mm) at different depths (cm) at the Göttinger Wald site

	0–5	5–10	10–20	20–30	30–40
Quartz	300	300	200	220	120
K-Feldspar	20	20	10	30	30
Albite	20	20	10	10	10
Calcite	0	10	20	390	500
Fe Minerals	10	10	10	10	10
Anatas	4	4	4	4	4
Kaolinite	71	64	67	20	13
Mg Chlorite	0	0	7	4	0
Illite	260	230	200	140	230
Vermiculite	170	170	230	140	78
Al Vermiculite	140	170	240	30	7

Table 3.3 Mineral composition of the soils (g kg^{-1}) (fine earth <2 mm) at different depths (cm) at the Zierenberg site (Butz-Braun, personal communication)

	0–5	5–10	10–20	20–40	40–80
Quartz	260	260	270	270	200
Feldspar	90	90	130	100	100
Hornblende	10	10	20	20	20
Pyroxene	100	100	100	80	80
Olivine	10	20	40	20	0
Kaolinite	60	40	50	60	30
Chlorite	0	10	10	20	0
Vermiculite	0	50	90	40	40
Illite/Muscovite	300	340	190	260	170
Smectite	170	80	140	130	360

smectite. The high content of Mg-rich minerals marks an important difference to the soil at the Göttinger Wald site. Moreover, basalt is known to contain between 7 and 30 g kg^{-1} of the phosphorus mineral apatite (Wedepohl 1968).

The bedrock at *Solling* is Triassic sandstone (>140 cm depth) which is covered with loess. The gravel content of the loess is low in the upper 60 cm (Table 3.1) and steadily increases between 60 and 140 cm depth indicating an admixture with weathered sandstone fragments by solifluction (Benecke 1984). The predominant minerals of the Triassic sandstone are quartz (450–680 g kg^{-1}), illite and mica (100–250 g kg^{-1}) and K-feldspar (80–200 g kg^{-1}) (Deutschmann 1994). The sandstone layer below the solifluction layer has higher contents of illite and mica (660–740 g kg^{-1}) and a lower content of quartz (170–260 g kg^{-1}). Therefore, the mineral composition of the soil shows an increasing illite content with depth from about 60–90 g kg^{-1} in the surface 30 cm to about 284 g kg^{-1} between 150 and 200 cm depth (Table 3.4). The quartz content decreases in the same direction from about 740 to 470 g kg^{-1}.

Table 3.4 Mineral composition of the soils (g kg^{-1}) (fine earth <2 mm) at different depths (cm) at the Solling site

	0–5	5–10	10–20	20–30	40–50	80–100	150–200
Quartz	740	720	700	690	680	640	470
K-Feldspar	80	120	130	130	120	120	160
Albite	20	20	15	15	15	0	0
Fe Minerals	13	15	17	17	17	17	30
Anatas	5	5	5	5	5	5	5
Kaolinite	33	26	27	25	34	65	34
Mg Chlorite	0	0	1	1	2	0	0
Mixed layer Chlorite/Vermiculite	0	0	0	0	0	0	10
Illite	81	62	69	89	109	147	284
Mixed layer Illite/Vermiculite	5	6	8	4	5	0	0
Vermiculite	6	10	11	3	0	0	0
Al Vermiculite	0	5	6	21	13	03	0
Vermiculite de-Al	17	11	11	0	0	0	0
Smectite	0	0	0	0	0	3	7

3.3 Soil Types, Soil pH, and Buffer Systems of Soils

The study sites differ with respect to the buffer systems indicated by the pH of the soils. The pH increases with depth and shows marked differences among the sites (Fig. 3.1). At the *Göttinger Wald* site, the relatively high pH of the soil is stabilized by a high buffer capacity and a high buffer rate (carbonate buffer range, pH(H_2O) >6.2, Ulrich 1981, 1987), except of the carbonate free parts of the topsoil and at some places at deeper soil depths where admixture with loess material occurs. About 26% of the soils at the Göttinger Wald research site are shallow (Ah–Cv soils, soil depth <20 cm) and are classified as *Rendzic Leptosols* according to the FAO/UNESCO classification, or as *Rendzina* in the German classification (Meiwes and Beese 1988). They are still rich in carbonate in the Ah-horizon and have high soil pH($CaCl_2$) values (6.1–7.0). The dominant soils at the Göttinger Wald research site are up to or >30 cm deep and have a Ah–Bv–Cv profile (*Calcaric Cambisol*, FAO/UNESCO classification, or *Terra fusca – Rendzina*, German classification). Occasional pockets of soils with a thicker loess deposit occur. The *Calcaric Cambisols* are often free of carbonate in the topsoil so that the pH($CaCl_2$) may drop down to 4.7. The prevailing buffer material in carbonate free soil layers are silicates (silicate buffer range, pH(H_2O) 5–6.2). The dominant soil type *Terra fusca–Rendzina* has a pH(H_2O) between 5.5 (min) and 6.9 (max) in the 0–10 cm depth depending on the content of carbonate (Table 3.5). This soil type is located between the carbonate and silicate buffer range. With increasing depth, soil pH(H_2O) may increase up to 8 indicating that the deeper soil is buffered by carbonate.

The soil at *Zierenberg* is carbonate free except in the 80–100 cm depth where some admixture from the underlying limestone was found (Jochheim 1992). Soil pH(H_2O) increased from 5.3 in the 0–20 cm depth to 6 in the 40–80 cm depth

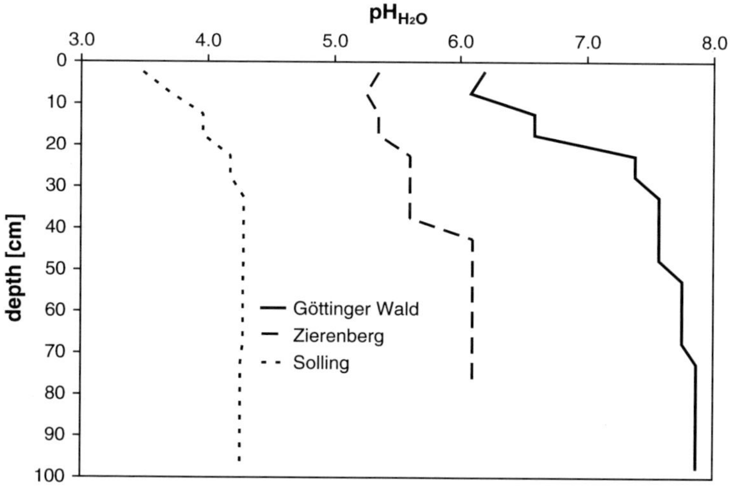

Fig 3.1 Depth functions of pH (H_2O) for Göttinger Wald, Zierenberg, and Solling

(Table 3.5). These values are thus lower than those found in the Göttinger Wald soil, and the soils are primarily assigned to the silicate buffer range. Because of high acidity in stemflow fluxes (Chap. 15, this volume), the soils in the vicinity of stems have been more acidified than those in the inter-stem areas. At the Zierenberg site, the soil pH(H_2O) is 1.2 units lower near the stems (20 cm distance) than within the stems in the 0–10 cm depth, and by 0.4 units in the 40–60 cm depth (Jochheim 1991). The soil type is classified as *Eutric Cambisol* (FAO/UNESCO classification) (*Braunerde*, *Ranker-Braunerde*, German classification).

The mineral soil at the *Solling* site is free of carbonates and has low pH (pH(H_2O) which slightly increased with depth from 3.4 to 4.3) (Table 3.5). The low pH is primarily related to low amounts of fast weathering minerals. The relatively high silicate content consists mainly of resistant phyllosilicates (mica, illite) and K-feldspar (Table 3.4). A low buffer rate and high acid load has resulted in a soil pH of the aluminum buffer range within the top 30 cm (Al-buffer range, pH (H_2O) < 4.2). Slightly higher soil pH(H_2O) exists below 30 cm depth (exchange buffer range, pH(H_2O) 5–4.2). Values below pH($CaCl_2$) 4.0 were found down to a depth of 6 m (Meiwes et al. 1994). The soil is classified as *Dystric Cambisol* (FAO/UNESCO classification) and *podsolige Braunerde* (German classification).

3.4 Cation Exchange Capacity, and Exchangeable Cations

The composition of a cation exchange complex is an expression of the acid–base status of a soil. The three study sites show big differences in their cation exchange composition (Table 3.6). At the *Göttinger Wald* site, more than 95% of the

Table 3.5 Soil chemical characteristics (mean and standard deviation (in brackets)) except for pH (min and max values are given) of the soils at Göttinger Wald (soil inventory in 1995), Zierenberg (soil inventory in 1996), and Solling (soil inventory in 1993), (bulk density, BD) (Göttinger Wald and Solling: $n = 6$, each is a mixture of four samples) (Zierenberg: $n = 4$)

	Depth cm	BD g cm^{-3}	pH_{H_2O}	pH_{KCl}	$CaCO_3$ g kg^{-1}	C_{org} g kg^{-1}	N g kg^{-1}	P µg g^{-1}	C/N g g^{-1}	C/P g g^{-1}
Göttinger Wald	L/F	–	5.52 (5.33–5.73)	5.16 (4.94–5.44)	0	445 (14)	13 (1)	527 (54)	34.5 (2.3)	854 (13.8)
	0–5	0.94	6.19 (5.66–6.76)	5.57 (4.99–6.22)	23.9 (44.5)	81.7 (8.7)	6.1 (0.6)	696 (110)	13.3 (0.3)	118 (9.3)
	5–10	0.96	6.08 (5.47–6.91)	5.19 (4.49–6.48)	5.4 (2.5)	56.9 (10.0)	4.9 (0.8)	637 (126)	11.7 (0.5)	89.7 (6.4)
	10–20	1.06	6.59 (6.12–7.35)	5.86 (5.27–6.96)	85.3 (88.4)	42.9 (8.2)	3.7 (0.7)	554 (95)	11.6 (0.4)	77.4 (6.6)
	20–30	1.27	7.38 (7.13–7.59)	6.77 (6.38–7.22)	127 (77.1)	29.6 (4.7)	2.6 (0.3)	462 (45)	11.4 (0.4)	63.8 (5.2)
	30–40	1.33	7.57 (7.31–7.72)	7.07 (6.75–7.59)	255 (98.4)	17.4 (3.0)	1.6 (0.4)	340 (66)	11.0 (1.7)	52.5 (14.4)
	40–50	1.33	7.75 (7.58–7.86)	7.11 (6.79–7.65)	283 (98.4)	16.7 (3.0)	1.4 (0.4)	349 (66)	12.1 (1.7)	46.4 (14.4)
	50–70	1.84	7.86 (7.69–7.97)	7.29 (6.97–7.66)	343 (90)	9.9 (1.6)	0.7 (0.3)	269 (72)	17.3 (7.0)	37.5 (6.1)
	70–80	1.82	7.81 (7.78–7.84)	7.31 (7.25–7.38)	186 (10.7)	11.3 (1.0)	1.1 (0.1)	439 (23)	10.8 (0.6)	25.8 (3.6)
Zierenberg	F	–	5,40 (4,60–5,80)	4,95 (4,10–5,30)	0	397 (16)	20.0 (1.3)	1145 (161)	20.0 (1.0)	351 (44)
	0–5	1.10	5.43 (5.3–5.7)	4.8 (4.7–5.0)	0	48.0 (3.2)	4.1 (0.4)	1218 (185)	11.9 (1.0)	40.1 (6.1)
	5–10	1.20	5.28 (5.2–5.4)	4.58 (4.5–4.7)	0	35.8 (3.3)	3.1 (0.4)	1178 (192)	11.7 (0.7)	30.8 (4.0)
	10–20	1.30	5.38 (5.3–5.5)	4.68 (4.6–4.8)	0	27.0 (1.4)	2.5 (0.2)	1156 (206)	11.1 (0.8)	23.9 (4.5)
	20–40	1.40	5.63 (5.5–5.8)	4.83 (4.7–5.0)	0	15.3 (3.4)	1.4 (0.3)	1100 (194)	10.9 (1.1)	14.3 (4.6)
	40–80	1.45	6.1 (6.0–6.2)	5.3 (5.2–5.4)	0	5.3 (1.3)	0.6 (0.1)	1275 (288)	8.7 (0.9)	4.3 (1.5)
Solling	L	–	4.75 (4.44–4.90)	3.91 (3.57–4.05)	0	494 (13)	25 (1)	1238 (76)	19.6 (0.6)	400 (22.4)
	F	–	3.85 (3.69–3.98)	2.87 (2.83–2.92)	0	444 (12)	23 (1)	979 (30)	19.5 (0.4)	454 (20.4)
	H	–	3.48 (3.35–3.62)	2.76 (2.68–2.88)	0	285 (65)	15 (3)	964 (98)	18.8 (0.8)	298 (67.8)
	0–5	1.04	3.49 (3.44–3.55)	2.84 (2.76–2.89)	0	61.5 (19.4)	3.2 (0.9)	505 (19)	18.8 (1.8)	120.9 (34.5)
	5–10	1.23	3.70 (3.60–3.76)	3.12 (2.99–3.28)	0	36.0 (15.0)	1.9 (0.6)	475 (16)	18.7 (2.5)	80.6 (35.8)
	10–20	1.19	3.96 (3.90–4.04)	3.46 (3.32–3.62)	0	24.8 (3.7)	1.5 (0.2)	370 (25)	17.0 (1.2)	67.0 (8.5)
	20–30	1.30	4.18 (4.14–4.22)	3.80 (3.68–3.92)	0	17.5 (1.6)	1.2 (0.1)	345 (14)	15.2 (0.8)	50.8 (3.1)
	30–50	1.48	4.28 (4.24–4.48)	4.03 (3.94–4.13)	0	10.8 (0.5)	0.8 (0.1)	299 (26)	12.9 (0.5)	36.2 (3.0)
	50–70	1.55	4.28 (4.23–4.42)	3.98 (3.93–4.07)	0	4.0 (1.8)	0.4 (0.1)	234 (32)	11.0 (4.6)	16.9 (7.0)
	70–100	1.55	4.26 (4.17–4.37)	3.83 (3.77–3.90)	0	2.2 (0.8)	0.3 (0.0)	253 (52)	7.0 (2.6)	9.1 (4.1)

Table 3.6 Exchangeable cations, cation exchange capacity (CEC, NH_4Cl extraction for Solling and Zierenberg, $BaCl_2$ extraction for Göttinger Wald), and base saturation (BS) (mean and standard deviation (in brackets)) of the soils at Göttinger Wald (soil inventory 1995), Zierenberg (mineral soil inventory in 1996) and Solling (soil inventory 1993) (Göttinger Wald and Solling: $n = 6$, each is a mixture of four samples) (Zierenberg: $n = 4$)

	Depth cm	CEC $mmol_c\ kg^{-1}$	Ca %	Mg %	K %	Na %	Mn %	Al %	Fe %	H %	BS %
Göttinger Wald	0–5	480 (114)	95 (1.2)	3.1 (0.7)	1.6 (0.5)	0.2 (0.1)	–	–	–	–	–
	5–10	435 (142)	96 (1.2)	2.4 (0.6)	1.4 (0.5)	0.3 (0.1)	–	–	–	–	–
	10–20	439 (113)	97 (0.6)	1.8 (0.3)	1.1 (0.2)	0.3 (0.1)	–	–	–	–	–
	20–30	406 (38.6)	97 (0.3)	1.3 (0.2)	1.0 (0.2)	0.2 (0.1)	–	–	–	–	–
	30–40	310 (68.8)	97 (0.2)	1.2 (0.2)	1.2 (0.1)	0.2 (0.1)	–	–	–	–	–
	40–50	295 (54.3)	97 (0.2)	1.2 (0.2)	1.3 (0.2)	0.2 (0.0)	–	–	–	–	–
	50–70	130 (142)	97 (0.4)	1.1 (0.3)	1.6 (0.2)	0.3 (0.1)	–	–	–	–	–
	70–80	334 (27.9)	97 (0.1)	1.2 (0.2)	1.4 (0.0)	0.3 (0.0)	–	–	–	–	–
Zierenberg	0–5	154 (10)	64 (4)	29 (4)	3.9 (0.6)	0.2 (0.1)	1.6 (0.4)	1.2 (0.3)	0.1 (0.0)	0	97 (1)
	5–10	124 (18)	60 (4)	30 (6)	4.4 (0.7)	0.2 (0.1)	1.4 (0.3)	4.7 (2.7)	0.2 (0.0)	0.02 (0.05)	94 (3)
	10–20	111 (8)	58 (4)	31 (6)	3.7 (2.3)	0.3 (0.1)	1.2 (0.3)	5.2 (1.8)	0.2 (0.0)	0	93 (2)
	20–40	104 (18)	60 (3)	35 (5)	2.1 (1.9)	0.4 (0.1)	0.6 (0.2)	1.7 (1.0)	0.2 (0.0)	0	98 (1)
	40–80	146 (12)	59 (6)	38 (6)	1.0 (0.3)	0.6 (0.1)	0.1 (0.1)	0.3 (0.0)	0.2 (0.0)	0	99 (0)
Solling	0–5	120 (33.0)	3.4 (0.8)	1.4 (0.2)	1.5 (0.1)	0.5 (0.2)	0.4 (0.1)	64.5 (3.5)	8.7 (0.8)	19.7 (4.2)	6.8 (1.1)
	5–10	110 (20.2)	1.8 (0.3)	0.9 (0.1)	1.1 (0.2)	0.5 (0.1)	1.5 (0.6)	79.2 (3.5)	4.7 (1.5)	10.4 (2.3)	4.2 (0.4)
	10–20	86.7 (11.9)	1.5 (0.1)	1.0 (0.4)	1.2 (0.1)	0.7 (0.3)	2.1 (0.4)	89.4 (2.1)	1.3 (0.8)	2.9 (1.9)	4.3 (0.8)
	20–30	60.6 (11.6)	2.4 (3.0)	1.1 (0.9)	1.3 (0.1)	0.8 (0.5)	1.3 (0.6)	83.8 (25.7)	1.7 (3.4)	7.6 (18.4)	5.6 (4.4)
	30–50	45.7 (5.1)	1.3 (0.3)	0.9 (0.3)	1.7 (0.2)	0.7 (0.1)	1.2 (0.2)	94.0 (0.6)	0.1 (0.1)	0.0 (0.0)	4.7 (0.7)
	50–70	49.7 (4.2)	1.4 (0.5)	0.8 (0.2)	2.1 (0.1)	0.8 (0.4)	1.0 (0.3)	93.9 (0.9)	0.1 (0.1)	0.0 (0.0)	5.1 (0.6)
	70–100	56.0 (15.2)	0.9 (0.2)	1.0 (0.5)	2.4 (0.2)	0.6 (0.2)	0.9 (0.5)	94.2 (0.7)	0.1 (0.0)	0.0 (0.0)	4.8 (0.4)

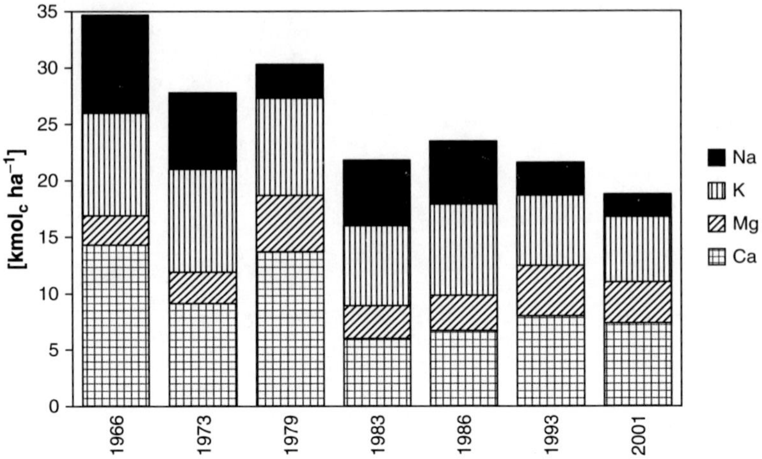

Fig. 3.2 Pools of exchangeable base cations in the mineral soil (0–50 cm) at the Solling site between 1966 and 2001 (for the calculation, an average bulk density was used which did not change significantly over the period) (Meesenburg et al. 1999)

exchange sites of the *Terra fusca – Rendzina* are occupied by Ca. Base saturation is mostly 100% for the whole soil profile. However, where loess has accumulated, pH values may be lower and exchangeable Al may appear in low concentrations in the 5–10 cm depth indicating the occurrence of silicate weathering. The cation exchange capacity (CEC) is about 3–4 times higher at this site than at the Solling and Zierenberg sites. At the *Zierenberg* site, the composition of the cation exchange complex is more balanced. About 60% of the exchange sites are occupied by Ca and 29–38% by Mg. Base saturation is less than 100% due to the presence of acid cations. At *Solling*, the base saturation is <7% in the upper 100 cm soil. Al is the dominating ion at the exchange sites with proportions up to 94%. Fe and H^+ account for 9–20% of exchange sites in the top soil, respectively, indicating a very strong acidification. Analysis of a soil core down to 30 m depth revealed a base saturation below 25% within the upper 6 m depth and 25–80% between 6 and 8 m depth (Meiwes et al. 1994). Below 8 m depth, base saturation is above 80%. Soil inventories conducted 7 times between 1966 and 2001 indicated that base cation pools in the Solling soil was decreasing during this period (Fig. 3.2). Nevertheless, a trend towards some re-translocation of base cations from the mineral soil into the organic layer was apparent at the Solling site, as was indicated by an increase in thickness of the organic layer during the last 30 years (Meiwes et al. Chap. 4, this volume).

3.5 Nutrient Status of Soils and Organic Layer Types

Soils are sources of nutrients for the plants as a result of processes such as mineral weathering, cation exchange and mineralization of organic matter. Exchangeable nutrients are highly available and Table 3.7 show large differences between the

Table 3.7 Total contents of exchangeable nutrients (kg ha^{-1}) in the mineral soil (0–100 cm depth) of Göttinger Wald, Zierenberg, and Solling sites

Site	K_{ex}	Ca_{ex}	Mg_{ex}
Göttinger Wald	1,100	49,400	430
Zierenberg	1,300	16,400	7,400
Solling	520	230	83

sites. Exchangeable K content decreases in the order Zierenberg, Göttinger Wald, Solling soils. Exchangeable Mg follows the same order among sites but the soils had much higher differences. Exchangeable Ca is about 3 times higher at Göttinger Wald than at Zierenberg, whereas it is very low at Solling. This indicates that the nutrient status at Zierenberg site is well balanced with respect to cations when compared to the Göttinger Wald site where the very high values of exchangeable Ca may result in an insufficient uptake of K and Mg by plants. The Solling site is very poor in exchangeable Ca and Mg which may induce deficiency of these elements in plants.

The organic layer at the base-rich Göttinger Wald and Zierenberg sites consists of unaltered dead remains of plants and animals (L or Oi layer) and a layer with fragmented, partly decomposed but still recognizable material underneath the L layer (F or Oe layer). The L/F mull humus indicates a rapid decomposition and disappearance of plant residues from the soil surface by earthworm activity and has resulted in relatively low nutrient pools in the organic layer (Table 3.8).

The organic layer of the acid Solling soil consists of L, F, and H layers and is classified as moder humus. The H layer (or Oa layer) consists of well-decomposed, amorphous organic matter with a thickness of 1.5 cm. The moder humus accumulate considerable amounts of nutrients of which Ca and Mg equal to about 90% and 100%, respectively, of the exchangeable pools within the upper 100 cm depth of the mineral soil. Heavy metal and Al content typically increased from L to F and then to H layers (Table 3.9) indicating increased enrichment of these elements during organic matter decomposition.

Mineral weathering of nutrients such as K, Ca, Mg, or P is a slow process and less important for plant nutrition in the short-term. In the long-term, it is the most important source for nutrients in terrestrial ecosystems except that of nitrogen. The juvenile soils at the study sites are relatively rich in primary silicates (see Sect. 3.2) which ensures a continuous release of base cations depending on mineral composition and content.

3.6 Additional Study Plots at the Solling Site

At the Solling site, additional plots were installed in 1982 adjacent to the permanent observation plot to study the effects of various treatments involving application of high doses of alkaline (liming, BK plot) or acidifying (ammonium sulphate, BN

Table 3.8 Thickness and contents of nutrients (mean and standard deviation (in brackets)) of the organic layer at Göttinger Wald, Zierenberg, and Solling (P, S, K, Ca, and Mg were estimated after HNO_3 pressure digestion (aqua regia extraction at Zierenberg) by ICP, N by combustion technique) (Göttinger Wald, soil inventory 1995; Zierenberg, soil inventory 1996; Solling, soil inventory 1993)

		Thickness	Amount	C	N	P	S	K	Na	Ca	Mg
		(cm)	Mg ha^{-1}	Mg ha^{-1}				kg ha^{-1}			
Göttinger Wald	L/F	1.3	17.7	7.8	200	9.5	19.4	71.2	2.6	369	35.6
			(9.0)	(3.7)	(100)	(5.3)	(9.6)	(54.6)	(1.8)	(199)	(24.9)
Zierenberg	F	–	11.2	4.4	223	12.8	–	15.1	–	161	31.7
			(–)	(0.2)	(15)	(1.8)		(0.8)		(72)	(16.2)
Solling	L	2.1	17.8	8.8	450	22.0	38.8	28.3	2.6	92.6	13.3
			(2.2)	(1.2)	(60)	(2.7)	(4.9)	(3.7)	(0.4)	(15.2)	(1.9)
	F	3.3	36.0	15.9	820	35.1	91.4	75.1	5.9	81.2	25.4
			(9.0)	(3.8)	(200)	(8.4)	(21.5)	(26.3)	(1.7)	(23.5)	(9.0)
	H	1.5	39.0	10.9	580	37.3	73.2	184	10.2	36.3	49.9
			(8.2)	(2.3)	(130)	(7.4)	(13.5)	(84.1)	(3.4)	(7.6)	(20.3)

Table 3.9 Element concentration of the organic layer at Solling, Göttinger Wald, and Zierenberg (Jochheim 1992) (HNO_3 pressure digestion at Solling and Göttinger Wald, aqua regia extraction at Zierenberg)

		Al	Fe	Mn	Zn	Cu	Cr	Ni	Cd	Pb
						mg kg^{-1}				
Göttinger Wald	L/F	12,523	5,893	481	69	9.4	14.3	7.9	0.30	28
		(2,967)	(1,594)	(70)	(9)	(0.9)	(3.7)	(1.6)	(0.09)	(6)
Zierenberg	F	1,880	3,200	1,030	86	15	–	0.65	25	
		(460)	(1,250)	(1,220)	(19)	(1.5)			(0.23)	(11)
Solling	L	2,051	1,959	1,868	95	19.1	–	–	0.63	66
		(799)	(636)	(385)	(6)	(2.9)			(0.05)	(20)
	F	7,581	6,096	371	115	31.3	–	–	0.62	284
		(814)	(600)	(37)	(8)	(3.8)			(0.14)	(41)
	H	19,356	12,107	148	87	30.8	–	–	0.28	367
		(5,667)	(2,465)	(45)	(14)	(6.3)			(0.08)	(54)

plot) substances. Parent material and physical soil characteristics are similar at the treated and untreated plots. Also the structure of the forest stands and silvicultural treatment are comparable with the main Solling plot since all plots belong to the same forest management unit. Tree cuttings were performed only for safety reasons, and the timber was retained on the plot.

At the BK plot, dolomitic lime was applied in 1982 with a dose of 30 Mg ha^{-1} (Beese and Stickan 1989). The BN plot was treated for 11 years between 1983 and 1993 with a dose of 1,000 mmol m^{-2} per year $(NH_4)_2SO_4$ (140 kg ha^{-1} per year N, 160 kg ha^{-1} per year S, Meesenburg et al. 2004).

Chemical characteristics of the soil solid phase at the BN plot showed only small differences compared to the untreated plot when measured 10 years after the start of the treatment, whereas the liming of the BK plot altered soil characteristics especially of the topsoil (Tables 3.10 and 3.11). Distributions of organic C within the soil profile as well as C pools are similar at the treated and untreated plots. N contents are enhanced over the whole profile at the BN plot resulting in an N pool (organic layer and mineral soil 0–100 cm) of 13.3 Mg ha^{-1}, which is 1.5 Mg ha^{-1} higher than the N pool of the untreated plot. 1.4 Mg ha^{-1} N was applied additionally at the BN plot between 1983 and 1992, but N output with soil solution was higher by 0.5 Mg ha^{-1} N during this period than from the untreated plot (Meesenburg et al. 2004). N pools at the limed plot (BK) are not different from the untreated plot.

The acid–base status of the soils at the BK plot has been markedly altered by liming. Contents of exchangeable Ca and Mg have increased manifold in the 0–20 cm soil, but also their concentrations in the subsoil have been raised significantly. Nevertheless, base saturation was above 10% in the 0–20 cm depth at the limed plot (Table 3.11). At the limed plot, pH($CaCl_2$) was higher than at the untreated plot in the soil to 100 cm depth. At the BN plot, base saturation is lower than in the untreated plot below 5 cm depth due to the application of acidity. A decrease of pH was not observed at the BN plot.

Table 3.10 Soil chemical characteristics (mean and standard deviation (in brackets)) except for pH (min and max values are given) of the limed (*BK*) and fertilized plot (*BN*) at the Solling site (soil inventory in 1993), (bulk density, *BD*) ($n = 5$, each is a mixture of four samples)

	Depth cm	BD $g\ cm^{-3}$	$pHCaCl_2$	Corg $g\ kg^{-1}$	N $g\ kg^{-1}$	P $\mu g\ g^{-1}$	C/N	C/P
BK	L	–	4.95 (4.84–5.08)	469 (20.5)	23.7 (0.9)	1256 (91)	19.8 (0.5)	375 (19.3)
	F	–	6.02 (5.65–6.21)	319 (20.6)	17.3 (1.0)	856 (45)	18.5 (0.5)	373 (18.1)
	H	–	5.14 (4.83–5.37)	339 (19.1)	19.1 (1.8)	1038 (39)	17.8 (0.6)	327 (32.6)
	0–2	1.05	4.61 (4.11–5.25)	82.2 (12.6)	4.5 (0.4)	587 (13)	18.1 (1.2)	140 (18.7)
	2–5	1.04	4.20 (4.09–4.37)	39.6 (06.4)	2.2 (0.3)	472 (20)	17.8 (0.6)	83.7 (10.1)
	5–10	1.23	3.95 (3.89–4.07)	27.0 (2.4)	1.5 (0.1)	409 (22)	17.5 (0.9)	66.1 (4.8)
	10–20	1.19	4.06 (3.96–4.36)	23.4 (3.4)	1.4 (0.2)	389 (23)	16.6 (0.8)	60.0 (6.8)
	20–30	1.30	4.14 (4.08–4.32)	16.0 (0.7)	1.1 (0.1)	362 (19)	14.9 (0.8)	44.3 (1.7)
	30–50	1.48	4.20 (4.16–4.24)	9.1 (0.5)	0.7 (0.1)	321 (35)	13.0 (1.3)	28.6 (1.7)
	50–70	1.55	4.14 (4.06–4.22)	3.6 (1.0)	0.3 (0.0)	268(21)	11.5 (3.2)	13.3 (3.3)
	70–100	1.55	4.07 (4.03–4.12)	1.9 (0.0)	0.3 (0.0)	282 (9)	6.0 (0.0)	6.7 (0.2)
BN	L	–	3.21 (2.75–3.74)	508 (20.5)	25.4 (1.0)	1043 (68)	20.0 (0.1)	488 (29.8)
	F	–	3.22 (2.84–3.75)	468 (25.5)	23.4 (1.5)	865 (62)	20.0 (0.6)	544 (45.3)
	H	–	2.82 (2.77–2.89)	352 (35.7)	19.2 (2.0)	1183 (76)	18.4 (0.3)	297 (24.9)
	0–2	1.05	3.21 (3.17–3.31)	101 (13.1)	6.1 (0.8)	708 (48)	16.5 (0.5)	142 (10.4)
	2–5	1.04	3.31 (3.25–3.44)	42.2 (8.3)	2.8 (0.5)	503 (48)	15.0 (1.5)	83.4 (11.6)
	5–10	1.23	3.45 (3.36–3.54)	28.5 (4.7)	2.0 (0.3)	410 (37)	14.6 (1.4)	69.4 (7.3)
	10–20	1.19	3.69 (3.61–3.78)	23.2 (3.2)	1.8 (0.2)	402 (32)	13.2 (1.3)	57.5 (5.8)
	20–30	1.30	3.94 (3.86–4.06)	15.3 (2.0)	1.4 (0.2)	371 (28)	11.3 (0.5)	41.4 (4.3)
	30–50	1.48	4.07 (3.99–4.21)	9.2 (2.0)	0.9 (0.1)	315 (30)	9.7 (1.1)	29.2 (5.3)
	50–70	1.55	4.02 (3.90–4.16)	7.1 (0.8)	0.8 (0.1)	324 (64)	9.2 (1.0)	22.6 (5.5)
	70–100	1.55	4.00 (3.94–4.06)	2.7 (1.8)	0.4 (0.2)	321 (55)	6.5 (1.1)	8.5 (5.5)

Table 3.11 Exchangeable cations, cation exchange capacity (*CEC*, NH_4Cl extraction), and base saturation (*BS*) (mean and standard deviation (in brackets)) of the limed (*BK*) and fertilized (BN) plots at the Solling site (soil inventory 1993) ($n = 5$, each is a mixture of four samples)

	Depth cm	CEC $mmol_c\ kg^{-1}$	Ca %	Mg %	K %	Na %	Mn %	Al %	Fe %	H %	BS %
BK	0–2	165 (33)	45 (16)	37 (11)	1.3 (0.4)	0.4 (0.1)	0.6 (0.2)	13 (23)	1.5 (1.4)	1.0 (2.0)	84 (26)
	2–5	129 (8.6)	24 (6.3)	23 (3.8)	1.5 (0.3)	0.4 (0.0)	0.4 (0.0)	42 (9.5)	4.2 (1.1)	4.8 (0.8)	49 (10)
	5–10	119 (8.0)	9.1 (2.1)	11 (1.2)	1.1 (0.1)	0.4 (0.0)	1.3 (0.4)	71 (3.5)	1.8 (0.7)	5.0 (0.6)	21 (3.1)
	10–20	100 (6.4)	8.1 (4.4)	8.6 (3.9)	1.2 (0.1)	0.4 (0.0)	2.0 (0.3)	78 (8.4)	0.6 (0.1)	1.5 (0.4)	18 (8.4)
	20–30	75 (3.4)	3.7 (1.2)	5.0 (1.3)	1.4 (0.2)	0.5 (0.1)	1.4 (0.2)	88 (2.3)	0.3 (0.0)	0.0 (0.0)	11 (2.5)
	30–50	58 (4.2)	2.1 (0.3)	3.5 (0.6)	1.9 (0.2)	0.6 (0.1)	1.0 (0.4)	91 (0.9)	0.1 (0.1)	0.0 (0.0)	8.1 (0.8)
	50–70	60 (4.0)	2.6 (1.2)	3.6 (0.8)	2.3 (0.2)	1.0 (0.6)	0.8 (0.5)	90 (2.3)	0.0 (0.0)	0.0 (0.0)	9.6 (2.1)
	70–100	60 (4.2)	1.3 (0.2)	2.1 (0.2)	2.9 (0.3)	0.8 (0.2)	0.9 (0.3)	92 (0.4)	0.0 (0.0)	0.0 (0.0)	7.1 (0.4)
BN	0–2	117 (56)	4.1 (0.4)	2.0 (0.1)	2.5 (0.3)	0.4 (0.1)	0.8 (0.2)	57 (4.9)	7.9 (1.2)	26 (5.6)	8.9 (0.3)
	2–5	123 (13)	1.8 (0.3)	1.2 (0.2)	1.9 (0.2)	0.4 (0.2)	0.9 (0.4)	70 (6.9)	7.6 (1.0)	16 (7.1)	5.3 (0.6)
	5–10	122 (7.7)	1.2 (0.2)	0.7 (0.1)	1.6 (0.2)	0.3 (0.1)	2.2 (1.0)	81 (3.3)	1.8 (0.9)	11.4 (3.4)	3.8 (0.3)
	10–20	130 (46)	1.1 (0.2)	0.7 (0.1)	1.5 (0.2)	0.4 (0.1)	2.8 (0.8)	89 (2.9)	0.5 (0.1)	3.7 (2.4)	3.7 (0.3)
	20–30	99 (39)	1.0 (0.1)	0.6 (0.1)	1.5 (0.2)	0.4 (0.1)	2.1 (0.4)	94 (0.3)	0.2 (0.1)	0.2 (0.3)	3.5 (0.3)
	30–50	63 (4.9)	0.9 (0.1)	0.6 (0.0)	1.7 (0.1)	0.5 (0.1)	1.3 (0.4)	95 (0.6)	0.1 (0.1)	0.0 (0.0)	3.6 (0.3)
	50–70	66 (2.6)	0.9 (0.2)	0.7 (0.1)	2.2 (0.3)	0.5 (0.1)	1.2 (0.3)	94 (0.7)	0.1 (0.1)	0.0 (0.0)	4.2 (0.5)
	70–100	67 (17)	0.7 (0.2)	0.6 (0.2)	2.3 (0.6)	0.5 (0.1)	1.2 (0.6)	87 (16)	0.1 (0.1)	7.7 (17)	4.2 (1.0)

3.7 Conclusions

- The soils at the three beech forest sites, Göttinger Wald, Zierenberg, and Solling, are derived from different parent material (limestone, basalt, loess over sandstone, respectively) and cover a wide range of soil types, buffer ranges and nutrient contents.
- The Göttinger Wald and Zierenberg sites show indications of soil acidification in the uppermost soil layers, whereas the soil at Solling is highly acid in the whole rooting zone.
- The base saturation of soils has values higher than 90% at the Göttinger Wald and Zierenberg sites. At the Solling site, exchangeable base cations in the mineral soil have decreased during 1966–2001 resulting in base saturation of less than 7%.
- The nutrient status of the sites differs considerably with the Solling site having lowest nutrient pools. Calcium dominates the exchange sites at the Göttinger Wald and Zierenberg sites while aluminium is the predominant cation at the Solling site. The Zierenberg site has the most balanced composition of exchangeable cations owing to a high content of exchangeable magnesium and potassium.
- Mineral soils of the additional study sites at Solling which were used for manipulation experiments, showed only small differences in soil properties after treatment with heavy doses of ammonium sulphate, but on liming the soil showed markedly altered acid–base status of the top 20 cm depth.

References

Beese F (1986) Parameter des Stickstoffumsatzes in Ökosystemen mit Böden unterschiedlicher Acidität. Göttinger Bodenkundl Ber 90

Beese F, Stickan W (1989) Das Buchenprojekt im Solling – ein interdisziplinärer Ansatz zur Erforschung der Ursachen neuartiger Waldschäden. Ber Forschungszentrum Waldökosysteme A 49:9–26

Benecke P (1984) Der Wasserumsatz eines Buchen- und eines Fichtenwaldökosystems im Hochsolling. Schr Forstl Fak Univ Göttingen u Nieders Forstl Versuchsanst 77

Brumme R (1986) Modelluntersuchungen zum Stofftransport und Stoffumsatz in einer Terra fusca – Rendzina auf Muschelkalk. Ber Forschungsz Waldökosysteme/Waldsterben A 24

Deutschmann G (1994) Zustand und Entwicklung der Versauerung des Bodens und des oberflächennahen Buntsandsteinuntergrundes eines Waldökosystems im Solling. Forschungszentrum Waldökosysteme der Universität Göttigen A 118

Jochheim H (1991) Chemische Bodeneigenschaften der Fest- und Lösungsphase in einem Buchenwald-Ökosystem in der Phase der Humusdisintegration. In: Eichhorn J (ed) Fallstudie Zierenberg: Stress in einem Buchenwaldökosystem in der Phase einer Stickstoffübersättigung. Forschungsber Hessische Forstliche Versuchsanstalt 13:20–25

Jochheim H (1992) Raum/zeitliche Variabilität des chemischen Bodenzustandes in der Phase des Stickstoff-Vorratsabbaus. Abschlussber BMFT-Project, Unpublished

Meesenburg H, Meiwes KJ, Schulze A, Rademacher P (1997) Bodendauerbeobachtungsflächen auf forstlich genutzten Böden (BDFF). Arb-H Boden 2:77–94

Meesenburg H, Meiwes KJ, Bartens H (1999) Veränderungen der Elementvorräte in Buchen- und Fichtenökosystemen des Sollings. Ber Freiburger Forstl Forschung 7:77–94

Meesenburg H, Merino A, Meiwes KJ, Beese FO (2004) Effects of long-term application of ammonium sulphate on nitrogen fluxes in a beech ecosystem at Solling, Germany. Water, Air, Soil Pollut 4:415–426

Meiwes KJ, Beese F (1988) Ergebnisse der Untersuchung des Stoffhaushaltes eines Buchenwaldökosystems auf Kalkgestein. Ber Forschungsz Waldökosysteme B 9

Meiwes KJ, Merino A, Fortmann H (1994) Untersuchungen der Versauerung in Bohrprofilen von Meßstellen des Grundwassergütemeßnetzes (GÜN) des Landes Niedersachsen. Ber Forschungsz Waldökosysteme B 34

Thöle R, Meyer B (1979) Bodengenetische und -ökologische Analyse eines Repräsentativ-Areals der Göttinger Muschelkalk-Scholle als Landschaftsökologische Planungsgrundlage. Göttinger Bodenkundliche Berichte 59

Ulrich B (1981) Ökologische Gruppierung von Böden nach ihrem chemischen Bodenzustand. Z Pflanzenernähr Bodenk 144:289–305

Ulrich B (1987) Stability, elasticity and resilience of terrestrial ecosystems with respect to matter balance. In: Schulze ED, Zwölfer H (eds) Ecological Studies 61. Springer, Berlin, pp 11–49

Wedepohl KH (1968) Die tertiären basaltischen Gesteine im nördlichen Hessen und südlichen Niedersachsen. In: Metz R (ed) Zur Mineralogie und Geologie der Umgebung von Göttingen. Sonderheft zur Jahrestagung 1968 der Vereinigung der Freunde der Mineralogie und Geologie (VFMG) eV, Springer, Heidelberg, pp 112–120

Wolff B, Riek W (1997) Deutscher Waldbodenbericht 1996, Band 1. Bundesministerium für Ernährung, Landwirtschaft und Forsten (BMELF), Bonn, Germany

Chapter 4
Changes in C and N Contents of Soils Under Beech Forests over a Period of 35 Years

K.J. Meiwes, H. Meesenburg, J. Eichhorn, C. Jacobsen, and P.K. Khanna

4.1 Introduction

Soil organic matter (SOM) contributes significantly to many vital ecosystem processes by affecting physical, chemical and biological properties of soils. Among the many ecosystem processes, SOM contributes significantly to the retention of water and nutrients, maintenance and functioning of soil biodiversity and cycling of nutrients and organic C. SOM is primarily a store of soil organic C (SOC) and plant nutrients and can therefore be considered as a suitable surrogate of fertility for many soils.

SOC occurs in many different forms in a soil and any change in its amount is the consequence of a balance between inputs and outputs of organic matter. A continuous increase in SOC is expected when the decomposition (output) pattern of the litter fall is impeded. This may occur due to a change in the type of litter fall (an increase in the proportion of more woody litter than leafy type) or to a change in its chemical composition (e.g. high C/N ratio) or a change in environmental and microbial factors (e.g. changes in the soil temperature and moisture conditions due to management practices or changes in the microbial activity due to population changes) (Schulze et al. 2000).

A number of theories have been postulated on the changes in SOC in forest ecosystems, which may occur during the development of a stand. Covington (1981) suggested that C in soils would undergo a decrease during the initial phases of stand development after clearfelling, which is then followed by a phase of C accumulation and finally reaching a phase in old growth stands of no further change. A decrease in soil C after clearfelling may, however, occur depending upon the type of the stand and status of stand structure and the amount of woody debris. For example, a breakdown of the stand structure will lead to a decrease in soil C in mature stands. The 'no change' or quasi-steady state situation will occur when the input of C balances the output despite further changes in stand growth. Berg and Matzner (1997) have described that various phases of decomposition are affected differently by climatic factors and N levels in the system. For example, an increase

R. Brumme and P.K. Khanna (eds.), *Functioning and Management of European Beech Ecosystems*, Ecological Studies 208,
DOI: 10.1007/978-3-642-00340-0_4, © Springer-Verlag Berlin Heidelberg 2009

in the N concentration of the litterfall from conifer stands will decrease the total amount of litterfall decomposed probably due to chemical changes in the lignin fraction in the presence of high N.

Nitrogen, phosphorus, sulphur and many other nutrients are intricately related with soil C, and any change in organic C in soils will affect their accumulation or release. Inputs of nitrogen through atmospheric depositions or fixation may allow a significant fraction of N to be retained as part of SOC at constant or near constant C: N ratios (De Vries et al. 2006). Any nutrient with limited supply or input will be depleted in the organic matter indicating a widening of its ratio with soil C and probably a reduction in its uptake and an increase in its retention (retranslocation processes) during litter fall. For example, Knoepp et al. (2005) showed in their study on the accumulation of litter layer in the southern Appalachian forests after 20 years that it was associated with changes of leaf litter quality. The leaf litter had decreased in P concentration and showed lower P release on decomposition.

This study provides two aspects of changes in soil C and N in beech stands at Solling, Zierenberg and Göttinger Wald. Firstly, a comparison will be made for the three mature beech stands for their soil C and N contents to find if there are common factors to explain the differences among the three sites. Secondly, the long-term data on soil C and N and other nutrients measured periodically over a period of 35 years at one of the sites will be used to assess the possible human effects in changing soil C, N and other nutrients and their consequences on ecosystem functioning. The long term analysis of data was restricted to the Solling site as similar long-term data from the other two sites were not available.

4.2 Content and Distribution of C and N in Soils from Three Beech Sites

Distribution of organic C and N in soils. Organic C concentration decreased with soil depth in all the three beech sites in an expected manner, when the depth from surface organic layer to 80 cm was considered (Table 4.1). C concentration of the LF layer ranged from 400 to 460 g kg^{-1} whereas it was about 300 g kg^{-1} in the H layer. L layer from the Zierenberg site had low C values indicating some mixing of L layer with the organic matter from deeper layers. In the mineral soil, C concentration ranged from 42–68 g kg^{-1} in the 0–10 cm depth to 5–21 g kg^{-1} in the 40–80 cm depth. Of the three sites, the highest values of organic C were observed at Göttinger Wald for all the comparable depths. For example, at 40–80 cm depth of the Solling and Zierenberg sites organic C concentration was 5 g kg^{-1} as compared to 21 g kg^{-1} in Göttinger Wald, which would be regarded as a high value for this depth of any soil.

Nitrogen concentration of soils decreased with depth in an expected way similar to that of organic C in these soils. Nitrogen values in the organic layer of 13–24 g kg^{-1} in the LF and of 15 g kg^{-1} in the H layers decreased to 2.5–5.4 g kg^{-1}

Table 4.1 Concentration of organic carbon and nitrogen in soils under European beech stands at Solling (*SO*) ($n = 6$; 1993), Göttinger Wald (*GW*) ($n = 6$; 1995) and Zierenberg (*ZB*) ($n = 4$; 1996); mean values and standard deviation (in parentheses) are given

Soil layer	Organic carbon (g kg^{-1})						Nitrogen (g kg^{-1})					
	SO		GW		ZB		SO		GW		ZB	
L + F	450.0	(9.0)	445.1	(14.0)	397.3	(16.0)	23.6	(0.3))	12.9	(0.5)	20.0	(1.3)
H	285.1	(65.0)	–	–	–	–	15.2	(0.3)	–	–	–	–
0–10	48.0	(17.0)	68.5	(9.0)	41.9	(3.0)	2.5	(0.7)	5.4	(0.7)	3.6	(0.4)
10–20	24.8	(4.0)	42.9	(8.0)	27.0	(1.0)	1.5	(0.2)	3.7	(0.7)	2.5	(0.2)
20–40	13.9	(1.0)	23.2	(4.0)	15.3	(3.0)	1.0	(0.1)	2.1	(0.3)	1.4	(0.3)
40–80	5.2	(1.0)	21.2	(3.0)	5.3	(1.0)	0.5	(0.1)	1.7	(0.3)	0.6	(0.1)

Table 4.2 Mean values of C:N ratio (g/g) in soils under European beech stands at Solling (*SO*) ($n = 6$, 1993), Göttinger Wald (*GW*) ($n = 6$, 1995) and Zierenberg (*ZB*) ($n = 4$, 1996); standard deviation in parentheses

Soil layer	C:N ratio (g/g)					
	SO		GW		ZB	
L + F	19.5	(0.39)	34.5	(2.28)	20.0	(0.99)
H	18.9	(0.82)	–	–	–	–
0–10	18.8	(2.05)	12.6	(0.40)	11.8	(0.85)
10–20	17.0	(1.21)	11.6	(0.44)	11.1	(0.83)
20–40	14.2	(0.52)	11.2	(0.94)	10.9	(1.07)
40–80	9.8	(1.33)	12.9	(1.70)	8.7	(0.90)

in the 0–10 cm depth and further to 0.5–1.7 g kg^{-1} in the 40–80 cm soil depth. Similar to C concentration, N concentration in Göttinger Wald soil was the highest of the three soils.

In the L + F layer, the C:N ratio was 35 at Göttinger Wald and about 20 at Solling and Zierenberg (Table 4.2). The C:N ratios decreased with soil depth of the mineral soil (0–10 cm compared with 40–80 cm), but the decrease was slight in Göttinger Wald when compared with the other two stands. A steeper gradient in the C:N ratio was observed in Solling. These differences in the C:N ratios indicate differences in the quality of SOM produced in these stands and probably in association with the differences in C and N cycling of these stands.

Total content of soil C and N in the three beech forests. Total organic C contents for 0–80 cm soil depths were similar at Solling (172 ± 31 Mg ha^{-1}) and Göttinger Wald (179 ± 22 Mg ha^{-1}) sites, whereas much lower contents were observed for Zierenberg (109 ± 9 Mg ha^{-1}) (Table 4.3). For the whole profile (0–80 cm), the amount of N was, however, about 20% higher at Göttinger Wald (14.2 ± 1.7 t Mg ha^{-1}) than at Solling (11.2 ± 1.5 Mg ha^{-1}) and Zierenberg (10.0 ± 1 Mg ha^{-1}). In order to compare organic C and N in these soils with similar soils in the area, soil survey data compiled for Lower Saxony (Bartens and Büttner 1997) and Hesse (Hocke, personal communication) were evaluated. The survey data

Table 4.3 Mean content of soil organic carbon and nitrogen in European beech stands at Solling (*SO*) ($n = 6$, 1993), Göttinger Wald (*GW*) ($n = 6$, 1995) and Zierenberg (*ZB*) ($n = 4$, 1996); standard deviation in parentheses

Soil layer	Organic carbon (Mg ha^{-1})						Nitrogen (Mg ha^{-1})					
	SO		GW		ZB		SO		GW		ZB	
L + F	24.9	(3.5)	7.8	(3.7)	4.4	(0.2)	1.3	(0.2)	0.2	(0.1)	0.2	(0.0)
H	11.0	(2.3)	–		–		0.6	(0.1)	–		–	
0–10	46.3	(16.4)	54.4	(7.1)	33.5	(2.3)	2.4	(0.7)	4.3	(0.6)	2.9	(0.3)
10–20	26.8	(4.0)	35.0	(6.7)	23.6	(1.2)	1.6	(0.2)	3.0	(0.5)	2.1	(0.2)
20–40	35.0	(2.1)	35.8	(5.8)	29.9	(6.7)	2.5	(0.2)	3.2	(0.4)	2.7	(0.6)
40–80	28.1	(4.6)	46.4	(6.2)	17.6	(4.2)	2.9	(0.4)	3.6	(0.6)	2.0	(0.3)
Sum	172.2	(31.2)	179.4	(21.8)	109.0	(9.5)	11.2	(1.5)	14.2	(1.7)	10.0	(1.0)

Table 4.4 Mean total content of organic C and total N in the soil survey collective from Lower Saxony and Hesse. The values are given for 40 cm soil depth. Data were obtained from Bartens and Büttner (1997) and Hocke (personal communication); standard deviation in parentheses

Sites	*n*	C (Mg ha^{-1})		N (Mg ha^{-1})	
Loess	9	95	(29)	5.6	(1.2)
Lime stone	6	102	(44)	8.3	(3.4)
Basalt	15	55	(19)	4.6	(1.6)

formed part of a study which was conducted on a 8 × 8 km grid to assess temporal changes in soils under forests. For comparing data with the present study, sites with similar parent materials (Solling – loess, Göttinger Wald – limestone and Zierenberg – basalt) were selected from the grid-based soil survey.

Amounts of C and N in those soils are shown in Table 4.4. Amount of C and N in all the three sites studied here showed 30–60% higher values than those found in soils of the same parent material in the region. This was probably due to differences in the altitude of the Solling and Göttinger Wald sites which are located 170 m and 130 m higher than the comparable grid-based soil survey sites. These differences in altitude are expected to affect the mean annual temperature on the sites and thus the stand productivity and the decomposition processes. Wirth et al. (2004) showed a clear increase of SOC with altitude in forest stands in the nearby region of Thuringia. The Zierenberg site has a north aspect and thus also has relatively lower temperatures than the comparable sites of the region. The forest stands of the Solling, Göttinger Wald and Zierenberg sites have not been managed since the investigation was started; this may also have influenced the content of C and N, at least in the top organic layer.

Distribution of C and N at various soil depths was different for the three beech sites depending upon the intensity of faunal activity and thus the presence or absence of H horizon of the top organic layer. In the Solling soil, where mixing

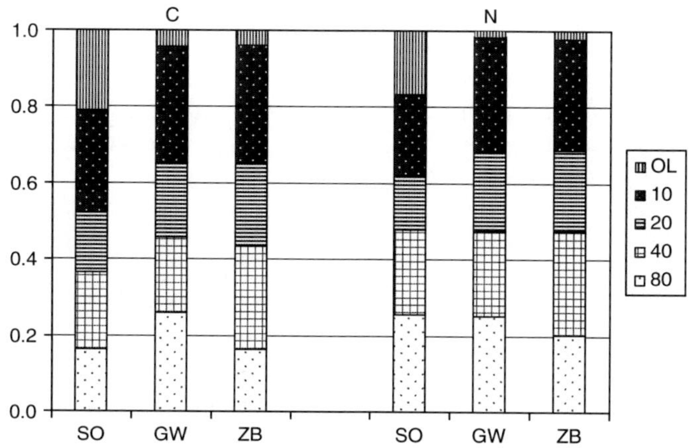

Fig. 4.1 Fractions of total organic C (*left side*) and N (*right side*) in different depths (*OL* organic layer, 10 = 0–10 cm, 20 = 10–20 cm, 40 = 20–40 cm and 80 = 40–80 cm). *SO*, *GW*, and *ZB* refer to three beech sites at Solling, Göttinger Wald and Zierenberg

and turnover of the soil was lacking due to low faunal activity, 36 Mg ha^{-1} of organic C occurred in the organic layer. This accounted for 20% of the total C in the surface 80 cm soil depth. At Göttinger Wald and Zierenberg, the organic C in the organic layer was about 4% of the total C when the 80 cm depth was considered (Fig. 4.1). The fraction of the total soil N in the organic layer of the three sites was less than that of C. At the Solling site, 12% of the total N in the soil to a depth of 80 cm was in the top organic layer whereas at the other two beech sites it was less than 2%.

In the 0–10 cm mineral soil, higher amounts of C and N occurred at Göttinger Wald and Zierenberg sites than at Solling, which was again related to the high earthworm activity in the former two soils. N content of the 0–10 cm mineral soil was 29% of the total at the Göttinger Wald and Zierenberg sites but only 21% at the Solling site when the whole soil depth of 80 cm was considered. Also, the subsoil (40–80 cm) at the Göttinger Wald site was richer in organic C and N than the subsoils at the Solling and Zierenberg sites. A quarter of the total soil C at the Göttinger Wald site was present in the subsoil which compared with 16% in soils at the Solling and Zierenberg sites, whereas the respective values for N in the subsoil were: Zierenberg 20%, Solling 26% and Göttinger Wald 25%.

Important factors determining the amount of organic C and N in forest soils are their inputs, the decomposition factors (substrate quality, temperature and moisture conditions) and the immobilisation factors (physical protection through clay content, chemical protection through Al content). Litterfall, including root litter, is the main input of C in these stands, whereas both litterfall and atmospheric depositions are important inputs for N. Annual litterfall values of these stands are, however, different, and are primarily related to the productivity of these stands. The annual litter production at Solling was 4.9 Mg ha^{-1} aboveground and 1.3 Mg ha^{-1}

fine roots. At Göttinger Wald, 5.1 Mg ha^{-1} originated from aboveground litter and 0.6 Mg ha^{-1} from fine roots. At Zierenberg, 5.8 Mg ha^{-1} came from aboveground litter and 0.6 Mg ha^{-1} from root litter.

The N inputs in these stands through litterfall and atmospheric deposition are (kg ha^{-1}): 68 and 36 (Solling), 64 and 25 (Göttinger Wald) and 75 and 23 (Zierenberg) (Meesenburg et al., Chap. 9, this volume, and Khanna et al., Chap. 5, this volume). Therefore, the differences in the inputs of C and N in the three sites can only partly explain the observed differences in the contents of C and N in these soils.

The decomposition factors, soil temperature and moisture, are similar on these three sites. The three sites are located at 400–500 m a.s.l., the mean annual air temperature at Solling is 7.0°C, at Göttinger Wald 7.7°C and at Zierenberg 7.6°C (1997–1998, Panferov et al., Chap. 1.2, this Volume). The annual precipitation values are, however, different: Solling 1168 mm, Göttinger Wald 525 mm and Zierenberg 644 mm (1990–1998). Despite the differences in precipitation, the climatic factors do not seem to be the main cause of different decomposition rates of organic C and N at these three sites.

Clay contents of the soils, as an indicator of physical protection to SOM decomposition, are different at these three sites with Göttinger Wald showing the highest value. Soils at the Solling site had 18–23% clay, Zierenberg 11–13% and Göttinger Wald 36–54% in the 0–30 cm depth and 23–30% in the 30–90 cm depth (the subsoil at the Göttinger Wald site had 49–61% stone content on a volume basis). Chemical protection as reflected by Al content or soil acidity in these soils is expected to be highest in Solling soil (lowest soil pH). It therefore seems that the physical protection of organic C by high clay concentration at the Göttinger Wald site and chemical protection at the Solling site are the major factors contributing to high SOC in these soils. Soil N contents of these soils follow soil C contents.

4.3 Long-term Periodic Measurements of C and N Contents in the Soil at Solling Site

Solling soil was periodically sampled and measured for organic C, N and other element contents during 1966–2001. Some of the data have been published previously (Ulrich et al. 1979; Matzner 1988; Tiktak et al. 1995; Meesenburg et al. 1999). In addition to the data from the beech site, some data from the spruce site at Solling will also be included as a comparison between the two sites. The combined data on soil C and N are presented as a sequence of temporal changes, the consequences of measured changes, and the methodological problems associated with such long-term measurements. Data on organic C and N for mineral soil and top organic layer are treated separately.

Methodological issues of long term field measurements. As is evident from Fig. 4.2, the C content of the top organic layer measured periodically for 35 years

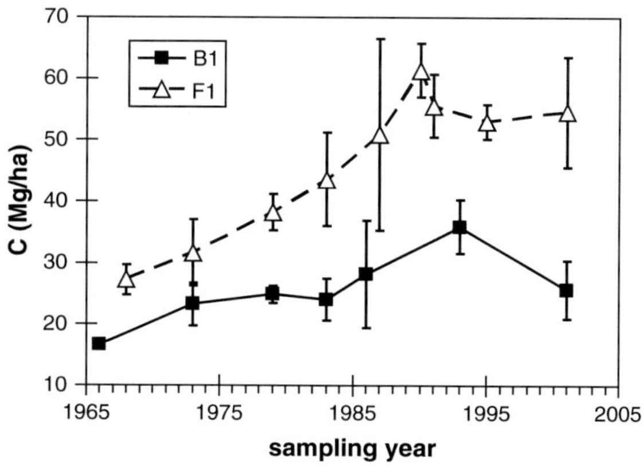

Fig. 4.2 Content of C in the organic layer of beech (B1) and spruce (F1) stand at the Solling site measured periodically since 1966; mean and standard deviation

showed high variability and some unexpected short-term trends. Many field and laboratory conditions influenced the type of data obtained, which are outlined below:

1. Changes in the study aims during the sampling period: Objectives of the initial soil sampling were not to set up a chronosequence study and therefore the number of replicates collected for each sample, preparation of soil and organic layer samples, and the use of field and laboratory methods were not the same nor standardised during the study period. In general, two to six mixed samples were collected and, for each mixed sample, three to six individual samples were combined. The diameter of soil corer varied from 8 to 27 cm. Some details of the field and laboratory methods are provided by Meesenburg et al. (1999).
2. The samples were collected by different persons and under different meteorological conditions which resulted in high variability associated especially in identifying the boundary between organic layer and mineral soil.
3. Chemical analysis of the soil changed over the period. Prior to 1979, C concentration of the organic layer was determined by subtracting the amount of oxides of elements in the ash (loss on ignition) from dry matter which was then divided by a factor of 1.81. This factor of 1.81 (standard deviation = 0.11) was obtained from a collective of samples ($n = 342$), which were analysed for C concentration by the dry combustion method, and these values were compared with the C concentration values calculated by the above method of using oxides in the ash. After 1979, C was determined by dry combustion using an organic carbon analyser.

Changes in the mineral soil. For the evaluation of time trends, the mean values of element contents were regressed against time (1966–2001) and a linear regression was fitted. Presence of any linear trend in time of the data was tested by using product moment correlations ($p < 10\%$) to find out if organic carbon contents measured at various times showed any linear trend. It was observed that soil C and N data from the mineral soil did not show any significant trend in time (Table 4.5).

Table 4.5 Contents of C and N (Mg ha^{-1}) of 0–50 cm mineral soil measured periodically at the Solling beech site; standard deviation in parentheses

Year	No. of samples	C		N	
1966	2	97	(3.6)	6.5	(0.3)
1973	3	98	(10.8)	6.7	(0.7)
1979	3	128	(11.9)	7.8	(0.5)
1981	4	66	(7.8)	4.4	(0.3)
1983	4	78	(8.8)	5.1	(0.8)
1986	3	103	(4.5)	6.5	(0.1)
1993	6	122	(22.0)	7.6	(1.1)
2001	6	127	(18.7)	7.1	(0.9)

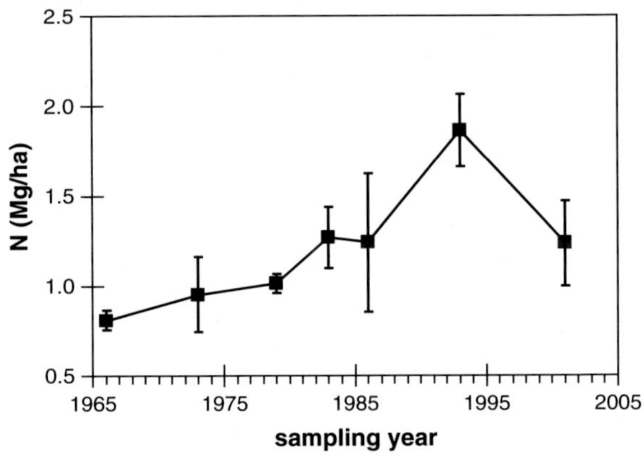

Fig. 4.3 Accumulation of N in the top organic layer of beech stand at Solling measured periodically since 1966; mean and standard deviation

The high content of C on this site and the associated variability were probably the reasons for the lack of any clear trend in the changes, if there were any to be detected.

Changes in the top organic layer. During the 35 years of the observation period, organic C content of the organic layer of the beech stand had increased from 20 to 32 Mg ha^{-1} showing a positive linear trend in time (Fig. 4.2). Comparative increase in the spruce stand showed a much higher increase from 29 to 62 Mg ha^{-1}. During the same period, the amount of N in the organic layer of the beech stand increased from 860 to 1600 kg ha^{-1} and that of Ca from 90 to 175 kg ha^{-1} (Figs. 4.3 and 4.4). No significant increase was observed for P. Despite the changes in the C and N contents, the C:N ratios did not show any significant change and probably remained constant during the study period. The annual increase for various elements was calculated to be 347 kg C ha^{-1}, 21 kg N ha^{-1} and 2.5 kg Ca ha^{-1} ($p <$ 0.10) (comparative values of annual increases for spruce stands were 999 kg C ha^{-1}, 42 kg N ha^{-1}, 3.2 kg Ca ha^{-1} and 1.8 kg P ha^{-1}). As the samples collected for the organic layer were not always divided into various horizons, it is not

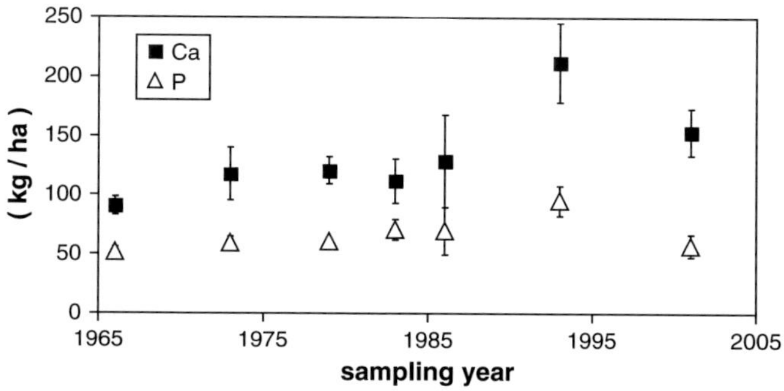

Fig. 4.4 Accumulation of P and Ca in the organic layer of beech stand at Solling measured periodically since 1966; mean and standard deviation

possible to assign this increase to a specific horizon of the organic layer. Moreover, there is a clear reduction in C and N values for the sampling period of 1993–2001 (Figs 4.2 and 4.3). However, the high level of variability in sampling as discussed above makes it mandatory to consider the changes during a single sampling period with caution. For example, a decrease in litter accumulation from 1993 to 2001 would need further verification in future years before any significance can be given to this decrease, which may have resulted from changes in the structure of canopy (removal of trees) or a decrease in atmospheric inputs of acidity and heavy metals.

4.4 Human Impacts and Management Issues at Solling

In order to assess human impacts on the rates of organic matter accumulation in the organic layer of the beech stand, one needs to consider changes in rates during the whole stand development. As no measurements of organic layer were available prior to 1966, the following method was used to calculate the annual accumulation rate.

It was assumed that the litter accumulated on the organic layer was essentially from the existing stand established in 1849 and that the initial 20 years of stand development contributed very little to the accumulated amount measured in 1966. The calculated amount therefore represents a high value for the accumulation of organic matter during the 1869–1966.

Table 4.6 gives the slopes of the accumulation of C, N, P and Ca for the measuring period of 1966–2001 (including r^2 and F significance) and the calculated values for the stand age of less that 20 years until 1966, assuming that at the beginning of this period there was no organic layer. The slopes indicating accumulation rate of C and N on the surface organic layer of the beech stand for 1966–2001 are much higher than those for the period 1869–1966 (Table 4.6). Together with C and N, calcium has also been retained. This accumulation of organic matter is not

Table 4.6 Coefficients (slopes) of the linear regression for C, N, P and Ca contents of the organic layer as a function of time for the period 1966–2001, and slope 1869–1966 based on the assumption of lacking organic layer during stand establishment (1869)

Variable	1869–1966	1966–2001	
	Slope	Slope	R^2 (*p* value)
C ($Mg\ ha^{-1}a^{-1}$)	0.124	0.347	0.50 (0.075)
N ($Mg\ ha^{-1}a^{-1}$)	0.006	0.021	0.51 (0.071)
P ($Kg\ ha^{-1}a^{-1}$)	0.386	0.562	0.21 (0.295)
Ca ($Kg\ ha^{-1}a^{-1}$)	0.6	2.461	0.55 (0.057)

related to any change in inputs as the inputs (leaf fall amount) have remained more or less constant as periodic measurements ($n = 13$) during the 35-year period indicated (2890 ± 510 kg ha^{-1} a^{-1}). Since forest management operations like thinning have not been conducted since 1966, there is no effect of forest management on this accumulation. The increase of the organic layer is presumed to be due to a reduction in the decomposition rate of the litter. A number of hypotheses can be formulated to describe the factors which may have influenced the conditions causing a decrease in litter decomposition on this site.

1. *A change in the micro-climatic conditions has occurred reducing the decomposition of litter*. It is difficult to test this hypothesis as the required long-term data are not available. However, the beech stand was already a mature one in 1966 and its structure did not change during the last 35 years, as is evident from a constant annual leaf litterfall on this site. Therefore, any micro-climatic change in soil temperature or moisture conditions is not expected to affect litter decomposition.
2. *Acid precipitation on the site has caused high acidity in the litter of the organic layer causing a reduction in decomposition rates*. High acidity in atmospheric deposition can reduce the rates of decomposition (CO_2 release) and N mineralisation (Wolters and Schaefer 1994; Schinner and Sonnleitner 1997). For example, Schaefer (1988) measured low CO_2 release from a number of soils which were collected close to the stem of the beech trees (high soil acidity due to stemflow) when compared with the soils collected from less acid areas. Similar results were obtained by Moloney et al. (1983) for conifer needles when studied under laboratory conditions. Wolters (1991) observed a decrease in CO_2 release when litter samples from the litter layer of the Göttinger Wald site were treated with acid rainwater. However, the effects of acidity on the decomposition of litter need to be considered with other accompanying factors such as high levels of soluble heavy metals and Al. For example, soluble Al may affect the litter decomposition directly by causing toxicity to micro-organisms (Jandl and Sletten 1999) or indirectly by complexing the soluble C and thereby reducing the amount of C available for microbial activity (Scheel et al. 2007).
3. *High levels of soluble heavy metals in the litter can slow down litter decomposition*. Contents of heavy metals in the organic layer are shown in Fig. 4.5.

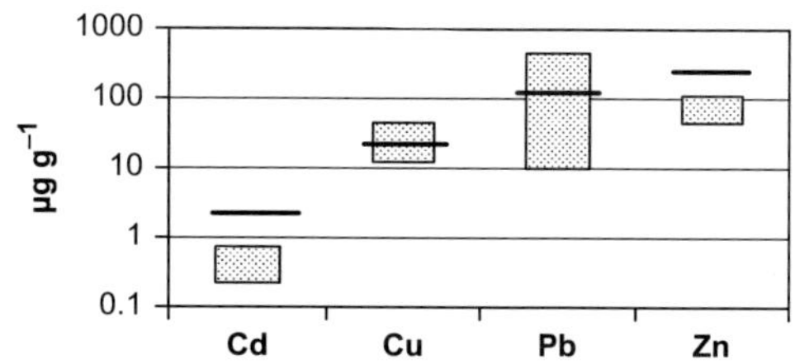

Fig. 4.5 Concentrations of heavy metals (10–90% percentile) in the top organic layer of the beech and spruce stand at Solling (*horizontal lines* indicate concentration levels where impedance of mineralisation is expected according to Tyler (1992)

The values presented here provide the data for the whole horizon, whereas it is possible that such values could be much higher when thinner layers of each horizon are considered. Those thin layers may form chemical barrier to microbial activity and C transfer in the soil. For example, Heil (2000) showed that contents of Zn, Cu and Pb, when measured for each cm depth of the organic layer under the adjoining spruce stand to that of beech stand at the Solling site, were highest in H horizons (mean values in mg kg^{-1} were: 500 for Pb, 40 for Cu and 120 for Zn). Considering the critical values of heavy metals indicating toxicity to soil fauna and microbes (Tyler 1992), the high levels of Cu and Pb of the organic layer under the beech stand may have affected its decomposition. Further evidence of the reduction in decomposition due to soil fauna activity was provided by Alberti et al. (1996). In their experiment at Wiesloch in Odenwald, Germany, decrease in litter decomposition due to heavy metal accumulation was primarily related to diminished activity of soil fauna rather than that of microbes. On their site, concentrations of Cd and Zn were much higher and of Pb somewhat higher than those measured in Solling. Schaefer (1988) reported that, in 11 different forest sites under beech located in the south of Lower Saxony, the Pb content of more than 200 mg kg^{-1} in Of horizons caused a decrease in CO_2 respiration. Pb concentration in Solling soils lie between 54 mg kg^{-1} and 460 mg kg^{-1}. The heavy metals relevant for toxicity to soil fauna and microbes are monomeric or cationic forms. In a water extract of a loess-derived upper soil (0–10 cm), the percentage of monomeric forms, determined by ion exchange and by the dialysis method, was 72–89% for Cd, 95–96% for Zn, 29–38% for Pb and 2–15% for Cu (König and Ulrich 1986). Similar results were obtained in water extracts of the top organic layer of an oak woodland ecosystem (Martin and Bullock 1994). At the Solling beech site there are no speciation data available, but it can be assumed that the distribution of the heavy metal species may be similar. Although the effects of heavy metal in reducing litter decomposition at Solling cannot be specified, their contribution cannot be ruled out (for details on litter decomposition refer to Schaefer et al., Chap. 6, this Volume).

4. *The site receives high N inputs which interact with litter components to make them less decomposable*. Nitrogen concentration in the litter may affect its decomposition. From a literature review, Berg and Matzner (1997) came to the conclusion that during initial decomposition phases high levels of N in the leaf

litter increased the weight loss, but in the later phases the decomposition rate decreased as the easily mineraliseable cellulose and other components were consumed and the residues were enriched with lignin and N. The decrease in decomposition was more in N-rich substrates. Berg and McClaugerthy (2003) postulated that this decrease in decomposition was related to the formation of less degradable lignin compounds which were formed under high initial N levels in the litter. This could result from two reaction mechanisms: (1) There may be a reduction in the formation of lignolytic enzymes, and (2) chemical changes may occur where lignin incorporates N and undergoes condensation reactions. Berg (2000) reported that a positive relationship existed between retardation in the decomposition of litter and its N concentration. An increase in N concentration from 4 to 13 mg g^{-1} increased the fraction of non-degradable litter component from about 10% to 50%. N concentration of freshly fallen leaves of beech at the Solling site was 13.3 ($\pm$1.8) mg g^{-1} sampled in 13 out of 30 years between 1968 and 1997, which lies in the range given by Berg (2000) where a significant decrease in litter decomposition would be expected. In samples collected on a transect from Sweden to Italy, Persson et al. (2000) showed that the rate of CO_2 development decreased with an increase in N content or with narrowing of C:N ratios of the litter. Yanai et al. (2001) and Hyvönen et al (2007) conclude in their literature reviews that in northern coniferous forests N fertilisation results in higher surface organic layer contents due to lower litter decomposition rates. This is in accordance with the observations of Olson et al. (2005), that fertilisation of Norway Spruce in boreal forests reduced heterotrophic soil respiration and increased the thickness of the mor layer. In northern temperate forests of Michigan, the annual addition of 30 kg ha^{-1} of NO_3–N for 10 years led to increased carbon content of the organic layer and the upper mineral soil (0–10 cm) (Pregitzer et al. 2007) as well as to reduced CO_2 respiration rates (Burton et al. 2004) and lignolytic activity (DeForest et al. 2004).

From the available data on the beech site at Solling, it is not easy to decide which of the four hypotheses described above can be accepted, including the factors considered individually or in combination which are involved in reducing litter decomposition. This study, being primarily a field-based one, was not designed to include the individual factors described above. But if the observation made on this site is a commonly occurring phenomenon on all German forests where acidity plays a major role in causing this reduction in litter decomposition, then one-third of the total forest area will show a similar type of reduction in litter decomposition. On the remaining areas, the top organic layers are expected to be less acid either due to base-rich soil parent material (BMVELF 2002) or due to recent application of lime materials (27% of the forest area in Germany have been limed). The effects of heavy metals will depend on a number of factors. For example, the possibility of Cd and Zn affecting decomposition processes may be small due to their low concentration in most forest soils (98% of sites showed low values according to the data summarised by BMELF 1997) when related to the critical values provided by Tyler (1992). Forest sites which may show values higher than the critical values given by

Tyler (1992) for Cu would lie around 40% of all sites. The effect of Pb in decreasing faunal activity may occur on 40% of the sites, but only 2% of the forest sites may show the effects of Pb on microbial activity. The effect of high levels of N can be evaluated by considering N values in the leaf litter. In Lower Saxony, Germany, on 12 monitoring sites measured during 1994–2004, N levels in the leaf litter were 14 ± 2 mg N g^{-1} indicating a non-degradable litter fraction of about 30–40% on these sites (Berg 2000). As the N inputs through atmospheric depositions are commonly high in forests of Germany, the hypothesis relating litter N concentration to litter decomposition has major significance and would need additional research efforts.

During litter accumulation, N and a number of other elements such as Ca and P would be retained in the surface organic layer. Retention of these elements may impact on the nutrition of forests in a number of ways. For example, the amount of N retained increases the nutrient potential of the site, which would require special attention at the time of harvesting. Management of such sites would require special efforts to keep the N losses by leaching and denitrification at a low level. Special management practices would still be required on forest sites, despite the high atmospheric N depositions. The amount of P and Ca removed from the cycling fraction by litter retention would remain unavailable for further uptake by plants decreasing their concentrations in leaves. For example, the P concentration of beech leaves at the Solling site has decreased to almost half the values observed 30 years ago. There may be other reasons for this decrease in foliar levels of P in this stand, but its retention in the surface organic layer also seems to be an important factor.

4.5 Summary

Comparison of C and N contents in the 80 cm soil depth of the three mature beech stands indicated that the Solling, Zierenberg and the Göttinger Wald sites differed in the total amount and the depthwise distribution. The important factors which differed on these sites were: the intensity of faunal activity for depthwise distribution of organic matter, clay content (for providing physical protection to decomposition of organic matter in Göttinger Wald soil) and high soil Al (for providing chemical protection to decomposition in Solling soil). In the highly acid Solling soil where faunal activity was low, an accumulation of C occurred in the organic layer accounting for 20% of the total C in the surface 80 cm soil depth. In high faunal activity soils at the Göttinger Wald and Zierenberg sites, only about 4% of the total C in the 80 cm depth was present in the organic layer.

Long-term data on soil C and N and other nutrients measured periodically over 35 years at the Solling site indicated a continual increase of organic layer expressed as C content from 20 to 32 Mg ha^{-1} showing a positive linear trend in time. During the same period, the amount of N in the organic layer of the beech stand increased from 860 to 1,600 kg ha^{-1}. This change in soil C, N and other nutrients was probably related to high atmospheric inputs of acidity, N and heavy metals causing a change in litter quality and decomposing conditions. This accumulation of organic matter in the surface organic layers may have significant effects on ecosystem functioning and on future management practices by affecting the cycling of nutrients, especially that of P.

References

Alberti G, Hauk B, Kammerer U, Köhler HR, Storch V (1996) Dekomposition – qualitative und quantitative Aspekte und deren Beeinflussung durch geogene und anthropogene Belastungsfaktoren. ECOMED Umweltforschung in Baden–Württemberg, 490S

Bartens H, Büttner G (1997) Bodenzustandserhebung in Niedersachsen 1990/91 – Datendokumentation. Ber Forschungszentrum Waldökosysteme Universität Göttingen C5:1–476

Berg B (2000) Litter decomposition and organic matter turnover in northern forest soils. Forest Ecol Manage 133:13–22

Berg B, Matzner E (1997) Effect of N deposition on decomposition of plant litter and soil organic matter in forest systems. Environ Rev 5:1–25

Berg B, McClaugerthy C (2003) Plant litter – decomposition, humus formation, carbon sequestration. Springer, Berlin

BMELF (Bundesministerium für Ernährung, Landwirtschaft und Forsten) (1997) Deutscher Waldbodenbericht 1996, Band 2, Bonn

BMVELF (Bundesministerium für Verbraucherschutz, Ernährung, Landwirtschaft und Forsten) (2002) Bericht über den Zustand des Waldes 2002, Bonn

Burton AJ, Pregitzer KS, Crawford JN, Zogg GP, Zak DR (2004) Chronic NO_3^- additions reduce soil respiration in northern hardwood forests. Glob Chang Biol 10:1080–1091

Covington WW (1981) Changes in forest floor organic matter and nutrient content following clearcutting in northern hardwoods. Ecology 62:41–48

DeForest JL, Zak DR, Pregitzer KS, Burton AJ (2004) Atmospheric nitrate deposition, microbial community composition, and enzyme activity in northern hardwood forests Soil Sci Soc Am J 68:132–138

De Vries W, Reinds GJ, Gundersen P, Sterba H (2006) The impact of nitrogen deposition on carbon sequestration in European forests and forest soils. Glob Chang Biol 12:1151–1173

Heil B (2000) C- und N-Dynamik sowie Elementgradienten in Böden temperater Waldökosysteme. Ber Forschungszentrum Waldökosysteme A167:1–127

Hyvönen R, Ågren GI, Linder S, Persson T, Cotrufo MF, Ekblad A, Freeman M, Grelle A, Janssens IA, Jarvis PG, Kellomäki S, Lindroth A, Loustau D, Lundmark T, Norby RJ, Oren R, Pilegaard K, Ryan MG, Sigurdsson BD, Strömgren M, van Oijen M, Wallin G (2007) The likely impact of elevated [CO_2], nitrogen deposition, increased temperature and management on carbon sequestration in temperate and boreal forest ecosystems: a literature review. New Phytol 173(3):463–480

Jandl R, Sletten RS (1999) Mineralization of forest soil carbon: interactions with metals. J Plant Nutr Soil Sci 162:623–629

Knoepp JD, Reynolds BC, Crossley DA, Swank WT (2005) Long term changes in forest floor processes in southern Appalachain forests. For Ecol Manage 220:300–312

König N, Ulrich B (1986) Molekülgrößenverteilung, Komplexierungs- und Adsorptionsverhalten natürlicher organischer Substanzen eines sauren Waldbodens in Lützelinde und ihr Einfluss auf die Schwermetallkonzentrationen in der Bodenlösung. Ber. d. Forschungszentrums Waldökosysteme/Waldsterben, Reihe B, 3:122–132

Martin MH, Bullock RJ (1994) The impact and fate of heavy metals in an oak woodland ecosystem. In: Ross SH (ed) Toxic metals in soil-plant systems. Wiley, Chichester, NY, Brinsbane, Toronto, Singapore, pp 325–365

Matzner E (1988) Der Stoffumsatz zweier Waldökosysteme im Solling. Ber Forschungszentrum Waldökosysteme/Waldsterben A40:1–217

Meesenburg H, Meiwes KJ, Bartens H (1999) Veränderungen der Elementvorräte im Boden von Buchen- und Fichtenökosystemen im Solling. Ber Freiburger Forstl Forschung 7:109–114

Moloney KA, Stratton LJ, Klein RM (1983) Effects of simulated acid rain, metal containing precipitation on coniferous litter decomposition. Can J Bot 61:3337–3342

Olson P, Linder S, Giesler R, Högberg P (2005) Fertilization of boreal forest reduces both autotrophic and heterotrophic soil respiration. Glob Chang Biol 11:1745–1753

Persson T, Karlsson PS, Seyferth U, Sjöberg RM, Rudebeck A (2000) Carbon mineralisation in European forest soils. In: Schulze ED (ed) Carbon and nitrogen cycling in European forest ecosystems. Ecol Stud 142:257–275

Pregitzer KS, Burton AJ, Zak DR, Talhelm AF (2007) Simulated chronic nitrogen deposition increases carbon storage in Northern temperate forests. Glob Chang Biol 14:1–12

Schaefer H (1988) Auswirkungen der Deposition von Luftschadstoffen auf die Streuzersetzung in Waldökosystemen – eine Fallstudie an den durch Stammabflußwasser stark säure- und schwermetallbelasteten Baumfuß-Bodenbereichen alter Buchen. Ber Forschungszentrum Waldökosysteme/Waldsterben A37:1–244

Scheel T, Dörfler C, Kalbitz K (2007) Precipitation of dissolved organic matter by aluminum stabilizes carbon in acidic forest soils. Soil Sci Soc Am J 71:64–74

Schinner F, Sonnleitner R (1997) Bodenökologie: Mikrobiologie und Bodenenzymatik. IV. Anorganische Schadstoffe, Springer, Berlin, pp 243–254

Schulze ED, Högberg P, van Oene H, Persson T, Harrison AF, Read D, Kjølleru A, Matteucci G (2000) Interactions between the carbon and nitrogen cycles and the role of biodiversity: a synopsis of a study along a north-south transect through Europe. In: Schulze ED (ed) Carbon and nitrogen cycling in European forest ecosystems. Ecol Stud 142:468–491

Tiktak A, Bredemeier M, van Heerden K (1995) The Solling dataset. Site characteristics, monitoring data and deposition scenarios. Ecol Modell 83:17–34

Tyler G (1992) Critical concentrations of heavy metals in the mor horizon of Swedish forests. Solna, Sweden, Swedish Environmental Protection Agency. Report 4078:1–38

Ulrich B, Mayer R, Khanna PK (1979) Deposition von Luftverunreinigungen und ihre Auswirkungen in Waldökosystemen im Solling. Schriften aus der Forstl Fak der Univ Göttingen und der Nieders Forstl Versuchsanstalt 58:1–291

Wirth C, Schulze ED, Schwalbe G, Tomczyk S, Weber G, Weller G (2004) Dynamik der Kohlenstoffvorräte in den Wäldern Thüringens. Mitteilungen der TLWJF 23:310

Wolters V (1991) Effects of acid rain on leaf litter decomposition in a beech forest on calcareous soil. Biol Fertil Soils 11:151–156

Wolters V, Schaefer M (1994) Effects of acid deposition on soil organisms and decomposition processes. In: Godbold DL, Hüttermann A (eds) Effects on acid rain on forest processes. Wiley, New York, pp 83–128

Yanai R, Stehman SV, Arthur MA, Prescott CE, Friedland AJ, Siccama TG, Binkley D (2001) Detecting change in forest floor. Soil Sci Soc Am J 67:1583–1593

Chapter 5
Vegetation

W. Schmidt

5.1 Introduction

Beech forests, i.e. deciduous forests dominated by *Fagus sylvatica*, form the potential and in part the current dominant climax vegetation over large parts of Central Europe (Ellenberg 1996; Bohn et al. 2003; Dierschke and Bohn 2004). *F. sylvatica* has a competitive advantage over most other Central European tree species due to its high growth performance, robust natural regeneration, broad ecological amplitude and its high degree of shade tolerance and canopy closure (Leuschner 1998). This results in different sub-communities and micro-habitats dependent on the canopy layer, and the processes that take place within them (Schaefer 1989, 2003; Leuschner 1999; Dierschke 2003; Dierschke and Bohn 2004; Schaefer and Schauermann, Chap. 7, this volume).

One of these canopy-dependent components is the ground vegetation (herb layer, under-storey vegetation), which in the strictest sense includes dwarf shrubs, herbs, grasses and terricolous mosses and lichens at various successional stages, and in the broadest sense includes also the tree regeneration itself (Schmidt 1991, 1999). The composition of the ground vegetation reacts very sensitively to edaphic and micro-climatic site factors. In beech forests, these factors include a low light regime and strong differences in the nutrient (base cations) and moisture regimes (Ellenberg 1996; Leuschner 1999; Schmidt 1999). These factors are in turn connected to the ecological function of the ground vegetation for the entire ecosystem, e.g. in its contribution to biomass production or nutrient cycling in the beech forests (Bolte et al. 2004; Schulze et al., Chap. 10, this volume).

Ecologically, three main beech forest ecosystems can be distinguished in Central Europe (Ellenberg 1996; Dierschke and Bohn 2004):

- Acidic, oligotrophic, predominantly species-poor beech forests
- Meso- to eutraphic, species-rich beech forests of mesic sites
- Meso- to eutraphic, species-rich beech forests of warm–dry sites

R. Brumme and P.K. Khanna (eds.), *Functioning and Management of European Beech Ecosystems*, Ecological Studies 208,
DOI: 10.1007/978-3-642-00340-0_5,

The Solling Project, beginning in 1966, (Ellenberg 1971; Ellenberg et al. 1986), marked for the first time in Central Europe the comprehensive investigation of the ecology of an acidic beech forest (*Luzulo-Fagetum*). Later on, in 1980, the acidic, species-poor beech forest in the Solling was compared to a species-rich beech forest on limestone (*Hordelymo-Fagetum*). The comparative study site was established near Göttingen (Schaefer 1989; Schmidt et al. 1989). Research on the causes of the new types of forest decline in the 1980s placed the stability of forests with respect to the increased input of nitrogen and acidity in the spotlight of forest ecosystem research. In the Zierenberg case study, a species-rich, mesic beech forest was likewise selected, which, unlike the Göttinger Wald site, had experienced the strong spread of nitrogen-indicators, especially stinging nettle (*Urtica dioica*), in recent decades (Eichhorn 1995; Mrotzek 1998).

The present contribution pursues three goals:

- To provide a floristic and ecological characterisation of the vegetation of the three beech forest ecosystems of the Solling, Zierenberg, and Göttinger Wald research sites.
- To describe patterns of vegetation dynamics during the last few decades.
- To include baseline data to describe the function of ground vegetation in the biomass production and nitrogen cycling of beech forest ecosystems (Schulze et al., Chap. 10, this volume).

For nomenclature of vascular plants Wisskirchen and Haeupler (1998) was followed, and for mosses, Koperski et al. (2000).

5.2 Vegetation Structure and Phyto-sociological Classification

5.2.1 Solling

This approximately 156-year-old beech stand (2003) was established through natural regeneration. Until its designation as an international biological programme (IBP) area in 1966, the stand received regular silvicultural treatments, with particularly intense logging during the Second World War (Ellenberg 1971; Ellenberg et al. 1986). Since 1966, no regular forest management practice has been employed. Windthrow patches, including those from the 100-year-storm of November 1972, have so far played a subordinate role in the dynamics of the stand. However, since the 1990s, white rot has infected some of the beech trees. Individual infected trees, mainly from the lower canopy, were removed between 1997 and 2000 wherever they possibly threatened existing research programmes. This has resulted in small gaps in the canopy of the beech forest. The main research plot (B1) has been continuously fenced to exclude hoofed game and has been neither limed nor fertilised, with the exception of a small subplot.

The beech forest site at Solling was assigned to the wood sorrel (*Oxalis*) variant of the typical woodrush-beech forest, *Luzulo-Fagetum typicum* (Gerlach 1970), the most commonly encountered close-to-nature forest community in the Solling area (Table 5.1). The tree layer is dominated exclusively by European beech (*F. sylvatica*), which in its optimal phase forms "cathedral forests" (Ellenberg 1996). A shrub layer is lacking, due either to deficiency of light (Eber 1972) or to the influence of the high game populations, which also prevented the establishment of natural beech regeneration outside the fenced areas in the 1960s (Gerlach 1970). Indicators of acidic soils are typical of the sparse, species-poor herb and moss layers, which according to Heinken (1995) and Ellenberg (1996) distinguish the acidic *Quercetalia* communities and the *Luzulo-Fagetum* from the nutrient-rich

Table 5.1 Vegetation of beech stands in the *Oxalis* variant of the *Luzulo-Fagetum typicum* in the Solling region 1966–1968 (Gerlach 1970) and 1999–2000 (Weckesser 2003)

	1966–1968 (n = 15) (Gerlach 1970)		1999–2000 (n = 25) (Weckesser 2003)	
Tree layer, cover (%)	89.0 ± 8.3		87.2 ± 6.8	
Shrub layer, cover (%)	0.0		**0.8 ± 2.9**	
Herb layer, cover (%)	**29.8 ± 26.9**		7.2 ± 8.5	
Moss layer, cover (%)	**4.5 ± 4.0**		0.7 ± 0.5	
Number of species	11.7 ± 3.1		14.6 ± 5.8	
	Frequency (%)	Cover (%)	Frequency (%)	Cover (%)
Species that increased				
Rubus idaeus	33	1.1	**76**	**2.0**
Dryopteris dilatata	7	+	**40**	+
Urtica dioica	–	–	**36**	+
Brachythecium rutabulum M	–	–	**32**	+
Cardamine flexuosa	–	–	**28**	**0.1**
Species that decreased				
Oxalis acetosella	**100**	**8.8**	72	2.9
Luzula luzuloides L, S	**100**	**6.7**	100	1.4
Polytrichum formosum L, S, M	**100**	**4.6**	96	0.4
Deschampsia flexuosa L, S	**87**	**3.0**	60	0.1
Carex pilulifera L, S	**87**	**1.1**	72	0.1
Calamagrostis arundiancea L				
Indifferent species				
Fagus sylvatica	80	2.0	100	2.7
Dicranella heteromalla L, S, M	60	0.4	68	0.1
Sorbus aucuparia	47	0.2	60	+
Dryopteris carthusiana	47	0.4	48	+
Agrostis capillaris L	40	1.3	56	0.1
Hypnum cupressiforme L, M	40	0.3	48	+
Mnium hornum L, S, M	33	0.2	20	+
Picea abies	20	0.1	56	+

Frequency (% of the relevés in which the species were recorded) and mean cover (%, +: $\leq$ 0.1%) of the most important species in the herb and moss layer. Significantly higher values are bolded ($p \leq 0.05$, Mann-Whitney U test). *L* Character or differential species (separating species) of the *Luzulo-Fagetum* (Heinken 1995, Härdtle et al. 1997), *S* soil acidity indicator (reaction figure $R \leq 3$ according to Ellenberg et al. 2001), *M* moss species beech forest (*Fagetalia*) communities: *Luzula luzuloides, Deschampsia flexuosa, Carex pilulifera, Dicranella heteromalla, Mnium hornum*, and *Polytrichum formosum*. Two species that are absent in the lowlands, *L. luzuloides* and *Calamagrostis arundinacea*, characterise the sub-montane and montane altitude of the *Luzulo-Fagetum typicum. Oxalis acetosella* and *Dryopteris carthusiana* are typical for the more mesic and somewhat more nutrient-rich sites on deeper soils in the low mountains.

5.2.2 Zierenberg

The approximately 159-year-old Zierenberg beech stand (2003) is on the north-eastern slope of a basalt knoll on the mountain Kleinen Gudenberg and has a strong gradient and pronounced heterogeneity in the soil parent material caused by the different thickness of the basalt cover over the adjacent layer of limestone. After about 1840, the former coppice with standards system was converted to a high forest system. In 1970, the stand was thinned to provide the appropriate conditions for natural regeneration (Eichhorn 1995; Mrotzek 1998; Schulze and Eichhorn 2000). Since 1972, the expansion of stinging nettle (*U. dioica*) was documented and was treated chemically in the thinned areas to further tree regeneration. In 1980, the soil was scarified, and in 1980 and 1983, the stand was further thinned, though still without the establishment of any successful natural beech regeneration. Low intensity shelterwood treatment was applied in 1996/97 and again in 1999 (Mrotzek 1998; Schulze and Eichhorn 2000). The thinned areas covered to a large extent the basalt-influenced upper slope. Although the plot has been fenced since 1989, roe deer have been sighted repeatedly within the experimental plot.

The ground vegetation of Zierenberg area has been determined on the one hand by the geological gradient from limestone- to basalt-influenced sites, and on the other by the different thinning operations done to the forest. The tree layer exhibits the highest degree of cover on the limestone-influenced lower slope, while the herb layer reaches its highest cover on the basalt-influenced upper slope (Table 5.2). The low tree cover on the upper slope resulted in greater light intensity for the herb layer (5.4% of relative light intensity) when compared to the lower slope (3.4%). Moreover, the basalt-influenced upper slope is characterised by lower pH values than the limestone-influenced lower slope (5.2 vs 5.7; see Meesenburg et al., Chap. 3 of this volume). The herb layer is dominated by either *Mercurialis perennis* or *U. dioica. M. perennis* cover increases from basalt (average 30% cover) to limestone (average 42% cover), whereas *U. dioica* occupies on average only a quarter of the area on limestone that it occupies on the basalt areas. This classification as well as the floristic composition as a whole (Table 5.2) reflects the ample nitrogen and moisture supply of the basalt-influenced sites on the upper slopes (Reif et al. 1985; Eichhorn 1995; Mrotzek 1998; Ellenberg et al. 2001). For *U. dioica*, however, basalt influence does not determine its herb layer cover as much as the relative light intensity on the surface organic layer as stratified sampling on the research plot

Table 5.2 Vegetation of 80 permanent plots (4 m^2 each) at the Zierenberg research site between 1993 and 2002. The stratification was based on the geological substrate (Schulze and Eichhorn 2000; *mixed* transitional zone between basalt and basalt-influenced limestone area)

	Total (*n* = 80)				Basalt (*n* = 35)				Mixed (*n* = 18)				Limestone (*n* = 27)			
Year	1993		2002		1993		2002		1993		2002		1993		2002	
Number of species 4 m^{-2}	7.2 ± 3.1		**7.9 ± 2.2**		7.9 ± 3.1		8.5 ± 2.3		6.7 ± 2.5		7.9 ± 2.3		6.6 ± 3.2		7.1 ± 1.9	
	F	C	F	C	F	C	F	C	F	C	F	C	F	C	F	C
Tree layer																
Fagus sylvatica	nd	nd	98	64.1	nd	nd	100	59.1	nd	nd	94	68.1	nd	nd	96	67.8
Shrub layer																
Fraxinus excelsior N	30	2.2	**61**	**20.3**	17	1.0	**43**	**12.1**	39	3.2	**67**	**27.6**	41	3.0	**81**	**25.9**
Acer pseudoplatanus N	3	+	**31**	**1.1**	–	–	**26**	**1.2**	–	–	**39**	**1.2**	7	+	**33**	**1.0**
Herb layer																
Urtica dioica F, N	**75**	**29.6**	53	21.6	97	46.8	80	39.9	67	23.7	44	11.3	**52**	**11.4**	22	4.9
Oxalis acetosella	**43**	**0.8**	23	0.2	40	0.8	29	0.2	39	0.8	17	+	**48**	**1.0**	19	0.2
Viola reichenbachiana	**25**	**0.5**	11	+	11	0.2	9	+	22	0.4	–	–	**44**	**0.9**	22	0.1
Vicia sepium	**20**	**0.5**	5	+	17	0.6	9	+	6	0.1	–	–	**33**	**0.7**	4	+
Galium odoratum	90	2.7	95	4.2	89	2.6	94	6.8	94	2.8	100	2.5	**89**	**2.6**	93	1.9
Fagus sylvatica	45	0.2	**61**	**1.4**	40	0.2	43	0.6	50	0.3	**78**	**2.1**	48	0.2	**74**	**2.0**
Circaea lutetiana F, N	14	0.3	**43**	**1.0**	17	0.4	**46**	**1.1**	17	0.3	56	1.0	7	0.2	**30**	**0.9**
Galium aparine N	15	0.3	**33**	**0.6**	20	0.5	34	0.7	11	0.4	39	0.3	11	0.1	26	0.8
Mercurialis perennis HF	99	34.3	98	42.9	100	29.5	**97**	**47.0**	94	31.3	94	36.4	100	42.4	100	41.9
Acer pseudoplatanus N	43	0.5	48	1.1	29	0.2	29	0.6	50	0.5	56	1.0	56	0.8	**67**	**1.8**
Lamiastrum galeobdolon	35	2.8	49	1.4	54	3.2	69	2.1	44	6.5	61	1.7	–	–	**15**	**0.2**
Fraxinus excelsior N	66	4.6	68	2.2	54	2.3	46	1.6	56	5.7	83	2.4	89	6.9	85	2.7
Milium effusum	25	0.7	20	0.2	40	1.1	29	0.4	28	0.8	22	0.1	4	+	7	+
Arum maculatum HF, N	21	0.6	26	0.2	29	1.0	20	0.2	11	0.2	28	0.2	19	0.4	33	0.2
Hordelymus europaeus HF	18	0.4	14	0.2	20	0.4	17	0.3	28	0.6	17	0.1	7	0.2	7	0.1
Anemone nemorosa	11	0.3	11	+	20	0.5	14	0.1	6	0.1	17	0.2	4	+	4	+
Athyrium filix-femina F	10	0.2	4	0.1	17	0.4	9	0.3	11	0.2	–	–	–	–	–	–
Alliaria petiolata N	9	0.2	14	0.2	17	0.4	17	0.3	–	–	6	+	4	+	15	0.1
Stachys sylvatica F, N	6	0.3	11	0.2	11	0.4	17	0.4	6	0.6	11	0.2	–	–	4	+

Frequency (*F* % of the plots in which the species were recorded) and mean cover (*C* %, +: ≤ 0.1%) of the most important species. *HF* Character or differential species (separating species) of the *Hordelymo-Fagetum*, *F* moisture indicator of the *Circaea*-Gruppe (Dierschke 1989a), *N* indicator of places rich in available nitrogen (nitrogen figure N ≥ 7 according to Ellenberg et al. 2001). Significantly higher values are in bold type ($p \leq 0.05$, Mann-Whitney U test), *nd* no data showed. In a typical beech forest light regime of less than 3.5% of the ambient light intensity (Eber 1972; Ellenberg 1996), *U. dioica* achieves less than 5% cover on average. However, if the relative light intensity is more than 7% of the ambient, *U. dioica* increases its cover to more than 50%. No other under-storey species reacts with similar vigour to the increased light supply. *M. perennis* is evenly represented along the light gradient with about 40% cover. Because the light values were measured above the herb layer, the true light availability for the lower-growing *M. perennis* under the taller-growing *U. dioica* is not reflected here. Mrotzek et al. (1996) were able to show that *M. perennis* reached its maximum cover where *U. dioica* occurred with less than 20% cover without, however, becoming completely excluded under almost closed patches of *U. dioica*.

This difference in behaviour of the two dominant species, *U. dioica* and *M. perennis*, with respect to the light regime and chemical soil factors, was demonstrated by a linear regression analysis using data from all 192 subplots (Mrotzek et al. 1996, Table 5.3). The respective correlation coefficient (r^2) values for the dependency of cover and height of *U. dioica* are consistently smaller for the soil factors pH, cation exchange capacity, and calcium content (each in the upper 0–15 cm of mineral soil) than for light intensity. For *M. perennis*, no linear correlation could be found between percent cover and the relative light intensity above the herb layer. However, for height, a weak correlation was found. Similarly low r^2 values were calculated with soil factors for combined growth of both *M. perennis* and *U. dioica*. According to the studies of Mrotzek (1998), there were no differences in the nitrogen supply between areas dominated by *U. dioica* or *M. perennis*. All studied plots showed high net nitrogen mineralisation rate of ca. 140 kg N ha^{-1} per year (Table 5.5), of which over 95% of the mineralised nitrogen was available in the form of nitrate. However, a clear difference in the soil nitrate content was observed during summer and autumn for the areas dominated by *U. dioica* and *M. perennis*. In the late summer and autumn, nitrate concentrations in the mineral soil were many times higher in areas where *M. perennis* had already died back and *U. dioica* was only weakly represented than in areas where *U. dioica* had dense cover.

Table 5.3 Correlation coefficients (r^2) of a linear regression between cover and growth height of *Urtica dioica* and *Mercurialis perennis* and various site factors at the Zierenberg research site (Mrotzek et al. 1996)

Independent factor	Dependent factor	*Urtica dioica*	*Mercurialis perennis*
Relative light intensity (% of full sun light intensity)	Cover	0.54*	0.00 n.s.
	Height	0.52*	0.18*
pH value (KCl) (0–15 cm)	Cover	0.11*	0.13*
	Height	0.15*	0.05*
Cation exchange capacity (mmol ieq l^{-1}) (0–15 cm)	Cover	0.08*	0.15*
	Height	0.14*	0.06*
Ca content (mval kg^{-1}) (0–15 cm)	Cover	0.20*	0.07*
	Height	0.28*	0.13*

n.s. Not significant, $*p \leq 0.001$; $n = 196$

The dominance of *U. dioica* and *M. perennis* overshadows the base floristic composition of the beech forest at Zierenberg and renders more difficult the already complex classification of species-rich beech forests on mesic sites in Central Europe (Dierschke 1989a, 2000a). Because of the presence of *M. perennis, Hordelymus europaeus, Arum maculatum*, and *Anemone ranunculoides* together with indicator species of moist soils with persistent, constant water supply species like *Carex remota, Circaea lutetiana, Ranunculus ficaria, Stachys sylvatica* and *U. dioica*, the *Hordelymo-Fagetum circaeetosum* (Dierschke 1989a) plant community type was assigned. Similarly, to the beech forests described by Seibert (1954) and Hakes (1991) from Hessian basalt sites, the *Hordelymo-Fagetum circaeetosum* with its dominance of *U. dioica* and/or *M. perennis* is species-poorer than the familiar *Hordelymo-Fagetum* on limestone sites (Dierschke 1989a; Hermy et al. 1993).

5.2.3 Göttinger Wald

The research site in the Göttingen Forest was established on a ca. 12-ha area in 1980. Following the management conversion from coppice with standards 100 years ago, the then-customary management method of thinning from below was implemented on the 120–130-year-old (2003) mixed deciduous beech forest stand until 1960. After the increased occurrence of beech slime flux (beech bark disease) at the beginning of the 1960s, silvicultural management was halted, and the stand remained unmanaged until today. However, the forest structure did not remain unchanged. Individual trees, especially those of *F. sylvatica* and *Acer platanoides*, were occasionally blown down or fell after being weakened by white rot. The resulting gaps have rapidly closed, however (Lambertz and Schmidt 1999).

F. sylvatica dominates the closed tall forest stand. The most common associated species is *Fraxinus excelsior*; in addition, *Acer pseudoplatanus, A. platanoides, Ulmus glabra, Quercus robur*, and *Q. petraea* can be found in the tree layer (Dierschke and Song 1982; Dierschke and Brünn 1993). Yield surveys carried out in 1983 at the beginning of studies indicated a younger stand age and smaller basal area in the NW than in the SE of the research plot (Pellinen 1986; Brünn 1992; Dierschke and Brünn 1993). In 1983, relatively high numbers of stems and higher standing volumes than expected in a managed beech forest were noted, which can be attributed to the lack of silvicultural treatment. A shrub layer was scarcely present at the beginning of the studies in 1980 (cover 5.2%; Table 5.4). Of the natural regeneration of *F. excelsior, A. pseudoplatanus, A. platanoides* and *F. sylvatica* which were richly represented in the herb layer, only the *F. excelsior* achieved the height of the shrub layer in scattered areas with better light conditions. The natural regeneration strongly suffered from roe deer browsing until the fencing of the experimental plot in the winter of 1980/81 (Dierschke and Song 1982; Eggert 1985).

Detailed analysis of the area showed a strongly horizontal structure of ground vegetation. The forest can be considered a sub-oceanic, sub-montane, mesic

Table 5.4 Vegetation of 281 permanent plots (100 m^2 each, east–west transect) at the Göttinger Wald research site from 1981 (vegetation relevés: Dierschke and Song 1982; Dierschke 1989b), 1991 (vegetation releves: Brünn 1992; Dierschke and Brünn 1993) and 2001 (vegetation releves: Rudolph 2002; Dierschke 2003, 2004, 2006)

	1981		1991		2001	
Tree layer, cover (%)	78.7		81.5		**89.0**	
Shrub layer, cover (%)	5.2		12.4		**17.1**	
Herb layer, spring, cover (%)	86.3		82.6		**97.0**	
Herb layer, summer, cover (%)	47.9		45.2		**53.8**	
Total number of species	76		72		56	
Number of herb layer species 100 m^{-2}	**24.0**		**22.4**		18.5	
	F	C	F	C	F	C
Woody species (shrub and herb layer, increase)						
Fraxinus excelsior	99	3.8	97	4.2	**100**	**12.1**
Fagus sylvatica	93	1.3	89	1.2	**99**	**12.4**
Acer platanoides	96	1.0	98	1.3	**98**	**3.9**
Acer pseudoplatanus	70	0.5	71	0.5	**90**	**2.3**
Ulmus glabra	48	0.3	48	0.3	**42**	**0.7**
Crataegus laevigata L	38	0.2	37	0.2	**43**	**0.4**
Species of the herb layer						
Species that increased						
Allium ursinum	91	37.6	91	54.2	**95**	**66.0**
Hedera helix	19	0.1	25	0.6	**25**	**2.0**
Species that decreased						
Mercurialis perennis DG	**90**	**28.7**	88	17.2	49	2.2
Lamiastrum galeobdololon DC	**92**	**9.0**	96	6.7	94	3.2
Oxalis acetosella DC	**84**	**7.0**	88	3.0	69	3.8
Galium odoratum	**96**	**6.7**	92	4.2	57	0.6
Hordelymus europaeus C	**100**	**3.0**	100	3.3	94	1.4
Melica uniflora	**64**	**2.1**	71	3.5	64	0.6
Primula elatior	**78**	**1.1**	54	0.5	39	0.2
Polygonatum verticillatum	**68**	**0.7**	37	0.7	53	0.4
Vicia sepium	**46**	**0.5**	33	0.4	12	+
Carex sylvatica DC	**60**	**0.4**	38	0.2	17	0.1
Euphorbia amygdaloides	**40**	**0.4**	11	+	8	+
Viola reichenbachiana	**62**	**0.3**	59	0.3	5	+
Ranunculus auricomus L	**50**	**0.3**	32	0.4	11	0.1
Lathyrus vernus L	**40**	**0.3**	40	0.3	34	0.2
Phyteuma spicatum	**50**	**0.2**	15	0.1	1	+
Indifferent species						
Anemone nemorosa	100	11.4	100	13.0	100	11.3
Asarum europaeum L	97	6.6	98	6.6	100	9.8
Anemone ranunculoides C	100	1.2	100	4.3	100	4.6
Arum maculatum C	94	2.7	93	1.7	100	1.8
Dryopteris filix-mas	74	1.6	72	1.8	82	1.0
Dentaria bulbifera	72	1.1	79	1.8	74	0.8

Frequency (*F* % of the plots in which the species were recorded) and mean cover (*C* %, +: $\leq 0.1\%$) of the most important species. *C* Character species of the *Hordelymo-Fagetum*, *DC* differential species separating from the *Carici-Fagetum*, *DG* differential species separating from the *Galio-Fagetum*, L differential species of the *Hordelymo-Fagetum lathyretosum* (Dierschke 1989a). Higher values are in bold type (only mean values published, therefore no statistics) limestone beech forest (*Hordelymo-Fagetum lathyretosum*, Dierschke 1989a), in which species such as *Anemone ranunculoides, Arum maculatum, Asarum europaeum, Hordelymus europaeus*, and *Lamiastrum galeobdolon* are equally represented. This basic floristic structure is overlain by small-scale spatial change of dominant species, conspicuous among which are *Allium ursinum, Mercurialis perennis*, and *Aconitum vulparia* with cover values of up to 100% in spring and early summer. In addition to dominance types (facies) there are also mixed types within the research plot, with different proportions of dominant species, resulting in greatly varied horizontal structure of the ground vegetation (Dierschke and Song 1982; Dierschke 1989a, 2003; Dierschke and Brünn 1993). In 1991, five of the most important communities within the association *Hordelymo-Fagetum lathyretosum* at the Göttinger Wald site were classified by Brünn (1992) to two variants: first, a central (typical) variant (percent cover of *Allium* less than 40%, often less than 5% or absent) and second, an *Allium ursinum* variant (percent cover of *Allium* more than 50% and often 70–100%; Dierschke 1989a) (Table 5.5). In the central variant, the *Anemone nemorosa* facies has relatively low cover in the herb layer. Only *A. nemorosa* and *Galium odoratum* achieved higher cover values. In 1991, the

Table 5.5 Distribution of small-scale vegetation types (facies of dominant herb layer species) within the total area (12 ha, 1,200 plots, 100 m^2 each) and along an East–west transect in 1981, 1991, and 2001 (2.81 ha, 281 plots, 100 m^2 each) of the *Hordelymo-Fagetum lathyretosum* on the Göttinger Wald research site (vegetation relevées: Dierschke and Song 1982, Brünn 1992, Dierschke and Brünn 1993, Rudolph 2002, Dierschke 2003, 2004, 2006)

Vegetation unit	Total area (12 ha)	East–west transect (2.81 ha)		
	1981	1981	1991	2001
Central (typical) variant				
***Anemone* facies**	**6.3%**	**4.4%**	**16.0%**	**19.6%**
Melica facies	0.8%	[a]	[a]	[a]
Anemone–Mercurialis facies	33.4%	[a]	[a]	[a]
Anemone–Aconitum facies	0.3%	[a]	[a]	[a]
***Mercurialis* facies**	**6.3%**	**24.5%**	**10.6%**	**2.5%**
Mercurialis–Melica facies	1.9%	[a]	[a]	[a]
Mercurialis–Aconitum facies	0.6%	[a]	[a]	[a]
Allium variant				
***Allium–Mercurialis* facies**	**9.1%**	**44.1%**	**26.5%**	**0.0%**
Anemone–Allium–Mercurialis facies	21.3%	[a]	[a]	[a]
Allium–Mercurialis–Melica facies	3.6%	[a]	[a]	[a]
***Allium* facies**	**5.1%**	**26.3%**	**45.8%**	**61.5%**
***Anemone–Allium* facies**	**8.7%**	**0.0%**	**0.0%**	**15.0%**
Allium–Melica facies	1.2%	[a]	[a]	[a]
***Aconitum* facies**	**1.2%**	**0.7%**	**1.1%**	**1.4%**
Stachys variant				
Ranunculus ficaria facies	0.2%	[a]	[a]	[a]
Sum total	100%	100%	100%	100%

[a]Denotes absence from the transect and/or not distinguished from the neighbouring (in bold type) units. Forms previously assigned to the *Galium* facies have been assigned to the *Anemone* units in their entirety

A. nemorosa facies was concentrated in the shadiest parts, in which the relative light intensity reached an average of only 3.4% (Brünn 1992; Dierschke and Brünn 1993). The *M. perennis* facies of the central variant had denser cover in spring and summer than the *Anemone* facies. Edaphically, the *Mercurialis* facies reached its optimum on rather shallow soils (mull rendzinas), which suffered slightly from moisture deficit in dry summers. *M. perennis* showed reduced vitality where soil pH dropped below 5.2, as was often the case in the *A. nemorosa* facies colonising relatively deep terra fusca, brown loam rendzina or para brown earth, where the upper horizons were more strongly acidified (Dierschke and Song 1982; Dierschke 1989b; Schmidt et al. 1989; Brünn 1992; Dierschke and Brünn 1993).

Within the *A. ursinum* variant, the *Allium* and *Allium-Mercurialis* facies harboured the highest cover values of *A. ursinum*, a spring geophyte which prefers to colonise nutrient-rich mull soils. The *A. ursinum* facies was optimally developed in areas of moderate light conditions (4.0% relative irradiance intensity) on deep, loamy and humus-rich terra fusca-rendzinas, which are often even moister in spring than the rather shallow mull-rendzina of the *Allium-Mercurialis* facies.

Based on the prevalence of *Allium ursinum, Corydalis cava*, and of other species of the same nutrient requirements, the *Aconitum vulparia* facies was assigned to the *Allium ursinum* variant of the *Hordelymo-Fagetum lathyretosum* (Dierschke and Song 1982; Brünn 1992; Dierschke and Brünn 1993). *Aconitum vulparia* occurred with high cover on a few small spots over the entire research plot; this may be an example of random colonisation of a highly competitive but dispersal-limited species.

5.3 Vegetation Ecology of the Beech Forest Ecosystems: Impact of Site Conditions

Floristic and structural differences of Central European forest communities can be attributed to a great extent to differences in soil chemistry (Tüxen 1954; Ellenberg 1996; Leuschner 1999). Following the mean indicator values of Ellenberg et al. (2001), the main site differences between the beech forest ecosystems of the Solling, Zierenberg, and Göttinger Wald research plots may be related to their differing degrees of acidity and their nutrient status, whereas differences in water supply and light intensity are comparably low (Table 5.6).

The acidic *Luzulo-Fagetum* of the Solling site clearly distinguishes itself from the more base-rich *Hordelymo-Fagetum* of the Zierenberg and Göttinger Wald sites in both cation exchange capacity and base saturation (Meesenburg et al., Chap. 3 of this volume). Species number and cover of the herb (and shrub) layer increases with increasing pH values, while the cover of the mosses decreases. The preponderance of acid-tolerant, or calcifuge, plants on the Solling site and the predominance of calcicolous plants (Runge and Rode 1991) on the Zierenberg and Göttinger Wald sites also reflect the differences in the acidity of the soils.

Table 5.6 Vegetation structure, diversity, and ecological indicator values of Ellenberg et al. (2001) for the three beech forest ecosystems of the Solling (based on 15 relevés from Gerlach 1970), Zierenberg (based on 192 relevés from Perona 1995) and the Göttinger Wald (based on 209 relevés from Brünn 1992) research sites

	Solling *Luzulo-Fagetum typicum*	Zierenberg *Hordelymo-Fagetum circaeetosum*	Göttinger Wald *Hordelymo-Fagetum lathyretosum*
Tree layer, cover (%)	**89.0 ± 8.3**	64.4 ± 22.5	**82.8 ± 14.5**
Shrub layer, cover (%)		**2.9 ± 3.7**	**12.4 ± 21.6**
Herb layer, cover (%)	29.8 ± 26.9	69.1 ± 23.4	**85.6 ± 18.6**
Moss layer, cover (%)	**4.5 ± 4.0**	0.1 ± 0.8	
Number of species (Herb layer)	7.9 ± 2.4	15.7 ± 4.9	**21.5 ± 4.2**
Mean indicator values of Ellenberg et al. (2001)			
Temperature	3.67 ± 0.37	5.29 ± 0.12	**5.33 ± 0.09**
Continentality	**3.44 ± 0.33**	3.09 ± 0.15	**3.28 ± 0.11**
Light	**4.51 ± 0.38**	3.44 ± 0.30	3.33 ± 0.21
Moisture	5.33 ± 0.27	**5.62 ± 0.19**	5.38 ± 0.09
pH reaction	3.05 ± 0.23	6.53 ± 0.19	**6.59 ± 0.15**
Nitrogen	4.14 ± 0.35	**6.56 ± 0.28**	6.26 ± 0.19
Site factors			
Relative light intensity (%)	4.4 ± 0.7	4.5 ± 2.2	4.2 ± 1.0
Organic layer (humus type)	Hemimor/humimor	Moder/vermimull	Vermimull
Thickness of organic layer (mm)	110	30–50	5
pH (0–5 cm, KCl, $CaCl_2$)	2.9–3.4	5.4 ± 0.5	**6.2 ± 0.7**
C/N (0–5 cm)	**18.8 ± 2.1**	11.8 ± 0.9	12.6 ± 0.4
N mineralisation (kg N ha^{-1} per year)	100 ± 11	**141 ± 23**	**137 ± 40**

The site data for the Solling (*SO*), Zierenberg (*ZI*) and the Göttinger Wald (*GW*) research sites were compiled based on studies by Eber (1972, SO), Ellenberg et al. (1986, SO), Brünn (1992, *GW*), Dierschke and Brünn (1993, *GW*), Perona (1995, *ZI*), Mrotzek et al. (1996, *ZI*), Mrotzek (1998, *ZI*), Leuschner (1999, *SO*, *ZI*, *GW*), Schmidt (2002b, *SO*, *GW*), Leuschner et al. (2004, *SO*, *GW*), and Meiwes et al. (Chapter 1.4, this volume, *SO*, *ZI*, *GW*). Significantly higher means (± standard deviation) are in bold type ($p \leq 0.05$, Mann-Whitney U test, not for thickness of organic layer and pH at *SO*)

Differences in the average nitrogen indicator value of Ellenberg et al. (2001) are not striking but still quite clear with values of 4.1 (Solling), 6.6 (Zierenberg) and 6.3 (Göttinger Wald). If one compares this with the annual nitrogen mineralisation, the Solling site with its average of 100 kg N ha^{-1} per year is about 40 kg N ha^{-1} per year lower than that of Zierenberg or Göttinger Wald (Mrotzek 1998; Schmidt 2002b). The differences in nitrogen availability are comparably low, such that this alone cannot explain the differences in the floristic composition and especially not the differences in diversity and cover values in the herb layers of the three beech ecosystems. The nitrogen mineralisation rate in the montane-influenced Solling is somewhat lower due to low average temperatures (Rehder 1970; Ellenberg 1977, 1996), but it is unlikely to be significantly limited by dryness at any of the three sites. The form of nitrogen may be equally as important for the species composition and productivity of the herb layer as the amount that is received. Since the studies of Bogner (1968), it has been known that a number of plants are physiologically not competitive in the presence of pure ammonium or nitrates, and thus are absent from certain forest communities. In the Solling site, nitrogen supply occurs both as ammonium and nitrates (Runge 1974; Ibrom and Runge 1989), while at the Zierenberg and the Göttinger Wald sites, it is primarily in the form of nitrate (Mrotzek 1998; Schmidt 2002b). Liming consistently enhances nitrification and is most likely the main reason why the ground vegetation of limed areas supports not only more indicators of base-rich sites but also of indicators of nitrogen-rich soils (Schmidt 2002a).

The high cover values of the herb layer with the dominance of *U. dioica* at Zierenberg is in part an expression of the high phosphorus availability of this basalt-influenced site (Mrotzek 1998, Meesenburg et al., Chap. 3 of this volume). According to Pigott and Taylor (1964), *U. dioica* is an indicator of soils with good phosphorus supply, and the species reacted in culture experiments more strongly to P-fertilisation than it did to N-fertilisation. Mrotzek (1998) found an unusually high P-uptake at Zierenberg with up to 10 kg P ha^{-1} per year through the *U. dioica*-dominated herb layer. The herb layer may have assimilated more phosphorus than the tree layer in the aboveground biomass (5.1 kg P ha^{-1} per year, Rademacher et al., Chap. 8, this volume). In the Solling and Göttinger Wald sites, the phosphorus supply is sufficient and not limiting for the development of beech forest communities according to the previous studies by Schmidt (1970). However, it would be necessary to investigate whether atmospheric inputs (Meesenberg et al., Chap. 15, this volume) have not strongly altered soil acidity and nitrogen supply such that phosphorus supply on these sites would require re-evaluation.

Differences in the nutrient supply are also reflected in differences of the humus form and the thickness of the organic litter layer. The thickness of the humus layer is 20 times greater in the *Luzulo-Fagetum typicum* in the Solling site than in the *Hordelymo-Fagetum lathyretosum* of the Göttinger Wald site. This is concomitant with a significant decrease in species number and cover of the herb layer (Table 5.6). Various litter accumulation experiments in forests have observed the negative relation between the amount of litter and species richness and biomass production of the herb layer (Grime 1979; Facelli and Pickett 1991; Xiong and Nilsson 1997; Xiong et al. 2001).

In addition to the nutrient supply and soil chemical conditions, fine root development is decisive for species richness and cover values of the herb layer in closed beech forests. In the *Luzulo-Fagetum* of the Solling site, 165 mg fine root mass per 100 ml soil volume was found in the upper soil horizons (O_f/O_h or A_h), while in the *Hordelymo-Fagetum* of the Zierenberg and the Göttinger Wald sites, only 110–140 mg 100 ml^{-1} was found. Together with three other beech sites, Leuschner

(1999) established a correlation between species number of the herb layer and a combination of the factors including soil pH value and density of fine roots (and thickness of topsoil). Herb layer cover showed a stronger relationship with exchangeable magnesium or base saturation, and with root density. However, no relationship with pH value was found. A high density of fine roots in the topsoil and upper mineral soil of poor, acidic beech forest sites can effectively exclude many species of the herb layer, because intense competition for nutrients and water hinders the establishment and growth of plants. However, when mean value of fine roots is expressed on an area basis there is hardly any difference between the Göttinger Wald and the Solling (320–350 g m^{-2} fine root biomass) sites, even if one includes the biologically very active deeper mineral soil horizons of the Göttinger Wald site (Leuschner et al. 2004). Area-based differences with higher values in the *Luzulo-Fagetum* were found only in the density of fine root tips and fine root necromass. Leuschner et al. (2004) therefore hypothesise that high root tip densities in acidic infertile soils compensate for low nutrient supply rates, and large necromasses are a consequence of adverse soil chemical conditions.

5.4 Vegetation Dynamics

5.4.1 Solling

In a comparison of 15 vegetation relevés (sample stands) of the *Oxalis* variant of the *Luzulo-Fagetum typicum* from Solling from 1966–1968 (Gerlach 1970) with 25 relevés from 1999–2000 (Weckesser 2003), the average cover values of the tree and shrub layers have scarcely changed (Table 5.1, Weckesser and Schmidt 2004) during that period. The cover values of the herb and moss layers, by contrast, have decreased significantly. However, the total number of herb and moss species per relevé has increased, mainly species such as *Rubus idaeus, Dryopteris dilatata, U. dioica*, and *Cardamine flexuosa* in the herb layer and *Brachythecium rutabulum* in the moss layer. These are not typical forest species of base- and nutrient-poor sites (Ellenberg et al. 2001; Schmidt et al. 2003) but rather species that react equally positively to increased nitrogen input, liming, and soil disturbance (Bobbink et al. 1998; Fischer 1999; Schmidt 1999, 2002a; Diekmann and Falkengren-Grerup 2002; Weckesser and Schmidt 2004). This contrasts with the decrease of the indicators of acidic and nutrient-poor sites (Ellenberg et al. 2001) such as *Luzula luzuloides, Deschampsia flexuosa, Carex pilulifera, Polytrichum formosum*, and *Oxalis acetosella*. Therefore, the disappearance of the typical species combination of the *Luzulo-Fagetum typicum* could be expected in future at the Solling site (Weckesser and Schmidt 2004), in agreement with the results of Wilmanns and Bogenrieder (1986) from the Kaiserstuhl region and of Röder et al. (1996) from the Spessart. This species-poor subunit of the *Luzulo-Fagetum*, which once covered large areas of the forest landscape of Central Europe (Ellenberg et al. 1986, Ellenberg 1996, Bohn et al. 2003), is increasing being displaced by more nutrient-demanding vegetation types, in which indicators of nitrogen, liming, and disturbance predominate (Fischer 1999, Schmidt 1999, 2002a, Weckesser and Schmidt 2004).

5.4.2 Zierenberg

Vegetation was monitored in the years of 1989 (Hubeny 1991) and 1995 (Perona 1995) on 40 subplots (2 × 2.5 m) along a transect from the basalt-influenced upper slope to the limestone-influenced lower slope. During this period, the percent cover of *U. dioica* has generally decreased, while that of *M. perennis* slightly increased. For both species, a close relationship to the relative light intensity on the sites was detected, which had markedly decreased by more than 2.5% during this period on over 40% of the subplots. This may be attributed to the increased canopy closure following the thinnings done in 1983 and 1985. The percent cover of *U. dioica* decreased especially in areas of low relative light intensity.

The change towards low light conditions continued in the following years, as shown by a comparison of 80 permanent plots that have been re-sampled since 1993 at intervals of 1–2 years (Table 5.2) in the context of the Europe-wide Level II biomonitoring programme (Schulze and Eichhorn 2000). From 1993 until 2002, the percent cover of *U. dioica* has significantly declined and at the same time the cover of *M. perennis* has increased (albeit not significantly). The decrease in frequency of occurrence of under-storey as documented by Schulze and Eichhorn (2000) until 1999 has thus continued. Whereas *U. dioica* was found on three-quarters of all permanent plots in 1993, it was observed on only around one-half of all plots in 2002. The low intensity shelterwood treatments that took place especially in edge areas in 1996/97 and 1999 (Mrotzek 1998; Schulze and Eichhorn 2000) did not improve the light situation for *U. dioica* in the central part of the research plot. Besides the increase in *M. perennis*, the increase of *Fraxinus excelsior* and *A. pseudoplatanus* in the shrub layer, and *F. sylvatica, Circaea lutetiana*, and *Galium aparine* in the herb layer was striking. Tree regeneration (*F. excelsior, A. pseudoplatanus, F. sylvatica*) has certainly benefitted directly from reduced game browsing and indirectly from the reduced competitiveness of *U. dioica*. The increase in the above-mentioned herbs, and the decrease of *Oxalis acetosella, Viola reichenbachiana* and *Vicia sepium*, are difficult to correlate with any changed habitat factor or competitive relationships.

5.4.3 Göttinger Wald

From 1981 to 2001, permanent plot studies along the east–west transect showed extraordinarily strong temporal dynamics of the limestone beech forest (Dierschke and Song 1982; Dierschke 1989b, 2003, 2006; Brünn 1992; Dierschke and Brünn 1993; Rudolph 2002, Table 5.4). The lack of silvicultural treatment was responsible for the increase in cover of the tree layer, and sufficient protection from game browsing was responsible for the strong development of the shrub layer. The shade-tolerant *F. sylvatica*, which at the beginning of the studies was less represented in the regeneration than *Fraxinus excelsior*, has now the highest cover in the understorey. Valuable broadleaved tree species such as *F. excelsior, A. pseudoplatanus*, and *A. platanoides* are only competitive against the beech on base-rich sites when they are thinned using group selection (Schmidt 1997, 2002b; Lambertz and Schmidt 1999; Wagner 1999) or when the stands are affected by large-scale wind-throw events (Schmidt 2002c, 2005; Kompa and Schmidt 2006).

Strong changes were observed in the herb layer (Table 5.4). The most prevalent trend was the loss in the number of species: since 1981, the total number of species decreased from 76 vascular plant species to 56 and the average number of species per relevé (100 m^2) decreased from 24.0 to 18.5. This result agrees with the observations of floristic and phyto-sociological studies in forest nature reserves of Lower Saxony and Hessia, in which the diversity of vascular plants was lower than in comparable, semi-natural managed forests. Greater amounts of light and soil disturbance are primarily responsible for this, allowing species that grow mainly in open areas to occur in woodlands (Schmidt 1999, 2003, 2005; Schmidt and Schmidt 2007).

Among the facies-forming species that decreased strongly were *M. perennis* and *Galium odoratum. Allium ursinum*, whereas *Anemone nemorosa* and *Aconitum vulparia* were among the dominant species whose frequency and cover had remained constant over the last 20 years. Changes in the facies-forming species resulted in a corresponding shift in the dominant vegetation of the research units (Table 5.5). For instance, while the *Mercurialis* facies of the central variant occupied almost a quarter of the east–west transect in 1981, it had nearly completely disappeared by 2001. In a reverse trend, the area occupied by the *Allium* facies in 1981 more than doubled to the current 60%. A more detailed analysis of the individual subplots, however, showed that *A. ursinum* in fact scarcely expanded its range, i.e. it colonised only few new plots. More commonly, the many small gaps present between existing patches of *A. ursinum* became covered with facies-forming species during the past two decades. This happened especially in the plots formerly mapped as the *Allium-Mercurialis* facies, but also in some areas earlier mapped as *Mercurialis* facies that contained small individual patches of *A. ursinum* in 1981. The majority of the *Mercurialis* facies has now been replaced by the *Anemone* facies, which increased its area from around 5% to almost 20%. A new map unit, the *Allium-Anemone* facies, was distinguished in 2001, which along with the *Allium* facies observed in this research study includes large parts of the former *Allium-Mercurialis* facies. The *Aconitum* facies remained largely constant over the study period.

There are various explanations for the observed changes. A tendency towards both a general spread and increased density of *A. ursinum* has been reported for more than a decade, for instance in southern German forests (Buck-Feucht 1989; Bücking 1989; Böhling 2003) and in Hainich National Park (Ahrns and Hofmann 1998). Strong decline of *A. ursinum* in the light gradient of group-selected shelterwoods (Schmidt 1997) and following the resumption of a coppice with standards system (Stegmann and Schmidt 2005), Winterhoff (1977) suggested that open, drier conditions were unfavourable for the hygromorphically constructed *A. ursinum* than closed, more densely shaded mature beech forest. The increase of layers of woody plants with a densely closed canopy and the absence of silvicultural intervention have benefitted *A. ursinum*, which with its large leaves can take advantage of the irradiation with high productivity photosynthesis before the leafing of the over-storey. This is less the case for species with leaves that remain green through the deep shade of the summer and persist on the surface organic layer, such as *M. perennis, Galium odoratum, Lamiastrum galeobdolon, Hordelymus europaeus* and *Melica uniflora* (Hellmold and Schmidt 1989; Kriebitzsch 1989). A growing season that starts early and lasts longer, which has been observed in the Göttinger Wald site (Dierschke 2000b) and in Central Europe for two decades (Chmielewski and Rötzer 2001), has had positive effects on *A. ursinum*; but it has not been experimentally proven nor the spread of *A. ursinum* through increased nitrogen inputs (Ahrns and Hofmann 1998). Ongoing fertilisation experiments at Göttinger Wald site since 1996 have so far not revealed any nitrogen effects, which may be due to the site being relatively rich in nitrogen (Schmidt 1999, 2002b, Table 5.6).

The decline of *M. perennis* can be in part attributed to reduced light conditions on the surface organic layer. The more densely closed tree layer and the rise in natural regeneration results in decreased available irradiation, especially in the M*ercurialis* facies, which was still widespread in 1981. Available light under a closed canopy and under the natural regeneration has decreased to less than 2% of full sun light intensity (Lambertz and Schmidt 1999). Under canopy gaps where the relative irradiance intensity was above 5%, however, dense patches of *M. perennis* were found (Dierschke and Brünn 1993). In addition to its decline due to light reduction, *M. perennis* reacted sensitively with strong losses of vitality in subsequent years to extreme summer dryness and fungal infection with *Synchytrichum mercurialis* and *Melampsora rostrupii* (Schmidt 1988, 1997; Dierschke and Brünn 1993; Dierschke 2003, 2006).

Browsing is another factor in the dynamics of under-storey species. According to Klötzli (1965) and Schmidt (1978), *A. ursinum* and *M. perennis* are not the preferred food species of the roe deer. *Anemone nemorosa* and *Hedera helix*, by contrast, are preferred food species and thus benefit immediately from the fencing of forested areas (Schmidt 1978; Morgenroth 1992; Gerber and Schmidt 1996; Gill 2006). *Hedera helix* in particular is favoured as a browse species especially during winters with high snowfall, as the evergreen liana can be reached at any time by the deer on the tree trunks. Corresponding effects were observed in late winter 2003 at the Göttinger Wald site, when two roe deer were accidentally trapped inside the enclosure for several weeks and within a short period of time browsed back all ivy within reach on the trunks of the beeches. Whereas severe winters limit the spread into the herb layer and ascent into the canopy of *Hedera helix*, a sub-Atlantic–sub-Mediterranean species, mild winters probably facilitate its spread (Oberdorfer 2001, Dierschke 2005a, b). Similar observations of the invasion of deciduous forests by evergreen plants (laurophyllisation) have been made in southern Switzerland

(Klötzli et al. 1996; Walther 1997, 1999; Walther and Grundmann 2001, Grund et al. 2005).

5.5 Conclusion

Tree and herb layer vegetation of the beech forest research sites, Solling, Zierenberg, and Göttinger Wald, resulted primarily from differences in soil fertility. The water supply and the light regime on the surface organic layer was found to be of minor importance. The nutrient-poor acid soil at the Solling site had species-poor, sparse ground vegetation (*Luzulo-Fagetum*) while the base-rich soils at the Zierenberg and Göttinger Wald sites supported species-rich, dense herb layers (*Hordelymo-Fagetum*). Although soil characteristics support similar herb layers on the Göttinger Wald and Zierenberg beech forests, the vegetation of the Zierenberg site is strongly influenced by anthropogenic disturbances (opening of the canopy, scarification of the soil, higher atmospheric nitrogen deposition).

Under-storey vegetation of all the three beech forests changed significantly during the past decades:

- Species richness and the proportion of non-forest plant species (indicators of disturbances, e.g. *U. dioica, Cardamine flexuosa*) depended on the intensity of forest management and declined significantly after forest harvesting had stopped.
- Atmospheric N deposition supported the increase of some herb layer species (e.g. *A. ursinum, U. dioica*); whereas acid-tolerant species of the nutrient-poor beech forests (e.g. *Oxalis acetosella, Luzula luzuloides*) have declined significantly.
- The protection or reduction from hoofed-game browsing was responsible for an overall increase of woody regeneration and of some few herb layer species (e.g. *Hedera helix*) which were particularly preferred food species of deer.
- The increase of evergreen broad-leaved (laurophyllous) species such as *Hedera helix* and spring geophytes such as *A. ursium* has profited from the trend of mild winters and the earlier start of the vegetation period, which can be attributed to global warming (global climate change).

These vegetational changes have been documented due to long-term permanent plot studies on the three beech forest research sites. For causal explanations of vegetation pattern and process, experiments with under-storey vegetation or important herb layer species in the field or under controlled condition in greenhouse or climatic chambers are urgently needed.

Acknowledgments In addition to the cited data sources, unpublished field data of Bernadett Lambertz, Ina–Maria Schulze, and Marcus Schmidt were evaluated for this publication. Andreas Parth supported data analysis and preparation of the figures. The manuscript benefited from comments by Andreas Bolte, Hartmut Dierschke, Partap K. Khanna, and Marcus Schmidt. Gina Gemma Lopez and Toby Spribille improved the English version. All this assistance is gratefully acknowledged.

References

Ahrns C, Hofmann G (1998) Vegetationsdynamik und Florenwandel im ehemaligen mitteldeutschen Waldschutzgebiet Hainich im Intervall 1963–1995. Hercynia 31:33–64

Bobbink R, Hornung M, Roelofs JG (1998) The effect of air-borne nitrogen pollutants on species diversity in natural and semi-natural European vegetation. J Ecol 86:717–738

Bogner W (1968) Experimentelle Prüfung von Waldbodenpflanzen auf ihre Ansprüche an die Form der Stickstoff-Ernährung. Mitt Ver Forstl Standortskd Forstpflzücht 18:3–45

Böhling N (2003) Dauerflächen-Untersuchungen in einem Eichen-Hainbuchenwald im Vorland der Schwäbischen Alb (Südwestdeutschland), 1978–2001: Der Niedergang von *Scilla bifolia* und die Invasion von *Allium ursinum*. Tuexenia 23:131–161

Bohn U, Neuhäusl R, Gollub G, Hettwer C, Neuhäuslova Z (2003) Karte der natürlichen Vegetation Europas 1:2500000. Erläuterungstext mit CD-ROM. Landwirtschaftsverlag, Münster

Bolte A, Lambertz B, Steinmeyer A, Kallweit R, Meesenburg H (2004) Zur Funktion der Bodenvegetation im Nährstoffhaushalt von Wäldern – Studien auf Dauerbeobachtungsflächen des EU Level II-Programms in Norddeutschland. Forstarchiv 75:207–220

Brünn S (1992) Kleinräumige Vegetations- und Standortsdifferenzierung in einem Kalkbuchenwald. DiplArb Univ Göttingen

Buck-Feucht G (1989) Vegetationskundliche Dauerbeobachtung in den Schonwäldern Hohe Reisach und Saulach bei Kirchheim unter Teck. Mitt Forstl Vers Forschungsanst Bad-Württ Waldschutzgebiete 4:267–306

Bücking W (1989) Bannwald Bechtaler Wald. Dauerbeobachtungen 1970–1988. Natur Landschaft 64:574–577

Chmielewski FM, Rötzer T (2001) Response of tree phenology to climate change across Europe. Agric For Meteorol 108:101–112

Diekmann M, Falkengren-Grerup U (2002) Prediction of species response to atmospheric nitrogen deposition by means of ecological measures and life history traits. J Ecol 90:108–120

Dierschke H (1989a) Artenreiche Buchenwald-Gesellschaften Nordwest-Deutschlands. Ber Reinh Tüxen Ges 1:107–148

Dierschke H (1989b) Kleinräumige Vegetationsstruktur und phänologischer Rhythmus eines Kalkbuchenwaldes. Verh Ges Ökol 17:131–143

Dierschke H (2000a) Entwicklung und Stand der Systematik mitteleuropäischer Buchenwälder. Forst Holz 55:467–470

Dierschke H (2004) Kleinräumige Dynamik in der Krautschicht eines Kalkbuchenwaldes. Ergebnisse von 20-jährigen Dauerflächen (1981–2001). Forst Holz 59:433–435

Dierschke H (2000b) Phenological phases and phenological species groups of mesic beech forests and their suitability for climatological monitoring. Phytocoenologia 30:469–476

Dierschke H (2003) Pflanzendiversität im Göttinger Kalkbuchenwald in Raum und Zeit. In: Gradstein SR, Willmann R, Zizka G. (Hrsg): Biodiversitätsforschung. Kleine Senkenberg-Reihe 45, pp 137–146

Dierschke H (2005a) Zur Lebensweise, Ausbreitung und aktuellen Verbreitung von *Hedera helix*, einer ungewöhnlichen Pflanze unserer Flora und Vegetation. Hoppea 66:187–206

Dierschke H (2005b) Laurophyllisation – auch eine Erscheinung im nördlichen Mitteleuropa? Zur aktuellen Ausbreitung von *Hedera helix* in sommergrünen Laubwäldern. Ber Reinh Tüxen-Ges 17:151–168

Dierschke H (2006) Long-term dynamics in the herb layer of a calcareous beech forest: investigations of permanent plots, 1981–2001. Pol Bot Stud 22:165–172

Dierschke H, Bohn U (2004) Eutraphente Rotbuchenwälder in Europa. Tuexenia 24:19–56
Dierschke H, Brünn S (1993) Raum-zeitliche Variabilität der Vegetation eines Kalkbuchenwaldes – Untersuchungen auf Dauerflächen 1981–1991. Scripta Geobot 20:105–151
Dierschke H, Song Y (1982) Vegetationsgliederung und kleinräumige Horizontalstruktur eines submontanen Kalkbuchenwaldes. In: Dierschke H (Red): Struktur und Dynamik von Wäldern. Ber Int Symp Int Ver Vegetationskde (Rinteln 1981), pp 513–539
Eber W (1972) Über das Lichtklima von Wäldern bei Göttingen und seinen Einfluß auf die Bodenvegetation. Scripta Geobot 3:1–150
Eggert A (1985) Zur Ökologie der Krautschichtvegetation in einem Bärlauch-Kalkbuchenwald. Diss Univ Göttingen
Eichhorn J (1995) Stickstoffsättigung und ihre Auswirkungen auf das Buchenwaldökosystem der Fallstudie Zierenberg. Ber Forschungszentrums Waldökosysteme A 124, University Göttingen
Ellenberg H (ed) (1971) Integrated experimental ecology. Methods and results of ecosystem research in the German Solling project. Ecological Studies 2, Springer, Berlin
Ellenberg H (1977) Stickstoff als Standortsfaktor, insbesondere für mitteleuropäische Pflanzengesellschaften. Oecol Plant 12:1–22
Ellenberg H (1996) Vegetation Mitteleuropas mit den Alpen in ökologischer, dynamischer und historischer Sicht. Ulmer, Stuttgart
Ellenberg H, Mayer R, Schauermann J (1986) Ökosystemforschung – Ergebnisse des Sollingprojekts 1966–1986. Ulmer, Stuttgart
Ellenberg H, Weber HE, Düll R, Wirth V, Werner W (2001) Zeigerwerte von Pflanzen in Mitteleuropa. 3. Aufl. Scripta Geobot 18:1–262
Facelli JM, Pickett STA (1991) Plant litter: its dynamics and effects on plant community structure. Bot Rev 57:1–32
Fischer A (1999) Floristical changes in central European forest ecosystems during the past decades as an expression of changing site conditions. In: Karjalainen T, Spieker H, Laroussinie O (eds) Causes and consequences of accelerating tree growth in Europe. EFI Proceedings 27, pp 53–64
Gerber R, Schmidt W (1996) Einfluß des Rehwildes auf die Vegetation von Eichen-Hainbuchenwäldern im südlichen Steigerwald. Verh Ges Ökol 26:345–353
Gerlach A (1970) Wald- und Forstgesellschaften im Solling. Schriftenr Vegetationskde 5:79–98
Gill R (2006) The influence of large herbivores on tree recruitment and forest dynamics. In: Danell K, Duncan P, Bergström R, Pastor J (eds) Large herbivore ecology, ecosystem dynamics and conservation. Cambridge University Press, Cambridge, pp 170–202
Grime JP (1979) Plant strategies and vegetation processes. Wiley, New York
Grund K, Conedera M, Schröder H, Walther GR (2005) The role of fire in the invasion process of evergreen broad-leaved species. Basic Appl Ecol 6:47–56
Hakes W (1991) Das Galio odorati-Fagenion im Habichtswald bei Kassel – Untersuchungen zur ökologischen Feingliederung. Tuexenia 11:381–406
Härdtle W, Heinken T, Pallas J, Welß W (1997) Synopsis der Pflanzengesellschaften Deutschlands. Heft 2 Querco-Fagetea (H5) Sommergrüne Laubwälder. Teil 1: Quercion roboris – Bodensaure Eichenmischwälder. Selbstverlag Floristisch-soziologische Arbeitsgemeinschaft, Göttingen, pp 1–51
Heinken T (1995) Naturnahe Laub- und Nadelwälder grundwasserferner Standorte im niedersächsischen Tiefland: Gliederung, Standortsbedingungen, Dynamik. Diss Bot 239:1–311
Hellmold C, Schmidt W (1989) Energiegehalt und Energiebilanz der Krautschicht eines Kalkbuchenwaldes. Verh Ges Ökol 17:177–188
Hermy M, Bremt Pvd, Tack G (1993) Effects of site history on woodland vegetation. In: Broekmeyer MEA, Vos W, Koop H (eds) European forest reserves. Pudoc Scientific, Wageningen, pp 219–232
Hubeny C (1991) Untersuchungen zur Waldbodenvegetation, insbesondere zum Wachstum von *Urtica dioica* L. im Rahmen der Fallstudie Zierenberg des Forschungszentrums Waldökosysteme. DiplArb Univ Göttingen

Ibrom A, Runge M (1989) Die Stickstoff-Mineralisation im Boden eines Sauerhumus-Buchenwaldes unter dem Einfluß von Kalkung oder Stickstoffdüngung. Ber Forschzent Waldökosysteme A49:129–140

Klötzli F (1965) Qualität und Quantität der Rehäsung. Veröff Geobot Inst ETH Zürich 38:1–128

Klötzli F, Walther GR, Carraro G, Grundmann A (1996) Anlaufender Biomwandel in Insubrien. Verh Ges Ökol 26:537–550

Kompa T, Schmidt W (2006) Zur Verjüngungssituation in südniedersächsischen Buchen-Windwurfgebieten nach einem lokalen Orkan von 1997. Forstarchiv 77:3–19

Koperski M, Sauer M, Braun W, Gradstein SR (2000) Referenzliste der Moose Deutschlands. Schriftenr Vegetationskde 34:1–519

Kriebitzsch WU (1989) CO_2- und H_2O-Gaswechsel von verschiedenen Krautschichtpflanzen in einem submontanen Kalkbuchenwald. Verh Ges Ökol 17:189–202

Lambertz B, Schmidt W (1999) Auflichtungen in Buchen- und Buchenmischbeständen auf Muschelkalk und Buntsandstein – Untersuchungen zur Verjüngungs- und Vegetationsstruktur. Verh Ges Ökol 29:81–88

Leuschner C (1998) Mechanismen der Konkurrenzüberlegenheit der Rotbuche. Ber Reinh Tüxen Ges 10:5–18

Leuschner C (1999) Zur Abhängigkeit der Baum- und Krautschicht mitteleuropäischer Waldgesellschaften von der Nährstoffversorgung des Bodens. Ber Reinh Tüxen Ges 11:109–131

Leuschner C, Hertel D, Schmid I, Koch O, Muhs A, Hölscher D (2004) Stand fine root biomass and fine root morphology in old-growth beech forests as a function of precipitation and soil fertility. Plant Soil 258:43–56

Morgenroth K (1992) Der Einfluß des Rehwildes auf die Vegetation des Göttinger Kalkbuchenwaldes. DiplArb Univ Göttingen

Mrotzek R (1998) Wuchsdynamik und Mineralstoffhaushalt der Krautschicht in einem Buchenwald auf Basalt. Ber Forschungszentrums Waldökosysteme A152, University Göttingen

Mrotzek R, Perona L, Schmidt W (1996) Einfluß von Licht und ausgewählten Bodenfaktoren auf die Verteilung von *Urtica dioica* L. und *Mercurialis perennis* L. in der Bodenvegetation des Buchenwaldökosystems der Fallstudie Zierenberg. Verh Ges Ökol 26:559–564

Oberdorfer E (2001) Pflanzensoziologische Exkursionsflora. Ulmer, Stuttgart

Pellinen P (1986) Biomasseuntersuchungen in einem Kalkbuchenwald. Diss Univ Göttingen

Perona L (1995) Standortbedingungen von *Urtica dioica* (L.) und *Mercurialis perennis* (L.) auf der Versuchsfläche der Fallstudie Zierenberg. DiplArb Univ Göttingen

Pigott CD, Taylor K (1964) The distribution of some woodland herbs in relation to the supply of nitrogen and phosphorus in the soil. J Ecol (Suppl) 52:175–185

Rehder H (1970) Zur Ökologie, insbesondere Stickstoffversorgung subalpiner und alpiner Pflanzengesellschaften im Naturschutzgebiet Schachen (Wettersteingebirge). Diss Bot 6:1–90

Reif A, Teckelmann M, Schulze ED (1985) Die Standortsamplitude der Großen Brennessel (*Urtica dioica* L.) – eine Auswertung vegetationskundlicher Aufnahmen auf der Grundlage der Ellenbergschen Zeigerwerte. Flora 176:365–382

Röder H, Fischer A, Klöck W (1996) Waldentwicklung auf Quasi-Dauerflächen im *Luzulo-Fagetum* der Buntsandsteinrhön (Forstamt Mittelsinn) zwischen 1950 und 1990. Forstw Cbl 115:321–335

Rudolph D (2002) Räumliche Vielfalt und zeitliche Variabilität in einem Kalkbuchenwald bei Göttingen. DiplArb Univ Göttingen

Runge M (1974) Die Stickstoff-Mineralisation im Boden eines Sauerhumus–Buchenwaldes. II. Die Nitratproduktion. Oecol Plant 9:219–230

Runge M, Rode M (1991) Effects of soil acidity on plant associations. In: Ulrich B, Sumner ME (eds) Soil acidity. Springer, Berlin, pp 182–202

Schaefer M (1989) Die Bodentiere eines Kalkbuchenwaldes: ein Ökosystemforschungsprojekt. Verh Ges Ökol 17:203–212

Schaefer M (2003) Diversität der Fauna in Wäldern. – Gibt es Gesetzmäßigkeiten? Ber Reinh Tüxen Ges 15:169–179

Schmidt W (1970) Untersuchungen über die Phosphorversorgung niedersächsischer Buchenwaldgesellschaften. Scripta Geobot 1:1–120
Schmidt W (1978) Einfluß einer Rehpopulation auf die Waldvegetation – Ergebnisse von Dauerflächenversuchen im Rehgatter Stammham 1972–1976. Phytocoenosis 7:43–59
Schmidt W (1988) Langjährige Veränderungen in der Krautschicht eines Kalkbuchenwaldes (Dauerflächenuntersuchungen). Tuexenia 8:327–338
Schmidt W (1991) Die Veränderung der Krautschicht in Wäldern und ihre Eignung als pflanzlicher Bioindikator. Schriftenr Vegetationskde 21:77–96
Schmidt W (1997) Zur Vegetationsdynamik von Lochhieben in einem Kalkbuchenwald. Forstwiss Centralbl 116:207–217
Schmidt W (1999) Bioindikation und Monitoring von Pflanzengesellschaften – Konzepte, Ergebnisse, Anwendungen, dargestellt an Beispielen aus Wäldern. Ber Reinh Tüxen Ges 11:133–155
Schmidt W (2002a) Einfluss der Bodenschutzkalkungen auf die Vegetation. Forstarchiv 73:43–54
Schmidt W (2002b) Stickstoffkreislauf in Schlaglücken eines Kalkbuchenwaldes. Allg Forst Jagd Z 173:67–76
Schmidt W (2002c) Die Naturschutzgebiete Hainholz und Staufenberg am Harzrand – Sukzessionsforschung in Buchenwäldern ohne Bewirtschaftung. Tuexenia 22:151–213
Schmidt W (2003) Vielfalt im Urwald – Einfalt im Wirtschaftswald? Untersuchungen zur Gefäßpflanzendiversität in Naturwaldreservaten. In: Gradstein SR, Willmann R, Zizka G (Hrsg): Biodiversitätforschung. Kleine Senkenberg-Reihe 45:185–204
Schmidt W (2005) Herb layer species as indicators of biodiversity of forest ecosystems – Examples from forest nature reserves and managed beech forests. For Snow Landsc Res 79:111–125
Schmidt M, Schmidt W (2007) Vegetationsökologisches Monitoring in Naturwaldreservaten. Forstarchiv 78:205–214
Schmidt W, Hartmann T, Kothe G, Schultz R (1989) Jahresrhythmus und Produktion der Krautschicht in einem Kalkbuchenwald. Verh Ges Ökol 17:145–157
Schmidt M, Ewald J, Fischer A, Oheimb Gv, Kriebitzsch WU, Ellenberg H, Schmidt W (2003) Liste der in Deutschland typischen Waldgefäßpflanzen. Mitt Bundesforschungsanst Forst-Holzwirtschaft 212:1–32, Anhang, CD
Schulze IM, Eichhorn J (2000) Veränderungen im Stickstoffhaushalt von Buchenwäldern auf Basalt: Die Ausbreitung der Großen Brennessel (*Urtica dioica* L.) und ihr Einfluss auf die natürliche Verjüngung der Buche. Forst Holz 55:435–441
Seibert P (1954) Die Wald- und Forstgesellschaften im Graf Görtzischen Forstbezirk Schlitz. Angew Pflanzensoz 9:1–63
Stegmann F, Schmidt W (2005) Der Northeimer Mittelwald – Wald- und vegetationskundliche Untersuchungen zu einem Naturschutzprojekt. Gött. Naturk. Ber 6:141–158
Tüxen R (1954) Über die räumliche, durch Relief und Gestein bedingte Ordnung der natürlichen Waldgesellschaften am nördlichen Rand des Harzes. Vegetatio 5/6:454–477
Wagner S (1999) Ökologische Untersuchungen zur Initialphase der Naturverjüngung in Eschen-Buchen-Mischbeständen. Schr Forstl Fak Univ Gött Niedersächs Forstl Versanst 129:1–262
Walther GR (1997) Longterm changes in species composition in Swiss beech forests. Ann Bot 55:77–84
Walther GR (1999) Distribution and limits of evergreen broad-leaved (laurophyllous) species in Switzerland. Bot Helv 109:153–167
Walther GR, Grundmann A (2001) Trends of vegetation change in colline and submontane climax forests in Switzerland. Bull Geobot Inst ETH 67:3–12
Weckesser M (2003) Die Bodenvegetation von Buchen-Fichten-Mischbeständen im Solling – Struktur, Diversität und Stoffhaushalt. Cuvillier, Göttingen
Weckesser M, Schmidt W (2004) Gehen dem *Luzulo-Fagetum* die Trennarten verloren? Veränderungen der Bodenvegetation in bodensauren Buchenwäldern und Fichtenbeständen des Solling in mehr als drei Jahrzehnten. Tuexenia 24:191–206

Wilmanns O, Bogenrieder A (1986) Veränderungen der Buchenwälder des Kaiserstuhls im Laufe von vier Jahrzehnten und ihre Interpretation – pflanzensoziologische Tabellen als Dokumente. Abh Landesmus Naturk Münster 48:55–79

Winterhoff W (1977) Über Verbreitungslücken einiger Arten im Göttinger Wald. Mitt FlorSoz Arbeitsgem NF 19/20:365–375

Wisskirchen R, Haeupler H (1998) Standardliste der Farn- und Blütenpflanzen Deutschlands. Ulmer, Stuttgart

Xiong S, Nilsson C (1997) Dynamics of leaf litter accumulation and its effects on riparian vegetation: a review. Bot Rev 63:240–264

Xiong S, Nilsson C, Johansson ME, Jansson R (2001) Responses of riparian plants to accumulation of silt and plant litter: the importance of plant traits. J Veg Sci 12:481–490

Chapter 6
Microbial Biomass

R. Brumme, M. Raubuch, J. Priess, C. P. Wang, and T.–H. Anderson

6.1 Introduction

Micro-organisms (bacteria, fungi) contribute to more than 90% of the carbon dioxide evolved during decomposition of forest litter (Schaefer 1991), indicating their role in the element cycling in forest ecosystems. Nitrogen is mostly cycled between primary producers and decomposer biota. The activity and growth of micro-organisms depend on carbon and nutrient supply and the physico-chemical environment because of their high surface area/volume ratio of microbial bodies (Hattori and Hattori 1976; Paul and Clark 1996). Soil acidity has often been shown to reduce microbial biomass (C_{mic}) and to increase metabolic respiration (qCO_2, ratio between microbial-respiration-C and microbial-biomass-C) in the surface mineral soil which was interpreted as an increased stress on micro-organisms living in acid soils (Anderson and Domsch 1993; Anderson, Chap. 20, this volume). In addition to the chemical stress, soil acidity may change the litter quality, litter amount and their distribution in the soil profile by acid sensitive earthworms. Quality and amount of litter determine the amount of micro-organisms that would live on the organic matter, and is indicated by the fraction of C_{mic} in organic carbon (C_{org}) (Anderson and Domsch 1986). Another effect of soil acidity relates to an increase in bacterial respiration with increasing pH. Less is known of microbial biomass and the mentioned relationships in the surface organic layer and deeper soil horizons. Here, we provide depthwise distribution of microbial carbon (C_{mic}), microbial nitrogen (N_{mic}), metabolic quotient (qCO_2), the fraction of C_{mic} in C_{org} (C_{mic}-to-C_{org} ratio) and the fraction of fungal respiration of soil respiration in three beech forest soils, the Göttinger Wald, Zierenberg, and the Solling sites (for site description see Part A). The following questions will be addressed. (1) Does soil acidity reduce the microbial biomass and the qCO_2 value in the acid soil profile at Solling site compared to the less acid soils at Göttinger Wald and Zierenberg sites? (2) Does better substrate quality at Göttinger Wald and Zierenberg sites increase the C_{mic}-to-C_{org} ratio? (3) What is the contribution of bacteria and fungi to soil respiration and how does soil acidity affect these contributions?

R. Brumme and P.K. Khanna (eds.), *Functioning and Management of European Beech Ecosystems*, Ecological Studies 208,
DOI: 10.1007/978-3-642-00340-0_6,

6.2 Microbial Carbon and Nitrogen

Microbial carbon (C_{mic}), measured by substrate-induced respiration method (SIR) (Anderson and Domsch 1978; Heilmann and Beese 1992) in May 1995, was the highest in the L layers (17–18.9 mg C_{mic} g^{-1}) and showed no difference among the three beech forest sites (Table 6.1). In the moder profile microbial biomass-C decreased from 18.9 mg g^{-1} in the L layer to 0.97 mg g^{-1} in the H layer at the Solling site. In the mineral soils, the lowest C_{mic} value was found at the Solling site (about 0.1 mg C_{mic} g^{-1}), and it did not change down to 20 cm depth. The microbial biomass values in the uppermost base-rich Göttinger Wald and Zierenberg soils were up to 10 times higher than those at the Solling site and decreased to about 0.2 mg C_{mic} g^{-1} in the 15–20 cm depth. These values are in the range observed for soils under beech and oak forests in northern Germany. Anderson (Chap. 20, this volume) reported values for microbial biomass in the range of 0.18–3.12 mg C_{mic} g^{-1} in 0–5 cm depth.

Similar microbial biomass values were found in the base-rich soils at the Göttinger Wald (988 kg C_{mic} ha^{-1}) and Zierenberg (919 kg C_{mic} ha^{-1}) sites for L to 20 cm soil depth out of which 11% and 36% were attributed to the surface organic layers of the two sites, respectively. The microbial biomass at the Solling site was 42% and 48% lower (532 kg C_{mic} ha^{-1}) than in the base-rich Zierenberg and Göttinger Wald soils. A high fraction of microbial biomass was present in the surface organic layer (67%) of the Solling site but it did not cover for the low values in the mineral soil. Similar values for the microbial biomass were reported for other European forest soils (148–619 kg C_{mic} ha^{-1} in L to 22 cm depth, micro-calorimeter method) by Raubuch and Beese (1995). They reported ratios of C_{mic} in the surface organic layer and the surface mineral soils (0–22 cm) between 0.3 in less acid to 1.9 in the highly acid soils, and emphasised the contribution of microbial decomposition in the surface organic layer.

Table 6.1 Microbial biomass carbon (substrate-induced respiration method, SIR) and pH(H_2O) values in the soil profile of three beech forests (standard error in parentheses) ($n = 5$) (adopted from Chodak et al. 2003)

	Göttinger Wald			Zierenberg			Solling		
	pH	C_{mic}		pH	C_{mic}		pH	C_{mic}	
		mg g^{-1}	kg ha^{-1}		mg g^{-1}	kg ha^{-1}		mg g^{-1}	kg ha^{-1}
L	–	17.0 (0.57)	104	–	17.9 (0.54)	329	–	18.9 (1.31)	209
F	–	–	–	–	–	–	–	2.75 (0.37)	94
H	–	–	–	–	–	–	3.4	0.97 (0.05)	53
0–5	5.0	1.18 (0.17)	370	5.6	0.83 (0.05)	250	3.4	0.11 (0.05)	37
5–10	4.7	0.51 (0.07)	233	5.4	0.51 (0.06)	202	3.7	0.13 (0.01)	52
10–15	4.8	0.30 (0.04)	146	5.5	0.28 (0.03)	85	4.0	0.12 (0.04)	49
15–20	5.1	0.26 (0.03)	135	5.7	0.18 (0.02)	53	4.2	0.09 (0.1)	38
$\sum_{L-20}$			988			919			532

Table 6.2 Microbial biomass nitrogen and C_{mic}/N_{mic} ratios (chloroform fumigation extraction method, CFE) in the soil profile of Göttinger Wald and Solling forests (standard deviation in parentheses) ($n = 3$)

	Göttinger Wald			Solling		
	N_{mic}			N_{mic}		
	mg g^{-1}	kg ha^{-1}	C_{mic}/N_{mic}	mg g^{-1}	kg ha^{-1}	C_{mic}/N_{mic}
L	2.06 (0.65)	1.26	8.6	1.92 (1.71)	2.12	10.8
F	–	–	–	0.98 (0.41)	3.35	7.4
H	–	–	–	0.40 (0.10)	2.20	8.6
0–5	0.30 (0.04)	9.47	5.3	0.06 (0.01)	1.87	10.5
5–10	0.18 (0.05)	8.12	6.1	0.03 (0.02)	1.07	10.2
10–20	0.10 (0.04)	10.4	7.0	0.02 (0.02)	1.91	8.8
$\sum_{L-20}$		29.3			12.5	

Microbial nitrogen, measured with the chloroform fumigation and extraction method (Brookes et al. 1985a, b), showed a similar distribution in the soil profiles at the Göttinger Wald and Solling sites as microbial carbon. However, microbial nitrogen was very low in the acid Solling soil as indicated by higher C_{mic}/N_{mic} ratios than those at the Göttinger Wald site in all soil layers (Table 6.2). Mostly, C_{mic}/N_{mic} ratios of 3–6 in agricultural and forest soils have been reported which are independent of land use pattern (Jenkinson 1988; Paul and Clark 1996). However, acid forest soils have much higher C_{mic}/N_{mic} ratios of up to 17, and a much larger variation, than the agricultural soils (Joergensen et al. 1995). The fungi, which are the primary decomposers of forest litter, could have a much higher C/N ratio (4–15, Paul and Clark 1996) which may explain the higher C_{mic}/N_{mic} ratio observed in the acid forest soil of the Solling site. In a study on ten acid and ten neutral surface mineral soils under beech, a high fungal-to-bacterial respiratory fraction ranging from 74 to −26 at high pH of ~6 and from 94 to 6 at low pH of ~3 were reported by selective inhibition of either fungal or bacterial metabolism (Blagodatskaya and Anderson 1998). The Zierenberg site showed a similar pH effect on the fungal-to-bacterial respiration from the basalt-dominated upper hill (main research area) to the lime dominated lower hill area. With increasing pH from the basalt to the lime-dominated soil, the fungal-to-bacterial respiration fraction decreased from 87 to 13 (pH(KCl) 4.1), 76 to 24, (pH(KCl) 5.5), and to 55 to 45 (pH(KCl) 6.4) (Fig. 6.1).

6.3 Metabolic Quotient and C_{mic}-to-C_{org} Relationship

Microbial carbon contributes between 0.3 and 7% of soil organic carbon (SOC) (C_{mic}-to-C_{org}) (Anderson and Domsch 1989) with mean values of 1.6% at Göttinger Wald, 1.9% at Zierenberg and 0.68% at Solling for L–20 cm depth in the soils. The L layers contained the highest content between 3.6% and 4.7% and low values of 0.6%–1.1% in the 15–20 cm depth (Fig. 6.2). The C_{mic}-to-C_{org} ratio is an indicator

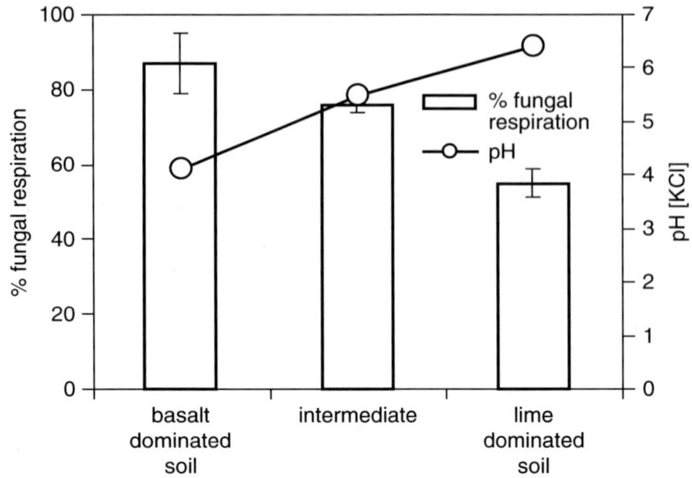

Fig. 6.1 Fraction of fungal respiration to total respiration (bacterial plus fungal) determined by selective inhibition with standard deviation and soil pH (KCl) at Zierenberg in three areas: from basalt-dominated (main research area), lime-dominated area and intermediate area containing both basalt and limestone ($n = 5{-}6$)

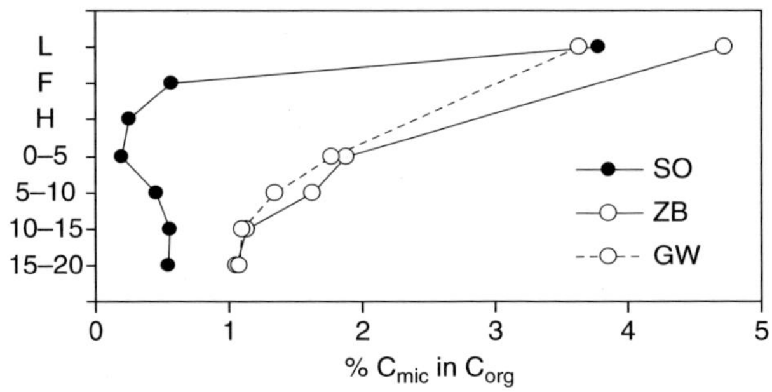

Fig. 6.2 %C_{mic} in C_{org} in the soil profiles at the Solling (*SO*), Zierenberg (*ZB*) and Göttinger Wald (*GW*) sites

of how much of the organic carbon is used for the growth of microbial cell mass (Anderson and Domsch 1986). The higher the C_{mic}-to-C_{org} fractions, the higher the amount of carbon that is present in available form and the higher the amount of micro-organisms that can live on the substrate. Thus, low C_{mic}-to-C_{org} ratio with increasing soil depth indicated low substrate availability. Fast incorporation of leaf litter by earthworms provided micro-organisms with easily available substrate in the base-rich Göttinger Wald and Zierenberg mineral soils (Schaefer and Schauermann, Chap. 7, this volume) especially from the luxuriant ground vegetation (Schmidt, Chap. 5, this volume), and explained the slow decrease of the C_{mic}-to-C_{org} ratio with soil depth in contrast to no change in the mineral layers of the acid Solling site

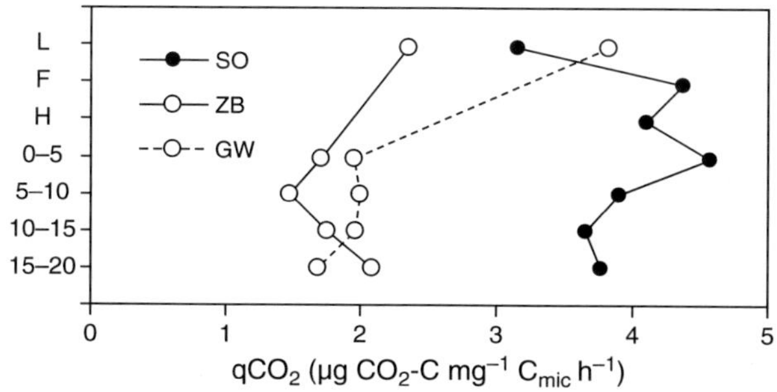

Fig. 6.3 Metabolic quotient (qCO_2, μg mg^{-1} C_{mic} h^{-1}) in the soil profiles at the Solling (*SO*), Zierenberg (*ZB*), and Göttinger Wald (*GW*) sites (adopted from Chodak et al. 2003)

(Fig. 6.2). The strong decrease in C_{mic}-to-C_{org} ratio from L to F layers at the Solling site was surprising because of the still discernible structure of the F layer suggesting that the substrate was highly available for micro-organisms. However, similar values for C_{mic}-to-C_{org} in the acid mineral soil and the F layer indicated that the microbial availability of the F substrate was similar to that of SOM in the mineral soil.

High metabolic quotient values (CO_2–C/C_{mic} or qCO_2) were found in the mineral soil layers of the acid Solling site (3.7–4.4 g CO_2–C mg^{-1} C_{mic} h^{-1}) (Fig. 6.3). The base-rich soils at the Göttinger Wald and Zierenberg sites had about 2 times lower qCO_2 values in the mineral soils indicating a much lower soil chemical stress to micro-organisms by aluminum, protons and heavy metals (Anderson and Domsch 1993; Duxbury 1995; Wood 1995).

6.4 Conclusions and Indications of Human Impacts

- The acid soil at the Solling site has the lowest microbial biomass, the lowest C_{mic}-to-C_{org} ratio and the highest qCO_2 value in all mineral soil layers of the three beech forest soils. High soil acidity, low bioturbation and sparse ground vegetation at the Solling site affected these parameters.
- Total microbial biomass was about 45% lower at the acid Solling site than at the base-rich Göttinger Wald and Zierenberg sites. About two-thirds of the total microbial biomass was present in the surface organic layer at Solling site whereas more than two-thirds were present in the base-rich mineral soils of the Göttinger Wald and Zierenberg sites.
- The luxuriant herb layer increased the substrate availability for micro-organisms (high C_{mic}-to-C_{org} ratio) in all mineral soil layers at the Göttinger Wald and Zierenberg sites by increasing earthworm activity. Low C_{mic}-to-C_{org} ratios indicated a low substrate availability in the acid mineral soil layers and the

strong decrease from L to the F/H layers suggested a rapid stabilisation of the remaining litter in the F layer.
- The fraction of fungal respiration decreased with increasing pH from a basalt-dominated soil (87% at pH(KCl) of 4.1) to a lime-dominated soil (55% at pH (KCl) of 6.4) along an acidity gradient at Zierenberg site.

References

Anderson JPE, Domsch KH (1978) A physiological method for the quantitative measurement of microbial biomass in soils. Soil Biol Biochem 10:215–221
Anderson T-H, Domsch KH (1986) Carbon link between microbial biomass and soil organic matter. In: Megusar F, Gantar M (eds) Proceedings of the 4th International Symposium of Microbial Ecology. Slovene Society for Microbiology, Lubljana, pp 467–471
Anderson T-H, Domsch KH (1989) Ratios of microbial biomass carbon to total organic carbon in arable soils. Soil Biol Biochem 21:471–479
Anderson T-H, Domsch KH (1993) The metabolic quotient for CO_2 (qCO_2) as a specific activity parameter to assess the effects of environmental conditions, such as pH, on the microbial biomass of forest soils. Soil Biol Biochem 25:393–395
Blagodatskaya EV, Anderson T-H (1998) Interactive effects of pH and substrate quality on the fungal-to-bacterial ratio and qCO_2 of microbial communities in forest soils. Soil Biol Biochem 30:1269–1274
Brookes PC, Kragt JF, Powlson DS, Jenkinson DS (1985a) Chloroform fumigation and the release of soil nitrogen: the effects of fumigation time and temperature. Soil Biol Biochem 17(6): 831–835
Brookes PC, Landman A, Pruden G, Jenkinson DS (1985b) Cloroform fumigation and the release of soil nitrogen: a rapid direct extraction method to measure microbial biomass nitrogen in soil. Soil Biol Biochem 17(6):837–842
Chodak M, Khanna P, Beese F (2003) Hot water extractable C and N in relation to microbiological properties of soils under beech forest. Biol Fert Soils 39:123–130
Duxbury T (1995) Ecological aspects of heavy metal responses in microorganisms. In: Marshall KC (ed) Advances in microbial ecology, vol. 8. Plenum, New York, pp 185–236
Hattori T, Hattori R (1976) The physical environment in soil microbiology: an attempt to extend principles of microbiology to soil microorganisms. CRC Crit Rev Microbiol 4:423–459
Heilmann B, Beese F (1992) Miniaturized method to measure carbon dioxide production and biomass of soil microorganisms. Soil Sci Soc Am J 56:596–598
Jenkinson DS (1988) The determination of microbial biomass carbon and nitrogen in soil. In: Wilson JR (ed) Advances in nitrogen cycling in agricultural ecosystems. CAB International, Wallingford, pp 368–386
Joergensen RG, Anderson T-H, Wolters V (1995) Carbon and nitrogen relationships in the microbial biomass of soils in beech (*Fagus sylvatica* L.) forests. Biol Fertil Soils 19:141–147
Paul EA, Clark FE (1996) Soil microbiology and biochemistry. Academic, San Diego
Raubuch M, Beese F (1995) Pattern of microbial indicators in forest soils along an European transect. Biol Fertil Soils 19:362–386
Schaefer M (1991) Secondary production and decomposition. In: Röhrig E and Ulrich B (eds) Ecosystems of the world 7 – Temperate deciduous forests. Elsevier, Amsterdam, pp 175–218
Wood M (1995) A mechanism of aluminium toxicity of soil bacteria and possible ecological implications. Plant Soil 171:63–69

Chapter 7
Soil Fauna

M. Schaefer and J. Schauermann

7.1 Introduction

The beech forests studied provide three different environments for the soil fauna and are located in a gradient from base-rich to acid soils. The forest "Göttinger Wald" is characterized by mull soil; soil conditions in the forest "Kleiner Gudenberg" (near the town of Zierenberg; hence denoted as forest "Zierenberg") are moder-like; and in the "Solling" forest moder soils prevail. In this chapter, the relationship between the fauna and the soil milieu is analysed, with the perspective to explain relational patterns with factors of a causal nature.

The beech forests are described in detail in several chapters of this volume (Meesenburg and Brumme; Panferov et al.; Meesenburg et al.; Schmidt). We highlight some habitat characteristics directly related to the soil fauna.

At the *Göttinger Wald* site, the soil is shallow and belongs to the soil series which has developed on the limestone, forming "terra fusca-rendzina" (about 50% of the area), rendzina (about 26%), terra fusca (about 14%) and some form of brown earths. The organic layer contains mainly leaf material (550 g dry mass m^{-2}) and the annual canopy leaf litter fall is 309 g dry mass m^{-2} (mean value for 1981–1991; Schmidt, unpublished results). The soil pH ranges from 6.8 to 4.3 with a mean value of about 5.8. The soil is base-rich with high cation exchange capacity. Further details are given by Schaefer (1990, 1991a) and Schaefer and Schauermann (1990).

The beech forest *Zierenberg* consists of three sites along the slope of a hill forming a gradient from basalt to limestone: it extends from the upper part of the hill (basalt site) via an intermediate site (transition zone) to the lower part of the hill (limestone site). The average mass of leaf litter is 1,900 g at the basalt site, 1,400 g at the intermediate site and 1,600 g dry mass m^{-2} at the limestone site. Annual beech leaf litter input (in 1989) was 509 g for the basalt, 445 g for the intermediate and 375 g dry mass m^{-2} for the limestone site (Hartwig et al. 1991). The input of litter of stinging nettles (*Urtica dioica*) is 70–90 g dry mass m^{-2} at the intermediate site (Scheu 1997). Soil pH of the basalt site is about 4–5, the values for the intermediate and limestone site are 5.7 and 6.7, respectively. Cation exchange

R. Brumme and P.K. Khanna (eds.), *Functioning and Management of European Beech Ecosystems*, Ecological Studies 208,
DOI: 10.1007/978-3-642-00340-0_7, © Springer-Verlag Berlin Heidelberg 2009

capacity is intermediate between that of the forests Göttinger Wald and Solling. Further details are presented in Eichhorn (1991) and Alphei (1995).

At the *Solling* site, the soil belongs to the series on carbonate-free silicate rocks: it is an acid brown earth with a moder humus form. The mass of organic layer on the soil increased from 2,960 g m^{-2} in 1966 to 4,460 g dry mass m^{-2} in 1979, while annual canopy leaf litter fall had a mean value of 290 g dry mass m^{-2} for the period from 1967 to 1976. Values of pH range from about 3 to 4. The soil is base-poor and cation exchange capacity is low. Further details are given in Ellenberg et al. (1986).

7.2 Fauna

The fauna of the two forests, Göttinger Wald and Solling, was sampled with different methods (soil and litter samples, photo-eclectors, pitfall traps) and different intensities over long periods, namely from 1980 to 1995 for Göttinger Wald and from 1967 to 1995 for Solling. For many (not all) animal groups, species numbers of the two forests can be compared because the intensity of sampling was similar. A synopsis of species numbers, population density and biomass is presented in Table 7.1, which is based on a synthesis of the data for the fauna of the two forests presented by Schaefer and Schauermann (1990). Additionally, more recent results are given for Nematoda (Solling) (Alphei 1995), Cryptostigmata, Gamasina and Uropodina (Göttinger Wald) (Schulz 1991a, b), Gamasina (Solling) (Buryn, unpublished results) and Diptera (Göttinger Wald) (Hövemeyer 1992). Some minor groups are omitted. For the Solling forest, several important and species-rich taxa were not studied.

The fauna of the Zierenberg forest was studied from 1990 to 1997 and is not completely known for the whole range of the soil fauna. The data for selected taxa of the soil fauna are presented for the gradient from basalt to limestone (Table 7.2). Data were obtained for Protozoa (Wellner, Coenen, Bonkowski, unpublished results), Nematoda (Alphei 1995), Enchytraeidae (Schlaghamerský 1998), Cryptostigmata (Heiligenstadt, unpublished results, Schulz 1991a), Collembola (Koch 1993), Gastropoda (Nieselt 1991), Lumbricidae (Bonkowski 1991), Isopoda (Nieselt 1991), Diplopoda (Nieselt 1991), Coleoptera (Nolte 1993), and Diptera (Markwardt 1993).

7.3 Macro-Gradient from Base-Rich to Acid Beech Forests

The series mull > mull-like moder > moder constitutes a macro-gradient with decreasing pH, decrease of the ratio C_{mic} to C_{org} and of the ratio bacteria to fungi, decreasing importance of the macrofauna, increasing significance of the meso-fauna and decreasing zoomass (cf. Fig. 12.7 in chapter 12). In the following sections, we compare the biota in the macrohabitats at Göttinger Wald, Zierenberg and Solling in more detail.

The mull-structured forest Göttinger Wald is characterised by high faunal biomass with earthworms as the dominant macrofauna group. This saprophagous

Table 7.1 Synopsis of species richness (*S*), mean annual population density (*N*, ind m^{-2}) and mean annual biomass (*B*, mg dry mass m^{-2}) in the beech forests Göttinger Wald (with mull soil) and Solling (with moder soil). Modified from Schaefer and Schauermann (1990). Additional data from Schulz (1991a, b), Hövemeyer (1992), Alphei (1995) and further unpublished sources

Animal group	Göttinger Wald			Solling		
	S	N	B	S	N	B
Microfauna						
Flagellata	?	2.7×10^9	54	?	?	?
Amoebina	?	3.5×10^9	1,133	?	?	?
Testacea	65	84×10^6	343	51	57×10^6	256
Turbellaria	3	859	8	3	1,882	4
Nematoda	110	732,000	146	90	3×10^6	65
Rotatoria	13	4,893	5	?	?	?
Tardigrada	4	4,207	4	?	41	9
Harpacticoida	?	3,873	2	1	3,300	0.6
Saprophagous and microphytophagous mesofauna						
Enchytraeidae	36	22,300	600	15	108,000	1,640
Cryptostigmata	75	22,445	241	72	101,810	195
Uropodina	11	1,971	19	4	1,525	?
Symphyla	2	57	?	1	?	?
Diplura	?	161	?	>1	277	?
Protura	?	2,481	?	>1	278	?
Collembola	48	37,835	153	50	63,000	246
Zoophagous mesofauna						
Gamasina	80	3,151	50	53	10,800	397
Saprophagous macrofauna						
Gastropoda	30	120	430	4	0	0
Lumbricidae	11	205	10,700	4	19	168
Isopoda	6	286	93	0	0	0
Diplopoda	6	55	618	1	0	0
Elateridae	11	37[a]	104[a]	4	332[a]	706[a]
Diptera	299	2,843[a]	161[a]	?	7,415[a]	628[a]
Zoophagous macrofauna						
Araneida	102	140	47	93	462	173
Pseudoscorpionida	3	35	16	2	89	10
Opilionida	8	19	11	4	20	6
Chilopoda	10	187	265	7	74	155
Carabidae	24	5	144	26	7	93
Staphylinidae	85	103	76	117	314	180

? = not studied; 0 = not present; [a]Larvae

guild is responsible for a high degree of bioturbation. In the moder-structured forest Solling, microphytophagous mesofaunal taxa (oribatid mites, collembolans) are prevalent. Generally, species diversity is lower in the Solling forest. A notable exception are staphylinid beetles. Total animal species number is 1,918 in the forest Göttinger Wald, among them 704 hymenopteran and 299 dipteran species (Schaefer 1996); the forest Solling contains 734 species (with the Hymenoptera and Diptera not studied), well below the corresponding 915 species of the mull-structured forest

Table 7.2 Synopsis of species richness (*S*), mean annual population density (*N*, ind m^{-2}) and mean annual biomass (*B*, mg dry mass m^{-2}) in the beech forest Kleiner Gudenberg (Zierenberg) for three sites along the slope of the hill: upper part of the hill (basalt site), intermediate site (transition zone), lower part of the hill (limestone site). From different sources

Animal group	Basalt site			Intermediate site			Limestone site			Total
	S	N	B	S	N	B	S	N	B	S
Microfauna										
Flagellata					5×10^6					
Amoebina			ca.800		5×10^6	ca.900		ca.400		
Ciliata	>53			>15			>29		>70	
Nematoda	71	1.8×10^6	High	85	1.87×10^6	medium	79	1.32×10^6	low	129
Saprophagous and microphytophagous mesofauna										
Enchytraeidae	18	23,000	900	19	22,000	1,000	11	13,500	500	23
Cryptostigmata	53	High	High	35	Medium	Medium	42	Low	Low	60
Uropodina	4	Low	Low	7	Medium	Medium	6	High	High	7
Collembola	38	44,800	140	45	44,200	167	38	30,400	123	49
Zoophagous mesofauna										
Gamasina	23	High	High	25	Medium	Medium	22	low	low	25
Saprophagous macrofauna										
Gastropoda	23	487	395[a]	28	881	809[a]	31	926	689[a]	33
Lumbricidae	8	114	15,000[b]	9	218	24,000[b]	9	67	23,000[b]	11
Isopoda	5	1,104	232	6	746	170	5	725	269	6
Diplopoda	6	272	646	8	200	365	8	355	639	8
Diptera		2,339[c]	288[c]		2,707[c]	384[c]		2,602[c]	493[c]	
Zoophagous macrofauna										
Carabidae	16		832[c]	17		1,823[c]	16		1,344[c]	21
Staphylinidae	34		147[c,d]	42		187[c,d]	29		122[c,d]	49

[a]Slugs not included. [b]Fresh wt. [c]Emerging adults per year (photo-eclectors). [d]Excluding Aleocharinae

Göttinger Wald. The fauna of the Zierenberg site appears to be intermediate between that of Göttinger Wald and Solling concerning species number, dominance of macro- and mesofauna and zoomass.

In a more detailed analysis, some general trends become apparent from the synopsis of the more important animal groups in the three forests (cf. Tables 7.1 and 7.2). Total soil zoomass amounts to about 15 g dry mass m^{-2} in the mull soil of the forest Göttinger Wald and about 5 g dry mass m^{-2} in the moder soil of Solling (with a value for Zierenberg in between). This difference is mainly due to the predominance of earthworms and, to a lesser degree, of saprophagous macroarthropods in the mull forest. The following faunal groups exhibit – partly striking – differences in the gradient Göttinger Wald → Zierenberg → Solling: Protozoa, Nematoda, Gastropoda, Enchytraeidae, Lumbricidae, Cryptostigmata, Isopoda, Diplopoda, Collembola, Diptera and some predatory macroarthropod groups (Araneida, Chilopoda and Staphylinidae). In the mull soil, many Protozoa (naked Amoebae, flagellates and ciliates), the nematodes, gastropods, lumbricids, isopods, diplopods and chilopods are favoured. The testaceans (Protozoa), enchytraeids, oribatid mites, collembolans, dipterans, araneids and staphylinids occur with higher population density and/or biomass in the moder soil. Thus, mainly microphytophagous mesofauna groups dominate in the moder environment. Zoophagous macroarthropods occur in comparable numbers in the three forest types. In the following paragraphs, some more detailed analyses of dominant species, abundance and biomass relations of edaphic key groups are given: microflora, Nematoda, Cryptostigmata, Collembola, earthworms and gastropods, and saprophagous macroarthropods (data mainly from Schaefer and Schauermann (1990), Schaefer (1991a)).

Microflora. In the series from mull to moder with a gradient of increasing acidity, the specific respiration of the microflora qCO_2 [unit CO_2–C released per unit microbial biomass, ($\mu g\ CO_2\text{–}C \times mg^{-1}\ C_{mic} \times h^{-1}$)] increases because more carbon is needed for the energy metabolism of the microorganisms (Fig. 7.1) (cf. Anderson, this volume). Less carbon is fixed in microbial biomass (decreasing ratio C_{mic} to C_{org}) in the moder soil. The species number of ectomycorrhizal fungi is twice as high on the base-rich site as compared to the acid site (Scheu, unpublished results).

Microfauna: Nematoda. Among the non-protozoan microfaunal groups, the Nematoda are clearly the most abundant (Alphei 1995). Species numbers are similar on the three sites; however, species composition is different. In the base-rich site, bacterial feeders are dominant whereas fungal or root feeders are prevalent in the forest on acid soil (Alphei 1998). Biomass was dominated by root hair feeders and the mycetophagous *Tylolaimophorus* in moder soil, whereas pantophagous taxa, mainly *Aporcelaimellus*, dominated in mull soil.

Saprophagous and microphytophagous mesofauna: Cryptostigmata, Collembola. Oribatid abundance is significantly higher in the moder than in the mull soil. However, biomass does not distinctly surpass values for the mull soil (mean body mass of mites in the Solling forest 0.002 mg, in the Göttinger Wald forest 0.007 mg dry mass). In Göttinger Wald, the oribatids *Steganacarus magnus*, *S. striculus* and *Nothrus palustris* dominate in biomass. In Solling *Platynothrus*

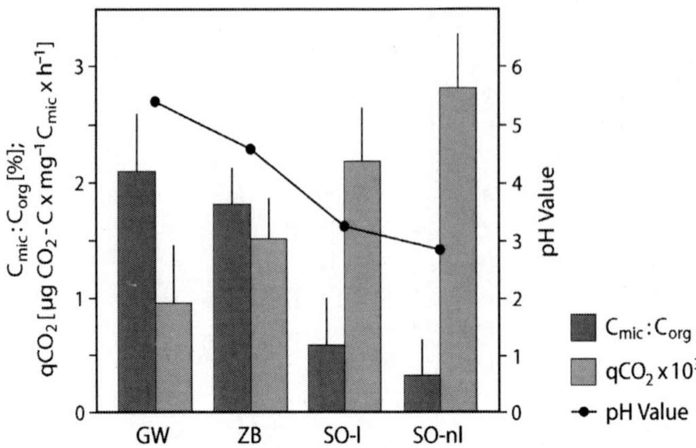

Fig. 7.1 Specific respiration qCO_2 and ratio C_{mic} to C_{org} in three different beech forests forming a gradient from base-rich to base-poor soil conditions. *GW* Göttinger Wald, *ZB* Kleiner Gudenberg (Zierenberg), *SO–l* Solling, limed, *SO-nl* Solling unlimed. Measurements in spring 1995 (means plus SD). (Scheu, unpublished results)

peltifer, *Nothrus silvestris*, *Nanhermannia coronata* and *Chamobates voigtsi* constituted over 50% of adult zoomass (Heiligenstadt 1988). In Göttinger Wald, mean annual population density and biomass of Collembola are about half of the corresponding values in the Solling soil. Thus, mean body mass is almost the same in both habitats (0.004 mg dry mass). In Göttinger Wald, several collembolan species dominate (measured in biomass): *Tomocerus flavescens*, *Lepidocyrtus lignorum, Folsomia quadrioculata*, *Isotomiella minor*, *Hypogastrura denticulata.* In Solling *Folsomia quadrioculata*, *Mesaphorura* sp., *I. minor*, and *Isotoma notabilis*, are the most abundant species. The oribatid and collembolan fauna of the Zierenberg site is more akin to that of Göttinger Wald (Koch 1993; Schulz 1991b).

Saprophagous macrofauna: Lumbricidae, Gastropoda. The Göttinger Wald mull soil is characterised by high zoomass of Lumbricidae (dominant in terms of biomass are *Lumbricus terrestris*, *Aporrectodea caliginosa*, *Octolasion* spp.) and of Gastropoda (dominants are *Perforatella incarnata*, *Arianta arbustorum*, *Aegopinella nitidula*, *Arion rufus*). In the Solling forest, only a few slug species, with negligible population density, and no snail species occur. In the moder soil, the earthworms are represented by the epigeic *Dendrobaena octaedra* and other species near tree stumps (e.g. *Lumbricus rubellus*); the mean body size of lumbricids is 8.8 mg dry mass which compares with 52.2 mg dry mass of the Göttinger Wald population. In the Zierenberg forest, endogeic species dominate with 15–24 g fresh mass m^{-2} (Bonkowski 1991). In this forest, Gastropods reach high values of density and biomass (Nieselt 1991).

Saprophagous macroarthropods. Diplopoda (dominant are *Glomeris marginata*, *G. conspersa*, and Isopoda (dominant is *Trichoniscus pusillus*) are typical faunal elements of the mull-structured forests of Göttinger Wald and Zierenberg (Schaefer 1991a; Nieselt 1991). Larvae of elaterid beetles are typical moder-soil inhabitants, consisting almost solely of *Athous subfuscus* in the Solling forest. The community in the Göttinger Wald forest is more diverse, with *A. subfuscus*, *A. vittatus* and *A. haemorrhoidalis* as the dominant species. In the Solling forest, population density and biomass of dipteran larvae are considerably higher than in the mull soil of Göttinger Wald. Mean average individual mass of the larvae in Göttinger Wald is 0.057 mg dry mass, while it is 0.085 mg dry mass in Solling. Dominant families in the Göttinger Wald (in terms of biomass) are Sciaridae, Empididae, Rhagionidae, Lestremiidae, Limoniidae and Tipulidae. In the Solling forest, Sciaridae and Sciophilidae have distinctly higher zoomass values than other dipteran groups. The dipteran community in the Zierenberg forest is rather similar to that of Göttinger Wald (Markwardt 1993).

Some presumptions about the reasons for different habitat preferences of the fauna are possible (Schaefer 1991a, b, c). Protozoa and Nematoda probably utilise the rich bacterial populations in mull soils. The reason for the suppression of shell-bearing gastropods in moder soils appears to be the low Ca availability; this element is required for the formation of the shell. Enchytraeids need high moisture, which is secured more in moder soils than in mull soils. Density and biomass of Enchytraeidae are inversely related to density and biomass of earthworms and are perhaps directly influenced by earthworm occurrence. The susceptibility of individuals to low pH values (Edwards and Bohlen 1996) may be one reason for the absence of most earthworm species in acid soils; another reason may be the restricted substrate feeding in the A and B horizons of moder soils. Oribatid mites and collembolans might be favoured in moder soils by the high quality of available fungal food. Fungi would provide enough Ca which is accumulated by these microfloral populations. Many collembolan species are preferably fungal feeders and may be favoured by high population densities of microfungi (Hopkin 1997). The reason for the absence of isopods and diplopods in acidic moder soils may be the low Ca availability; this element being an important constituent of the skeleton. Generally, the absence of herbs will have a detrimental effect on some groups of the saprophagous macrofauna (see Chap. 12). The voluminous organic layers in the moder ground floor may favour the establishment of predatory macrofauna populations, such as spiders, chilopods, carabids and staphylinids. According to Wallwork (1970), zoophagous arthropods (such as centipedes, spiders, pseudoscorpions, opilionids and predatory beetles) are generally better represented in moder profiles than in mull, probably in parallel to the high numbers of microarthropods as potential prey.

7.4 Meso-Gradient from Basalt to Limestone at the Zierenberg Site

Some of the animal groups are distributed in a characteristic pattern along the hill gradient of the Zierenberg site (cf. Table 7.2). There is a tendency for high faunal diversity in the intermediate site; however, species numbers were rather similar in all three sites covering the gradient. Nematodes and the mesofauna reach high numbers in the basalt site. Groups of the macrofauna (saprophagous and zoophagous) attain high densities in the intermediate site and are less pronounced in the limestone site. The distribution pattern of nematodes and earthworms is discussed below in more detail.

Nematoda. The limestone site is characterised by lower nematode numbers and biomass (Alphei 1995). In this case, lower soil moisture appears to be a decisive factor. Herbivorous populations are well developed in the intermediate and basalt sites because of the high availability of aboveground and belowground herb litter. The root ectoparasites *Paratylenchus*, *Helicotylenchus* and *Rotylenchus* had their highest density at the intermediate site with abundant *Urtica* litter. It appears that *Helicotylenchus* is favoured by this nitrogen-rich plant material.

Lumbricidae. The earthworms had their highest density and biomass at the intermediate site (Bonkowski 1991). Obviously, they are favoured by abundant high-quality herb litter material (mainly stinging nettle) and by the balanced moisture regime of this site. Some endogeic lumbricids prefer soil rich in Protozoa and the rhizosphere of herbs and/or grasses. Adverse factors in the basalt site might be frost in winter, and in the limestone site, low soil water content in summer. However, the low numbers of *L. terrestris* at the Zierenberg site are difficult to explain (Bonkowski 1991; Schaefer et al. 1993).

Obviously, the presence of stinging nettle in the intermediate site as a rich high-quality resource favours the saprophagous fauna and offers high structural diversity for carabid and staphylinid beetles. High availability of basic cations in the limestone site is a key factor for the development of the macrofauna. More acid conditions are responsible for the dominance of the mesofauna in the basalt site. Here, bioturbation is less pronounced, which explains the fact that the fauna does not occur deeper in the soil than at the limestone site.

7.5 Conclusions

Some distinct patterns emerge for the distribution of the animal taxa in the macro- and meso-gradient from mull to moder. The change of abiotic soil conditions and of the availability of specific food resources from base-rich to acid soils leads to faunal and microfloral change. The patterns are diverse and may concern the composition of the fauna on higher taxonomic levels and on species levels, population densities, faunal and microfloral biomass, life-forms and trophic types and mean body size.

Some of the trends are interpretable, as was demonstrated for the macro-gradient and the meso-gradient. As factors responsible for the observed patterns, the following emerge: soil pH, cation exchange capacity, depth and structure of the litter layer, bioturbation by earthworms, composition and biomass of the microflora and quality of litter resources (herb, canopy litter). However, often causal relationships are cryptic. The spatial sequence of woodland sites with different pH conditions can be regarded as a false time-series for human impact via acid rain (see Chap. 12).

Acknowledgements We thank David H. Wise (Chicago, Illinois) for reading and commenting on the manuscript.

References

Alphei J (1995) Die freilebenden Nematoden von Buchenwäldern mit unterschiedlicher Humusform: Struktur der Gemeinschaften und Funktion in der Rhizosphäre der Krautvegetation. Berichte des Forschungszentrums Waldökosysteme Göttingen (A) 125:1–165

Alphei J (1998) Differences in soil nematode community structure of beech forests: comparison between a mull and moder soil. Appl Soil Ecol 9:9–15

Bonkowski M (1991) Untersuchungen zum Verteilungsmuster der Regenwürmer (Lumbricidae) in einem Buchenwald auf Basalt und Kalk. Diplomarbeit, Göttingen

Edwards CA, Bohlen PJ (1996) Biology and ecology of earthworms. Chapman and Hall, London

Eichhorn J (ed) (1991) Fallstudie Zierenberg: Stress in einem Buchenwaldökosystem in der Phase einer Stickstoffübersättigung. Forschungsberichte der Hessischen Forstlichen Versuchsanstalt 13, Münden

Ellenberg H, Mayer R, Schauermann J (eds) (1986) Ökosystemforschung: Ergebnisse des Sollingprojektes. Ulmer, Stuttgart

Hartwig J, Heiligenstadt B, Judas M, Karst H, Schauermann J (1991) Abiotische Parameter der tierökologischen Untersuchungsflächen im Basalt-Kalk-Gradienten des Kleinen Gudenbergs (Zierenberg, Nordhessen). Forschungsberichte der Hessischen Forstlichen Versuchsanstalt 13:45–52

Heiligenstadt B (1988) Zur Ökologie von Hornmilben (Acari: Oribatida) in einem Moder-Buchenwald: Phänologie und räumliches Verteilungsmuster. Diplomarbeit, Göttingen

Hövemeyer K (1992) Die Dipterengemeinschaft eines Kalkbuchenwaldes: eine siebenjährige Untersuchung. Zoologische Jahrbücher, Abteilung Systematik 119:225–260

Hopkin SP (1997) Biology of the springtails. Oxford University Press, Oxford

Koch M (1993) Zur Ökologie der Springschwänze (Insecta: Collembola) eines Buchenwaldes in einem Gradienten von Basalt zu Kalk. Diplomarbeit, Göttingen

Markwardt D (1993) Zur Ökologie der Zweiflügler (Insecta: Diptera) eines Buchenwaldes in einem Gradienten Basalt–Kalk. Diplomarbeit, Göttingen

Nieselt A (1991) Untersuchungen zum Verteilungsmuster saprophager Makroarthropoden (Isopoda, Diplopoda) und Gastropoden in einem Buchenwald auf Basalt und Kalk. Diplomarbeit, Göttingen

Nolte S (1993) Das Verteilungsmuster der Käfer (Coleoptera) eines Buchenwaldes: Die Bedeutung von Brennesselbeständen in einem Gradienten von Basalt zu Kalk. Diplomarbeit, Göttingen

Schaefer M (1990) The soil fauna of a beech forest on limestone: trophic structure and energy budget. Oecologia 82:128–136

Schaefer M (1991a) Animals in European temperate deciduous forests. In: Röhrig E, Ulrich B (eds) Temperate deciduous forests (Ecosystems of the World). Elsevier, Amsterdam, pp 503–525

Schaefer M (1991b) Ecosystem processes: secondary production and decomposition. In: Röhrig E, Ulrich B (eds) Temperate deciduous forests (Ecosystems of the World). Elsevier, Amsterdam, pp 175–218

Schaefer M (1991c) The animal community: diversity and resources. In: Röhrig E, Ulrich B (eds) Temperate deciduous forests (Ecosystems of the World). Elsevier, Amsterdam, pp 51–120

Schaefer M (1996) Die Artenzahl von Waldinsekten: Muster und mögliche Ursachen der Diversität. Mitteilungen der Deutschen Gesellschaft für allgemeine und angewandte Entomologie 10:387–395

Schaefer M, Schauermann J (1990) The soil fauna of beech forests: comparison between a mull and moder forest. Pedobiologia 34:299–314

Schaefer M, Bonkowski M, Judas M, Schauermann, J (1993) Die Habitatbindung von Bodentieren: das Problem einer Kausalanalyse am Beispiel des Verteilungsmusters der Regenwurmfauna eines Buchenwaldes in einem Gradienten von Basalt zu Kalk. Mitteilungen der Deutschen Bodenkundlichen Gesellschaft 69:67–74

Scheu S (1997) Effect of litter (beech and stinging nettle) and earthworms (*Octolasion lacteum*) on carbon and nutrient cycling in beech forests on a basalt-limestone gradient: a laboratory experiment. Biol Fertil Soils 24:384–393

Schlaghamerský J (1998) The enchytraeids of a beech forest on a basalt-limestone gradient. In: Pižl V, Tajovský K (eds.). Soil zoological problems in Central Europe. Academy of Sciences of the Czech Republic, České Budějovice, pp 179–189

Schulz E (1991a) Die Milbenfauna (Acari: Mesostigmata und Cryptostigmata) in Lebensräumen auf Kalkgestein: Populationsökologie, Sukzession und Beziehungen zum Lebensraum. Berichte des Forschungszentrums Waldökosysteme Göttingen (A) 79:1–245

Schulz E (1991b) Die Milbenfauna (Acari: Gamasina, Uropodina, Oribatei) in verschiedenen Buchenwäldern. Berichte des Forschungszentrums Waldökosysteme Göttingen (B) 22:373–376

Wallwork JA (1970) Ecology of soil animals. London, McGraw-Hill

Part B
Ecosystem Processes

Chapter 8
Tree Growth, Biomass, and Elements in Tree Components of Three Beech Sites

P. Rademacher, P.K. Khanna, J. Eichhorn, and M. Guericke

8.1 Introduction

The role of site factors in determining tree growth, especially those relating to nutrient availability, is well recognised. For example, it is well known that high inputs of atmospheric N in recent decades have increased the growth of a number of tree species in Europe (Oren and Schulze 1989; Spiecker 1999; Dieter and Elsasser 2002). Various nutrition-related plant physiological reasons are given for this increase in forest growth. One is the improvement in widespread N deficiency that had occurred in European forests prior to high atmospheric N inputs. However, it is not known how widespread this growth increase is and for how long this would extend in the future (Oren et al. 1988; Scarascia-Mugnozza et al. 2000).

Among many site parameters, those affecting the extent of net allocation of photosynthetic resources to belowground components are considered important for the aboveground tree growth. For example, the allocation of photosynthates to the production and respiration of fine roots, which may be high in soils of high acidity and low base content, may lead to decreased allocation to the stem (woody components) (Alexander and Fairley 1983; Hofmann 1985; Santantonio and Hermann 1985; Dyckmans et al. 2000). Similar to soil acidity and nutrient availability, soil moisture is another important site factor which is expected to affect wood growth through changes in photosynthesis, canopy size and allocation patterns (Jurgensen and Leaf 1965; Wittwer et al. 1975).

It is sometimes hypothesised that the growth of beech trees may differ in relation to differences in climatic and latitude conditions. But the recent studies involving long transects in Europe by Scarascia-Mugnozza et al. (2000) indicated that stand age was the main factor affecting the growth of beech, and probably the local management and site specific factors also affected the pattern of growth. The present study aims to describe the role of chemical soil factors as they affect beech tree growth and biomass distribution of various tree components on the three long-term experimental sites.

R. Brumme and P.K. Khanna (eds.), *Functioning and Management of European Beech Ecosystems*, Ecological Studies 208,
DOI: 10.1007/978-3-642-00340-0_8,

Nutrients are withdrawn from the cycling fractions through accumulation in the woody components including bark which may affect levels in leaves. Any severe decrease in nutrient levels would affect growth of the existing stand. Moreover, nutrient removal through wood harvesting may remove a significant proportion of available soil nutrients affecting growth and sustainability of future stands. In mature beech stands, up to 1,000 kg N, 100 kg P, 500 kg K, 2,000 kg Ca and 150 kg Mg per hectare can be incorporated in the above- and belowground biomass, depending on the site and tree species (Rademacher 2002). A comparison between the amount of nutrients removed from the nutrient cycling fraction and accumulated in the woody biomass with the inputs from atmospheric depositions and also the rate of mineral weathering in soils (Ulrich et al. 1975) would provide an indication of the sustainable nature of the nutrient cycle in these stands. For example, the amount of calcium in the litter layer and in the mineral (0–50 cm depth) soil layer in oak stands on very poor heathland sites was found to be one-third of the amount of Ca accumulated in the stand which appears unsustainable (Bredemeier 1987). In spruce and beech sites on poor to average sites in the Central Mountain ranges in Germany, the amount of Ca in the topsoil and in the stand was found to be of similar magnitude (Ulrich 1972; Ulrich et al. 1986; Matzner 1988). However, the nutrient rich soils or calcareous soils supporting beech contained 5–30 times the amount of available nutrients than were present in the stand, indicating a high buffer of stored nutrients in the soil for future use by forest stands (Boyle and Ek 1972; Ulrich et al. 1975; Meiwes and Beese 1988; Hochbichler et al. 1994; Rademacher 2002).

Biomass production and nutrient retention rates are commonly calculated by using conventional inventory methods (Kreutzer 1976; Weetmann et al. 1979; Madgwick and Tamm 1987), while allometric methods are employed for assessing the biomass of various tree components. However, due to lack of long-term measurement of various stand parameters required for inventory purposes, computer growth models are also used (Montès et al. 2000; Augusto et al. 2000; Joosten and Schulte 2003). The influence of site nutrition and atmospheric pollution is evaluated by long-term trends in tree growth and by using the 'stability' indicators of forest ecosystems (Young 1967; Ulrich et al. 1975; Weetman et al. 1979; Binkley 1986; Lindberg et al. 1986; Likens et al. 1996; Meesenburg et al. 1998).

This chapter provides the growth data from the three experimental sites which in some cases were measured over three decades. The three sites, Solling, Göttinger Wald, and Zierenberg, carry beech stands of almost similar age that grow under similar climatic conditions and are located close to each other, but differ in their soil properties (see Chaps. 1–7, this volume). They represent a wide range of soil chemical properties with major differences in soil acidity, providing a unique opportunity to study the differences in stand growth. We hypothesised that allocation of resources to roots will differ on these sites as a significant amount of resources are involved in the production and maintenance of fine root system under acid soil conditions. In less acid soils, the allocation of biomass to aboveground may result in higher growth rates of merchantable wood. The allocation of biomass resources to different tree components were analysed to identify relationships between tree growth and factors affecting the growth of different tree components, particularly of stem-wood. In addition, the nutrient accumulations by different tree components at these sites were analysed with respect to the amount of removal during harvesting depending upon the intensity of harvesting. The sustainability of such practices in terms of depletion of soil nutrients is also discussed.

8.2 Sites and Methods of Data Collection

The history of stand development differs at the three sites. The 153-year-old beech stand (2000) at the Solling site has developed from natural regeneration, undergoing periodical silvicultural thinnings until 1966 after which a few trees affected by disease or windthrow were felled (Table 8.1). The main experimental plot was fenced and was not limed or fertilised. The understorey at this site is sparser than at the other sites (Ellenberg et al. 1986).

The beech site of Göttinger Wald (132-year-old stand in 2000) is located in a beech regeneration forest following the commonly employed system of beech management. It was managed under a regime of thinning from below until 1960 (Schmidt, Chap. 5, this volume). This experimental site was fenced upon establishment in 1980 and left untreated. It has lush understorey vegetation (Pellinen 1986; Meiwes and Beese 1988).

The beech stand at Zierenberg was regenerated 156 years ago (2000) (Eichhorn 1995). Due to a dense understorey of stinging nettle (*Urtica dioica*), it was frequently thinned during the 1980s to create better conditions for regeneration. The soil was cultivated in 1980 to enhance natural regeneration, which was followed by low intensity shelterwood harvesting in 1996/97 and 1999 (Mrotzek 1998; Schulze and Eichhorn 2000).

8.2.1 Stem Growth

Forest inventory data at the intensively monitored Solling, Göttinger Wald and Zierenberg sites were collected as part of research programs by the Universities of Göttingen and Kassel, and the State Forest Research Centres of Lower Saxony and Hessen. At the Solling research site, 244 beech trees at the B1 plot (1 ha area) representing different diameter classes were measured initially in 1968 (Ellenberg et al. 1986). For the measurements taken in 2000, there were only 199 trees ha^{-1} left on the site. At the Göttinger Wald site (0.31 ha area), 283 trees ha^{-1} were measured in 1995 which had decreased to 245 trees in the inventory taken in 2000. At the Zierenberg site (0.77 ha area), there were 134 trees ha^{-1} in 1990 and decreased to 132 trees in 2000.

The diameters at breast height of all trees at all sites have been measured at least every 5 years. The height was measured periodically, and the height data were plotted against diameter measurements each time. Diameters at breast height values

Table 8.1 Periodically measured stand data for the Solling, Göttinger Wald and Zierenberg intensively monitored sites (*DBH* mean diameter in breast height, YC yield classes according to Schober 1987) *n.d.* Not determined

Year	Age	Living trees					Removed trees			Cumulative growth	
		Tree density	Height	DBH	Basal Area	Wood volume[a]	Tree density	Basal Area	Wood volume[a]	Basal Area[c]	Wood volume[a]
	Years	$N\ ha^{-1}$	m	cm	$m^2\ ha^{-1}$	$m^3\ ha^{-1}$	$N\ ha^{-1}$	$m^2\ ha^{-1}$	$m^3\ ha^{-1}$	$m^2\ ha^{-1}$	$m^3\ ha^{-1}$
Solling											
1968	121	244	25.3	37.9	27.6	351	n.d.	n.d.	n.d.	27.6	351
1972	125	239	27.3	39.6	29.4	404	5	0.5	5	29.9	409
1977	130	231	27.7	41.2	30.7	429	8	0.7	10	31.9	444
1980	134	226	28.2	42.3	31.7	452	5	0.3	4	33.2	471
1985	138	223	28.6	43.4	33.0	475	3	0.3	4	34.8	498
1990	144	217	28.8	44.9	34.4	499	6	0.4	6	36.6	528
1996	149	212	29.6	46.1	35.4	529	5	0.7	11	38.3	569
2000	153	199	30.0	46.8	34.2	519	13	2.5	38	39.6	597
YC3	150	152	29.0	45.4	24.6	364					
Göttinger Wald											
1980	113	283	31.2	38.9	32.3	504	n.d.	n.d.	n.d.	32.3	504
1983	116	262	31.5	40.8	33.1	522	21	2.6[b]	41[b]	35.7	562
1996	129	252	33.6	45.0	39.2	678	10	1.6[b]	28[b]	43.4	747
2000	132	245	33.7	45.7	39.5	686	7	1.2	21	44.9	776
YC1	130	196	36.1	46.1	32.8	586					
Zierenberg											
1990	145	134	35.3	56.3	28.3	502	n.d.	n.d.	n.d.	28.3	502
1996	151	133	36.0	58.3	30.6	555	6	0.23[b]	4[b]	30.8	559
2000	156	132	36.7	59.9	32.0	595	5	0.24	5	32.5	604
YC1	150	145	38.3	54.4	33.8	647					

[a]Wood volume = wood >7 cm diameter in breast hight plus bark (merchantable wood)
[b]Estimated using calculated values as no data were available for this period
[c]The cumulative values include those of all trees removed during previous measurement periods

were used to obtain basal area (m^2 ha^{-1}). The mean tree basal area was calculated for each site and converted to mean tree diameter. From the diameter/height relationships, the mean tree height was calculated for each site.

The wood volume (wood >7 cm diameter at breast height) was then calculated by allocating the stem a truncated cone form (Kramer and Akça 1982). Wood volume was computed by using the total basal area and the mean tree height for each of the measuring periods. The removal or death of a tree during the measuring periods was recorded and derivations of its basal area and wood volume were obtained in the same way as for living trees. Cumulative values of total basal area and tree volumes were then obtained for the measuring period by including standing and harvested trees. These cumulative values were then used to calculate the periodic increments of basal area and wood volumes.

8.2.2 *Biomass Measurements of Different Tree Components on Two Sites*

8.2.2.1 Solling Site

Thirty-six trees belonging to different diameter and age classes (40, 80 and 120 years) were harvested from six plots in the Solling area in 2000 to determine the biomass of tree components. Biomass was calculated by determining the volume (length and circumference) of woody components along the main stem (first 30 cm and subsequently in 1-m lengths) and for all the branches of diameters greater than 7 cm. The bark volume was calculated from bark thickness measurements taken at 1- to 2-m distances along the stem and branches (Rademacher and Müller-Using 2004). A reduction in wood volume arising from moisture loss during drying of 16.3% was considered in the calculation while such a factor was not applied for bark. To calculate wood and bark biomass, density values of wood 0.66 g cm^{-3} and of bark 0.42 g cm^{-3} were used. These values were taken from Kollmann (1951) and Kramer and Akça (1982).

After harvesting the crown, branches <7 cm in diameter including bark were allocated to different diameter classes and their fresh weights were recorded. Individual samples were dried in the laboratory at 103°C. After determining the moisture content, total dry weight was calculated. Samples of the wood and bark were also taken for analysis. They were dried at 60°C, ground and analysed for different nutrients.

All root material with a diameter of 2 mm or more was defined as coarse roots. Coarse roots from 36 harvested trees were measured by manual excavation under water-saturated soil conditions as described in detail by Bolte et al (2004). After cleaning the roots, the root system was divided into three-dimensional sections comprising concentric rings (Nielsen 1995; Kuhr 2000) by recording three parameters: (1) distance from the stem base in 50-cm sections up to a maximum distance of 400 cm along the horizontal coarse roots, (2) soil depth using 25-cm sections (top soil) and 50-cm sections to a maximum depth of 200 cm, and (3) orientation in four discrete directions: NE, SE, SW and NW. All root radii entering and leaving a section were recorded to calculate volumes for each root section to derive the total coarse root volume for each tree. Oven-dry mass of coarse roots was derived from the dry weight of a random sample of different root parts ($n = 313$); coarse root parts were oven-dried for at least 48 h at 103°C to a constant weight. Average dry root densities were calculated using a linear regression between coarse root volume and coarse root biomass without intercept. Dry root density value was 0.536 t m^{-3} for beech ($r^2 = 0.94$; Bolte et al. 2004).

The biomass allocation ratio of the below (including fine roots) and aboveground beech tree compartments was 0.10 ± 0.03. These results are similar to those in other studies on comparable sites in the region. However, the ratio for pure stands in other regions growing under different site conditions is somewhat higher. Suppressed beech trees (diameter at breast height <10 cm) commonly have the lowest values.

Foliage biomass was considered to be same (2.7 Mg ha^{-1}) as the mean litter fall value at the Solling site and the data were converted to per tree basis by considering the number of trees on a site and the individual tree biomass. Fine root biomass was taken as 6 Mg ha^{-1} (Murach et al. Chap. 9, this volume).

8.2.2.2 Göttinger Wald Site

Inventory data were available for 19 beech trees from four plots at the Göttinger Wald site from an earlier investigation by Pellinen (1986), where seven different tree components were investigated. However, as for the Solling site, these were reduced to six main components (see Tables 8.4 and 8.7). For this study, the biomass data for each component as collated by Pellinen (1986) and Murach et al. (Chap. 9, this volume) were used. No sample trees were harvested at Zierenberg.

8.2.3 Biomass Modelling at Different Experimental Sites

Linear regression equations were calculated from the basal area of the harvested trees and the measured biomass of various tree components. The following tree components were separated: leaves, branches <7 cm diameter including bark, wood >7 cm diameter in breast height, bark of wood >7 cm, coarse roots >0.2 cm diameter including bark and fine roots <0.2 cm diameter (Table 8.3, Figs. 8.2–8.5). These regressions were then used to calculate the biomass for each stand using the basal area of individual trees at the Solling and Göttinger Wald sites. The use of linear regression equations instead of the commonly used log transformations meant that a correction factor for bias in log transformation equations was not required for biomass calculations. As the basal area of the largest harvested tree was smaller than some of the trees found at the sites, the linear regressions were extrapolated with the assumption that they remained valid for large-sized trees. The variation in the estimated value is expected to be high for the larger trees, potentially resulting in large errors in the estimations.

8.2.4 Nutrient Concentration of Various Tree Components

Samples of wood, bark, branches and coarse and fine roots were collected from the harvested trees to measure nutrient concentrations. Samples of fully grown leaves were collected in August 1999 at the Solling and Göttinger Wald sites from the trees adjoining those harvested for biomass sampling in the following year, and preceding the tree harvest. Green leaves were collected from various parts of the canopy (lower, middle and upper). Samples were dried, ground and digested using HNO_3 (under pressure) prior to analysis using ICP- emission spectroscopy (Meiwes and Beese 1988; König and Fortmann 1998; Rademacher et al. 2001; Rademacher and Müller-Using 2004).

8.3 Stand Growth

8.3.1 Solling Site

In 1968, stand density of the 121-year-old beech stand at the Solling site was 244 trees per ha. During the 32 years of measurements, 45 trees were removed because they died or had disease or constituted a safety risk. The harvesting regime at the site did not conform to conventional management practices as outlined by the yield tables. According to the yield tables, only 152 trees with smaller basal area and lower stem volume per ha should have been retained at the stand age of 150 years, instead of the 199 trees per ha presently on the site. This high density is expected to affect tree growth, as well as the density of the beech natural regeneration and understorey growth (Table 8.1).

Table 8.1 shows that trees at this site are still growing in height and diameter at the age of more than 150 years. Mean height increased by 4.7 m in the last 32 years (from 25.3 to 30.0 m, 19% increase in 32 years) and diameter at breast height increased by about 9 cm (from 37.9 to 46.8 cm, 23.5% increment).

In 2000, the living basal area was 34.2 $m^2\ ha^{-1}$ (24% more than in 1968), and the standing wood volume was 519 $m^3\ ha^{-1}$. A small loss (38 $m^3\ ha^{-1}$ since 1996) of wood volume due to tree deaths was recorded. On average, the annual increment in basal area was 0.38 $m^2\ ha^{-1}$ per year and merchantable wood including bark (wood +bark >7 cm) volume over the 32-year period was 7.7 $m^3\ ha^{-1}$ per year (Table 8.2). The values measured for various periods, usually about 5 years, were highly variable and did not reveal a clear trend over time. Annual growth for these parameters during the last 10 years is comparable with the previous 20 years, both higher than those predicted by the yield tables. The annual growth of merchantable wood (wood >7 cm

Table 8.2 Annual growth parameters for different periods and yield class values (YC 1 at age 130 years and 150 years for Göttinger Wald and Zierenberg, respectively, YC 3 at age 150 years for Solling) according to Schober (1987) for the Solling, Göttinger Wald, and Zierenberg intensively monitored sites (tree density refers to values at the end of the period; BA basal area, YC yield classes after Schober 1987)

Tree age Year	Tree density $N\ ha^{-1}$	BA increment $m^2\ ha^{-1}$ per year	Wood volume[a] increment $m^3\ ha^{-1}$ per year
Solling			
121–134	226	0.43	9.2
134–144	217	0.34	5.7
144–153	199	0.33	7.7
121–153	244–199	0.38	7.7
YC	152	0.36	6.5
Göttinger Wald			
113–129	252	0.69	15.2
129–132	245	0.50	9.7
113–132	283–245	0.66	14.3
YC	196	0.45	10.1
Zierenberg			
145–151	133	0.41	9.5
151–156	132	0.34	9.0
145–156	134–132	0.38	9.3
YC	145	0.41	9.5

[a]Wood volume = wood >7 cm diameter in breast height plus bark

diameter) including bark is about 17% higher than expected in yield tables during the last 32 years. Although the number of trees at the Solling site is 25% higher than yield tables predict, mean tree height and diameter are actually slightly higher than the yield table values. The annual basal area and volume increment values fall within the ranges given in yield tables. However, the aboveground biomass and basal area are about 40% higher, indicating higher growth rates or lower tree removals in the past.

Guericke (2002) reported that maximum annual diameter at breast height increments of up to 7 mm occurred in the largest diameter class, 60–70 cm, in thinned mature beech stands. At the Solling B1 site, diameter increments in the diameter class 60–70 cm of up to 5 mm were recorded and up to 8 mm in the 50–60 cm diameter class. In contrast, the annual diameter at breast height increment in the lower diameter classes at the B1 site declined considerably, from 3.1 to 1.1 mm, due to the decline in light availability with increasing stand density.

Despite the less favourable soil characteristics found at Solling, the volume and growing stock levels are high compared to other stands on similar sites, with more than 500 m^3 of merchantable wood volume including bark (Nihlgård 1972; Kreutzer 1976; Khanna and Ulrich 1991; Röhrig 1991; Jacobsen et al. 2003). However, the annual growth increment rate at Solling is comparable to the average growth on similar sites under best conditions. The percentage of trees with diameter at breast height >50 cm increased from 14% in 1980 to 27% in 2000 in the stand at the Solling site. The target diameter of >65 cm for one tree per hectare was first reached at the stand age of around 120 years. Thereafter, there were two stems at 130 years, three stems at 145 years, five stems at 149 and six stems at 153 years which had reached the target diameter. In a ca. 150-year-old beech stand in Lower Saxony, Guericke (2002) measured 20 trees with diameter at breast height >65 cm, whereas the majority of trees belonged to the 45–60 cm diameter class. At Solling, the 45- to 60-cm-diameter class comprised of only about half that number of trees. Over 100 years observation period, the basal area of the remaining stand after thinning in Guericke's study ranged between 25 and 30 $m^2\ ha^{-1}$, depending on tree number and tree diameter, whereas the basal area at Solling ranged between 28 and 35 $m^2\ ha^{-1}$ during the last 30 years.

8.3.2 *Göttinger Wald Site*

Beech trees at the Göttinger Wald site were 132 years old in 2000, with 245 trees ha^{-1} (Table 8.2). Over a 20-year measurement period, the mean tree height (2000: 33.7 m) has increased by 8%, whereas the mean diameter at breast height (in 2000: 45.7 cm) has increased by 17%.

In 2000, the basal area of the standing trees amounted to 39.5 $m^2\ ha^{-1}$, and the volume of merchantable wood (wood >7 cm diameter including bark) was 686 $m^3\ ha^{-1}$. The annual productivity calculated at this site is highly variable from period to period, indicating the importance of annual variation in certain factors affecting tree growth. In the last 20 years, the basal area increment was 0.66 $m^2\ ha^{-1}$ per year and volume increment 14.3 $m^3\ ha^{-1}$ per year. Thus, annual growth is much higher here than at the Solling site (1.7 times basal area and 1.9 times wood volume). The high productivity of this site may be related to many factors. The mean annual growth rates for basal area and for wood are 44% and 39% higher, respectively, than expected from the yield tables. Similar to these results, high increment and yield values have been reported by others on only a few sites with favourable nutrient and growth conditions (Khanna and Ulrich 1991; Röhrig 1991; Hochbichler et al. 1994; Persson et al. 2000).

8.3.3 *Zierenberg Site*

The Zierenberg site was included as a long-term experimental site in 1990 (Eichhorn 1995). At that time the trees were 145 years old and had been subjected to standard thinning and regeneration felling regimes (more details in the chapter by Schmidt,Chap. 5, this volume). These practices left a relatively small number of trees on the site (132) compared to those on the other two sites (Table 8.1).

Mean tree diameter at breast height value at this site was 59.9 cm in 2000 representing an increase of 3.6 cm over the 10-year measuring period. The average tree height at Zierenberg was 36.7 m. In 2000, the basal area at Zierenberg was

32 m^2 ha^{-1}, and the merchantable wood volume (wood >7 cm diameter including bark) 595 m^3 ha^{-1}. For the 10 years preceding 2000, the mean annual growth rates for basal area was 0.38 m^2 ha^{-1} and the volume was 9.3 m^3 ha^{-1} (Table 8.2). The basal area values were similar to those on the Solling site although fewer trees were present, indicating that annual diameter growth rates of single trees at Zierenberg site were much higher (0.070 m^3 ha^{-1} per year compared to 0.054 m^3 ha^{-1} per year at Göttinger Wald and 0.035 m^3 ha^{-1} per year at Solling).

8.4 Comparison of Stand Growth at the Three Experimental Sites

The relationship between tree height and stem diameter followed an exponential pattern on all the three sites (Fig. 8.1), indicating that height growth slowed down once trees reached 30–35 cm in diameter at breast height whereas diameter growth continued to increase at the usual rate. However, an increase in height growth was recorded at all the three sites despite the age of the stands and the lack of intensive management of the sites (Ellenberg et al. 1986; Ulrich et al. 1986; Matzner 1988). The annual basal area and wood volume production during the last 10 years followed the order: Solling < Zierenberg < Göttinger Wald (Table 8.2). Due to low tree density at the Zierenberg site, the annual growth of the single trees was higher at this site than at the other two sites following the order: Zierenberg > Göttinger Wald > Solling.

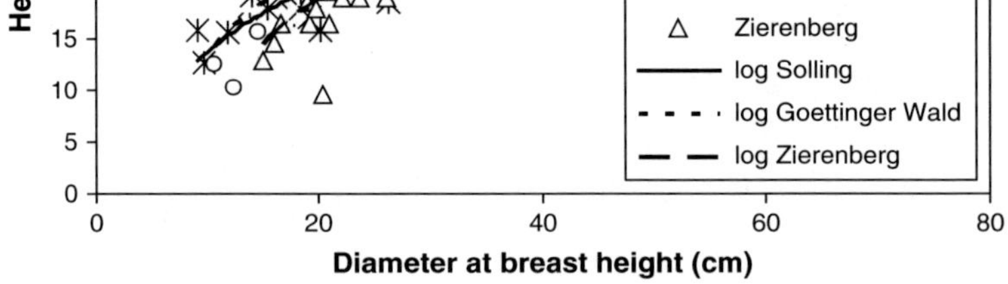

Fig. 8.1 Tree height in relation to diameter at breast height in beech trees at Göttinger Wald, Zierenberg and Solling. Data: Göttinger Wald (Pellinen 1986), Zierenberg (Hess. Forstl. Versuchsanstalt, unpublished), Solling (Rademacher and Müller-Using 2004). Logarithmic functions: Solling, $y = 9.3609\ln(x) - 7.9090$, $r^2 = 0.872$; Göttinger Wald, $y = 12.0523\ln(x) - 14.3155$, $r^2 = 0.816$; Zierenberg, $y = 15.5912\ln(x) - 27.8508$, $r^2 = 0.851$

The annual values of wood volume increments observed at these sites were 7.5 m^3 for Solling and 9 m^3 for Zierenberg which fall within the yield classes 2 and 1. At Göttinger Wald, the annual increments in wood volume ranged between 10 and 15 m^3, corresponding to higher levels than given for yield class 1. At all the three sites, the stem growth rates are much higher than predicted by the yield tables which may be due to high atmospheric nutrient inputs, longer growing periods and better spectral conditions for photosynthetic activities. Other factors such as the high frequency of mast years may have adversely affected the stem growth in individual years (Burschel 1966). Diversion of photosynthates to the production of fruiting bodies has been shown to affect stem growth (Ellenberg et al. 1986). Following the very dry year 2003 in Germany, a reduced stem growth (height and diameter) was accompanied by an increased fructification in 2004 at seven beech forests including the Zierenberg site (Eichhorn et al 2008).

The basal area at the three study sites in 2000 followed: Zierenberg (32 m^2 ha^{-1}) < Solling (34 m^2 ha^{-1}) < Göttinger Wald (40 m^2 ha^{-1}). The mean annual basal area increment for the total observation period of 0.38 m^2 at both Solling and Zierenberg lie within the values given in the yield table, whereas the 0.66 m^2 value at Göttinger Wald (yield table value 0.45 m^2) indicated much higher tree growth.

The order of sites for mean annual volume increment per hectare was Solling (7.7 m^3 ha^{-1} per year) <Zierenberg (9.3 m^3 ha^{-1} per year) <Göttinger Wald (14.3 m^3 ha^{-1} per year). These site differences were primarily due to tree density arising from forest management practices. Persson et al. (2000) reported the values of wood volume increment in 80- to 160-year-old beech stands along the north–south gradient from Sweden to Italy. The distribution of wood volume increments ranged from 2.4 to 5 m^3 ha^{-1} per year in France and Italy, and from 9 to 12 m^3 ha^{-1} per year in Denmark and the Czech Republic. The corresponding values for the Solling and Zierenberg sites were in the middle to upper range whereas the Göttinger Wald stand with 14 m^3 ha^{-1} per year showed the maximum annual volume increment in Europe. This value was similar to the value obtained for spruce stands in Denmark by Persson et al. (2000).

However, for values relating to basal area distribution, Persson et al. (2000) found that maximum values occurred most frequently in the Central European region (72 m^2 ha^{-1}) whereas minimum values occurred in northern Sweden and the south of Italy. The temperature and precipitation conditions in Central Europe are more favourable during the growing season, and there are high atmospheric depositions of N and S causing these high growth rates, particularly in the Czech Republic, Germany and France despite the low soil pH values (Persson et al. 2000).

A number of soil chemical characteristics such as pH values, exchange capacity, base saturation, Ca/Al ratio in the soil solution and percentage of Al at exchange sites were more favourable on the Göttinger Wald and Zierenberg sites where high growth rates were recorded. In addition, other site parameters such as elevation, mean annual temperature, the mineral content of the geological substrate, soil type, the buffer region of the mineral soil and humus type are more favourable for tree growth at the Göttinger Wald and Zierenberg sites than at the Solling site (Ulrich and Summer 1991; Eichhorn 1995; Becker et al. 2000; Persson et al. 2000; Jacobsen et al. 2001; Meesenburg et al. see Chap. 3, this volume).

8.5 Biomass of Harvested Trees and Annual Biomass Increments

The aboveground and belowground biomass of a single beech tree ranged from a few kilograms for trees of smaller diameters to about 3–4 Mg for trees of the largest diameters harvested at Solling and Göttinger Wald sites (Figs. 8.2 and 8.3) indicating the close relationship between tree biomass and tree basal area (Table 8.3).

Relationships between basal area of single trees to total biomass and aboveground biomass were described as linear regressions (Table 8.3) because the r^2 values obtained for linear regressions were higher than those obtained when

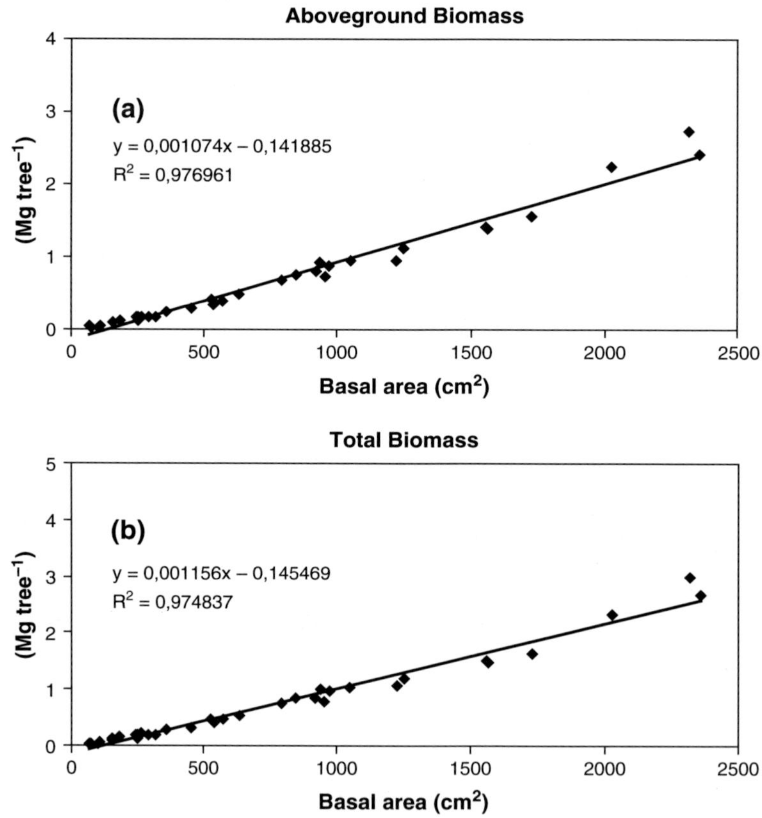

Fig. 8.2 Aboveground (**a**) and total biomass (**b**) of harvested trees in relation to basal area at breast height for the Solling area

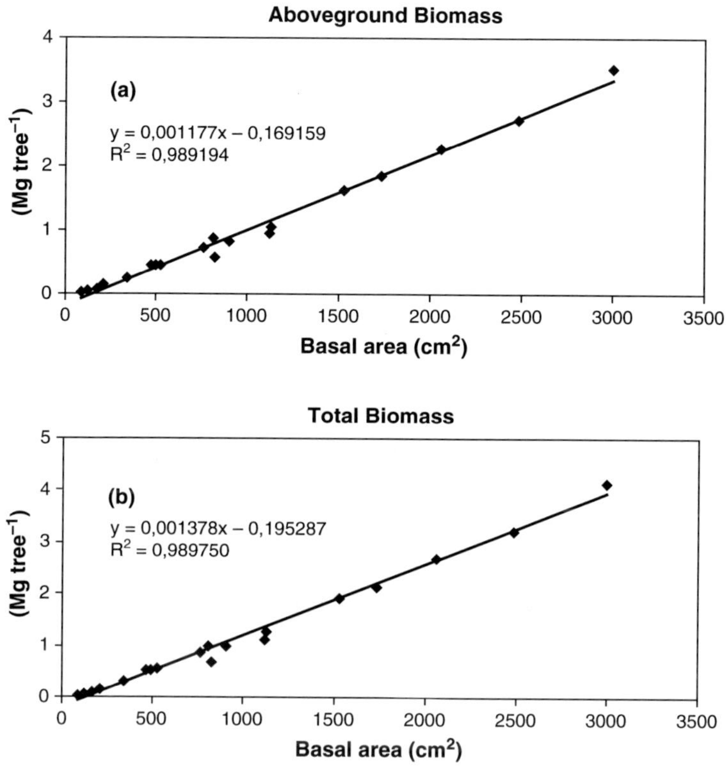

Fig. 8.3 Aboveground (**a**) and total biomass (**b**) of harvested trees in relation to basal area at breast height for the Göttinger Wald area

exponential functions were fitted to the data (not presented here). Stem-wood is the main component of biomass, particularly when tree diameter at breast height is greater than 20 cm (Figs. 8.4 and 8.5). The r^2 values of linear regressions between basal area and stem-wood is about 0.97–0.99 (for exponential regressions, 0.73–0.81), and slightly lower for other tree components, such as bark, branches, leaves and coarse roots (Table 8.3).

Despite the higher tree number at the Göttinger Wald site the biomass of a single tree was considerably higher than at the Solling site (Table 8.1, Figs. 8.2–8.5). The total biomass values of aboveground and belowground components were calculated to be ca. 394 Mg ha^{-1} at Solling and 508 Mg ha^{-1} at Göttinger Wald (Table 8.4).

Merchantable wood (>7 cm diameter) was 268 Mg at Solling and 371 Mg at Göttinger Wald. Whereas trees at the Solling site had higher biomass of branches <7 cm diameter (71 Mg ha^{-1}) than at the Göttinger Wald site (35 Mg ha^{-1}), that of roots >0.2 cm diameter (74 Mg ha^{-1}) and bark (22 Mg ha^{-1}) were higher at the Göttinger Wald site (Table 8.4). The reasons for these high biomass values of coarse roots and bark at the Göttinger Wald site are not known but may be related to the calcareous nature of the soil. In addition to high plasticity of beech in allocating

Table 8.3 Linear regression parameters (y = ax−b) between biomass of various tree components (y, Mg per tree) and tree basal area, (x = basal area in cm^2) for 36 beech trees from the Solling area (Rademacher and Müller-Using 2004) and 20 beech trees from the Göttinger Wald area (Pellinen 1986)

Component	r^2	a	b
Solling			
Wood (>7 cm)	0.974	0.000839	0.12140
Bark (>7 cm)	0.943	0.000046	0.00639
Branches[a] (<7 cm)	0.909	0.000179	0.01327
Leaves	0.919	0.000010	0.00083
Coarse roots[a]	0.746	0.000081	0.00358
Aboveground biomass	0.977	0.001074	0.14189
Total biomass	0.975	0.001156	0.14547
Göttinger Wald			
Wood (>7 cm)	0.990	0.000961	0.13423
Bark (>7 cm)	0.962	0.000063	0.01069
Branches[a] (<7 cm)	0.968	0.000143	0.02288
Leaves	0.939	0.000010	0.00135
Coarse roots[a]	0.987	0.000201	0.02613
Aboveground biomass	0.989	0.001177	0.16916
Total biomass	0.990	0.001378	0.19529

[a]Including bark

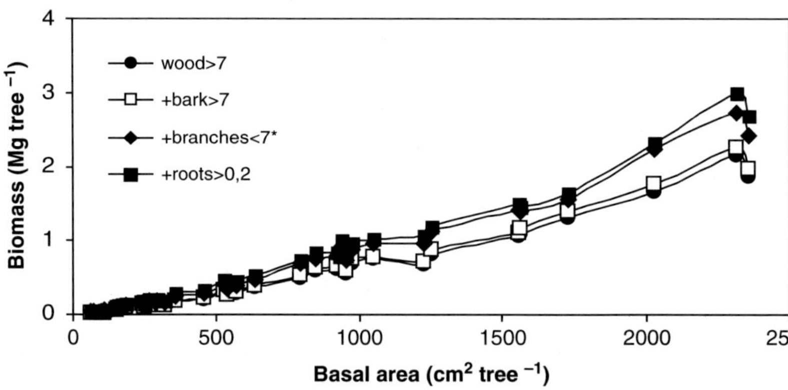

Fig. 8.4 Biomass in various components of harvested beech trees in relation to their basal area for the Solling area (Rademacher and Müller-Using 2004). *Asterisk* indicates branches <7 cm diameter including bark and leaves

biomass to coarse roots depending on the site conditions, Bolte et al. (2004) observed that the method of assessing coarse root biomass might be the cause of the major variations observed. There are primarily two factors involved: (1) definition of root crown and problem associated with clear separation of root from the stem stump, which in most cases may overestimate the root biomass, and (2) assessment of biomass of missing roots due to broken root-ends during excavation.

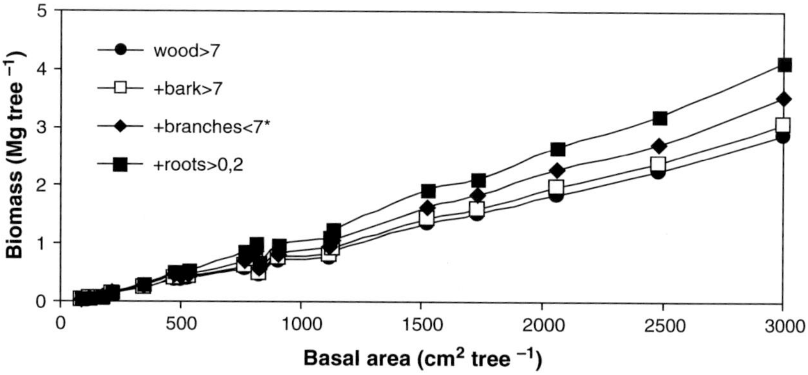

Fig. 8.5 Biomass in various components of harvested beech trees in relation to their basal area for the Göttinger Wald area (data: Pellinen 1986). *Asterisk* indicates branches <7 cm diameter including bark and leaves

The aboveground biomass in a forest stand depends on stand age, tree species composition and site characteristics, and may range from 100 to 330 Mg ha^{-1} in temperate mixed broadleaved forests aged 30–200 years (Khanna and Ulrich 1991). Hochbichler et al. (1994) found that 300 Mg ha^{-1} of aboveground biomass had accumulated in a 40-year-old beech forest in Austria, of which the merchantable wood >7 cm diameter including bark comprised 80%. The total net aboveground production of all woody components in forests during a period of 100 years commonly ranges from 300 to 700 Mg ha^{-1}. However, in managed forests of that age, about 50% of this total woody biomass is commonly removed during intermittent thinnings. Thus, the aboveground standing biomass at the average age of 100 years for ten beech sites was found to range from 60 to 350 Mg ha^{-1} (Rademacher 2002). Among others, Jacobsen et al. (2003) calculated 290 ± 100 Mg ha^{-1}; Kreutzer (1976) 115–300 Mg ha^{-1} in 130–140-year-old beech; Nihlgård (1972) 324 Mg ha^{-1}in 80-year-old beech; and Scarascia-Mugnozza et al. (2000) 210–330 Mg ha^{-1}. Based on eight studies in northwestern European beech sites, Röhrig (1991) found the aboveground biomass in 90- to 150-year-old stands to be about 200–400 Mg ha^{-1}.

From the literature review provided by Rademacher (2002), the annual total biomass production rate (aboveground and belowground components for a mean rotation length) was found to range from 4 to 7 Mg ha^{-1} per year for beech, which were similar to the values obtained for other tree species with values of 2–6 Mg ha^{-1} per year for pine and 4–8 Mg ha^{-1} per year for spruce. For beech stands, the most productive growth phase occurs at an age of more than 125 years with values of 9.2 for the Solling site and 11.4 Mg ha^{-1} per year for the Göttinger Wald site (Table 8.4). This difference in total biomass production on the two sites reflects both the better growing conditions and the high tree density on the Göttinger Wald site.

Leaves constituted a similar fraction of about 38% and 33% of the total biomass produced annually at Solling and Göttinger Wald, respectively. However, the annual production of fine roots formed a major difference between the two sites

Table 8.4 Stand biomass (Mg ha^{-1}) and biomass increment (Mg ha^{-1} per year) of various tree components calculated by using regression equations given in Table 8.3 and the basal area of each tree (Table 8.1). The literature values are taken from the list given in Tables 8.5 and 8.6

Site	Leaves	Branches <7 cm	Bark >7 cm	Wood >7 cm	Root >0.2 cm	Root[b] <0.2 cm	Aboveground biomass	Total biomass
	Stand biomass (Mg ha^{-1})							
Solling	3.5	70.6	17.0	267.8	28.6	6.2	359	394
Göttinger Wald	3.7	34.8	21.6	370.6	74.3	2.9	431	508
	Annual biomass production (Mg ha^{-1} per year)							
Solling	3.5	0.4	0.1	1.7	0.2	3.3	5.7	9.2
Göttinger Wald	3.7	0.5	0.3	4.9	1.0	1.0	9.4	11.4
	Literature data – Stand biomass (Mg ha^{-1})							
Min-max	2–5	26–122		98–602[a]	26–84	2–6	128–660	220–705
	Literature – Annual biomass production (Mg ha^{-1} per year)							
Min-max	2.8–4.7	0.4–4.6		2.0–6.7[a]	0.2–1.3	1.5–3.8	2.9–12.5	3.5–16.6

[a]Values include wood plus bark
[b]Fine root biomass and annual fine root production from Chap. 9

because 36% of the total biomass produced was invested in fine roots at Solling and only 9% at the Göttinger Wald (Table 8.4). Thus, 74% of the annual biomass produced each year was invested in the turnover fractions (fine roots and leaves) at the Solling site whereas only 42% at the Göttinger Wald site. This difference in the allocation of total biomass resulted in much higher wood production in the Göttinger Wald stand. It appears that beech trees growing under acid conditions need to invest high resources for the growth and maintenance of fine root biomass, affecting the aboveground production.

Biomass content of aboveground and belowground components of trees on these two sites has been previously reported in a number of studies which are provided here for comparison purposes (see also Tables 8.5 and 8.6). For Göttinger Wald, Pellinen (1986) calculated that the value of an annual production of merchantable wood >7 cm diameter including bark was about 4.7 Mg ha^{-1} per year, with 0.7 Mg each of branches <7 cm and of roots with bark. Meiwes and Beese (1988) reported an annual production of 2.8 Mg leaves and 2.8 Mg fine roots in the first inventory at Göttinger Wald. For the Solling site, Ellenberg et al. (1986) calculated, for the period 1967–1977, the annual production of merchantable wood including bark to be 5–7 Mg ha^{-1}, of branches <7 cm 0.6–0.9 Mg ha^{-1}, and for coarse roots, 0.5–0.7 Mg.

8.6 Nutrients in Various Tree Components

Due to the calcareous nature of soils at the Göttinger Wald site (Table 8.7) the Ca concentrations of all the different components of beech trees at Göttinger Wald were much higher than those for the Solling stand and were about two times higher than the mean values reported in literature (Table 8.7; also Bauer et al. 2000; Jacobsen et al. 2003). Similar concentrations of Ca have been reported in the beech sapwood on nutrient-rich sites in the Czech Republic and Italy (Bauer et al. 2000). Phosphorus concentration in all the tree components except the leaves was lower at the Göttinger Wald site than at the Solling site. Concentration of Mg was similar on the two sites for almost all the tree components. Leaves from the Göttinger Wald site had higher concentration of K than those from Solling. C concentration of tree components ranged from about 470 to 530 g kg^{-1} at both the Solling and Göttinger Wald sites.

Comparing nutrient concentrations of various components, leaves had the highest concentrations of almost all the elements except for Ca which was higher in bark. Bark was, next to leaves, the element-rich component of the trees on each site. For most elements, fine roots of <0.2 cm thickness had similar concentrations as bark on both the experimental sites. These differences in element concentrations have nutrient management implications if bark is removed from the site with other woody components during tree harvesting, as leaves and fine roots are commonly left on the site for decomposition and mineralisation.

Table 8.5 Biomass content (Mg ha^{-1}) of beech tree components in Europe: *wood* >7 cm diameter plus bark, *branches* <7 cm diameter plus bark, FR fine roots, *CR* coarse roots plus bark

Publication	Site	Age (yr)	Leaf	Wood	Branches	Sum	FR	CR	FR + CR	Total
Rademacher and Müller-Using (2004)	Solling/D (Goseplack 27)	40	3	98	27	128				
Rademacher and Müller-Using (2004)	Solling/D (Steinhoff 129)	40	3	111	37	151				
Hochbichler et al. (1994)	Flyschwienerwald/A	40	4	240	57	301		54		355
Ellenberg et al. (1986)	Solling/D (B4)	67	3	151	35	189		28		220
Scarascia-Mugnozza et al. (2000)	Jezeri/CZ	79	4	161	40	205	3	48	51	255
Rademacher and Müller-Using (2004)	Solling/D (Goseplack 16)	80	3	194	47	243				
Rademacher and Müller-Using (2004)	Solling/D (Verliehausen 325)	80	3	232	52	287				
Ellenberg et al. (1986)	Solling/D (B3)	88	3	165	26	194		26		223
Holm and Jensen (1981); in: Röhrig (1991)	Danmark	90	2	170	43	215			43	258
Pellinen (1986)	Göttinger Wald/D (site D)	100	3	268	35	306				
Garelkov (1973); in: Röhrig (1991)	Belgien	100	3	280	32	315			38	353
Pellinen (1986)	Göttinger Wald/D (site C)	100	3	296	37	336				
Garelkov (1973); in: Röhrig (1991)	Belgium	100	5	365	49	419			50	469
Scarascia-Mugnozza et al. (2000)	Collelongo/I	104	3	174	32	209	4	55	59	268
Pellinen (1986)	Göttinger Wald/D (site B)[a]	115	3	277	36	316				
Meiwes and Beese (1988)	Göttinger Wald/D	115	3	316	37	357	3	26	29	385
Pellinen (1986)	Göttinger Wald/D (site A)[b]	115	3	316	41	360				
Pellinen (1986); in: Röhrig (1991)	Göttinger Wald/D	116	3	349	38	390				
Scarascia-Mugnozza et al. (2000)	Gribskov/DK	118	5	266	60	331	3	71	74	405
Scarascia-Mugnozza et al. (2000)	Schacht/D	120	4	223	46	272	3	54	57	329
Rademacher and Müller-Using (2004)	Solling/D (Goseplack 570)	120	3	303	38	343				
Rademacher and Müller-Using (2004)	Solling/D (Rottmünde 47)	120	3	316	35	354				
Ellenberg et al. (1986); in: Röhrig (1991)	Solling/D	126	3	255	34	292			32	324

Kreutzer (1976)	Modelling/YC IV	130	2	178	34	214		48		262
Ellenberg et al. (1986)	Solling/D (B1)	130	3	273	36	313		33		348
Duvigneaud (1971); in: Röhrig (1991)	Belgium	130	3	214	122	339			68	407
Rademacher et al. (this investigation)	Göttinger Wald/D (model A-D)	130	4	392	35	431	3	74	77	508
Kreutzer (1976)	Modelling/YC I	140	3	366	51	420		84		504
Kestemont (1975); in: Röhrig (1991)	Belgium	144	3	319	51	373			74	447
Lemée (1978); in: Röhrig (1991)	France	150	3	195	48	246			40	286
Rademacher et al. (this investigation)	Solling/D (modelling B1)	150	4	285	71	359	6	29	35	394
Rademacher (2002)	Solling/D (modelling B1)	155	5	361	42	408	2	36	38	446
Rademacher (2002)	Göttinger Wald/D (modelling)	155	3	602	55	660	3	42	45	705
Scarascia-Mugnozza et al. (2000)	Aubure/F	161	3	204	48	255	2	54	56	311
All investigations (min – max)	All sites	40–161	2–5	98–602	26–122	128–660	2–6	26–84	29–74	220–705

[a]Include 2% *Acer*
[b]Include 14% *Acer*

Table 8.6 Annual biomass accumulation (Mg ha^{-1} a^{-1}) of beech tree components in Europe: *wood* >7 cm diameter plus bark, *branches* <7 cm diameter plus bark, *FR* fine roots, *CR* coarse roots plus bark

Publication	Site	Age (yr)	Leaf	Wood	Branches	Sum	FR	CR	FR + CR	Total
Ellenberg et al. (1986)	Solling/D (B4)	67	3.1	6.7	0.7	10.6	2.5	1.1	3.6	14.2
Scarascia-Mugnozza et al. (2000)	Jezeri/CZ	79	3.7	4.2	4.6	12.5	2.9	1.2	4.1	16.6
Ellenberg et al. (1986)	Solling/D (B3)	88	3.4	5.4	0.4	9.2	2.5	0.7	3.2	12.4
Pellinen (1986)	Göttinger Wald/D (site C)	100	2.9	4.8	0.6	8.3				
Pellinen (1986)	Göttinger Wald/D (site D)	100	3.1	4.7	0.6	8.4				
Scarascia-Mugnozza et al. (2000)	Collelongo/I	104	2.8	2.8	3.1	8.7	3.8	1.0	4.8	13.5
Pellinen (1986)	Göttinger Wald/D (site B)[a]	115	3.0	4.8	0.7	8.5				
Pellinen (1986)	Göttinger Wald/D (site A)[b]	115	3.3	4.7	0.6	8.6				
Scarascia-Mugnozza et al. (2000)	Gribskov/DK	118	4.7	2.0	2.2	8.9	2.9	0.5	3.4	12.3
Scarascia-Mugnozza et al. (2000)	Schacht/D	120	3.5	2.0	2.2	7.7	2.6	0.5	3.1	10.8
Ellenberg et al. (1986)	Solling/D (B1)	130	3.0	5.9	0.7	9.7	2.5	0.6	3.1	12.8
Kreutzer (1976)	Modelling/YC IV	130		2.3	0.5	2.9		0.7		3.5
Rademacher et al. (this investigation)	Göttinger Wald/ model site A-D	130	3.7	5.2	0.5	9.4	1.0	1.0	2.0	11.4
Kreutzer (1976)	Modelling/YC I	140		5.0	0.9	6.0		1.3		7.2
Rademacher et al. (this investigation)	Solling/D, modelling B1	150	3.5	1.8	0.4	5.7	3.3	0.2	3.5	9.2
Rademacher (2002)	Solling/D (B1/modelling)	155	4.6	3.7	0.6	8.9	2.0	0.4	2.4	11.3
Rademacher (2002)	Göttinger Wald/D (modelling)	155	2.8	5.7	0.7	9.2	2.8	0.4	3.2	12.4
Scarascia-Mugnozza et al. (2000)	Aubure/F	161	2.8	2.0	2.2	7.0	1.5	0.5	2.0	9.0
All investigations (min–max)	All sites	67–161	2.8–4.7	2.0–6.7	0.4–4.6	2.9–12.5	1.5–3.8	0.2–1.3	2.0–4.8	3.5–16.6

[a]Include 2% *Acer*
[b]Include 14% *Acer*

Table 8.7 Nutrient and carbon concentrations (mean and standard deviation in brackets (values in g kg^{-1}) in various components of trees harvested in the Solling area (Rademacher and Müller-Using 2004) (n = 36) and the Göttinger Wald area (Meiwes and Beese 1988) (n = 19), and for up to 11 literature studies (for roots, n = 3–5)

Element	Leaves	Branches <7 cm	Bark >7 cm	Wood >7 cm	Roots >0.2 cm	Roots <0.2 cm
Solling						
C	510 (14)	488 (18)	510 (10)	497 (7)	501 (9)	528 (16)
N	23.9 (2.4)	4.3 (1.6)	8.1 (1.8)	1.7 (1.8)	4.4 (1.1)	7.5 (2.3)
P	1.32 (0.19)	0.41 (0.14)	0.43 (0.18)	0.13 (0.13)	0.28 (0.09)	0.51 (0.28)
K	6.6 (1.5)	1.6 (0.4)	2.3 (0.5)	1.3 (0.3)	1.8 (0.6)	3.0 (1.0)
Ca	6.8 (2.0)	2.5 (1.1)	11.8 (3.0)	1.1 (1.0)	1.1 (0.2)	1.4 (0.5)
Mg	1.54 (0.71)	0.40 (0.15)	0.61 (0.19)	0.33 (0.11)	0.29 (0.10)	0.60 (0.16)
Göttinger Wald						
C	500 (9)	496 (9)	474 (10)	489 (10)	488 (7)	529 (1)
N	25.1 (5.0)	4.0 (0.6)	6.3 (1.0)	1.8 (0.3)	5.0 (0.5)	7.9 (1.2)
P	1.34 (0.25)	0.29 (0.05)	0.36 (0.05)	0.09 (0.04)	0.21 (0.05)	0.27 (0.06)
K	10.6 (1.4)	1.7 (0.3)	2.6 (0.6)	1.1 (0.2)	1.5 (0.5)	1.2 (0.4)
Ca	17.5 (3.2)	8.6 (1.4)	40.9 (5.3)	1.4 (0.4)	7.1 (1.9)	12.0 (2.2)
Mg	1.52 (0.43)	0.38 (0.08)	0.46 (0.11)	0.29 (0.13)	0.35 (0.08)	0.66 (0.12)
Literature*						
C	490 (15)	474 (7)	468	477 (9)	466	507 (17)
N	26.0 (3.2)	4.3 (1.4)	7.4 (1.6)	1.2 (0.3)	3.0 (0.9)	7.2 (0.7)
P	1.46 (0.33)	0.48 (0.21)	0.50 (0.16)	0.10 (0.05)	0.35 (0.20)	0.60 (0.28)
K	8.7 (2.5)	1.5 (0.4)	2.3 (0.5)	0.9 (0.2)	1.3 (0.3)	2.2 (1.0)
Ca	8.9 (5.7)	4.0 (1.9)	20.5 (12.4)	1.0 (0.5)	2.7 (2.2)	5.3 (5.8)
Mg	1.25 (0.58)	0.36 (0.13)	0.59 (0.21)	0.25 (0.11)	0.43 (0.30)	0.74 (0.19)

*Scarascia-Mugnozza et al. 2000; Rademacher 2002; Jacobsen et al. 2003

Among the woody components, branches <7 cm and roots of >0.2 cm thickness have similar values of higher concentrations of all elements than of the wood. Despite the low values of all the elements in these three components, they still reflected the differences in the amount of the elements available on the two sites. For example, the concentrations of Ca were higher in these three components for Göttinger Wald than those for Solling, whereas those of P were lower at Göttinger Wald.

In addition to the site factors affecting nutrient availability, age and size of trees are important factors which may affect the concentrations of tree components especially those of a woody nature. Common nutrient concentrations in the above-ground beech components of smaller trees are high due to the higher proportion of nutrient-rich non-woody components (Jacobsen et al. 2003). This was also observed in our study of different sized trees harvested in the Solling area (data not shown). However, Ca concentrations in the bark formed an exception, where higher values were observed in the bark of larger trees due to its accumulation in the bark.

Nutrient concentrations in the mature leaves of beech trees depend upon a number of soil- and site-specific factors and may vary depending on their atmospheric inputs. For example, the continuing high inputs of N indicated only small differences among sites with 80% of the sites in Germany recording 21–26 g N kg^{-1} in beech foliage (BMELF 1997). The concentration of P in leaves varied more with values from 1 to 1.5 g kg^{-1}. Concentrations of cations K, Ca and Mg in beech leaves depended very much on the soil conditions. The values for base cations in the beech leaves in stands from the Central Mountain Ranges in Germany (alkaline or calcium-rich soils) ranged from about 4 to 10 g K kg^{-1}, 5 to 15 g Ca kg^{-1} and 0.7 to 2 g Mg kg^{-1} (Glavac 1987; Büttner et al. 1993; Asche 1997; Jacobsen et al. 2003). Similar values were reported at all the survey sites included in the national tree health assessment (BMELF 1997). The long-term studies at the Solling and Göttinger Wald sites indicated significant temporal changes in the base content of leaves. Leaf Ca concentration was reported by Meiwes and Beese (1988) to be high in the 1980s at the Göttinger Wald site, when compared to the values reported for the beech sites of a north–south gradient across Europe (Bauer et al. 2000). Recently observed values of Ca in leaves are somewhat lower, but around 17 g Ca kg^{-1} (Table 8.7). In contrast, the high Mg concentration in the beech foliage of the Solling region as reported in this study (Table 8.7) was very high when compared with low values ($\leq$0.7 g kg^{-1}) found in other investigations in former years, when acid deposition was much higher (Ellenberg et al. 1986; Büttner et al. 1993; NFV, unpublished), and Mg was not applied together with liming. From various studies on nutrient concentrations in various tree species, the following order is commonly reported: Beech $\geq$ Oak $>$ Spruce $>$ Pine (Kreutzer 1976; Augusto et al. 2000; Rademacher 2002; Jacobsen et al. 2003).

For the calculation of the total amount of elements in the tree components at the Solling and Göttinger Wald sites, the measured values for different elements (Table 8.7) and the biomass data for different components (Table 8.4) were used. Leaves and fine roots contributed very little to the total content of nutrients in the trees of a mature stand (Table 8.8 and 8.9). However, when the total amount retained annually is considered, the amount recycled through leaves and fine roots is much higher than the amount retained by the woody components. This small fraction of the total amount of nutrients retained annually when accumulated over a long period of 100–150 years on these sites forms a significant amount of nutrients in the woody components of trees. Wood and bark components retained about 6 kg N annually at the Solling site and 18 kg ha^{-1} per year at the Göttinger Wald site (Table 8.10).

The total amount of N retained by the aboveground biomass was about 1,000 kg per ha at the Solling and Göttinger Wald sites and more than 70% of that in the wood and branches (Table 8.8). The total woody biomass (which also included bark) accumulated about 2,200 kg ha^{-1} (in Göttinger Wald) of Ca, which was three times higher value than that for the Solling site. Amounts of phosphorus, potassium and magnesium retained did not differ greatly between the two stands. The values given here fall in the ranges given in literature for the contents of P, K, Ca and Mg in tree components (Table 8.9, Jacobsen et al. 2003).

Mean annual retention of nutrients by trees (woody components) during the last 10 years was calculated to be 0.9 kg P, 9 kg K and 2 kg Mg ha^{-1} at Göttinger Wald

Table 8.8 Carbon (Mg ha^{-1}) and element contents (kg ha^{-1}) of various tree components calculated by element concentrations (Table 8.7) multiplied with stand biomass (Table 8.4) using regression equations given in Table 8.3 for the Solling (SO) and Göttinger Wald (GW) areas

Site	Leaves	Branches <7 cm	Bark >7 cm	Wood >7 cm	Coarse roots >0.2 cm	Fine roots <0.2 cm	Above ground biomass	Woody biomass[a]	Total biomass
Carbon									
SO	1.78	34.5	8.7	133	14.3	3.27	178	191	196
GW	1.85	17.3	10.2	181	36.3	1.53	211	245	248
Nitrogen									
SO	84	305	138	451	126	47	978	1,020	1,151
GW	93	141	137	663	374	23	1,034	1,315	1,441
Phosphorus									
SO	4.6	28.9	7.3	33.9	8.0	3.2	74.7	78.1	85.9
GW	5.0	10.1	7.8	33.4	15.6	0.8	56.3	67.9	72.7
Potassium									
SO	23.0	113.4	38.3	334	51	18.7	509	537	578
GW	39.2	58.2	55.1	408	110	3.5	561	631	674
Calcium									
SO	24	176	200	291	32	9	691	699	731
GW	65	301	884	504	528	35	1,754	2,217	2,317
Magnesium									
SO	5.4	28.4	10.4	88.6	8.4	3.7	133	136	145
GW	5.6	13.2	9.9	107.5	26.0	1.9	136	157	164

[a]Including values of branches, bark, wood and coarse roots

Table 8.9 Range (minimum–maximum) values of element content (kg ha^{-1}) of various components of beech trees for 11 literature values (for roots $n = 3–5$; Rademacher 2002; Jacobsen et al. 2003)

Element	Leaves	Branches <7 cm	Bark >7 cm	Wood >7 cm	Coarse roots >0.2 cm	Fine roots <0.2 cm	Above ground biomass	Woody biomass[a]	Total biomass
N	61–156	80–460	47–150	127–446	77–282	14–45	315–1,212	331–1,338	406–1,539
P	3.1–9.5	7.5–55.7	2.8–11.0	7.0–44.6	5.5–39.8	0.7–5.0	20–121	23–151	27–166
K	17–59	31–153	15–47	99–327	37–116	3–18	160–586	180–643	200–720
Ca	9–77	58–476	66–549	70–446	18–354	3–63	203–1,549	213–1,826	222–1,966
Mg	1.8–9.7	6.4–39.5	3.1–13.4	20–107	4.8–52.8	1.2–5.3	31–170	34–213	37–228

[a]Including values of branches, bark, wood and coarse roots

Table 8.10 Annual values of carbon (Mg ha^{-1}) and element contents (kg ha^{-1}) retained by various tree components for the Solling (*SO*) and Göttinger Wald (*GW*) sites

Site	Leaves	Branches <7 cm	Bark >7 cm	Wood >7 cm	Root >0.2 cm	Root[b] <0.2 cm	Above ground biomass	Woody biomass[a]	Total biomass
Carbon									
SO	1.81	0.19	0.05	0.83	0.08	1.3	2.88	1.15	4.26
GW	1.86	0.24	0.14	2.42	0.48	0.6	4.66	3.28	5.74
Nitrogen									
SO	85	1.7	0.8	2.8	0.7	38	90.3	6.0	129
GW	93	2.0	1.9	8.8	5.0	22	106.1	17.7	133
Phosphorus									
SO	4.68	0.16	0.04	0.21	0.04	1.9	5.1	0.5	7.0
GW	4.96	0.14	0.11	0.45	0.21	0.9	5.7	0.9	6.8
Potassium									
SO	23.3	0.6	0.2	2.1	0.3	8.3	26.2	3.2	34.8
GW	39.5	0.8	0.8	5.4	1.5	1.7	46.5	8.5	49.7
Calcium									
SO	24.0	1.0	1.2	1.8	0.2	1.7	28.0	4.2	30
GW	65.1	4.3	12.1	6.7	7.0	2.8	88.2	30.1	98
Magnesium									
SO	5.5	0.16	0.06	0.55	0.05	1.7	6.3	0.8	8.1
GW	5.7	0.19	0.14	1.4	0.35	0.7	7.4	2.1	8.5

[a]Including values of branches, bark, wood and coarse roots

[b]From Chap. 9 (this volume), fine root litter production (mean of element concentrations in living and dead fine roots) (*SO*: 0–50 cm depth; *GW*: 0–20 cm depth)

which were twice as large as those measured at Solling (Table 8.10). The amount of N retained by the woody components was three times more at Göttinger Wald. The value of 30 kg ha^{-1} Ca at Göttinger Wald was about seven times higher than at Solling. The distinct differences in annual rates of nutrients retained at different sites may primarily be the result of a reduction in growth and nutrient accumulation in recent years at Solling, and the ongoing high nutrient uptake and biomass production at Göttinger Wald.

Low values of nutrients retained annually in woody biomass components compared to those turned over through leaves and fine root production indicated the presence of highly efficient systems of element uptake and element re-translocation in these trees. However, the two sites differed in the annual amount required for the production of fine roots which was higher for N, P, K and Mg at the Solling site.

8.7 Stand Harvesting and Nutrient Export

Continuous growth of woody components in beech trees of more than 120 years old at these sites indicates the continuing removal of nutrients from the turnover components and soil to those which do not turnover (woody components), which would have implications for the current and future sustainability of a forest site, especially that relating to its nutrient levels. This would thus depend upon the amount of nutrients withdrawn annually from the turnover fraction and that exported in future harvesting operations in relation to the amount replenished through soil weathering and atmospheric inputs (Ulrich 1972; Weetman et al. 1979; Federer et al. 1989; Adams et al. 2000; Scarascia-Mugnozza et al. 2000). For two of the beech stands investigated in this study, the following harvesting options involving the removal of different tree components can be considered: merchantable wood >7 cm diameter without bark, merchantable wood with bark, aboveground tree components and total tree (above- and belowground stand).

There are two approaches which can be used to consider the effects of nutrient withdrawal from the turnover fractions and of the amount of nutrients removed during harvesting operations. The first approach is the annual nutrient input/output analysis where input refers to nutrient supply (atmospheric inputs plus mineral weathering rate) and the output refers to the amount retained by woody non-turnover plant components. The implicit assumption in this analysis is that the rest of the amount taken up by the vegetation is cycled through the soil plant system in due course of the considered time period. For any element where input is lower than output, a continuous decrease in soil supply is expected indicating a possible effect on tree growth and sustainability of nutrient supply for future.

The annual amounts of nutrients retained presently by woody biomass if continued until the harvest would be 6 and 18 kg N, 0.5 and 0.9 kg P, 3 and 9 kg K, 4 and 30 kg Ca and 1 and 2 kg Mg ha^{-1} per year on the Solling and Göttinger Wald sites, respectively (Table 8.10). These output values fall within the ranges reported for comparable beech stands by Rennie (1955), Kreutzer (1976) and other literature studies (Khanna and Ulrich 1991; Nihlgård 1972; Weiner and Grodzinski 1984; Horn et al. 1989). The atmospheric input values for N of 20–23 kg ha^{-1} per year at the two sites fall within the distribution range reported by Persson et al. (2000) for Central Europe. Thus, the N balance on the two sites is positive and some of it may be accumulated in the soil. The atmospheric inputs of K (2–3 kg ha^{-1} per year), Ca (4–5 kg) and Mg (1 kg) differ only slightly between the two study sites (Meesenburg et al. see Chap. 15, this volume). However, the two sites differed in the amount of cations released on mineral weathering. Values of K released through weathering at the Solling site was notably high as compared with Göttinger Wald (20 kg vs 8 kg ha^{-1} per year) whereas those of Ca (1.6 kg and 235 kg ha^{-1} per year) were high at the Göttinger Wald site, whereas the values for Mg were similar (3.5 kg ha^{-1} per year) on both the sites (Brumme et al. see Chap. 16, this volume). However, the interpretation will change when the leaching losses and soil and litter retention processes are considered as discussed in other chapters of this volume.

The second approach is to compare the total amount of nutrients removed with the tree harvest with those present in the available forms in the soil where two options can be considered: (1) final harvest only and (2) final harvest plus all the commercial thinnings (removal of biomass) carried out until final harvest. Here, the issue is to assess the nutrient cost of biomass harvest for future management and sustainability. For example, total tree harvest where maximum amount of nutrients would be removed (Table 8.8) may not affect the supply of K, Ca and Mg by the soil at the Göttinger Wald site. This soil contains many times higher amounts than that exported in the harvest (amounts available in the soil are given as: about 1,200 kg K ha^{-1}, 50,000 kg Ca ha^{-1} and 500 kg Mg ha^{-1}, when calculated as exchangeable elements in mineral soil, 0–100 cm, plus the total amount stored in the forest floor, Chap. 3). However, the amount of exchangeable Ca at the Solling site lies below the amount removed upon harvest. The Mg available in the Solling soil is only about 1.3–1.8 times that of the total biomass. The available soil K in the Solling soil is 1.5–2.3 times the amount in the tree components. When additional export of the biomass of thinnings is included, there are expected to be significant losses of cations from the Solling site which would need replenishment for future stands. Table 8.8 shows that for most nutrients the amount present in the woody components of the tree would dominate the fraction exported, whereas for some nutrients, especially N and P, leaving the crown and roots would make a significant contribution to the total amount removed from the site. Similarly, Boyle and Ek (1972) showed a decrease of the soil to plant ratio in nutrient supply of 3–1.5 in case of intensified forest use, e.g. the additional use of crown material.

The range of nutrient accumulation in biomass depends on the site characteristics, tree age and biomass production. A range of values were found in the compiled list for beech stands (Table 8.9). When compared with the values of the two stands studied here, the one at Göttinger Wald had much higher values than those in the literature for nutrient accumulation in the biomass, which followed its high growth rates and high biomass accumulation. Nutrient accumulation in tree biomass also varies with the tree species. Rademacher (2002) compiled such data for comparison purposes of 100-year-old stands of pine, spruce and beech. Beech stands showed high quantities of nutrients accumulated in the aboveground biomass per hectare indicating a high element demand, whereas pine stands accumulated the lowest amount of all tree species.

8.8 Summary

For the growth of beech stands and the sustainability of sites, appropriate tree nutrition is an important factor. In this study, tree growth measurements at three beech forest sites were recorded for different periods, and biomass and nutrient inventories were obtained by harvesting individual trees on two of those sites. The following important conclusions can be drawn from this study:

- The mature beech stands had been subjected to different management options in the past and represented three highly variable soil conditions (highly acid to less acid surface mineral soil with Ca-rich to Ca-poor conditions, and from high P to low P soils).
- The three sites are mature stands of beech and have accumulated huge amounts of merchantable wood and woody biomass and still continue to grow in height and stem diameter.
- The three sites show the following order in their annual wood volume increments: Göttinger Wald > Zierenberg > Solling.
- The differences in the biomass content of different tree components indicated that the fast turnover fraction (leaves and fine roots) was lower at the Ca-rich Göttinger Wald site, causing differences in the allocation of biomass to woody components or the merchantable wood.
- Except for P, elements in different tree components of the Göttinger Wald stand had higher or equal concentrations compared to those of the Solling stand.
- Harvesting of woody components may export only a fraction of the amounts of K, Ca and Mg present in the soil at the Göttinger Wald site and thus may not affect the future supply of these nutrients. However, the amount of these elements removed with wood harvest may constitute a significant fraction of the amounts present in the soil at the Solling site, indicating the need for additional inputs to achieve sustained supply for the growth of future stands.

Acknowledgements The authors thank the North West German Forestry Research Institute, headed by Professor. Dr. B. Müller-Using, and now by Professor Dr. H. Spellmann, and previous employer of the first author for conducting biomass and element analysis at forestry sites in the Solling. We thank Dr. Ulrich (NW-FVA) for biomass inventory data of Zierenberg and Dr. Pellinen as well as Dr. Meiwes (NW-FVA) for biomass data of the Göttinger Wald site. Professor Dr. Bolte (FH Eberswalde) kindly supplied inventory data of the root biomass in the Soling site. We thank the Federal Ministry of Education and Research for sponsoring research in the biomass and element inventories in the Solling as part of the case study "Forest Landscape Solling" (FKZ 0339474C/3).

References

Adams MB, Burger JA, Jenkins AB, Zelazny L (2000) Impact of harvesting and atmospheric pollution on nutrient depletion of eastern US hardwood forests. For Ecol Manage 138:301–319

Alexander IJ, Fairley RI (1983) Effects of N fertilisation on populations of fine roots and mycorrhizas in Spruce humus. Plant Soil 71:49–53

Asche N (1997) Nährelementgehalt in Buchenblättern unter besonderer Berücksichtigung der zeitlichen Variation auf basenarmen Standorten in Nordrhein-Westfalen. Forstw Cbl 116: 394–402

Augusto L, Ranger J, Ponette Q, Rapp M (2000) Relationships between forest tree species, stand production and stand nutrient amount. Ann For Sci 57:313–324

Bauer GA, Persson H, Persson T, Mund M, Hein M, Kummetz E, Matteuch G, vg Oene H, Scarascia-Mugnozza G, Schulze E-D (2000): Linking plant nutrition and ecosystem processes. In: Schulze E-D (ed) Carbon and nitrogen cycling in European forest ecosystems. Ecological Studies 142. Springer, New York, pp 64–98

Becker R, Block J, Schimming CG, Spranger T, Wellbroock N (2000) Critical loads für Waldökosysteme – Methoden und Ergebnisse für Standorte des Level II-Programms. Arbeitskreis A der Bund-Länder Arbeitsgruppe Level II, Bundesministerium für Ernährung, Landwirtschaft und Forsten, Berlin

Binkley D (1986) Forest nutrition management. Wiley, New York

BMELF (1997) Deutscher Waldbodenbericht 1996. Ergebnisse der bundesweiten Bodenzustandserhebung im Wald von 1987–1993 (BZE), Bd 1 und 2. Bundesministerium für Ernährung, Landwirtschaft und Forsten, Berlin

Bolte A, Rahmann T, Kuhr M, Pogoda P, Murach D, Gadow Kv (2004) Relationships between tree dimension and coarse root biomass in mixed stands of European beech (*Fagus sylvatica L.*) and Norway spruce (*Picea abies* [*L.*] Karst.). Plant Soil 264:1–11

Boyle JR, Ek AR (1972) An evaluation of some effects of bole and branch pulpwood harvesting on site macronutrients. Can J For Res 2:407–412

Bredemeier M (1987) Stoffbilanzen, interne Protonenproduktion und Gesamtsäurebelastung des Bodens in verschiedenen Waldökosystemen Norddeutschlands. Berichte des Forschungszentrums Waldökosysteme/Waldsterben, Reihe A, Bd 33, 183 S

Burschel P (1966) Untersuchungen in Buchenmastjahren. Forstw Cbl 85:204–219

Büttner G, Hartmann G, Thomas FM (1993) Vorzeitige Vergilbung und Nährstoffgehalte des Buchenlaubes in Südniedersachsen. Forst und Holz 48:627–630

Dieter M, Elsasser P (2002) Carbon stocks and carbon stocks changes in the tree biomass of Germany's forests. Forstw Cbl 121:195–210

Dyckmans J, Flessa H, Shangguan Z, Beese F (2000) A dual ^{13}C and ^{15}N long term labelling technique to investigate uptake and translocation of C and N in Beech (*Fagus sylvatica L.*). Isot Environ Health Stud 36:63–78

Eichhorn J (1995) Stickstoffsättigung und ihre Auswirkungen auf das Buchenwaldökosystem der Fallstudie Zierenberg. Berichte des Forschungszentrums Waldökosysteme. Reihe A 124

Eichhorn J, Dammann I, Schönfelder E, Albrecht M, Beck W, Paar U (2008) Untersuchungen zur Trockenheitstoleranz der Buche am Beispiel des witterungsextremen Jahres 2003. In: Ergebnisse angewandter Forschung zur Buche. Beiträge aus der Nordwestdeutschen Forstlichen Versuchsanstalt, Band 3. Universitätsverlag, Göttingen, Germany

Ellenberg H, Mayer R, Schauermann J (1986) Ökosystemforschung –Ergebnisse des Sollingprojektes. Verlag Eugen Ulmer/Stuttgart

Federer CA, Hornbeck JW, Tritton LM, Martin CW, Smith CT (1989) Long-term depletion of calcium and other nutrients in eastern US forests. Environ Manage 13:593–601

Glavac V (1987) Calcium-, Magnesium-, Kalium- und Zink-Gehalte in Blättern eines immissionsgeschädigten Rendzina-Buchenwaldes. Allgemeine Forst Zeitschrift 42:303–305

Guericke M (2002) Untersuchungen zur Wuchsdynamkik der Buche. Wachstumskundliche Beobachtungen und waldbauliche Konsequenzen auf Grundlage langfristig beobachteter Durchforstungsversuche. Forst und Holz 27:331–337

Hochbichler E, Putzgruber N, Krapfenbauer A (1994) Biomassen- und Nährstoffuntersuchungen in einem 40jährigen Buchenbestand (*Fagus sylvatica L.*). Centralblatt für das gesamte Forstwesen 111:1–22

Hofmann G (1985) Die potentielle natürliche Nettoprimärproduktion an oberirdischer Pflanzentrockenmasse – ihre Quantifizierung und Kartierung für das Gebiet der DDR. Beiträge für die Forstwirtschaft 19:110–116

Horn R, Schulze E-D, Hantschel R (1989) Nutritional balance and element cycling in healthy and declining Norway spruce stands. In: E-D Schulze, OL Lange, R Oren (eds) Ecological Studies 77, Springer, Berlin, pp 444–455

Jacobsen C, Schönfelder E, Paar U, Eichhorn J (2001) Die Bodenlösungschemie in einem stickstoffgesättigten Buchenwaldökosystem in Nordhessen (Fallstudie Zierenberg). Allgemeine Forst- und Jagdzeitung 172:117–126

Jacobsen C, Rademacher P, Meesenburg H, Meiwes KJ (2003) Gehalte und chemische Elemente in Baumkompartimenten – Literaturstudie und Datensammlung. Studie im Auftrage des BMELF, Berichte des Forschungszentrums Waldökosysteme Reihe B, Bd 69

Joosten R, Schulte A (2003) Schätzung der Nährstoffexporte bei einer intensiven Holznutzung in Buchenwäldern. Allg Forst- u J-Ztg 174:157–168

Jurgensen MF, Leaf AL (1965) Soil moisture- fertility interactions related to growth and nutrient uptake of Red Pine. Soil Sci Soc Am Proc 29:294–299

Khanna PK, Ulrich B (1991) Ecochemistry of temperate deciduous forests. In: Röhrig E, Ulrich B (eds) Ecosystems of the World 7. Elsevier, Amsterdam, pp 121–163

Kollman F (1951) Technologie des Holzes und der Holzwerkstoffe. Springer, Berlin

König N, Fortmann H (1998) Probenvorbereitungs-, Untersuchungs- und Elementbestimmungs-Methoden des Umweltanalytik-Labors der Niedersächsischen Forstlichen Versuchsanstalt und des Zentrallabors II des Forschungszentrums Waldökosysteme. Berichte des Forschungszentrums Waldökosysteme Reihe B, Bd 53

Kramer H, Akça A (1982) Leitfaden für Dendrometrie und Bestandesinventur. JD Sauerländer's, Frankfurt

Kreutzer K (1976) Effects on growth in next rotation (regeneration). Symposium on the harvesting of a larger part of the forest biomass. ECE/FAO-Symposium in Hyvinkää/Finland, pp 78–90

Kuhr M (2000) Grobwurzelarchitektur in Abhängigkeit von Baumart, Alter, Standort und sozialer Stellung. Dissertation Fakultät für Forstwissenschaften und Waldökologie der Georg-August-Universität Göttingen

Likens GE, Driscoll CT, Busco DC (1996) Long-term effects of acid rain: response and recovery of a forest ecosystem. Science 272:244–246

Lindberg SE, Lovett GM, Richter DD, Johnson DW (1986) Atmospheric deposition and canopy interactions of major ions in a forest. Science 231:141–145

Madgwick HA, Tamm CO (1987) Allocation of dry weight increment in crowns of *Picea abies* as affected by stand nutrition. OIKOS 48:99–195

Matzner E (1988) Der Stoffumsatz zweier Waldökosysteme im Solling. Berichte des Forschungszentrums Waldökosysteme/Waldsterben, Reihe A, Bd 40

Meesenburg H, Rademacher P, Meiwes KJ (1998) Stoffeintrag über atmogene Depositionen in verschiedene Ökosysteme Niedersachsens und deren Auswirkungen. Arb-H Boden 1/1998: 67–77

Meiwes KJ, Beese F (1988) Ergebnisse der Untersuchung des Stoffhaushaltes eines Buchenwaldökosystems auf Kalkgestein. Berichte des Forschungszentrums Waldökosysteme, Reihe B, Bd 9

Montès N, Gauquelin T, Badri W, Bertaudiere V, Zaoui EH (2000) A non-destructive method für estimating above-ground forest biomass in threatened woodlands. For Ecol Manage 130:37–46

Mrotzek R (1998): Wuchsdynamik und Mineralstoffhaushalt der Krautschicht in einem Buchenwald auf Basalt. Berichte des Forschungszentrums Waldökosysteme. Reihe A 152:164–181

Nielsen CCN (1995) Detailed instructions for root architecture assessments with the Rootarch method. Arboretum, Royal Vet and Agriculture University, Int Rep No 7, Hoersholm, DK

Nihlgård B (1972) Plant biomass, primary production and distribution of chemical elements in a beech and a plant spruce forest in South Sweden. OIKOS 23:69–81

Oren R, Schulze E-D (1989) Nutritional disharmony and forest decline: a conceptual model. In: Schulze E-D, Lange OL, Oren R (eds) Forest decline and air pollution, Ecological Studies 77. Springer, Berlin, pp 425–443

Oren R, Schulze E-D, Werk KS, Meyer J (1988) Performance of two *Picea abies* (*L.*) Karst. stands at different stages of decline. VII. Nutrient relations and growth. Oecologia 77:163–173

Pellinen P (1986) Biomasseuntersuchungen im Kalkbuchenwald. Diss Forstl Fak Göttingen

Persson T, v Oene H, Harrison AF, Karlsson PS, Bauer GA, Cerny J, Coûteaux M-M, Dambrine E, Höberg P, Kjøller A, Matteuch G, Rudebeck A, Schulze E-D, Pages T (2000) Experimental Sites in the NIPHYS/CANIF-Projekt. In: Schulze E-D (ed) Carbon and nitrogen cycling in European forest ecosystems. Ecological Studies 142, Springer, Berlin, pp 14–46

Rademacher P (2002) Ermittlung der Ernährungssituation, der Biomasseproduktion und der Nährelementakkumulation mit Hilfe von Inventurverfahren sowie Quantifizierung der Entzugsgrößen auf Umtriebsebene in forstlich genutzten Beständen. Manuskript der Habilitationsschrift am Forstlichen Fachbereich der Universität Göttingen

Rademacher P, Müller-Using B (2004) Bioelemententzug bei der Holznutzung in Rein- und Mischbeständen aus Buche und Fichte. Abschlussbericht im BMBF-Vorhaben ÖK-2.3.3: Indikatoren und Strategien für eine nachhaltige, multifunktionelle Waldnutzung – Fallstudie Waldlandschaft Solling, Forschungszentrum Waldökosysteme Reine B 71:81–109

Rademacher P, Müller-Using B, Meesenburg H, Meiwes KJ (2001) Inventory and budget of nutrients in managed German forest stands – a method for sustainable nutrient management. In: Horst WJ, Bürkert A, Claassen N, Flessa H, Frommer WB, Goldbach H, Merbach W, Olfs HW, Römheld V, Sattelmacher B, Schmidhalter U, Schenk MK, v. Wirén N (eds) Plant nutrition – Food security and sustainability of agro-ecosystems, Kluwer Academic Publishers/ Netherlands, pp 914–915

Rennie PJ (1955) The uptake of nutrients by mature forest growth. Plant Soil 7:49–95

Röhrig E (1991) Biomass and productivity. In: Röhrig E and Ulrich B (eds) Ecosystems of the World 7. Elsevier, Amsterdam, pp 165–174

Santantonio D, Hermann RK (1985) Standing crop, production, and turnover of fine roots on dry, moderate, and wet sites of mature Douglas-fir in Western Oregon. Ann Sci For 42:113–142

Scarascia-Mugnozza G, Bauer GA, Persson T, Matteuch G, Masci A (2000) Tree Biomass, Growth and Nutrient Pools. In: Schulze E-D (ed) Carbon and nitrogen cycling in European forest ecosystems. Ecological Studies 142, Springer, Berlin, pp 49–62

Schober R (1987) Ertragstafeln wichtiger Baumarten bei verschiedener Durchforstung. Sauerländer, Frankfurt

Schulze E-D, Eichhorn J (2000) Veränderungen im Stickstoffhaushalt von Buchenwäldern auf Basalt: Die Ausbreitung der Großen Brennessel (*Urtica dioica L.*) und ihr Einfluss auf die natürliche Verjüngung der Buche. Forst und Holz 55:435–441

Spiecker H (1999) Overview of recent growth trends in European forests. Water Air Soil Pollut 116:33–46

Ulrich B (1972) Betriebswirtschaftliche Konsequenzen von Eingriffen in den Naturhaushalt von Wald-Ökosystemen. Forstarchiv 7:129–132

Ulrich B, Summer ME (1991) Soil acidity. Springer, Berlin

Ulrich B, Mayer R, Sommer U (1975) Rückwirkungen der Wirtschaftsführung über den Nährstoffhaushalt auf die Leistungsfähigkeit der Standorte. Forstarchiv 46:5–8

Ulrich B, Mayer R, Matzner E (1986) Vorräte und Flüsse der chemischen Elemente. In: Ellenberg H, Mayer R, Schauermann J (eds) Ökosystemforschung – Ergebnisse des Sollingprojektes, Ulmer, Stuttgart, pp 375–417

Weetmann GF, Lemmer EC, Algar D (1979) Boreal forests nutrient balance after logging. Volunteer paper for the Symposium on Impact of Intensive Harvesting on Forest Nutrient Cycling, Syracuse, August 13–16:1–60

Weiner J, Grodzinski W (1984) Energy, nutrient, and pollutant budgets of the forest ecosystems. In: Grodzinski W, Weiner J, Maycock PF (eds) Forest ecosystems in industrial regions. Springer, Berlin, pp 203–230

Wittwer RF, Leaf AL, Bickelhaupt DH (1975) Biomass and chemical composition of fertilizes and/or irrigated *Pinus resinosa Ait* plantations. Plant Soil 42:629–651

Young HE (1967) Symposium on primary productivity and mineral cycling in natural ecosystems. Young HE (ed) Symposium-paper: Ecological Society of America/American Association for the Advancement of Science, NY

Chapter 9
Fine Root Biomass, Turnover and Litter Production

D. Murach, A. Horn, W. Ke-Hong, and C. Rapp

9.1 Introduction

Fine roots are part of the regenerative system of trees. The annual renewal of fine roots can represent an important sink for assimilates in forest ecosystems (Agren et al. 1980). Estimates for the production of fine roots range between 5 and 70% of the total net primary production of trees (Santantonio 1990). Thus, fine root turnover may equal or even exceed the amount of carbon that is required for foliar litter production, making the quantification of fine root turnover an essential component of carbon budgets in forest ecosystems (Bartelink 1998).

However, only limited data are available on belowground C allocation as it is not possible to directly measure root growth, mortality and decomposition in an undisturbed forest soil. Even less is known about the factors controlling fine root turnover. Because of the insufficient database, reliable hypotheses about the regulation of belowground processes in forest ecosystems have great uncertainty (Vogt et al. 1996). However, due to their importance in global climate change, more information on the factors controlling carbon allocation within plants is required to assess the ability of forest ecosystems to sequester carbon (Kroon and Visser 2003).

The emphasis of the present chapter is to assess if soil acidification affects the fine root production or turnover of European beech. In addition to the quantification of carbon cycling via fine root dynamics, the importance of fine root litter production in element recycling and thus the soil organic matter pool will be evaluated.

9.2 Methods

The chapter is based on case studies of the main plots and fertilisation experiments in beech stands at the Solling and Göttinger Wald sites, two experimental sites of the forest ecosystem research centre at Göttingen with long-term inventories of fine root biomass and necromass. In addition to the data from untreated plots at the

R. Brumme and P.K. Khanna (eds.), *Functioning and Management of European Beech Ecosystems*, Ecological Studies 208,
DOI: 10.1007/978-3-642-00340-0_9, © Springer-Verlag Berlin Heidelberg 2009

Solling and Göttinger Wald sites, data from a fertilised and a limed plot adjacent to the Solling plot were used in our analysis.

The soils at the Solling sites are acid brown earths derived from sandstone, and at Göttinger Wald brown earths and rendzinas derived from limestone (for detailed description of the sites, refer to Chap. 3, this volume). The limed plot at Solling was limed in 1982 with 30 Mg ha^{-1} dolomite (Beese and Prenzel 1985). The dolomite was not worked into the soil but was applied on the surface organic layer and represented about 6 Mg ha^{-1} Ca and 3.5 Mg ha^{-1} Mg. The fertilised plot at the Solling site had been fertilised annually with ammonium sulphate (ca. 140 kg ha^{-1} NH_4–N) from 1983 until 1993. A detailed description of the treated sites was given by Rapp (1991) and Meiwes et al. (1998).

Fine root turnover was determined by using data of sequential coring. In the present investigation, rootlets with a diameter of less than 2 mm were defined as fine roots, and those with a diameter between 2 and 5 mm as small roots. The years of sampling were 1986/1987 for the Solling site and 1995–1997 for Göttinger Wald. Normally, the samples were taken 6–7 times per year. To determine standing amount of fine roots, at each sampling date 10–12 soil cores were taken randomly from each plot with a root auger 8 cm in diameter. For the Solling plots. the aboveground organic layer was separated from the mineral soil, and the mineral soil was divided into segments of 5 cm depth for the first 10 cm and of 10 cm depth for the remaining depths. For the Göttinger Wald plot, where an aboveground organic layer was not present, the mineral soil was separated into depth segments of 0–5 cm and 5–20 cm. For the Solling site, the maximum coring depth was 50 cm. Rapp (1991) has shown that more than 80% of the total fine root biomass can be found within 50 cm depth in the area between the tree stems at this site. For Göttinger Wald, coring was restricted by the stony bedrock to a depth of about 20 cm.

After sampling, the soil segments were soaked in de-ionized water and placed in a vacuum chamber at −700 hPa to improve the disintegration of the soil. Next day the roots were washed free of soil using de-ionised water on 1.5-mm mesh size nylon net. All root segments longer than 1 cm were collected from the sieve and any remaining soil particles were removed with a soft brush. Root fragments of the O layers were separated from the leaf litter manually without using water. However, this treatment did not recover completely the fine root necromass from the O layer. The roots were stored at 2–4°C in de-ionized water until they were further sorted.

To distinguish living from dead rootlets, the degree of discolouration, tissue structure of the central stele and resistance of rootlets against separation of extra-xylary tissues from the central stele by pulling longitudinally were used as criteria. The assessment was carried out under a binocular microscope. Further criteria were included which have been described in detail by Murach (1984) and were similar to those suggested by Kohmann (1972), Roberts (1976) and Persson (1979).

The compartment model (Fig. 9.1) was used to calculate the temporal changes of the standing amount of fine root biomass and necromass. Fine root biomass describes the live root material, and necromass the dead root content.

Relative extreme values during the temporal development of living biomass and necromass were considered by comparing the rank sums of the sampling dates derived

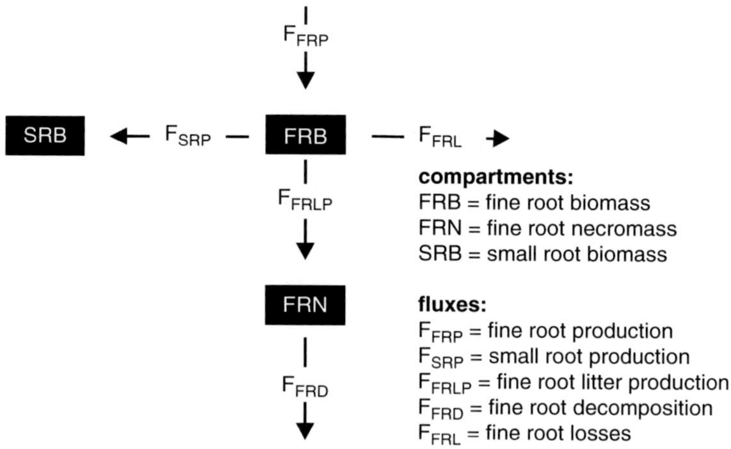

Fig. 9.1 Compartment model for the calculation of fine root turnover

from the Kruskal–Wallis-test (H-test) (Lienert 1973). Successive sampling dates were grouped for the Dunns-test (Lienert 1973), when the rank sums were similar.

Only significant differences between grouped sampling dates were considered as relative extreme values and were used for the subsequent formation of the time series providing the fine root standing crops. This series was based on the means of the starting date and the significant relative extreme values with series values between the relative extreme values calculated as a linear function of time.

To estimate the fluxes F_{FRP}, F_{FRLP} and F_{FRD} (for definition see Fig. 9.1), the method of balancing transfers developed by Santantonio (1978) was applied. This method balances the observed changes in standing crops of fine root biomass and necromass between two sampling dates by providing flux values from one root compartment to another. The decision matrix used for balancing transfers is described in Table 9.1. The matrix is based on Fairley and Alexander (1985) and modified by Murach (1991). It describes the equations used to estimate the fluxes in the standing crop of fine root biomass and necromass between two sampling dates. An example of calculation of fluxes based on the decision matrix of Table 9.1 is given in Table 9.2.

The annual fluxes of fine roots (described by the suffix 'a' below) for each soil depth were calculated by adding up the fluxes calculated for the single sampling dates:

$$\text{Production}\, F_{\mathrm{FRP}_a} = \sum_{i=1}^{i=n} F_{\mathrm{FRP}_i} \quad (\mathrm{kg\,ha^{-1}\;per\;year}), \tag{9.1}$$

$$\text{Litter production}\, F_{\mathrm{FRLP}_a} = \sum_{i=1}^{i=n} F_{\mathrm{FRLP}_i} \quad (\mathrm{kg\,ha^{-1}\;per\;year}), \tag{9.2}$$

Table 9.1 Decision matrix for the selection of equations to calculate the fluxes F_{FRP} fine root production, F_{FRLP} fine root litter production and F_{FRD} fine root decomposition between two sampling dates

Differences in necromass between two sampling dates	Differences in biomass between two sampling dates: ΔFRB > 0	ΔFRB < 0: ΔFRN > ΔFRB	ΔFRB < 0: ΔFRN ≤ ΔFRB	ΔFRB = 0
ΔFRN > 0	$F_{FRP} = \Delta FRB + \Delta FRN$ $F_{FRLP} = \Delta FRN$ $F_{FRD} = 0$	$F_{FRP} = \Delta FRB + \Delta FRN$ $F_{FRLP} = \Delta FRN$ $F_{FRD} = 0$	$F_{FRP} = 0$ $F_{FRLP} = -\Delta FRB$ $F_{FRD} = -\Delta FRB - \Delta FRN$	$F_{FRP} = \Delta FWN$ $F_{FRLP} = \Delta FWN$ $F_{FRD} = 0$
ΔFRN < 0	$F_{FRP} = \Delta FRB$ $F_{FRLP} = 0$ $F_{FRD} = -\Delta FRN$	$F_{FRP} = 0$ $F_{FRLP} = -\Delta FRB$ $F_{FRD} = -\Delta FRB - \Delta FRN$		$F_{FRP} = 0$ $F_{FRLP} = 0$ $F_{FRD} = -\Delta FRN$
ΔFRN = 0	$F_{FRP} = \Delta FRB$ $F_{FRLP} = 0$ $F_{FRD} = 0$	$F_{FRP} = 0$ $F_{FRLP} = -\Delta FRB$ $F_{FRD} = -\Delta FRB$		$F_{FRP} = 0$ $F_{FRLP} = 0$ $F_{FRD} = 0$

Table 9.2 Example for calculation of fluxes based on the decision matrix of Table 9.1

Given	Data of date 1→	$FRB_1 = 2{,}300$ kg ha^{-1}	$FRN_1 = 1{,}500$ kg ha^{-1}
	Data of date 2→	$FRB_2 = 2{,}100$ kg ha^{-1}	$FRN_2 = 1{,}800$ kg ha^{-1}
Calculated		$\Delta FRB = -200$ kg ha^{-1}	$\Delta FRN = 300$ kg ha^{-1}
		$F_{FRP} = 100$ kg ha^{-1}	
		$F_{FRLP} = 300$ kg ha^{-1}	
		$F_{FRD} = 0$ kg ha^{-1}	

$$\text{Decomposition}\, F_{\text{FRD}_a} = \sum_{i=1}^{i=n} F_{\text{FRD}_i} \quad (\text{kg ha}^{-1}\ \text{per year}), \tag{9.3}$$

$$\text{Turnover}\, F_{\text{FRT}_a} = F_{\text{FRP}_a} - (\text{FRB}_n - \text{FRB}_1) \quad (\text{kg ha}^{-1}\ \text{per year}); \tag{9.4}$$

for $(FRB_n - FRB_1) \geq 0$.
FRB_n fine root biomass at the end of the investigation period (here: after 1 year),
FRB_1 fine root biomass at the beginning of the investigation period.

By summing up the data for the different soil depths values of the fluxes for the whole profile were obtained.

According to (equation 9.4), the calculation of fine root turnover is based on two terms: F_{FRP} that describes the amount of fine roots produced during the investigation period from the first ($i = 1$) to the last ($i = n$) sampling, and (2) ($FRB_n - FRB_1) \geq 0$ that describes an increase of the standing crop of fine root biomass at the end of the investigation period. Such a potential increase of the standing crop of fine root biomass is not included in the calculation of fine root turnover. Thus, fine root turnover characterises the flux of assimilates into the fine root system which is required for the regeneration of the fine roots to maintain a constant amount of fine root biomass.

F_{FRLP} describes the mortality rate of the fine roots within the investigation period. This flux is an estimate of the amount of fine roots potentially available for decomposition.

The calculation of the fine root production (F_{FRP}) neither includes the ingrowth of fine roots into the small root class (F_{SRP}) nor any kind of loss of fine roots (F_{FRL}), which for example may be caused by herbivores. These fluxes cannot be easily quantified using the coring method. Although rarely estimated, consumption of fine roots by herbivores in general does not seem to be important in temperate forests where estimates usually do not exceed 10% of the fine root standing crop (Fogel 1985).

Estimation of fine root decomposition (F_{FRD}) requires a differentiation between dead roots and decomposed roots or other organic matter. In the present investigation, only rootlets longer than 0.5 cm were included for assessing necromass.

It is not yet possible to define confidence intervals for the calculated fluxes of the compartment model in Fig. 9.1, as the sources of error are not fully known. One such error is the differentiation of live roots from dead roots, another one is the unknown temporal changes of root mass measured in arbitrary time intervals (Singh et al. 1984), and the third one is the uncertainty in the estimation of fine root decomposition.

Despite these shortcomings, Vogt et al. (1998) recommended the use of sequential coring in combination with the decision matrix as a useful approach in estimating fine root production and turnover. Universally agreed methods for estimating root biomass, fine root production and turnover do not exist in the literature; there has been controversial discussion for more than a decade (Grier et al. 1981; Singh et al. 1984; Lauenroth et al. 1986; Vogt et al. 1986a, 1998; Nadelhoffer and Raich 1992).

Release of elements from roots into the soil via decomposition and mineralisation can not be measured directly. Based on the mortality rate of the fine roots (F_{FRLP}), an estimate of the amount of fine root mass that is potentially available for decomposition can be calculated. The element flux based on F_{FRLP} may be considered as input into the plant available element pool in the soil as these elements belong to the organic matter pool of the soil.

The potential element release via fine root litter production (F_{FRER}) was calculated by using the element concentrations of the living and dead fine roots:

$$F_{FRERL}\ (\text{kg ha}^{-1}\ \text{per year}) = F_{FRLP}\ (\text{kg ha}^{-1}\ \text{per year})\ c_{FRB}\ (\text{mg g}^{-1})\ 10^{-3}, \quad (9.5)$$

$$F_{\mathrm{FRERD}}\ (\mathrm{kg\,ha^{-1}\,per\,year}) = F_{\mathrm{FRLP}}\ (\mathrm{kg\,ha^{-1}\,per\,year}) \times c_{\mathrm{FRN}}\ (\mathrm{mg\,g^{-1}})\ 10^{-3}. \quad (9.6)$$

F_{FRERL} refers to element release via fine root litter production based on element concentrations in living fine roots, F_{FRERD} element release via fine root litter production based on element concentrations in dead fine roots, c element concentration, FRB fine root biomass, FRN fine root necromass.

Data for the element concentrations were taken from Murach and Wiedemann (1988). The equation using the element contents of the living fine roots (F_{FRERL}) describes the upper limit of the recyclable element pool (F_{FRER}). F_{FRERD} characterises the lower limit by applying the element contents of the dead fine roots. It is not known whether the difference between (F_{FRERL}) and F_{FRERD} values may be attributed to the re-translocation of elements during root senescence. In contrast to leaves, the re-translocation of the elements can not be easily calculated by comparing living and dead fine roots as the age of the dead fine roots remains unknown, and the fine root necromass includes dead roots in varying degrees of decomposition.

9.3 Fine Root Production and Turnover

Fine root production and mortality values showed high level of annual fluctuations in all plots (Table 9.3). This was indicated by the differences of the average and the maximum values of the annual fine root fluxes. The calculated data corresponded well with the limited amount of data published from other field studies. For example, Scarascia-Mugnozza et al. (2000) estimated that fine root production rates in European beech stands ranged from 1,500 to 3,800 kg ha^{-1} per year for different sites in Europe. The Göttinger Wald site showed the lowest values for all calculated fluxes to a depth of 20 cm (Table 9.4). They were two to three times higher at Solling than at the Göttinger Wald site. Fluxes below 20 cm could not be

Table 9.3 Annual fine root fluxes (kg ha^{-1} per year) as mean and maximum values and specific fine root turnover in the total soil profiles (0–50 cm) (1986–1987)

	F_{FRP}	F_{FRLP}	F_{FRD}	F_{FRT}	F_{FRT}/ FRB	F_{FRP} (max)	F_{FRLP} (max)	F_{FRD} (max)	F_{FRT} (max)
Solling, control	3,300	2,600	1,400	2,000	0.32	5,300	2,700	1,800	2,700
Solling, fertilised	3,300	2,400	2,000	2,000	0.36	4,500	2,700	2,900	2,100
Solling, limed	2,700	2,300	1,500	1,500	0.27	3,700	2,700	2,400	1,700

F_{FRP} fine root production; F_{FRLP} fine root litter production; F_{FRD} fine root decomposition; F_{FRT} fine root turnover; F_{FRT}/ FRB specific fine root turnover

Table 9.4 Annual mean values of fine root fluxes for different soil depths (kg ha^{-1} per year) (Solling: 1986–1987; Göttinger Wald: 1995–1997)

	O-layer				0–20 cm				20–50 cm			
	F_{FRP}	F_{FRLP}	F_{FRD}	F_{FRT}	F_{FRP}	F_{FRLP}	F_{FRD}	F_{FRT}	F_{FRP}	F_{FRLP}	F_{FRD}	F_{FRT}
Solling	700	800	800	500	2,000	1,500	500	1,200	600	300	100	300
Solling, fertilised	300	200	300	200	1,500	1,100	700	800	1,500	1,100	1,000	1,000
Solling, limed	400	600	600	300	1,700	900	200	800	600	800	700	400
Göttinger Wald					1,000	800	700	800				

F_{FRP} fine root production; F_{FRLP} fine root litter production; F_{FRD} fine root decomposition; F_{FRT} fine root turnover calculated at the Göttinger Wald site as roots growing between the weathered limestone could not be sampled with the auger, but one can assume similar relations between the two sites for the total profile. According to Leuschner et al. (1998) who determined fine root mass using the profile-wall-method, root mass in a soil depth of 40–100 cm was 25% of the root mass in 0–40 cm in comparable beech stands at the Göttinger Wald site. Thus, there is no indication that the soil depth from 20 to 50 cm in Göttinger Wald could balance the calculated differences in fine root fluxes between the Göttinger Wald and Solling sites (Table 9.3). Even if maximum values found at the different plots for 20–50 cm soil depth (Table 9.4) were added to the profile values of Göttinger Wald, the fine root production and the mortality values would remain about twice as high at Solling as at the Göttinger Wald site.

Fine root turnover was lower at the Göttinger Wald site than at all the Solling plots, but the specific turnover (for definition see Table 9.3) tended to be higher indicating a shorter life-span of the fine roots at Göttinger Wald site. The average life-span of the fine roots can be estimated by the reciprocal of the F_{FRT}/FRB-ratios (Table 9.3). Thus, the fine root turnover depended more on the standing crop of fine roots than on their life-span.

The life-span of the fine roots was not influenced by the ammonium or liming treatment as indicated by similar F_{FRT}/FRB-ratios for all Solling plots. However, fine roots from the ammonium fertilised plot showed slightly higher values of the ratios indicating a shorter life-span, and for the limed plot, these values were low indicating a long life-span for the fine roots.

Different treatments changed the proportion of the annual root turnover fluxes in the different soil depths of the whole profile (Table 9.4). At the fertilised plot at the Solling site, the soil depth from 20 to 50 cm had about 50% of the total fine root production (F_{FRP}) and turnover (F_{FRT}). At the limed and control plots of the Solling site, the proportions of the sub-soil contributing to total F_{FRT} were low, about 15 and 20%, respectively.

Despite the high values of fine root turnover in the sub-soil of the fertilised Solling plot, the fine root turnover for the whole soil profile was not significantly different to that of the control plot (Table 9.3). Thus, the root turnover was at this site low in the O layer and in the surface mineral soil (Table 9.4). However, in the liming treatment, the fine root turnover (Table 9.3) of the whole soil profile was reduced.

Mechanisms controlling the growth and life-span of roots in forest ecosystems are poorly understood because of the small database available for making an assessment. The number of investigated stands is about 200 worldwide (Vogt et al. 1996), comprising more than 40 different species and a wide range of environmental conditions. There is very little quantitative information available on the interactions among fine roots, ecto-mycorrhiza and micro-faunal activities to formulate useful hypotheses about the control mechanisms of fine root turnover in forests.

Vogt et al. (1996) showed that on the global scale predictive environmental variables for fine root biomass or turnover change depended on the grouping of ecosystems, for example, those including species, climatic factors, forest types or soil orders. An obvious pattern on the global scale was that sites with the highest accumulation of soil organic matter had the highest standing crops of fine root biomass, and showed the highest amount of carbon allocation for belowground production. Such soils were acid with high levels of Al and Fe. Although some environmental variables showed significant correlations to fine root biomass and turnover, they explained a very small part of the variation on a global scale, illustrating the complexity of interactions of environment factors and species.

Based on the limited amount of available data, the supply of nutrients (Vogt et al. 1986b; Santantonio 1989; Gower et al. 1992), particularly of N and P (Nadelhoffer et al. 1985; Vogt et al. 1991) and water (Santantonio and Hermann 1985) are the potential factors determining root turnover. Other important factors are temperature, soil acidity and Al stress, mycorrhizas, root respiration and C-accumulation in the soil (Vogt et al. 1986b, 1990, 1991; Vogt and Bloomfield 1991).

In a spruce stand located close to the beech site at Solling, surface applied lime reduced the fine root turnover from 3,300 to 2,000 kg ha^{-1} per year especially in the O layer and 0–10 cm soil depth (Murach 1991). This was related to the higher base saturation of the soil and the lower H^+ -concentration of the soil solution in these layers than in those of the control plot. At the control plot, the temporal pattern of the fine root biomass showed a higher variation with minimum values occurring during acidification pulses. During such acidification pulses, the pH in soil solution was around 3.0 and Ca/H ratios in solution were less than 0.2 mol mol^{-1}, and they occurred in soils during periods of rewetting. Murach (1991) assumed that these short-term H^+-induced stresses resulted in an increase of fine root turnover. Fine root turnover in the sub-soil was low due to low amounts of fine root biomass which did not change on liming. Even during periods when fine root concentrations decreased in the surface soil due to acidification pulses, no compensatory root growth was observed in the sub-soil. Permanent Al stress in acid sub-soils possibly restricted the growth and turnover of fine roots of spruce in the surface soil where Al stress was low due to soil organic matter complexation (Murach and Matzner 1987), and H^+-induced stress was a short-term duration event.

Deeper soil layers in beech stands at the Solling site contributed significantly to the total fine root turnover (Table 9.4). This was in contrast to the results obtained on the adjacent spruce stands. Murach (1991) found no significant fine root fluxes at soil depths below 20 cm either in the limed or the unlimed spruce plots.

The highest production and turnover rates of fine roots in the sub-soils of all plots included in this study were observed at the fertilised Solling plot where ammonium fertilisation reduced the root growth in the O layer and the surface upper mineral soil layer presumably because of increased acid stress (Rapp 1991; Murach 1994; Paar 1994). The fine root growth was high in the deeper mineral soil (Table 9.5) together with the associated higher rates of fine root turnover.

This contrasting fine root growth pattern between beech and spruce was in accordance with results obtained from hydroponic experiments: The lower sensitivity of beech seedlings towards Al stress, i.e. low Ca/Al ratios in nutrient solution, and a higher sensitivity towards H^+-stress, i.e. low Ca/H ratios, than for spruce was demonstrated by Rost–Siebert (1985).

Table 9.5 Average standing crop of total fine root biomass and necromass, and per cent depthwise distribution of fine roots at the experimental plots

Depth	Solling control		Solling limed		Solling fertilised		Göttinger Wald		Zierenberg[a]	
	FRB	FRN	FRB	FRN	FRB	FRN	FRB[c]	FRN	FRB[b]	FRN
O-horizon (%)	23	11	21	12	9	11	0	n.d.	1	n.d.
0–5 cm (%)	8	23	7	16	7	17	30	n.d.	17	n.d.
5–10 cm (%)	16	17	11	14	15	14	15	n.d.	17	n.d.
10–20 cm (%)	24	23	26	24	31	25	15	n.d.	25	n.d.
20–30 cm (%)	17	16	22	19	21	18	14	n.d.	18	n.d.
30–50 cm (%)	12	10	13	15	17	15	24	n.d.	22	n.d.
Total profile (kg ha^{-1})	6,200	2,600	5,600	2,600	5,600	2,300	2,900[d]		2,800	n.d.

FRB fine root biomass; *FRN* fine root necromass

[a]Data for Zierenberg calculated from Paar (1994)

[b]Paar (1994) did not differentiate soil depth 0–5 cm and 5–10 cm, only 0–10 cm

[c]Values calculated as relative proportions of 0–20 cm depthwise gradients at Göttinger Wald site from Brumme (1986) and total fine root data from Leuschner et al. (1998)

[d]Data from Leuschner et al. (1998)

Similar to the results obtained for the whole soil depth (Table 9.3). high fine root turnover in the sub-soil of the fertilised Solling plot was not related to a shorter life-span of the roots but to the high fine root concentrations in these soil layers as compared to those of the control plot. There was no evident correlation between the life-span of beech fine roots and soil acidification when all plots and soil depths were considered.

However, liming reduced the fine root turnover of beech (Table 9.3) as was also observed for spruce stands (Murach 1991). The O layer and the surface mineral soil of the limed Solling plot were affected most, where the supply of base cations increased five-fold compared to the control plot (Rapp 1991), indicating that an increase in base saturation reduced fine root turnover in beech stands.

This pattern of fine root growth at the Solling site was similar to those observed at other beech stands in Germany. In Fig. 9.2, available data on fine root production in field studies of beech stands in Germany are pooled. The regression analysis gives some evidence to the hypothesis that fine root production decreases with increasing base saturation of the soil. Thus, the pattern described by Vogt et al. (1996) where the highest accumulation of fine root biomass and the highest carbon allocation in belowground production occurred in soils with high levels of Al and Fe at the global scale was supported by the data presented here.

9.4 Element Recycling with Litter of Fine Root

Element release from the fine roots to the soil (F_{FRERL} and F_{FRERD}-values) when calculated by using the element concentrations either in living fine roots or in the dead fine roots (Table 9.6) showed no differences for major elements. However, some differences between F_{FRERL} and F_{FRERD} values were apparent for Al and Fe, as these elements accumulated in dead fine roots (Murach 1984). The potential annual return of nutrients to the soil via F_{FRLP} was about 26–40 kg ha^{-1} per year for N and about 1–5 kg ha^{-1} per year for the other nutrients, except K for the Solling site where values of about 10 kg ha^{-1} per year were calculated.

At the Solling site about two to six times more Mg and K were recycled than at the Göttinger Wald site. This is caused by the higher fine root litter production at Solling which has higher element concentrations. However, there were major differences in the depthwise distribution of F_{FRERL} values among the three plots at the Solling site. Fine root litter production and element concentrations in the sub-soils of the fertilised and limed Solling plots were high resulting in high elements inputs through litter. These were comparable in magnitude with those of 0–20 cm depths and also from the control plot at Solling.

Some literature data is given in Table 9.7 as comparative values of nutrient transfer via fine roots production in other deciduous forests. Notable differences were low K value for the northern hardwood forest (Burke and Raynal 1994) and high Ca value for the Belgian beech forest (Van Praag et al. 1988). The high Ca value for the Belgian beech forest is site-specific as it is reflected in high Ca

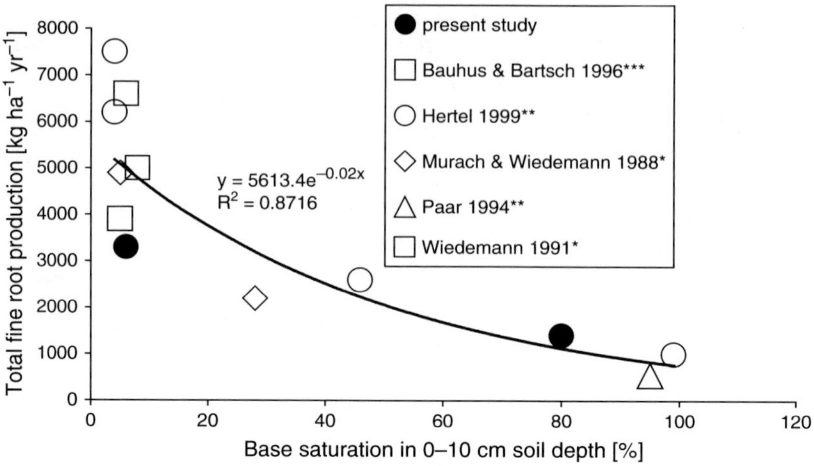

*	FFRP calculated as in the present study based on published data
**	FFRP calculated with minimum/maximum method
***	FFRP calculated with ingrowth core method

Fig. 9.2 Total fine root production in the soil profiles of beech stands in Germany in relation to increasing base saturation (fraction of exchangeable K, Ca and Mg in effective cation exchange capacity (CEC_e)). *Asterisk* F_{FRP} calculated as in the present study based on published data; *Double asterisk* F_{FRP} calculated with minimum/maximum method; *Triple asterisk* F_{FRP} calculated with ingrowth core method

concentrations of the fine roots (van Praag et al. 1988). The element turnover data for fine roots in this study were compared with the amounts in the leaf litterfall (Table 9.8). The belowground plant components contributed in significant amounts to the total elemental cycling via litterfall, especially those of N, P, K and Mg. For the cycling of Ca, the fine roots are distinctly less important.

9.5 Summary

The present study had the focus of assessing the role of fine roots in the carbon and nutrient dynamics on two beech sites. In order to consider changes due to soil parameters, two additional sites at Solling were included. The sites were an acid soil at Solling, a limed plot at Solling and an ammonium sulphate fertilised plot at Solling, and a calcareous soil at Göttinger Wald.

Fine root production (F_{FRP}), fine root litter production (F_{FRLP}), fine root decomposition (F_{FRD}) and fine root turnover (F_{FRT}) were determined using statistically significant differences in the temporal development of fine root biomass and necromass derived from data of sequential coring for which the method of balancing transfers based on a decision matrix was applied.

Table 9.6 Potential values of annual element release to the soil via fine root litter production (F_{FRLP}) (kg ha^{-1} per year). F_{FRERL} refers to element release via F_{FRLP} based on element concentrations in living fine roots, F_{FRERD} element release via F_{FRLP} based on element concentrations in dead fine roots

	O-layer		0–20 cm		20–50 cm		Total profile	
	F_{FRERL}	F_{FRERD}	F_{FRERL}	F_{FRERD}	F_{FRERL}	F_{FRERD}	F_{FRERL}	F_{FRERD}
Solling, control plot								
N	15.6	13.7	19.5	19.2	3.5	3.5	38.6	36.4
P	0.7	0.6	1.1	1.0	0.2	0.2	1.9	1.8
K	1.9	2.2	6.3	4.7	0.8	0.8	9.0	7.6
Ca	1.0	0.9	0.9	0.3	0.2	0.1	2.1	1.3
Mg	0.3	0.5	0.9	1.2	0.2	0.2	1.4	1.9
Mn	0.2	0.2	0.3	0.2	0.1	0.0	0.7	0.4
Fe	0.7	3.7	9.6	15.6	0.6	2.5	10.9	21.8
Al	1.5	6.4	12.6	26.1	2.3	6.5	16.5	39.0
C	400		750		150		1,300	
Solling, ammonium sulphate fertilised plot								
N	3.8	3.6	15.2	13.0	11.8	10.3	30.8	26.9
P	nd	nd	nd	nd	nd	nd	nd	nd
K	0.2	0.4	4.6	5.3	3.7	3.5	8.6	9.2
Ca	0.2	0.3	0.7	0.3	0.6	0.5	1.6	1.1
Mg	0.1	0.2	1.1	1.4	0.9	1.0	2.1	2.6
Mn	0.1	0.0	0.3	0.2	0.4	0.2	0.7	0.4
Fe	0.1	1.1	8.6	12.1	4.1	5.6	12.8	18.8
Al	0.2	1.8	15.5	22.0	16.6	26.5	32.4	50.3
C	100		550		550		1,200	
Solling, limed plot								
N	11.5	11.0	10.6	10.5	6.9	8.2	29.0	29.7
P	0.5	0.5	0.5	0.6	0.4	0.4	1.4	1.4
K	1.3	1.1	4.4	3.8	3.0	2.6	8.8	7.4
Ca	5.3	5.8	0.7	0.3	0.6	0.3	6.5	6.4
Mg	2.3	1.7	1.0	1.1	0.8	0.7	4.1	3.6
Mn	0.1	0.1	0.2	0.1	0.2	0.1	0.5	0.3
Fe	0.6	2.6	6.9	11.6	2.6	4.8	10.1	19.1
Al	0.6	4.3	10.8	16.3	11.2	19.2	22.6	39.8
C	300		450		400		1,150	
Göttinger Wald, control plot								
N			25.6	18.4			25.6	18.4
P			0.9	0.8			0.9	0.8
K			2.4	1.0			2.4	1.0
Ca			3.2	2.4			3.2	2.4
Mg			0.7	0.7			0.7	0.7
Mn			0.5	0.5			0.5	0.5
Fe			3.6	6.5			3.6	6.5
Al			6.0	10.9			6.0	10.9
C			600				600	

Table 9.7 Fine root litter production (as organic matter) and potential annual nutrient release to the soil via F_{FRLP} (kg ha^{-1} per year)

	F_{FRLP}	N	P	K	Ca	Mg
North Hardwood[a]	2,300	23	1.4	0.3	5.2	1.3
White oak[b]	1,100	15	0.7	4.1	10.3	2.0
European Beech[c]	1,900	32	1.9	1.0	36	3.9
Solling, control[d]	2,600	39	1.9	9.0	2.1	1.4
Solling, fertilised[d]	2,400	31	n.d.	8.6	1.6	2.1
Solling, limed[d]	2,300	29	1.4	8.8	6.5	4.1
Göttinger Wald[e]	1,200	26	0.9	2.4	3.2	0.7

F_{FRLP} fine root litter production
[a]Burke and Raynal (1994). Roots <2 mm in diameter; Adirondack State Park, New York, USA. Mainly Acer and Fagus. Data for F_{FRP}, but $F_{FRP} \approx F_{FRLP}$ in this study
[b]Joslin and Henderson (1987), cited in Burke and Raynal (1994). Roots <2 mm in diameter. Missouri, US
[c]Van Praag et al. (1988). Roots <1 mm in diameter; Belgian Ardennes
[d]This study. Roots <2 mm in diameter, 0–50 cm depth
[e]This study. Roots <2 mm in diameter, 0–20 cm depth

Table 9.8 Mean annual organic matter and element recycling via litterfall (kg ha^{-1} per year) (data from Beese et al. 1991)

	F_{LLP}[a]	N	P	K	Ca	Mg
Solling, control	3,000	44	3	12	18	1
Solling, fertilised	3,000	45	3	13	17	1
Solling, limed	3,100	38	2	12	28	2
Göttinger Wald	2,800	35	1	14	53	3

[a]F_{LLP} Leaf litter production; means for years 1981–1985

For the Solling site, the maximum coring depth was 50 cm. For Göttinger Wald, coring was restricted by the stony bedrock to a depth of about 20 cm. The following results were obtained:

- Carbon allocation into the fine roots was distinctly higher for all Solling plots than for the Göttinger Wald site. Production and turnover of fine roots at Solling were estimated to be 3,300 kg and 2,000 kg ha^{-1} per year for a depth of 50 cm. For the Göttinger Wald site, these values were 1,000 kg for production and 800 kg ha^{-1} per year for turnover for a depth of 20 cm (Tables 9.3 and 9.4) indicating that fine root production was low in high base saturation soil (Fig. 9.2).
- At the Solling site, beech stands had high amounts of fine root biomass in the highly acid sub-soil layers (Table 9.5), which contributed significantly to the total fine root production and fine root turnover (Table 9.4). This contrasted with the fine root distribution at the spruce site, where fine root turnover was restricted to the O layer and the surface mineral soil. These differences in the fine root

biomass and dynamics were similar to those observed under laboratory experiments, where beech seedlings were found to be less sensitive to Al stress.

- The potential annual return of nutrients to the soil via fine root litter production contributed about 25–40 kg ha^{-1} of N and about 1–5 kg ha^{-1} of most other nutrients except that of K at the Solling site. In the acid soils of the Solling site about two to six times the amount of Mg and K were recycled via F_{FRLP} than in the calcareous soil of the Göttinger Wald site.
- Liming and fertilisation with ammonium sulphate increased fine root litter production and nutrient return via fine root litter in the sub-soil of the Solling site indicating the possibility of changing root distribution in highly acid soils by ameliorative measures.
- The amount of nutrients cycled through fine roots may form a significant fraction of total litter content including that of foliage, except for Ca.

Acknowledgement This work was supported by the BMBF.

References

Agren GI, Axelsson B, Flower–Ellis JGK, Linder S, Persson H, Staaf H, Troeng E (1980) Annual carbon budget for a young scots pine. In: Persson T (ed) Structure and function of northern coniferous forests – an ecosystem study. Ecological Bulletin, Stockholm, pp 307–313

Bartelink HH (1998) A model of dry matter partitioning in trees. Tree Physiol 18:91–101

Bauhus J, Bartsch N (1996) Fine root growth in beech (*Fagus silvatica*) forest gaps. Can J For Res 26:2153–2159

Beese F, Prenzel J (1985) Das Verhalten von Ionen in Buchenwald-Oekosystemen auf podsoliger Braunerde mit und ohne Kalkung. AFZ 43:1162–1164

Beese F, Waraghai A, Woehler I, Stickan W, Meiwes KJ (1991) Phaenologie und Inhaltsstoffe von Buchenblaettern in Relation zur Aciditaet von Boeden. Ber d Forschungszentr Waldoekosysteme d Univ Goettingen, B25:1–178

Brumme R (1986) Modelluntersuchungen zum Stofftransport und Stoffumsatz in einer Terra fusca-Rendzina auf Muschelkalk. Ber d Forschungszentr Waldoekosysteme d Univ Goettingen, A24:1–206

Burke MK, Raynal DJ (1994) Fine root growth phenology, production, and turnover in a northern hardwood forest ecosystem. Plant Soil 162:135–146

Fairley RI, Alexander IJ (1985) Methods of calculating fine root production in forests. In: Fitter AH et-al (eds) Ecological interactions in soils. Special publication No. 4 of the British Ecological Society. Blackwell, Oxford, pp 37–42

Fogel R (1985) Roots as primary producers in below-ground ecosystems. In: Fitter AH et-al (eds) Ecological interactions in soil. Special Publication No. 4 of the British Ecological Society, Blackwell, Oxford, pp 23–36

Gower ST, Vogt KA, Grier CC (1992) Carbon dynamics of rocky mountain Douglas-fir: influences of water and nutrient availability. Ecol Monogr 62:43–65

Grier CC, Vogt KA, Keyes MR, Edmonds RL (1981) Biomass distribution and above– and below-ground production in young and mature Abies amabilis zone ecosystems in the Washington Cascades. Can J For Res 11:155–167

Joslin JD, Henderson GS (1987) Organic matter and nutrients associated with fine root turnover in a white oak stand. For Sci 33:330–346

Kohmann K (1972) Root ecological investigations on pine (*Pinus silvestris*) – I. Problems of methodology and general root relationships. Medd Norsk Skogfors Ves 30:325–357

Kroon H, de Visser EJW (2003) Root ecology. Ecological Studies 168. Springer, Berlin

Lauenroth WK, Hunt HW, Swift DM, Singh JS (1986) Reply to Vogt et al. Ecology 67:580–582

Leuschner C, Hertel D, Muhs A, Schmid I (1998) Feinwurzel-Bestandesmassen der Rotbuche an verschiedenen Standorten innerhalb ihrer oekologischen Amplitude in Nordwest- und Mitteldeutschland. Verh Ges Oekologie 28:429–434

Lienert GA (1973) Verteilungsfreie Methoden in der Biostatistik, Vol 1, Anton Hain, Meisenheim

Meiwes KJ, Merino A, Beese FO (1998) Chemical composition of throughfall, soil water, leaves and leaf litter in a beech forest receiving long term application of ammonium sulphate. Plant Soil 201:217–230

Murach D (1984) Die Reaktion der Feinwurzeln von Fichten (*Picea abies* Karst) auf zunehmende Bodenversauerung. Göttinger Bodenkundl Berichte 77:1–126

Murach D (1991) Feinwurzelumsaetze auf bodensauren Fichtenstandorten. Forstarchiv 62: 12–17

Murach D (1994) Einfluß verschiedener Bodenfaktoren auf Wurzelgradienten bei Fichte und Buche – Abschlußbericht zum BMFT-Forschungsvorhaben Stabilitaet von Waldoekosystemen. Ber d Forschungszentr Waldoekosysteme d Univ Göttingen B37:69–78

Murach D, Matzner E (1987) The influence of soil acidification on root growth of Norway spruce (*Picea abies* Karst) and European beech (*Fagus silvatica* L). In: Lavender DP (ed) Woody plant growth in a changing chemical and physical environment. IUFRO workshop Vancouver, pp 171–188

Murach D, Wiedemann H (1988) Forschungsbericht Dynamik und chemische Zusammensetzung der Feinwurzeln von Waldbaeumen als Maß fuer die Gefaehrdung von Waldoekosystemen durch toxische Luftverunreinigungen. Ber d Forschungszentr Waldoekosysteme d Univ Göttingen B10:1–287

Nadelhoffer KJ, Raich JW (1992) Fine root production estimates and belowground carbon allocation in forest ecosystems. Ecology 73:1139–1147

Nadelhoffer KJ, Aber JD, Melillo JM (1985) Fine roots, net primary production and nitrogen availability: a new hypothesis. Ecology 66:1377–1390

Paar U (1994) Untersuchungen zum Einfluß von Ammonium und Nitrat auf wurzelphysiologische Reaktionsmuster der Buche. Ber d Forschungszentr Waldoekosysteme d Univ Goettingen A115:1–124

Persson H (1979) Fine-root production, mortality and decomposition in forest ecosystems. Vegetation 41:101–109

Rapp C (1991) Untersuchungen zum Einfluß von Kalkung und Ammoniumsulfat-Duengung auf Feinwurzeln und Ektomykorrhizen eines Buchenaltbestandes im Solling. Ber d Forschungszentr Waldoekosysteme d Univ Goettingen A72:1–293

Roberts J (1976) A study of root distribution and growth in a *Pinus sylvestris* L (Scots pine) plantation in East Anglia. Plant Soil 44:607–621

Rost-Siebert K (1985) Untersuchungen zur H- und Al-Ionentoxizitaet an Keimpflanzen von Fichte (*Picea abies* Karst) und Buche (*Fagus silvatica* L) in Loesungskultur. Ber d Forschungszentr Waldoekosysteme d Univ Göttingen A12:1–219

Santantonio D (1978) Seasonal dynamics of fine roots in mature stands of Douglas fir of different water regimes – a preliminary report. In: Riedacker A, Gagnaire-Michard J (eds) Symposium Physiologie des Racines et Symbioses. IUFRO Proceedings, Nancy, pp 190–203

Santantonio D (1989) Dry matter partitioning and fine root production in forests – new approaches to a difficult problem. In: Pereira JS, Landsberg JJ (eds) Biomass production by fast growing trees. Kluwer, Dorderecht, pp 57–72

Santantonio D (1990) Modelling growth and production of tree roots. In: Dixon RK et-al (eds) Process modelling of forest growth responses to environmental stress. Timber, Portland, Oregon, USA, pp 124–141

Santantonio D, Hermann RK (1985) Standing crop, production, and turnover of fine roots on dry, moderate, and wet sites of mature Douglas-fir in Western Oregon. Ann Sci For 42:113–142

Scarascia-Mugnozza G, Bauer GA, Persson H, Matteucci G (2000) Tree biomass, growth and nutrient pools. In: Schulze ED (ed) Carbon and nitrogen cycling in European forest ecosystems. Ecological Studies 142. Springer, Berlin, pp 49–62

Singh JS, Lauenroth WK, Hunt HW, Swift DM (1984) Bias and random errors in estimators of net root production: a simulation approach. Ecology 65:1760–1764

Van Praag HJ, Sougnez-Remy S, Weissen F, Carletti G, (1988) Root turnover in a beech and a spruce stand of the Belgian Ardennes. Plant Soil 105:87–103

Vogt KA, Bloomfield J (1991) Tree root turnover and senescence. In: Waisel Y, Eshel A, Kafkafi U (eds) Plant Roots. The Hidden Half. Marcel Dekker, New York, pp 287–306

Vogt KA, Grier CC, Gower ST, Sprugel DG, Vogt DJ (1986a) Overestimation of net root production: a real or imaginary problem? Ecology 67:577–579

Vogt KA, Grier CC, Vogt DJ (1986b) Production, turnover, and nutrient dynamics of above- and belowground detrius of world forests. Adv Ecol Res 15:303–377

Vogt KA, Vogt DJ, Gower ST, Grier CC (1990) Carbon and nitrogen interactions for forest ecosystems. In: Persson H (ed) Above and below-ground interactions in forest trees in acidified soils. Air Pollution Research Report 32. CEC, Brussels, pp 203–235

Vogt KA, Vogt DJ, Bloomfield J (1991) Input of organic matter to the soil by tree roots. In: McMichael BL, Persson H (eds) Plant roots and their environment. Developments in agricultural and managed-forest ecology 24. Elsevier, Amsterdam, pp 171–190

Vogt KA, Vogt DJ, Palmiotto PA, Boon P, O'Hara J, Asbjornsen H (1996) Review of root dynamics in forest ecosystems grouped by climate, climatic forest type and species. Plant Soil 187:159–219

Vogt KA, Vogt DJ, Bloomfield J (1998) Analysis of some direct and indirect methods for estimating root biomass and production of forests at an ecosystem level. Plant Soil 200:71–89

Chapter 10
Phytomass, Litter and Net Primary Production of Herbaceous Layer

I.-M. Schulze, A. Bolte*, W. Schmidt, and J. Eichhorn

10.1 Introduction

Understorey vegetation may form an important component of mineral cycling pathways in forest ecosystems. The important cycling processes are: nutrient uptake, storage, translocation and return by vegetation, litterfall and nutrient release from residues by mineralisation of litter and soil organic matter for further plant uptake (Duvigneaud and Denaeyer-DeSmet 1973; Charley and Richards 1983; Ellenberg et al. 1986; Eichhorn 1995). Although the biomass of herbaceous plants is low compared to that of the tree layer (Yarie 1980; Muller and Bormann 1976; Ellenberg et al. 1986; Giese et al. 2003; Peichl and Arain 2006), high element concentrations of herbaceous plants usually increase their importance in element cycling (Scott 1955; Turner et al. 1978; Blank et al. 1980; Rodenkirchen 1995; Bolte et al. 2004). Through short-term changes in species composition, element concentration and storage (Höhne 1963; Likens and Bormann 1970; Linke and Roloff 1995), the herbaceous vegetation forms a variable element pool that may reduce nutrient losses to external sinks of forest ecosystems (Eichhorn 1995; Mellert et al. 1998; Bartsch 2000; Schmidt 2002; Olsson and Falkengren-Grerup 2003; Bolte 2006).

Litter production of herbaceous layer may sometimes equal that of the tree layer (Höhne 1962). The amount and higher quality (particularly C/N ratio; Wittich 1944, 1961; Wolters 1999) of readily decomposable substrates produced by the herbaceous layer may alter the community of decomposers in a soil (Jandl et al. 1997) changing the element turnover rates (Wise and Schaefer 1994) that may affect C and N retention by forest soil (Bauhus et al. 2004).

The objective of this chapter is to describe the role of herbaceous vegetation in the nutrient cycling processes at the three beech forest sites of Göttinger Wald, Zierenberg and Solling through differences in the mass and quality of annually produced litter. The internal and external nitrogen balances in relation to the ground vegetation will be included in the discussion. For a valid assessment of any

*Corresponding author

R. Brumme and P.K. Khanna (eds.), *Functioning and Management of European Beech Ecosystems*, Ecological Studies 208,
DOI: 10.1007/978-3-642-00340-0_10, © Springer-Verlag Berlin Heidelberg 2009

subsequent comparisons among sites involving quantitative parameters of the biomass and nitrogen, temporal and spatial dynamics in herbaceous biomass and N-cycling will be included, as well as the different methods used to assess changes in biomass and N-cycling.

10.2 Definitions and Methods

The *biomass* is defined as all the above- and belowground parts of live plant material. The *necromass* at the time of the inventory is the dead plant material, which includes the standing dead plant parts. The *phytomass* describes the combined amounts of biomass and necromass. Plant material shed to the surface organic layer is termed as *litter fall* and is not part of the phytomass. All parameters are expressed as dry matter (kg ha^{-1}, g m^{-2}). These definitions follow the terminology of the International Biological Program (I.B.P; Newbould 1967; Milner and Hughes 1968).

Two methods were used to assess various components of the herbaceous vegetation: (1) biomass harvesting to estimate net primary production (NPP) (Sect. 10.2.1), and (2) models using the mean shoot length and coverage of species or morphological groups for estimates of aboveground phytomass (Sect. 10.2.2).

10.2.1 Biomass and NPP Estimates using Harvesting Methods

Estimation of biomass by the harvest method is the basic procedure of assessing production components (Newbould 1967; Lieth and Whittaker 1975). *Net primary productivity* (NPP) (Whittaker 1975; Schütt et al. 1992) can be described by the following equation (Milner and Hughes 1968; Larcher 1994) (all values in weight per unit area and time):

$$\mathrm{NPP} = \Delta B + \mathrm{Ld} + \mathrm{Lc},$$

where
$\triangle B$: biomass increase
Ld: biomass losses by death (necromass) and shedding (litter fall)
Lc: biomass losses through consumption (e.g. by herbivorous animals or insects)

Increase in biomass ($\triangle B$) of trees remains positive for a long period and leads to high storage values whereas for herbaceous plant communities it is usually assumed that $\triangle B$ value for a given stand when considered at the same time every year becomes close to 0 in the long-term (Larcher 1994). Biomass accumulation and decomposition then approach a ‘quasi-equilibrium’ state. Consequently, the annual NPP of the herbaceous communities is primarily given by the *necromass* plus *litterfall* values (Ld). However, Ld can only partially be determined by repeated inventories due to variations in the phenology of plant functional groups and the highly dynamic biomass turnover of herbaceous plants, in particular of their root systems (Eggert 1985).

The proportion of NPP utilised by consumers (Lc) in the investigated stands is very low (less than 1% of NPP according to estimates by Schaefer 1989, 1996) and is thus not further considered here. For *Allium ursinum* in Göttinger Wald, Eggert (1985) and Hövemeyer (1995) determined an annual loss of 1–3% dry leaf matter consumed by hoverfly larvae (*Cheilosia fasciata*). These results were obtained from set of 120 uniform harvest plots with 1 m^2 surface. For *Mercurialis perennis*, Kothe (1983) reported a loss of 5% of the leaf area due to insect damage which was measured in 40 uniform harvest plots of 1 m^2 surface in a *Mercurialis* facies.

For the assessment of the annual NPP, the "maximum–minimum method" (Newbould 1967) is commonly used where the difference between the maximum and the minimum amount of phytomass is calculated. Usually, this annual NPP value equals the annual amount of litterfall plus necromass whereas Lc is neglected and $\triangle B$ equals zero. The assumption of $\triangle B = 0$ is not valid in all cases; therefore, modified calculation methods which include long-term changes in $\triangle B$ are used (Mrotzek 1998; Brünn 1999). The annual amount of nitrogen uptake by herbaceous vegetation was deduced by using a similar model as for the assessment of the biomass and litter production where the amounts of nitrogen in the litter fall, necromass and the storage ($\triangle B$) were used.

Harvest methods provide good estimates of the standing biomass for a given plot. However, these assessments are destructive and cannot be repeated at the same location. Thus, the accuracy of estimates of biomass increase ($\triangle B$), and consequently net primary production (NPP), depends on the structural homogeneity of the plots which are selected for the repeated harvesting. For vegetation monitoring purposes, e.g. within the EU Level II program (De Vries et al. 2002; Bolte et al. 2004), destructive harvest methods are often not applicable (Bolte et al. 2002).

10.2.2 Aboveground Phytomass Estimates by Dimension Analysis

Maximum aboveground storage of the herbaceous phytomass can be estimated by models which are based on biomass assessment of 31 widespread species and 13 morphological groups of the ground vegetation in forest stands of North Germany (Bolte 1999; Bolte et al. 2002; Tables 10.1, 10.2). These models were developed from ground vegetation structure and biomass data which originated from studies of 1,602 plots in 61 stands on various sites throughout northeast Germany and were sampled during 1993–1996. The selected plots covered the abundance range (coverage: 1–100%) for all selected species in forests predominantly dominated by European beech, oak and Scots pine. Based on plant species, shoot length and plant density, plot area selected for sampling varied from 0.05 m^2 (mosses) to 1.28 m^2 (tall ferns) which allowed the survey of sufficient number of shoots per plot (Donita 1972). In each plot, percentage coverage and the maximum length of 15 randomly

Table 10.1 Assessment functions for aboveground phytomass of different plant species according to Bolte (1999, modified)

No	MG	Species[b]	Empirical parameters[a] (eq. 10.1)			Cd_{nlin}
			A	B	C	
1	sh	*Adoxa moschatellina*	0.2919	1.0535	–	0.94
2		*Oxalis acetosella*	0.1520	1.1090	–	0.92
3	mh	*Anemone nemorosa, A. ranuncoloides*	0.1853	1.2731	–	0.92
4		*Corydalis cava*	0.2027	1.2000	–	0.81
5		*Galium odoratum*	0.0038	1.0308	1.6429	0.86
6		*Lamium galeobdolon*	0.1618	1.2832	–	0.87
7		*Maianthemum bifolium*	0.0800	1.4286	–	0.87
8		*Mercurialis perennis*	0.0258	1.3338	0.6425	0.92
9		*Ranunculus ficaria*	0.3930	1.0699	–	0.90
10		*Stellaria holostea*	0.9375	0.9602	–	0.92
11	th	*Aegopodium podagraria*	0.0050	0.9901	1.4418	0.95
12		*Impatiens parviflora*	0.0013	1.5500	0.9015	0.88
13		*Urtica dioica*	0.0130	1.1055	0.9013	0.91
14	sg	*Deschampsia flexuosa*	0.1095	0.8900	0.7698	0.89
15		*Melica uniflora*	0.0289	0.9317	1.0126	0.89
16	mg	*Agrostis capillaris, Poa nemoralis*	0.0104	1.0786	1.2639	0.91
17		*Brachypodium pinnatum, B. sylvaticum*	0.0076	1.2545	1.0600	0.88
18		*Dactylis polygama*	0.0001	0.9228	2.5215	0.96
19	tg	*Calamagrostis epigejos*	0.0001	0.9927	2.2550	0.87
20		*Milium effusum*	0.1096	1.0918	0.4147	0.91
21		*Molinia caerulea*	0.0045	1.3120	1.1230	0.92
22	f	*Dryopteris carthusiana*	0.0039	1.2887	1.1811	0.85
23		*Dryopteris filix-mas*	0.0017	1.3997	1.0831	0.93
24	tf	*Pteridium aquilinum*	0.0003	1.2634	1.5391	0.96
25	dsh	*Calluna vulgaris*	0.1620	0.9808	1.0187	0.95
26		*Vaccinium myrtillus*	0.0373	0.9656	1.4117	0.93
27		*Vaccinium vitis-idaea*	0.1363	1.1427	0.8627	0.98
28	sh	*Rubus idaeus*	0.0003	0.9658	2.2250	0.92
29	pm	*Dicranella heteromalla, Dicranum polysetum Dicranum scoparium, Hypnum cupressiforme, Pohlia nutans*	0.8161	1.1543	0.4306	0.90
30	mm	*Brachythecium* spec., *Pleurozium schreberi, Scleropodium purum*	0.6937	0.8291	0.6726	0.87
31	dm	*Leucobryum glaucum*	1.8028	1.2476	0.4231	0.83

MG refers to morphological groups of species (for further description, refer to Table 10.2), Cd_{nlin}: non-linear coefficient of determination according to Sachs (1976) and (eq. 10.1)

[a]Values are rounded from 12 to 4 decimal points

[b]Nomenclature: vascular plants: Wisskirchen and Haeupler (1998), mosses: Koperski et al. (2000)

Table 10.2 Assessment functions for above ground phytomass of morphological groups of the forest understorey vegetation according to Bolte (1999, modified)

No	Morphological groups	Mean shoot length (cm)	Empirical parameters[a]			Cd_{nlin}
			A	B	C	
1	Small herb (sh)	5–**10**–15	0.1187	0.9663	0.4311	0.86
2	Middle herb (mh)	10–**15**–40	0.0687	1.2594	0.3624	0.81
3	Tall herb (th)	20–**55**–160	0.0037	1.5158	0.8057	0.80
4	Small grass (sg)	20–**35**–60	0.0426	0.9779	0.9083	0.88
5	Middle grass (mg)	30–**50**–80	0.0047	1.0748	1.4185	0.83
6	Tall grass (tg)	30–**70**–120	0.0001	1.2050	2.0590	0.88
7	Fern (f)	30–**55**–90	0.0445	1.4812	0.2582	0.81
8	Tall fern (tf)	50–**100**–215	0.0003	1.2634	1.5391	0.96
9	Dwarf shrub (dsh)	10–**20**–40	0.1641	1.0080	0.9028	0.86
10	Shrub (sh)	25–**55**–120	0.0003	0.9658	2.2250	0.92
11	Pad moss[b] (pm)	1–**3**–10	0.8161	1.1543	0.4306	0.90
12	Mat moss[c] (mm)	5–**7**–10	0.6937	0.8291	0.6726	0.87
13	Dense moss[d] (dm)	4–**5**–8	1.8028	1.2476	0.4231	0.83

Mean shoot length refers to the range (min, max) for the 129 plots, the bold value represents the mean for all plots; Cd_{nlin}: refers to non-linear coefficient of determination according to Sachs (1976) and (eq. 10.1)
[a]Values are rounded from 12 to 4 decimals
[b]Mostly acrocarpous moss species, growing in singular pads
[c]Pleurocarpous moss species, growing in more or less extended mats (carpets)
[d]Very dense pads of *Leucobryum glaucum*

selected shoots were recorded separately for all species analysed. Dry weight of species phytomass was measured after harvesting and drying aboveground shoots for 48 h at 105°C in the laboratory. Sampling was timed between blooming and fruiting of each species and ranged from April to September. During this period, aboveground plant growth of most herbaceous plant species occurred (Höhne and Fiedler 1963; Persson 1975; Rodenkirchen 1982).

A non-linear regression analysis was performed (Proc Nlin, iteration: Gauss-Newton, SAS 6.12; SAS Inst. 1989). Allometric functions of the following formula (10.1) were used to predict aboveground biomass of ground vegetation species or plant morphological groups relative to percentage cover (PC) and mean shoot length (MS):

$$P = a\,\mathrm{PC}^{b}\,\mathrm{MS}^{c}, \qquad (10.1)$$

P: Aboveground phytomass (dry matter, 105°C, g m^{-2})
PC: Percentage cover (species, morphological groups)
MS: Mean shoot length (above ground, cm)
$a \ldots c$: Empirical parameters obtained by fitting the measured data

Fitting criteria for the optimal regression models were maximum values of the non-linear coefficient of determination (Cd_{nlin}, Sachs 1976) and the independence of the residuals from the dependent variable (Hartung 1989). Model derivation for both species (SP) and morphological groups (MG) allowed the estimation of those species that were not sampled by attributing them to morphological groups with regard to morphology (fern, herb, grass, moss, shrub) and shoot length (Table 10.2). Standard deviation of estimations (standard deviation of the residuals) ranged between 18.8 and 20.7% of the predicted values for species models (SP) and for morphological group models (MG). Both model types were successfully validated comparing phytomass values of predicted (x) and measured (y) values of 50 independent samples ($y = x$, SP: $R^2 = 0.85$, MG: $R^2 = 0.85$; Bolte 1999).

Based on the allometric models in Tables 10.1 and 10.2, aboveground phytomass of the ground vegetation was calculated with MS Access 2002 using the macro package PhytoCalc 1.31 (Bolte and Parth 2004).

For the beech stands at Göttinger Wald and Solling, percent coverage and mean shoot length for each species with coverage of more than 1% were recorded in 2002 during the Level II survey for which four and six square subplots (each of 100 m^2) were measured. At Göttinger Wald, these subplots were not systematically distributed in the stand, but covered characteristic facies of ground vegetation (see Chap. 5, Table 5.9). At Zierenberg, percent coverage of all and upper shoot length (UL: mean length of the dominant shoots, 90th percentile) of the most abundant species, stinging nettle (*Urtica dioica)* and dog's mercury (*M. perennis)* were assessed in 2002 in 80 small square subplots (4 m^2) that were systematically distributed using a grid approach within ca. 4-ha area. For the calculation of mean shoot length ML from upper shoot length values UL (see (10.2)), conversion factors were deduced for each morphological group by linear regression analyses using the model samples (n = 1,602; Table 10.3). For less abundant species for which shoot lengths were not recorded, mean values (displayed bold values in Table 10.2) of the corresponding morphological species group were used.

$$\mathrm{ML} = a + b\,\mathrm{UL}, \qquad (10.2)$$

where
ML: Mean shoot length (cm)
UL: Upper shoot length, 90th percentile (cm)
$a \ldots c$: Empirical parameters

Estimation of the above ground herbaceous phytomass was obtained for the beech stands by calculating: (1) the arithmetic mean of all six subplot values at Solling with a homogenous herbaceous structure, (2) the frequency weighted mean phytomass of all 80 subplots stratified according to bedrock type and relative irradiance intensity at Zierenberg, and (3) the area weighted phytomass mean of the four subplots attributed to different vegetation facies at Göttinger Wald (Dierschke 2006).

In contrast to the harvest methods, this non-destructive dimension analysis enables repeated estimations of plant phytomass for the same plot. Thus, it can be used for vegetation monitoring without any restriction. However, to date, the assessment models apply only to the aboveground parts of the understorey vegetation. The accuracy of the model predictions depends on data quality of the input parameters and the model errors described (18.8% for species models and 20.7% for morphological group models).

Table 10.3 Conversion functions for upper shoot length (90-%) to mean shoot length of morphological groups of the forest understorey vegetation (Bolte and Parth 2004)

No	Morphological groups of species	Sample number (n)	Empirical parameters (10.2)		R^2
			A	B	
1	Small herb (sh)	90	1.0471	0.6767	0.85
2	Middle herb (mh)	455	0.9175	0.6930	0.90
3	Tall herb (th)	149	−0.6697	0.7647	0.95
4	Small grass (sg)	158	7.9168	0.4758	0.83
5	Middle grass (mg)	158	5.3827	0.6557	0.84
6	Tall grass (tg)	173	−3.3267	0.7747	0.89
7	Fern (f)	52	−5.0700	0.8481	0.90
8	Tall fern (tf)	48	0.7862	0.8659	0.96
9	Dwarf shrub (dsh)	138	2.1576	0.6595	0.88
10	Shrub (sh)	50	4.9191	0.5994	0.96
11	Pad moss[a] (pm)	74	−0.0658	0.8326	0.97
12	Mat moss[b] (mm)	35	0.4573	0.6966	0.88
13	Dense moss[c] (dm)	22	−0.1319	0.8625	0.96

[a]Mostly acrocarpous moss species, growing in singular pads
[b]Pleurocarpous moss species, growing in more or less extended mats (carpets)
[c]Very dense pads of *Leucobryum glaucum*

10.3 Phytomass Estimates

The maximum phytomass determines the pool for matter and nutrient storage and turnover of the herbaceous vegetation. Thus, it is an important parameter for production studies. Biomass and necromass, both phytomass components, provide the basis for net primary production estimates.

10.3.1 Total Biomass Estimated from Harvest Studies

The maximum biomass values for below- and aboveground components, which were derived from harvest studies, indicated that markedly high amount of biomass in the herbaceous layer occurred at the nutrient-rich sites of Göttinger Wald and Zierenberg (Table 10.4). At each of these sites, high values (peak biomass aboveground plus belowground; see Table 10.4) between 1,865 kg in Zierenberg and 2,100 kg in Göttinger Wald dry matter per hectare were observed during two separate studies undertaken in 1981/1982 and 1995.

At the Solling site, the total phytomass of the herbaceous and moss layer ranged from 27 to 56 kg ha^{-1} which was only 1–3% of the maximum values obtained at the Göttinger Wald and Zierenberg sites. The calculated maximum aboveground biomass values were comparable to the modelled phytomass values shown in Table 10.5. It is, however, worth noting that necromass of herbaceous plants had already occurred

Table 10.4 Maximum biomass of the herbaceous layer (kg ha^{-1}, ± SD; *estimated from SD of *A. ursinum* populations, only) for different years derived from harvest studies

	Göttinger Wald[a]				Zierenberg[b]			Solling[c]
	1982	1981	1981	1981	1995	1995	1995	1968–1972
	Allium- facies	*Mercurialis*-facies	*Anemone*-facies	**Mean**	*Urtica*-facies	*Mercurialis*-facies	**Mean**	**B 1, mean**
aboveground	1,253 ± n.a.	976 ± 426	281 ± 103	**795** ± n.a.	1,145 ± n.a.	469 ± n.a.	**560** ± n.a.	**13–22 ± n.a.**
Belowground	1,932 ± n.a.	2,301 ± 563	623 ± 198	**1,548** ± n.a.	1,765 ± n.a.	1562 ± n.a.	**1,589** ± n.a.	**18–34 ± n.a.**
Peak biomass (aboveground plus belowground)[d]	2,801 ± 608*	2,952 ± 611	837 ± 224	**2,098 ± 500**	2,541 ± n.a.	1760 ± n.a.	**1,865** ± n.a.	**27–56 ± n.a.**

n.a. Not available

[a]Hartmann (1982), Schultz (1982), Eggert (1985), Kothe (1988), Schmidt et al. (1989). Mean biomass values are area-weighted means of biomass values using proportion of the area under following facies according to Schmidt, Chap. 5, Table 5.5: *Anemone* facies: 6.3%, *Anemone-Mercurialis* facies (mean): 33.4%, *Mercurialis* facies: 6.3%, *Allium-Mercurialis* facies (mean): 9.1%, *Anemone-Allium-Mercurialis* facies (mean): 21.3%, *Allium* facies: 5.1%, *Anemone-Allium* facies (mean): 8.7%, other facies (mixture of previous facies according to their area proportion): 9.8%

[b]Mrotzek (1998). Mean biomass are the area-weighted means of biomass values considering the plot proportion of more than 7% relative irradiance intensity (*Urtica* facies, proportion of area: 13.5%) and less than 7% relative irradiance intensity (*Mercurialis* facies, proportion of area: 86.5%). The proportions were calculated from data that displays the classification of 192 plots into irradiance intensity classes according to Perona (1995) and Mrotzek (1998) (cf. Schmidt, this volume, Chap. 5)

[c]Eber (1972, 1982), Bennert (1973, 1980), Ellenberg et al. (1986). Vegetation type *Luzulo-Fagetum typicum*, *Oxalis acetosella*-facies

[d]Maximum total biomass is less than the sum of maximum above and belowground biomass because of different sampling dates

Table 10.5 Estimates of the aboveground phytomass of the herbaceous layer (dominant species and total sum) for spring (April–May) and summer 2002 (June–July) calculated with PhytoCalc 1.31 (Bolte and Parth 2004) (kg ha^{-1}, ± SD) at the Göttinger Wald, Zierenberg, and Solling sites (tree regeneration is not included)

	Göttinger Wald					Zierenberg				Solling
	Allium ursinum	*Anemone nemorosa*	*Asarum europaeum*	Other species	Total[a]	*Urtica dioica*	*Mercurialis perennis*	Other species	Total[b]	B 1, total[c]
Spring 2002	508 ± 280	111 ± 140	6 ± 9	28 ± 26	**653 ± 123**	118 ± 201	235 ± 221	26 ± 27	**379 ± 306**	–
Summer 2002	3 ± 2	<1 ± <1	20 ± 20	14 ± 12	**37 ± 25**	238 ± 398	238 ± 229	39 ± 104	**515 ± 474**	**52 ± 47**

[a]Göttinger Wald: Phytomass values are area-weighted means considering the proportion of area of the central floristic variant in the east–west transect (2.81 ha) in 2001 with 22.1% (*Anemone* facies: 19.6%, *Mercurialis* facies: 2.5%) and the *Allium* variant with 77.9% (*Allium* facies: 61.5%, *Anemone-Alium* facies: 15.0%, *Aconitum* facies: 1.4%) according to Schmidt (this volume, Chap. 5, Table 5.5)

[b]Zierenberg: Frequency weighted-means of 80 regularly distributed plots (4 m^2 area, basalt n = 35, Limestone n = 18, Mixed basalt-limestone n = 27) according to Schmidt (this volume, Chap. 5, Table 5.2)

[c]Solling: Arithmetic means of six plots (100 m^2 area)

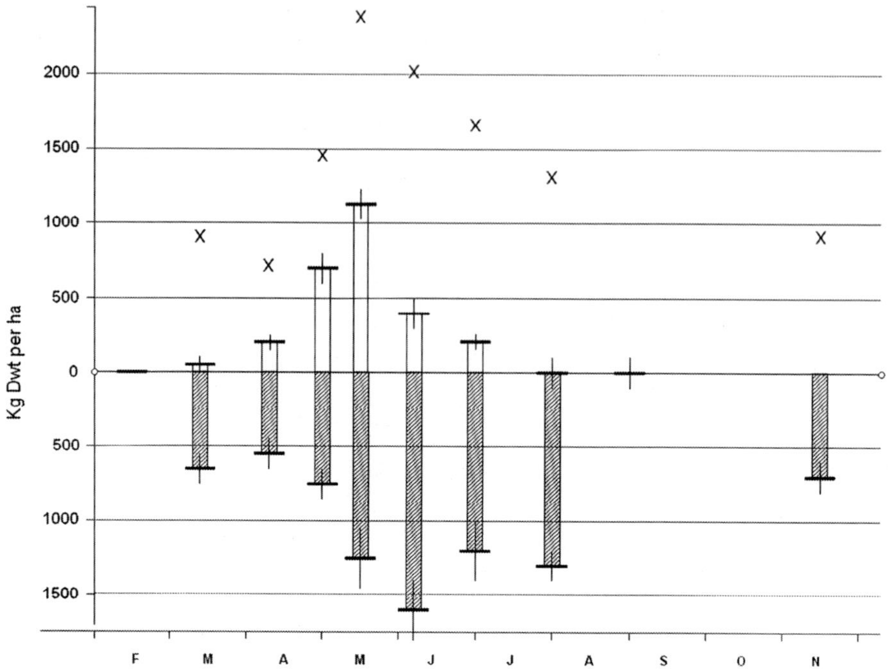

Fig. 10.1 Biomass and necromass of the herbaceous layer of *Allium* facies at Göttinger Wald in 1982 (according to Eggert 1985). *x* sum of aboveground and belowground biomass, *hatched boxes* belowground, *blank boxes* aboveground biomass

before maximum production of phytomass as shown for *A. ursinum* (Fig. 10.1, Eggert 1985).

At the Göttinger Wald site, the beech forest on limestone is characterised by a wide range of total biomass production according to the different species composition and small-scale heterogeneity of vegetation density ("patchiness", Rodenkirchen 1995). In the *Anemone nemorosa* facies with 837 kg ha^{-1}, considerably less biomass occurred compared to the *Allium* and *Mercurialis* facies. These differences reflected the morphological and growing properties of the dominant species and the low surface cover (about 50%) of the herb layer at this site.

At the Zierenberg site, *U. dioica* and *M. perennis* occurred in combinations where *Urtica* facies were characterised by a high coverage and a high biomass (2,541 kg dry weight per ha; peak biomass aboveground plus belowground). The differences in the aboveground biomass between the *U. dioica* facies and the *M. perennis* facies may be explained by the different morphology of the dominating species. *M. perennis* reaches a maximum height of 40 cm, whereas *U. dioica* can reach a height of up to 240 cm at the Zierenberg site.

In beech forests which are poor in understorey species, as at Solling, the biomass of the moss layer may be larger than that of the herbaceous layer. This was obvious in a single inventory of the quantitatively relevant moss *Polytrichum formosum* at Solling in 1970 that attained an aboveground biomass dry weigth of 18.5 kg ha^{-1}

and a similar value for the belowground biomass (Ellenberg et al. 1986). Total biomass of the herbaceous and the moss layer was about 62 kg ha^{-1}. Among the values reported in literature for beech forests throughout central Europe, the herb layer biomass in the nutrient poor beech forest at Solling (27–56 kg ha^{-1}) represented the lower end of the range. The mean values for aboveground biomass ranged from 10 to 460 kg ha^{-1}, and a maximum value of 1,700 kg ha^{-1} was reported (Hughes 1975; Leibundgut 1977; Kubiček 1980; Moldenhauer 1998; Anders et al. 2002; Weckesser 2003).

The nutrient-rich sites of Göttinger Wald and Zierenberg are close to the maximum value recorded in closed forests (Hughes 1975; Leibundgut 1977; Kubiček 1980; Moldenhauer 1998; Anders et al. 2002; Weckesser 2003). High amounts of herbaceous biomass of more than 4,000 kg ha^{-1} have been reported only for oak-dominated hardwood forests with geophytes (e.g. *A. ursinum, A. nemorosa*) in Southern Sweden (Olsson and Falkengren-Grerup 2003).

10.3.2 Aboveground Phytomass Estimated by Plant Dimension Analysis

Aboveground phytomass of the herbaceous layer on the three beech sites was calculated from percentage coverage and average shoot length using the model PhytoCalc 1.31 (Bolte and Parth 2004). Phytomass values given in Table 10.5 do not include the contribution of the tree regeneration.

Maximum total aboveground phytomass of the herbaceous layer were highest on the nutrient-rich soils at Göttinger Wald with 653 kg ha^{-1} and at Zierenberg with 515 kg ha^{-1} (Table 10.5). Low values occur at the nutrient-poor Solling beech site.

At Zierenberg, the dominant hemicryptophyte *U. dioica* accumulates the highest phytomass in summer. Its growth is stimulated by high irradiance intensity during the whole growing season. High light conditions were observed at the Zierenberg site mostly in the area of upper slope of basalt bedrock (Mrotzek 1998, Table 10.6). In contrast to *U. dioica*, phytomass of *M. perennis* is more evenly distributed within the different hillside situations and bedrock units on the site. An early increase of above ground phytomass was observed in spring (see Sect. 10.4) which was maintained until summer.

10.3.3 Comparison of Methods of Biomass Assessment

The comparison of harvest and plant dimension analysis (modelling) methods (Tables 10.4, 10.5) showed some differences. Modelled values were lower, 82% at the Göttinger Wald site and 68–92% at Zierenberg site, but of comparable size. However, at the Solling site, the modelled value of biomass was more than double

Table 10.6 Aboveground phytomass of the herbaceous layer for the dominant species and total values at Zierenberg for different hillside areas and bedrock units (means in kg ha^{-1}, ± SD) in 2002 estimates using PhytoCalc 1.31, Bolte and Parth (2004)

Zierenberg[1]	Upper side (n = 35 plots) Basalt			Middle side (n = 18 plots) Basalt/limestone			Lower side (n = 27 plots) Limestone		
	Urtica dioica	*Mercurialis Perennis*	**Weighted mean for the site**	*Urtica dioica*	*Mercurialis Perennis*	**Weighted mean for the site**	*Urtica dioica*	*Mercurialis Perennis*	**Weighted mean for the site**
Spring 2002	$234^a \pm 252$	$250^a \pm 180$	$\mathbf{518^a \pm 288}$	$38^b \pm 76$	$190^a \pm 236$	$\mathbf{257^b \pm 296}$	$20^b \pm 61$	$245^a \pm 259$	$\mathbf{275^b \pm 266}$
Summer 2002	$466^a \pm 459$	$251^a \pm 182$	$\mathbf{786^a \pm 513}$	$86^b \pm 160$	$204^a \pm 231$	$\mathbf{303^b \pm 334}$	$45^b \pm 129$	$244^a \pm 283$	$\mathbf{303^b \pm 299}$

n is the number of plots; species' and weighted means of the same season followed by different letters are significantly different at $P < 0.05$ (Kruskal–Wallace H test). Upper hillside site represents the main experimental area

[1]Zierenberg: Stratification of 80 plots of different hillside and bedrock sections according to Schulze and Eichhorn (2000), plot size 4 m^2

the value measured by the harvest method. Besides the methodological differences of assessing the biomass, the different values may have resulted from the temporal change of the vegetation structure and the different locations of the plots measured within the stands. At Zierenberg, the abundance of highly productive *U. dioica* (Mrotzek 1998) has decreased between 1993 (harvest studies) and 2002 (PhytoCalc study; Table 5.2). At Solling, a windthrow of one beech tree adjacent to the monitoring plots in 1990 increased the ground vegetation cover values. The high small-scale heterogeneity of ground vegetation structure may have caused major differences at the Göttinger Wald site reflected in the plots used for harvest and those used for monitoring for modelling purposes. Moreover, the more detailed description of the vegetation facies in 1981 (Dierschke and Song 1982) was used for mean biomass calculation from the harvest study made during the same year. For the following considerations, only values based on the harvest method are used.

10.4 Phenological Patterns of Annual Development of Herbaceous Biomass

The herbaceous vegetation showed a strong seasonal dynamic of increasing and decreasing phytomass that allowed the separation of time intervals between biomass and litter production of the aboveground components (Fig. 10.1; Jandl et al. 1997). Through regular measurement of above- and belowground biomass production, it was possible to recognise internal translocation processes related to species-specific patterns and strategies.

The herbaceous layer at Göttinger Wald is dominated by spring geophytes like *A. usinum* (Schmidt et al. 1989; Ellenberg 1996; Schmidt, Chap. 5, this volume) that are most competitive in the shade of forests with closed canopies (Ernst 1979, 1983; Eggert 1985; Hellmold and Schmidt 1989; Kriebitzsch 1989; Böhling 2003; Schmidt, Chap. 5, this volume). Aboveground growth of *A. ursinum* at Göttinger Wald (Fig. 10.1), started early in the year in March before leaf flushing of the beech trees and reached its maximum in mid-May (1,253 kg ha^{-1}). Under leafed canopy from mid-May, the relative irradiance intensity declined to 2–5% compared to open field conditions (Dierschke and Brünn 1993; Lambertz and Schmidt 1999). During this time, *Allium*'s biomass decreased and at the beginning of June aboveground biomass was only 1/3 (418 kg ha^{-1}) of the amount measured in May. Root growth started later than growth of the aboveground parts in mid-April, but lasted for about 14 days longer, leading to an increase in the biomass proportion belowground during the second half of the year. This specific temporal course of aboveground biomass also reflected the phenological adaptation of other geophytes like *A. nemorosa* to changes in light and soil temperature conditions under beech stands (Hellmold and Schmidt 1989; Heine-Koch 1993; Dierschke 2000). An extended period of belowground development, when compared to the shoot growth, pointed to processes of element mobilisation in the geophytes, which survive through belowground storage components (Eggert 1985; Demars and Boerner 1997).

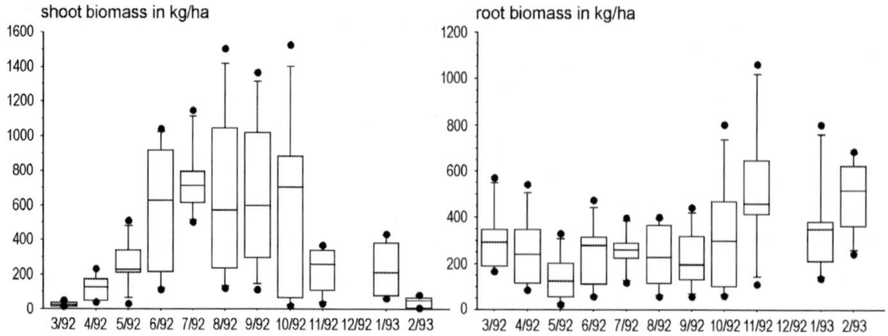

Fig. 10.2 Aboveground and belowground biomass of *U. dioica* at Zierenberg 1992/1993 (Eichhorn 1995). Box-whisker-plots with range (*points*), 10/90th percentiles (whisker plot), 25/75th percentiles (*box plot*), median (*middle line*)

At the Zierenberg site, with higher below-canopy irradiance intensity compared to Göttinger Wald site (Perona 1995; Mrotzek 1998), the dominant species were *U. dioica* and *M. perennis* of which first *M. perennis* and then *U. dioica* became the dominating fraction of the understorey. The shoot biomass of the nettle *U. dioica*, e.g., reached high values during June to October with a median value of about 650 kg ha^{-1} (Fig. 10.2). During these months, the aboveground biomass showed a large spatial variability and, for some areas with high and densely growing nettle, shoot biomass of up to 1,520 kg ha^{-1} was calculated for October (Eichhorn 1995). *U. dioica* may prolong its photosynthetic activity during summer when aboveground parts of the geophytes dominating at the Göttinger Wald site have already died (Teckelmann 1987). There is an annual course of the aboveground and belowground biomass with low amounts of shoot biomass and high values of the root biomass occurring after the growing period. In February, the highest value of 510 kg ha^{-1} (median value) of root biomass was determined at Zierenberg.

The shoot/root ratios are indications of the ecological adaptation of the species to different resources including water, light and nutrients. In the *Allium* facies of Göttinger Wald, the dominating spring and summer geophytes *A. ursinum*, *M. perennis* and *A. nemorosa* showed a shoot/root ratio < 1 over the entire year (Table 10.7). Their belowground components acted as storage for survival and therefore dominated over the aboveground plant parts most of the year with peak in May. However, the shoot/root ratios of *U. dioica* at Zierenberg ranged from 2.37 to 3.07 during periods when aboveground values peaked (Table 10.7) pointing to a strategy of aboveground competitive advantage (Mrotzek 1998).

At the Solling site, *Luzula luzuloides* and *Avenella flexuosa* were the dominant species, besides tree seedlings of *Fagus sylvatica*, and showed a distinct seasonal development with increasing biomass until summer and decreasing towards autumn (Ellenberg et al. 1986). At the Solling site, the total biomass ranged from 11 to 27 kg (1969) and 14 to 25 kg (1970) per hectare (mosses excluded). The biomass in 1969 did not peak until early July and in 1970 until the beginning of August. The

Table 10.7 Annual course of shoot/root ratio of the *Allium* facies at Göttinger Wald[a] and of *Urtica dioica* at Zierenberg[b]

Harvest month	Feb	Mar	Apr	May	Jun	Jul	Aug	Sep	Oct	Nov
Allium-facies[a]	0.06	0.13	0.40	0.82	0.18	–	0.04	–	–	0.04
Urtica dioica[b]	0.10	0.09	0.54	1.84	2.48	2.74	2.52	3.07	2.37	0.57

[a]Eggert (1985)
[b]Eichhorn (1995)

reason for these annual fluctuations may be related to climatic conditions, masting and competitive effects of tree seedlings, herbivory and the influence of fungal infections on the herbaceous vegetation (Ellenberg et al. 1986; Schmidt 1988).

10.5 Annual Production of Biomass and Litterfall of the Herbaceous Layer

Mean annual biomass production (NPP) was high at Göttinger Wald with a mean value of 1,035 and at Zierenberg with 1,069 kg ha^{-1} (Table 10.8). However, the herbaceous layer at the Solling site produced per year only 1–3% of the biomass produced annually at both other beech forests. Depending on the model used (Eggert 1985; Mrotzek 1998), the annual maximum litter production at Göttinger Wald was only half of the amount at Zierenberg.

At the Göttinger Wald site, the amount of biomass produced annually by different facies showed that *Allium* facies was the most productive (1,907 kg ha^{-1}) followed by the *Mercurialis* facies (1,139 kg ha^{-1}) and the *A. nemorosa* facies (256 kg ha^{-1}). At Zierenberg, high annual NPP values of 2,117 kg dry matter per ha were determined for *U. dioica* in the *Urtica* facies. These exceeded the values for the *Allium* facies at Göttinger Wald.

The *Mercurialis* facies at Zierenberg and at Göttinger Wald showed comparable biomass production values. The *Anemone* facies, however, produced at Göttinger Wald very low biomass during the same phenological period (Hellmold and Schmidt 1989). A lower base saturation (Schmidt et al. 1989) and reduced light conditions (Dierschke and Brünn 1993) in the *Anemone* facies may result in differences in NPP values.

Net primary production was markedly lower than the maximum biomass accumulated (compare Table 10.6), especially in the belowground components (Mrotzek 1998). Belowground values underestimate the actual production, because it was not possible to accurately determine the fine root turnover (Eggert 1985; Mrotzek 1998).

The highest annual litter production was observed under *U. dioica* in the *Urtica* facies at Zierenberg with a mean value of 1,950 kg ha^{-1}, and under *M. perennis* in the *Mercurialis* facies, 853 kg ha^{-1} was observed (Table 10.8). In *Urtica* facies, the mean litter production of the dominating *U. dioica* slightly exceeded the biomass

Table 10.8 Maximum biomass (NPP) and litter production of the herbaceous layer at the Göttinger Wald, Zierenberg and Solling sites expressed as kg ha^{-1} a^{-1}. Value ranges (in parentheses)

	Göttinger Wald[a]				Zierenberg[b]			Solling[c]
	1982	1981	1981	1981	1995	1995	1995	1968–1972
	Allium- facies	*Mercurialis*-*facies*	*Anemone*-facies	**Weighted mean**	*Urtica*-facies[e]	*Mercurialis*- facies[f]	**Weighted mean**	**B 1, mean**
Max. biomass production (NPP)	1,907	1,139	256	**1,035**	1,915 (1,713–2,117)	938 (703–1,172)	**1,069**	**14–30**
Max. litter production (=Ld)	970 (370–1,570)	479[d]	94	**481**	1,950 (1,748–2,152)	853 (618–1,087)	**1,001**	**16**

[a]Hartmann 1982; Schultz 1982; Eggert 1985; Kothe 1988; Schmidt et al. 1989. Mean biomass values are the area-weighted means of biomass values referred to the area proportion of following facies according to Schmidt, Chap. 5, Table 5.9: *Anemone*-facies: 6.3%, *Anemone-Mercurialis* facies (mean): 33.4%, *Mercurialis* facies: 6.3%, *Allium-Mercurialis* facies (mean): 9.1%, *Anemone-Allium-Mercurialis* facies (mean): 21.3%, *Allium* facies: 5.1%, *Anemone-Allium* facies (mean): 8,7%, other facies (mixture of previous facies according to their area proportion): 9.8%

[b]Mrotzek (1998). Mean biomasses are the area-weighted means of biomass values considering the plot proportion with more than 7% open field irradiance (*Urtica* facies, area proportion: 13.5%) and less than 7% open field irradiance (*Mercurialis* facies, area proportion: 86.5%). The proportion was calculated from data in Table 5.4 (Schmidt, this volume, Chap. 5) that includes the classification of 192 plots based on irradiance intensity classes according to Perona (1995) and Mrotzek (1998)

[c]Eber (1972, 1982), Bennert (1973, 1980), Ellenberg et al. (1986). Vegetation type *Luzulo-Fagetum typicum*, *Oxalis acetosella* facies (relevé no. 2/1991)

[d]Minimum value according to Hartmann (1982)

[e]Value only for *U. dioica*

[f]Value only for *M. perennis*

production (Table 10.8). These results supported the observations carried out on permanent plots at Zierenberg over several years. The cover and growth of *U. dioica* has reduced in some parts of the site during the last few years. This was mainly caused by reduced light conditions after closing of the beech forest canopy. This process became more obvious in those parts of the research areas where nettle growth was poor and was less evident where it was vigorous. In the formation dominated by the shade tolerant plant *M. perennis*, a continuous increase was observed (Mrotzek 1998; Schulze et al. 2000).

10.6 N-storage, Dynamic and Internal N-Cycling

Herbaceous vegetation in forests may constitute a sink for nitrogen when annual net balance is considered. However, ground vegetation may also be a source for nitrogen when herbaceous phytomass decreases due to a shortage of growth resources as may occur in summer droughts (Bolte et al. 2004). Once a stable amount of understorey has been established which is in balance with the conditions of the stand, no further annual biomass increment is expected. However, the seasonal changes in herbaceous growth may influence the internal element budget. Olssen and Falkengren-Grerup (2003) found that, in spring, excessive nitrate (more than 90%) was assimilated by the understorey which was dominated by geophytes in an oak forest, when tree canopy showed little activity and nitrogen availability was high. During the growing period, the uptake of nitrate and ammonium by the ground vegetation may contribute to a decrease in the N-concentration in the soil solution (Eichhorn 1995).

The amount of N retained by the herbaceous vegetation (Table 10.9) was determined by the biomass and the high N concentrations of the dominant species *A. nemorosa*, *M. perennis*, *A. ursinum* and *U. dioica*. A comparison of the three stands showed that the highest mean N storages occurred at Zierenberg with 52 kg N ha^{-1}. Similar values of 48 kg N ha^{-1} were determined for the *Allium* facies in Göttinger Wald and 51 kg N ha^{-1} for the *Mercurialis* facies at Zierenberg. These amounts are high compared to the value of 11 kg N ha^{-1} that was recorded by Moldenhauer (1998) at a 131-year-old beech stand on a comparable nutrient-rich site at Knüll (Hessia). The nitrogen retained by the herbaceous vegetation at Solling was comparatively low with 1.2 kg N ha^{-1}. For a mature beech stand on Pleistocene sandy sediments of lower nutrition, Anders et al. (2002) reported 4.6 kg N ha^{-1} in aerial parts of the ground vegetation. In different deciduous forests, maximum values of nitrogen in the herbaceous layer ranged from 1 to 55 kg ha^{-1} (Höhne 1962; Eggert 1985; Ellenberg et al. 1986; Anders et al. 2002). Thus, the Zierenberg site and the Solling site represented the upper and the lower ends of the range.

In the *Allium* facies of Göttinger Wald (Table 10.9), the annual nitrogen returned via the litter fall and decomposition ranged between 21 and 31 kg ha^{-1}. In comparison, the nitrogen in all facies of the Zierenberg site is higher. The highest amounts of N-uptake by *U. dioica* at Zierenberg was a maximum of 77 kg ha^{-1} per year. The stinging nettle contributed 65 kg ha^{-1}per year and *M. perennis* 12 kg ha^{-1}

Table 10.9 Maximum N-storage (kg ha^{-1}), amount of N-uptake (kg ha^{-1} a^{-1}), and amount of N-flux with the litterfall (kg ha^{-1} a^{-1}) of the herbaceous vegetation at the Göttinger Wald, Zierenberg, and Solling sites

	Göttinger Wald[a]		Zierenberg[b]			Solling[c]
	1982	1982	1995	1995	1995	1968–1972
	Allium-facies	Weighted mean	*Urtica*-facies	*Mercurialis*-facies	Weighted mean	B 1, mean
N-storage	48	**42**	68	51	**52**	**1.2**
N-uptake	–	**21**	77	30	**36**	**0.4**
N-litterfall	–	**21–31**	78	30	**36**	–

[a]Hartmann (1982), Schultz (1982), Eggert (1985; 1989), Kothe (1988), Schmidt et al. (1989, unpublished data) and Sah (1990). Mean values are calculated from biomass shown in Table 10.4 or production values of Table 10.8 and nitrogen concentrations of the dominating species *A. nemorosa, M. perennis* and *A. ursinum* (living plant material and shed dead material, compare also Sah 1990)

[b]Mrotzek (1998). Mean values are the area-weighted means considering the plot proportion with more than 7% relative light intensity (*Urtica* facies, proportion of area: 13.5%) and less than 7% relative light intensity (*Mercurialis* facies, proportion of area: 86.5%) according to Perona (1995) and Mrotzek (1998)

[c]Eber (1972, 1982), Bennert (1973, 1980), Ellenberg et al. (1986). Vegetation type *Luzulo-Fagetum typicum, Oxalis acetosella* facies (Table 5.1)

per year to the total amount. The annual amount of N-uptake of 30 kg ha^{-1}was determined for the facies dominated by *M. perennis*. On average, similar amounts of 36 kg ha^{-1} per year are taken up as released with the litterfall in the annual course. This is lower than those values between 48 and 55 kg ha^{-1} per year reported by Höhne (1962) for comparable stands. With an annual N-uptake of only 0.4 kg ha^{-1} by the herbaceous vegetation, the Solling site represents a base-poor site with low turnover rates for N.

The low values of nitrogen storage and uptake at the Solling site are supported by Anders et al. (2002) who estimated 3.3 kg ha^{-1} annual nitrogen uptake by the herbaceous plants aboveground of a closed beech forest on a base-poor sandy site in northeastern Germany. This is contrasted by Höhne (1962) who reported 19 kg ha^{-1} nitrogen uptake for the ground vegetation under the canopy of a mixed beech-pine stand on a site with comparably poorer nutrition.

10.7 Annual Changes in N-Contents

The N-contents in the shoots and leaves of stinging nettles at the Zierenberg site were about 70 g kg^{-1} at the time of bud break in March (Fig. 10.3, Eichhorn 1995; Mrotzek 1998). During the growing period from April to September, the N-content decreased to 60 g kg^{-1} and showed markedly lower N-content in the shoots than in the leaves. This phenological variance of element concentration in herbaceous

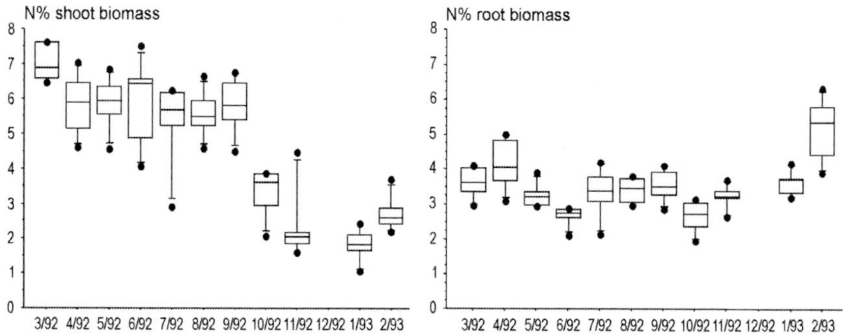

Fig. 10.3 N-contents in the above and belowground biomass of *U. diocia* expressed as % dry matter (Eichhorn 1995)

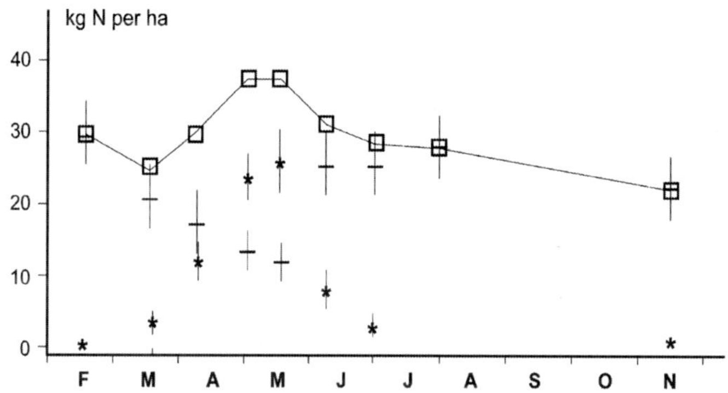

Fig. 10.4 Nitrogen storages (kg ha^{-1}) of the herbaceous layer at Göttinger Wald, *Allium* facies 1982 (Eggert 1985). *Star* N in aboveground biomass, *line* N in belowground biomass, *box* total N in biomass

plants is also described by Höhne and Fiedler (1963), Bennert (1973) and Chapin (1980). Plant tissue is diluted for N due to fast spreading of leaves at the beginning of the growing period (Eggert 1985). From October onwards, the N-content decreased below 40 g kg^{-1}. Due to a parallel accumulation of N in the throughfall below the densely growing nettles and very little increase in the N-concentration in the roots, it was assumed that the shoots of the nettles release nitrogen (Eichhorn 1995). N-content of roots varied very little from March to November (median values around 30–40 g N kg^{-1}). The highest N-content of roots occurred in February and reached 60 g N kg^{-1}. Fine roots were always richer in nitrogen than the coarse roots and rhizomes. This result leads to concepts of whole plant perspectives as proposed by Givnish (1988).

In the *Allium* facies at Göttinger Wald (Fig. 10.4; Eggert 1985), total N-content of 48 kg ha^{-1} was determined during May and June. In November, there was still found 27 kg N ha^{-1}. Re-translocation and mobilisation of nitrogen become obvious as development of the above- and belowground components takes place. From

February to May, nitrogen was transported from the belowground organs into the aboveground ones. In mid-May, after a sudden reduction of the aboveground biomass, N increased in roots and continued until August when it began to decrease again.

The extent of N translocation usually depends on the plant species and may be considered an adaptation to the nitrogen supply, especially when nitrogen is limiting growth. Species growing at nutrient poor sites often show a higher internal N re-translocation than plants at nutrient-rich sites. The amount of internal N-mobilisation depends on plant parts (Werner 1983) and can constitute up to two-thirds of the total N required by the plant (Eggert 1985). For *A. ursinum*, translocation of 25 kg N ha^{-1} was observed (Fig. 10.4) indicating high translocation ability. Eichhorn (1995) calculated for *U. dioica* an annual re-translocation of nitrogen at the end of the growing period from the aboveground to the belowground organs of 10–45 kg N ha^{-1}, and Mrotzek (1998) reported for *M. perennis* 2–5 kg N ha^{-1}. Compilation of data on element re-translocation by Aerts (1996) indicates values between 25 and 80% (mean: 41%) for most herbaceous species. A comparably low re-translocation of nitrogen is typical for species on nutrient-rich sites (Chapin 1980; Chapin et al. 1990).

10.8 C/N Ratios

The C/N ratio is used as a measure of litter quality determining decomposition (Wittich 1961); and herbaceous plants usually have low C/N ratios. They ranged between 8.1 (*M. perennis*) and 18.8 (*A. ursinum*) in the leaf components of understorey plants on these three sites (Table 10.10). Aboveground litter of almost all species had higher C/N ratios. A comparison of these values suggests that Re-translocation of nitrogen occur at the time of leaf senescence to varying extents in plants.

The C/N ratio in the leaf litter of beech is about 40 (Rademacher et al., Chap. 8, this volume). The C/N ratios of the herbaceous vegetation may be related to the humus form. Bolte (1999) showed that with an improvement of the humus form the

Table 10.10 C/N-ratios of the leaf biomass at the time of maximum aboveground biomass and the aboveground leaf litter of herbaceous species

	C/N leaf biomass	C/N leaf litterfall
Allium ursinum[a]	18.8	19
Anemone nemorosa[a]	10.1–12.8	24.3
Arum maculatum[a]	13.2	27
Mercurialis perennis[a]	8.1	6.9
Asarum europaeum[a]	10.7	15.5
Lamiastrum galeobdolon[a]	8.7	27.3
Urtica dioica[b]	11.2–14.6	–

[a]Eggert (1985): Göttinger Wald
[b]Teckelmann (1987)

C/N ratios of the Oh/Ah-horizon decreased whereas the N-contents of herbaceous plants growing on these sites increased and the C/N ratio in the aboveground plant material was reduced from 31 (raw humus) to 12 (mull). Plant litter on nutrient-rich sites provides readily decomposable substrates and thus more favourable conditions for the activity of decomposers.

10.9 Ground Vegetation as a Part of Internal N-Cycling in Beech Stands

Contribution of the herbaceous vegetation to the annual nitrogen turnover may be significant in some cases (Alriksson and Eriksson 1996; Buchmann et al. 1996). For example, herbaceous vegetation at Göttinger Wald contribute with 25% and at Solling with less than 1% to the total annual N-uptake by vegetation (Table 10.11). The herbaceous layer in forest ecosystems thus forms an intermediate nutrient store for cycling (Muller and Bormann 1976; Eichhorn 1995; Bolte 1996; Olsson 2002). This was of particular significance at the beginning of the growing period or after tree removals when the element uptake of the actively growing herbaceous vegetation may reduce nutrient losses and nitrate leaching from forests (Mellert et al. 1998; Olsson and Falkengren-Grerup 2003). However, due to the seasonal nature of the understorey growth, the release of N during mineralisation in autumn or early spring may not always synchronise with the demand for element uptake by trees (Jandl et al. 1997) and may cause significant N losses in drainage water during

Table 10.11 Component of vegetation in the nitrogen balance of beech stands at Göttinger Wald and Solling (storage and uptake by trees adopted from Chap. 8, belowground N-litter from Chap. 9 and aboveground litter from Chap. 11)

	Göttinger Wald	Solling
N-storage [kg ha^{-1}]		
Tree layer		
Aboveground[a]	941	894
Belowground	397	173
Total tree layer	1,338	1,067
Herb layer	42	1.2
N-uptake [kg ha^{-1} a^{-1}]		
Tree layer		
Aboveground[b]	81	74
Herb layer	21	0.4
N-litter [kg ha^{-1} a^{-1}]		
Tree layer		
Aboveground	63	68
Belowground	18–26	36–39
Herb layer	21–31	–

[a]Leaf biomass not considered
[b]Litter fall plus increment

winter and early spring (Eichhorn 1995). High nitrate uptake by vegetation dominated by geophytes during summer may reduce nitrate losses from the ecosystem even when their biomass is low due to a high nitrate reductase activity (NRA) per unit biomass (Olsson and Falkengren-Grerup 2003).

Differences in the quality of litter between herbaceous and woody vegetation are important for the supply of readily decomposable substrates for mineralisation (Wittich 1944; Scott 1955; Wise and Schaefer 1994). For example, at the Zierenberg site, the N-content in the herbaceous litter for decomposition was as high as N-content in the beech leaf litter. Moreover, the litter of the herbaceous vegetation has lower C/N ratios than that of beech trees indicating better litter quality (Jandl et al. 1997). Positive effects of *A. ursinum* ground cover on the soil fauna, particularly collembolans and earthworms, have been reported for a beech forest which may enhance decomposition processes (Kopeszki and Jandl 1994; Wolters 1999). Herbaceous species like *Melica uniflora* and *Epilobium angustifolium* may facilitate decomposition (Ponge and Delhaye 1995; Bauhus et al. 2004). Wise and Schaefer (1994) hypothesised that for an accelerated decomposition of beech litter the contribution of high quality herbaceous litter is required. For the biotic incorporation by earthworms, high quality herbaceous litter plays an important role by triggering decomposition processes after liming (Bauhus et al. 2004). Thus, the retention of C and N by mineral soil may be positively affected by species-rich and highly productive herbaceous vegetation.

10.10 Conclusion

Important issues and implications of the study are:

- The three beech stands have a similar age, but differ in site conditions, forest history and management. Herb layer vegetation is mainly affected by the chemical soil site conditions and irradiance intensity which may vary in time and space.
- The Solling site has a poorly developed herbaceous layer, while the Göttinger Wald site is rich in number of species with high variability at a small spatial scale. The dominant species at the Zierenberg site include *U. dioica* and *M. perennis*.
- Two methods were used to assess the maximum aboveground biomass of the herbaceous layer at the three sites. The first method involved biomass harvests conducted in different years from 1968 to 1995. With this method, the total above- and belowground biomass was 56 kg ha^{-1} at Solling, 2,100 kg ha^{-1} at Göttinger Wald and 1,865 kg ha^{-1} at Zierenberg. The amount of mean biomass production (NPP: "maximum–minimum") was 1,035 kg ha^{-1} per year at Göttinger Wald, 1,069 kg ha^{-1} per year at Zierenberg and 30 kg ha^{-1} per year at Solling. The second method made use of the PhytoCalc 1.31 model which involved percent coverage and average shoot length. The calculated maximum

aboveground phytomass of the herbaceous layer from relevés in 2002 were 653 kg ha^{-1} at Göttinger Wald and 515 kg ha^{-1} at Zierenberg, and 52 kg ha^{-1} at the nutrient-poor Solling site.

- Amount of N retained by the understorey vegetation depends on the biomass and N concentrations in the plant parts. The herbaceous vegetation at Zierenberg (*U. dioica* and *M. perennis*) stored a significant amount (mean: 52 kg ha^{-1}, *Urtica*-facies: 68 kg ha^{-1}) when compared to Göttinger Wald (mean: 42 kg ha^{-1}, *Allium* facies: 48 kg ha^{-1}) and particularly the Solling site (1.2 kg ha^{-1}). Different values were recorded also for mean annual N-uptake (Göttingen Wald: 21 kg ha^{-1} per year, Zierenberg: 36 kg ha^{-1} per year, Solling: 0.4 kg ha^{-1} per year) and mean annual N-losses with the litterfall (Göttingen Wald 21–31 kg ha^{-1} per year, Zierenberg: 36 kg ha^{-1} per year) of the understorey vegetation indicating high proportion of total turnover of N at the Zierenberg and Göttingen Wald sites.
- Intermediate store of nutrients is probably the most important function of the herbaceous layer in forest ecosystems. At times of low tree uptake at the beginning of the growing period or after tree removals, the element uptake by the ground vegetation may reduce leaching losses. However, due to the seasonality in the growth of understorey plants, the release of N during mineralisation of herbaceous litter may not always synchronise with the demand for plant uptake and may cause significant N losses in drainage water.
- High quality litter of the understorey vegetation plays an important role in the decomposition processes of overstorey components by enhancing the activity of the soil fauna (collembolans, earthworms) and triggering litter decomposition.

Acknowledgments The vegetation studies that provided the basis for the present analyses were kindly funded by the German Research Council DFG (SFB 135), the Federal Ministry of Science and Education (BMBF FKZ 0339474B, 0339474C/3, 0339500B/C) and the European Union (EU Level II programme). The authors are grateful to Andreas Parth (Goettingen University, Institute of Silviculture) for his skillful contributions to the "PhytoCalc" calculations and the editors, Partap Khanna and Rainer Brumme, for helpful comments.

References

Aerts R (1996) Nutrient resorption from senescing leaves of perennials: are there general patterns? J Ecol 84:597–608

Alriksson A, Eriksson HM (1996) Carbon and nutrient distribution in the Sävar tree species experiment. Silvestra 2:1–19

Anders S, Beck W, Bolte A, Hofmann G, Jenssen M, Krakau U-K, Müller J (2002) Ökologie und Vegetation der Wälder Nordostdeutschlands. Dr. Kessel, Remagen-Ow

Bartsch N (2000) Element release in beech (*Fagus sylvatica* L.) forest gaps. Water Air Soil Pollut 122:3–16

Bauhus J, Vor T, Bartsch N, Cowling A (2004) The effects of gaps and liming on forest floor decomposition and soil C and N dynamics in a *Fagus sylvatica* forest. Can J For Res 34:509–518

Bennert HW (1973) Chemische-ökologische Untersuchungen an Arten der Krautschicht eines montanen Hainsimsen-Buchenwaldes (*Luzulo-Fagetum*). Diss FU Berlin

Bennert H-W (1980) Über die stoffliche Zusammensetzung von Waldbodenpflanzen eines montanen Hainsimsen-Buchenwaldes. 1. Asche-, Stickstoff- und Proteingehalt. Flora 170:433–455

Blank JL, Olson RK, Vitousek PM (1980) Nutrient uptake by a diverse spring ephemeral community. Oecologia 47:96–98

Böhling N (2003) Dauerflächen-Untersuchungen in einem Eichen-Hainbuchenwald im Vorland der Schwäbischen Alb (Südwestdeutschland), 1978–2001: Der Niedergang von *Scilla bifolia* und die Invasion von *Allium ursinum*. Tuexenia 23:131–161

Bolte A (1996) Die Bodenvegetation in Kiefernökosystemen – eine Steuergröße für den Wasser- und Stoffhaushalt. Mitt Bundesforschungsanst Forst- Holzwirtschaft 185:97–111

Bolte A (1999) Abschätzung von Trockensubstanz-, Kohlenstoff- und Nährelementvorräten der Waldbodenflora – Verfahren, Anwendung, Schätztafeln. Forstwiss Beiträge Tharandt, Ulmer, Stuttgart

Bolte A (2006) Biomasse- und Elementvorräte der Bodenvegetation auf Flächen des forstlichen Umweltmonitorings in Rheinland-Pfalz. Ber Forschungszentr Waldökosysteme B72

Bolte A, Parth A (2004) Programmbeschreibung PhytoCalc 1.3. Univ Göttingen, Inst für Waldbau, Abt I

Bolte A, Anders S, Roloff A (2002) Schätzmodelle zum oberirdischen Vorrat der Waldbodenflora an Trockensubstanz, Kohlenstoff und Makronährelementen. Allg Forst- u Jagdztg 173:57–66

Bolte A, Lambertz B, Steinmeyer A, Kallweit R, Meesenburg H (2004) Zur Funktion der Bodenvegetation im Nährstoffhaushalt von Wäldern – Studien auf Dauerbeobachtungsflächen des EU Level II-Programms in Norddeutschland. Forstarchiv 75:207–220

Brünn S (1999) Untersuchungen zum Mineralstoffhaushalt von *Calamagrostis epigeios* (L.) Roth in stickstoffbelasteten Kiefernwäldern. Ber Forschungszentr Waldökosysteme A160

Buchmann N, Schulze ED, Gebauer G (1996) Partitioning of ^{15}N-labelled ammonium and nitrate among soil, litter, below and above-ground biomass of trees an understorey in a 15-year-old *Picea abies* plantation. Oecologia 102:361–370

Chapin III FS (1980) The mineral nutrition of wild plants. Ann Rev Ecol Syst 11:233–260

Chapin III FS, Schulze ED, Mooney HA (1990) The ecology and economics of storage in plants. Ann Rev Ecol Syst 21:423–447

Charley JL, Richards BN (1983) Nutrient allocation in plant communities: Mineral cycling in terrestrial ecosystems. In: Lange OL, Nobel PS, Osmond CB, Ziegler H (eds) Physiol Plant Ecol 4:5–45

Demars BG, Boerner REJ (1997) Foliar phosphorus and nitrogen resorption in three woodland herbs of contrasting phenology. Castanea 62:695–701

De Vries W, Reinds GJ, De Zwart D, Aamlid D, Neville P, Posch M, Auee J, Voogd JCH, Vel EM (2002) Intensive Monitoring of Forest Ecosystems in Europe – Technical Report 2002. EC-UN/ECE, Brussels, Geneva

Dierschke H (2000) Phenological phases and phenological species groups of mesic beech forests and their suitability for climatological monitoring. Phytocoenologia 30:469–476

Dierschke H (2006) Long-term dynamics in the herb layer of a calcareous beech forest: investigations on permanent plots, 1981–2001. Pol Bot Stud 22:165–172

Dierschke H, Brünn S (1993) Raum-zeitliche Variabilität der Vegetation eines Kalkbuchenwaldes – Untersuchungen auf Dauerflächen 1981–1991. Scripta Geobot 20:105–151

Dierschke H, Song Y (1982) Vegetationsgliederung und kleinräumige Horizontalstruktur eines submontanen Kalkbuchenwaldes. In: Dierschke H (Red.): Struktur und Dynamik von Wäldern. Ber Int Symp Int Ver Vegetationskde (Rinteln 1981), pp 513–539

Donita N (1972) Zahl und Größe der Probeflächen bei der direkten Bestimmung der Krautschicht-Biomasse im Walde. Oecol Plant 7:85–94

Duvigneaud P, Denaeyer-De Smet S (1973) Biological cycling of minerals in temperate deciduous forests. In: Reichle DE (ed) Analysis of temperate forest ecosystems. Ecological Studies 1, 1st corrected reprint, Springer, Berlin

Eber W (1972) Über das Lichtklima von Wäldern bei Göttingen und seinen Einfluß auf die Bodenvegetation. Scripta Geobot 3

Eber W (1982) Struktur und Dynamik der Bodenvegetation im Luzulo-Fagetum. In: Dierschke H (Red): Struktur und Dynamik von Wäldern. Ber Int Symp Int Ver Vegetationskde (Rinteln 1981), pp 495–511

Eggert A (1985) Zur Ökologie der Krautschichtvegetation in einem Bärlauch-Kalkbuchenwald. Diss Univ Göttingen

Eggert A (1989) Zur saisonalen Kohlenstoff- und Stickstoffdynamik der Krautschichtvegetation eines submontanen Kalkbuchenwaldes. Verh Ges Ökol 17:167–176

Eichhorn J (1995) Stickstoffsättigung und ihre Auswirkungen auf das Buchenwaldökosystem der Fallstudie Zierenberg. Ber Forschungszentr Waldökosysteme A124, University Göttingen

Ellenberg H (1996) Vegetation Mitteleuropas mit den Alpen. 5th edn. Ulmer, Stuttgart

Ellenberg H, Mayer R, Schauermann J (1986) Ökosystemforschung – Ergebnisse des Sollingprojektes 1966 – 1986. Ulmer, Stuttgart

Ernst WHO (1979) Population biology of *Allium ursinum* in Northern Germany. J Ecol 67: 347–362

Ernst WHO (1983) Population biology and mineral nutrition of Anemone nemorosa with emphasis on its parasitic fungi. Flora 173:335–348

Giese LAB, Aust WM, Kolka RK, Trettin CC (2003) Biomass and carbon allocation of disturbed riparian forests. Forest Ecol Manage 180:493–508

Givnish T (1988) Adaption to sun and shade: a whole-plant perspective Aust. J Plant Physiol 15:63–92

Hartmann T (1982) Zur Biomasseentwicklung von *Mercurialis perennis* – Beständen in einem Kalkbuchenwald. Dipl Arbeit Univ Göttingen

Hartung J (1989) Statistik. 7th ed R Oldenbourg, München and Wien

Heine-Koch J (1993) Beziehungen zwischen phänologischer Entwicklung von Pflanzen und Frühjahrserwärmung in Laubwäldern bei Göttingen. Dipl Arb Univ Göttingen

Hellmold C, Schmidt W (1989) Energiegehalt und Energiebilanz der Krautschicht eines Kalkbuchenwaldes. Verh Ges Ökol 17:177–188

Höhne H (1962) Vergleichende Untersuchungen über Mineralstoff- und Stickstoffgehalt sowie Trockensubstanzproduktion von Waldbodenpflanzen. Arch Forstwes 11:1085–1141

Höhne H (1963) Der Mineral- und Stickstoffgehalt von Waldbodenpflanzen in Abhängigkeit vom Standort. Arch Forstwes 12:791–805

Höhne H, Fiedler HJ (1963) Über den Einfluß des Entwicklungszustandes von Waldgräsern auf ihren Gehalt an Mineralstoffen und Stickstoff. Arch Forstwes 12:676–696

Hövemeyer K (1995) Trophic links, nutrient fluxes, and natural history in the Allium ursinum food web, with particular reference to life history traits of two hoverfly herbivores (Diptera: Syrphidae). Oecologia 102:86–94

Hughes MK (1975) Ground vegetation net production in a Danish beech wood. Oecologia 18:251–258

Jandl R, Kopeszki H, Glatzel G (1997) Effect of a dense *Allium ursinum* (L) ground cover on nutrient dynamics and mesofauna of a *Fagus sylvatica* (L) woodland. Plant Soil 189:245–255

Koperski M, Sauer M, Braun W, Gradstein SR (2000) Referenzliste der Moose Deutschlands. Schriftenr Vegetationskde 34: 1–519

Kopeszki H, Jandl R (1994) Die Mesofauna, insbesondere Collembolenfauna, im Buchen-Wienerwald in Abhängigkeit von Streu-Akkumulation und -Depletion. Zool Anz 231:123–134

Kothe G (1983) Biometrische Methoden zur Bestimmung der Stoffproduktion von *Mercurialis perennis* L. in einem Kalkbuchenwald. Dipl Arb Univ Göttingen

Kothe G (1988) Biometrische Bestimmung der Biomasse und Blattfläche von *Mercurialis perennis*-Beständen. Flora 180:203–221

Kriebitzsch W-U (1989) CO_2- und H_2O-Gaswechsel von verschiedenen Krautschichtpflanzen. Verh Ges Ökol 17:189–202

Kubíček F (1980) Productivity of herb layer in natural spruce ecosystems in Slovakia. In: Klimo E (ed) Stability of spruce forest ecosystems – International symposium. pp 495–504

Lambertz B, Schmidt W (1999) Auflichtungen in Buchen- und Buchenmischbeständen auf Muschelkalk und Buntsandstein – Untersuchungen zur Verjüngungs- und Vegetationsstruktur. Verh Ges Ökol 29:81–88
Larcher W (1994) Ökophysiologie der Pflanzen, 5th edn. Ulmer, Stuttgart
Leibundgut H (1977) Über die Phytomasse der Bodenvegetation einiger Waldgesellschaften. Schweiz Zeitschr Forstwes 128:730–740
Lieth H, Whittaker R (eds) (1975) Primary production of the biosphere. Springer, Berlin
Likens GE, Bormann FH (1970) Chemical analysis of plant tissues from the Hubbard Brook Ecosystem in New Hampshire. Yale University School of Forestry Bull 79, New Haven
Linke J, Roloff A (1995) Landwirtschaftliche Arten als Hilfspflanzen bei der Bestandesbegründung. Allg Forst- u J-Ztg 166:69–76
Mellert K-H, Kölling C, Rehfuess KE (1998) Vegetationsentwicklung und Nitrataustrag auf 13 Sturmkahlflächen in Bayern. Forstarchiv 69:3–11
Milner C, Hughes RE (1968) Methods for the measurement of the primary production of grassland. IBP Handbook 6, Blackwell, Oxford
Moldenhauer H (1998) Strukturbeschreibung der Vegetation in Buchenwäldern unterschiedlicher Altersstufen anhand der Artenzusammensetzung, der Biomasse und der Elementgehalte. Dipl Arb Univ Göttingen
Mrotzek R (1998) Wuchsdynamik und Mineralstoffhaushalt der Krautschicht in einem Buchenwald auf Basalt. Ber Forschungszentr Waldökosysteme A152
Muller RN, Bormann FH (1976) Role of *Erythronium americanum* Ker in energy flow and nutrient dynamics of a northern hardwood forest ecosystem. Science 193:1126–1128
Newbould PJ (1967) Methods for estimating the primary production of forests. IBP Handbook 2, Blackwell, Oxford
Olsson M (2002) Uptake of and preference for nitrate, ammonium and amino acids by understorey species in deciduous forests. Doctoral dissertation Lund Univ
Olsson MO, Falkengren-Grerup U (2003) Partitioning of nitrate uptake between trees and understorey in oak forests. For Ecol Manage 179:311–320
Perona L (1995) Standortbedingungen von *Urtica dioica* (L.) und *Mercurialis perennis* (L.) auf der Versuchsfläche der., Fallstudie Zierenberg. Dipl Arb Univ Göttingen
Peichl M, Arain AA (2006) Above- and belowground ecosystem biomass and carbon pools in an age-sequence of temperate pine plantation forests. Agric For Meteorol 140:51–63
Persson H (1975) Deciduous woodland at Andersby, Eastern Sweden: field-layer and belowground production. Acta Phytogeogr Suec 62
Ponge JF, Delhaye L (1995) The heterogeneity of humus profiles and earthworm communities in a virgin beech forest. Biol Fertil Soils 20:24–32
Rodenkirchen H (1982) Wirkung von Meliorationsmaßnahmen auf die Bodenvegetation in einem ehemals streugenutzten Kiefernstandort in der Oberpfalz. Schriftenr Forstw Fak Univ München 53
Rodenkirchen H (1995) Nutrient pools and fluxes of the ground vegetation in coniferous forests due to fertilizing, liming and amelioration. Plant Soil 168–169:383–390
Sachs L (1976) Angewandte Statistik, 5th edn. Springer, Berlin
Sah SP (1990) Vergleich des Stoffhaushaltes zweier Buchenwaldökosysteme auf Kalkgestein und Buntsandstein. Göttingen. Ber Forschungszentr Waldökosysteme A 59
SAS Inst. Inc. (1989) SAS/STAT User's Guide, Version 6, 4th ed, Vols. 1 and 2, Cary, NC
Schaefer M (1989) Die Bodentiere eines Kalkbuchenwaldes: ein Ökosystemforschungsprojekt (Zur Funktion der Fauna in einem Mullbuchenwald 1.). Verh Ges Ökol 17:203–212
Schaefer M (1996) Die Bodenfauna von Wäldern: Biodiversität in einem ökologischen System. Abh Math-Naturwiss Kl Akad Wiss Lit Mainz 2:1–81
Schmidt W (1988) Langjährige Veränderungen der Krautschicht eines Kalkbuchenwaldes (Daueruntersuchungen). Tuexenia, pp 327–338
Schmidt W (2002) Einfluss der Bodenschutzkalkung auf die Waldvegetation. Forstarchiv 73:43–54
Schmidt W, Hartmann T, Kothe-Heinrich G, Schultz R (1989): Jahresrhythmus und Produktion der Krautschicht in einem Kalkbuchenwald. Verh Ges Ökol 17:145–157

Schultz R (1982): Zur Biomasseentwicklung von *Anemone nemorosa*-Beständen in einem Kalkbuchenwald. Diplom Arb Univ Göttingen

Schulze I-M, Eichhorn J (2000) Veränderungen im Stickstoffhaushalt von Buchenwäldern auf Basalt: Die Ausbreitung der Großen Brennnessel und ihr Einfluss auf die natürliche Verjüngung der Buche. Forst und Holz 55:435–441

Schulze I-M, Bolte A, Seidling W, Stetzka K, Wellbrock N (2000) Vegetationskundliche Aufnahmen im Level II-Programm: Methoden, Auswertungen, erste Ergebnisse. Forstarchiv 71:76–83

Schütt P, Schuck HJ, Stimm B (1992): Lexikon der Forstbotanik. Ecomed, Landsberg (Lech)

Scott DRM (1955) Amount and chemical composition of the organic matter contributed by overstorey and understorey vegetation to forest soils. Yale University School of Forestry Bull 62, New Haven

Teckelmann M (1987) Kohlenstoff-, Wasser- und Stickstoffhaushalt von *Urtica dioica* L an natürlichen Standorten. Diss Univ Bayreuth

Turner J, Long JN, Backiel A (1978) Under-storey nutrient content in an age sequence of Douglas-fir stands. Ann Bot 42:1045–1055

Weckesser M (2003) Die Bodenvegetation von Buchen-Fichten-Mischbeständen im Solling – Struktur, Diversität und Stoffhaushalt. Cuvillier, Göttingen

Werner, W (1983) Untersuchungen zum Stickstoffhaushalt einiger Pflanzenbestände. Scripta Geobot:16

Whittaker RH (1975) Communities and ecosystems, 2nd edn. Macmillan, New York

Wise DH, Schaefer M (1994) Decomposition of leaf litter in a mull beech forest: comparison between canopy abd herbaceous species. Pedobiologia 38:269–288

Wisskirchen R, Haeupler H (1998) Standardliste der Farn- und Blütenpflanzen Deutschlands. Ulmer, Stuttgart

Wittich W (1944) Untersuchungen über den Verlauf der Streuzersetzung auf einem Boden mit Mullzustand IV. Forstarchiv 20:110–114

Wittich W (1961) Der Einfluss der Baumart auf den Bodenzustand. Allg Forstzeitschr 16:41–45

Wolters V (1999) *Allium ursinum* litter triggering decomposition on a beech forest floor – the effect of earthworms. Pedobiologia 43:528–536

Yarie J (1980) The role of understorey vegetation in the nutrient cycle of forested ecosystems in the mountain hemlock biogeoclimatic zone. Ecology 61:1498–1514

Chapter 11
Biomass and Element Content of Foliage and Aboveground Litterfall on the Three Long-Term Experimental Beech Sites: Dynamics and Significance

P.K. Khanna, H. Fortmann, H. Meesenburg, J. Eichhorn, and K.J. Meiwes

11.1 Introduction

The amount and dynamics of litterfall in a forest stand is recognised as an integrated response, reflecting biologically (hereditary defined) controlled processes of shedding of leaves, roots, bark, fruits and other plant components and the physiologically controlled factors of environmental nature (climate, water availability, nutritional supply and toxicity parameters). Changes in the dynamics of litterfall components (leaves, fruits, florescence, seeds, twigs, etc.), in the short-term, is primarily a reflection of climatic perturbations affecting water and nutrient supply. For example, drought conditions in the early summer preceding mast year (mast year−1) were found to be a very strong predictor of mast production in beech stands of Europe and eastern North America (Piovesan and Adams 2001). An unusually moist, cool summer the year before the drought (mast year−2) increased the predictability of mast production which was associated with carbohydrate build up within the trees which might prime them for floral induction the following year (mast year−1). There is, however, very little known of the accumulation and redistribution of nutrients in association with masting in beech.

Litterfall constitutes the transfer of energy between different trophic levels, where energy-rich plant materials are transferred for use by microbial and faunal populations in and on the soil. Despite the significance of litterfall in many ecosystem processes, the main emphasis on litterfall studies during the earlier periods had been to describe the fluctuations, distribution and composition of litterfall (Viro 1955; Kittredge 1948). Emphasis has now shifted to study the role of litterfall in ecosystem functioning especially its role in nutrient cycling processes in forest ecosystems (Khanna and Ulrich 1991). The aim of many studies in forest ecosystems is to develop management options where sustainability of ecosystem productivity and ecological functioning have been appropriately achieved, and for that purpose litterfall can act as a useful indicator. One such parameter is leaf biomass that is related to the health of forest ecosystems, which can be directly deduced from leaf litter, especially in deciduous stands. Another one is

R. Brumme and P.K. Khanna (eds.), *Functioning and Management of European Beech Ecosystems*, Ecological Studies 208,
DOI: 10.1007/978-3-642-00340-0_11,

the decomposition rate of litterfall components which depends on the quality of litter fall and is thus closely related to the functionality and activity of microbial and faunal populations in soils (Chaps. 12, 20, this volume). Thus, litter fall quality of overstorey and understorey vegetation is closely related to the biodiversity on a forest site. Some essential components of the multi-functions of a forest ecosystem that are closely related to the amount and quality of litterfall are: C source for microbial activity, food for animals (biodiversity) and the use of nutrients for the growth and activity of flora and fauna.

Due to continually changing input conditions and climatic factors, site factors and stand parameters strongly and differently affect forest productivity and nutrient cycling processes. Leuschner et al. (2006) evaluated 50 beech stands in Central Europe to assess the important soil factors which describe the niche breadth of their growth. The stands could be delineated into three main groups depending upon the exchangeable base cations, C/N ratio in the mineral soil and C/P of the surface organic layer. In order to manage such ecosystems in a landscape context, it is essential to develop suitable knowledge based on such interactions, e.g. on the role of litter fall parameters which may enable one to extend site-specific information to landscape level.

In this chapter, the relative amounts of leaf and fruiting bodies will be considered in relation to climatic and other variables which may affect their production. The question will be addressed as to how does the annual leaf litter relate to the productivity on these sites?

Mast production in beech is a significant event involving the transfer of huge amounts of C and nutrients. Both high plant productivity and high amount of atmospheric nutrient inputs may cause the high frequency and amount of mast produced (Gruber 2006). This has further consequences for the amount of nutrients retained by the trees for mast production, leading to low values of Mg, K, and P as cycling fractions in these stands. Mast production may also affect the production of foliage and/or wood. Thus, the consequences of changes in mast production (increase in the frequency of mast years has been observed during recent decades; Paar et al. 2000) for the growth and functioning of beech stands will be specifically discussed in this chapter.

All the three stands studied have been receiving high amounts of atmospheric inputs of elements and acidity (Chap. 15), which may have a long-term effect on their foliage chemistry either through direct effects on the amount of leaf nutrients leached from the leaves or indirectly through nutrient uptake. Foliage chemistry and leaf litter chemistry are intimately related, thus affecting the nutrient turnover processes in the soil and litter layer. There are many other factors which affect the concentration of nutrients in the total litterfall. For example, season of litterfall, moisture and other stress factors, relationships with other elements and nutrient withdrawal patterns during leaf senescence will determine the concentration and amount of nutrients transferred through litterfall. Some of these factors will be described in this chapter.

The patterns of element withdrawal at the time of leaf senescence form an important process which affects the nutrient status of the stand and may depend upon the absolute and annual variation in the element concentrations in the foliage. The following questions will be addressed. What types of annual patterns of element withdrawal are observed on different sites? How do they relate to the supply of nutrients by soils and to the stand productivity?

The amount of leaf litter and foliar mass are related; however, this relationship depends upon a number of factors (Stachurski and Zimka 1975). Usually, the amount of leaf litter collected at the surface organic layer does not equal the amount of leaf produced because of the unaccountable losses of leaf material. There are direct losses of leaf material where it is consumed by insects and undergoes microbial decomposition and indirect losses where leaf material is transferred to other plant components prior to or at the time of senescence. Is it possible to make an assessment of these losses, for example by using marker element such as Ca and then assess the foliar mass from the amount of leaf litter? As it is hard to directly measure changes in the amount of foliar mass, the possibility of assessing the amount through leaf litter will be tested in this chapter.

11.2 Site Description and Collection and Measurements of Aboveground Litterfall Components and Green Foliage

The three experimental sites at Solling, Göttinger Wald and Zierenberg differ in their stand, climatic and soil characteristics, which are expected to affect the amount and chemistry of litterfall. Some of the relevant stand and site parameters are given in Table 11.1. The stands contained mostly beech with very small fractions of other trees in the overstorey, but differed in their understorey which was a reflection of stand and soil parameters (Chaps. 5, 10). The stands were aged

Table 11.1 Some relevant parameters for site description of the three beech stands in relation to the amount and chemistry of litter fall. Range of soil properties are given for 0–50 cm soil depth

Site	Solling	Göttinger Wald	Zierenberg
Age (years) (2003)	165	120–130	160
Density (trees ha^{-1})	199	245	132
Mean tree height (m)	29.6	33.7	36.7
Mean tree diameter (cm)	46.8	45.7	59.9
Basal area (m^2)	34.2	39.5	32.0
Mean annual temp. °C (rainfall mm)	6.9 (1,193)	7.4 (709)	6.9 (754)
Understorey components	Luzulu-Fagetum	Hordelymo-Fagetum	Hordelymo-Fagetum
Soil pH (H_2O)	3.5–4.5	5.5–7.9	5.3–5.8
N (g kg^{-1})	0.8–3.2	1.4–6.1	1.4–4.1
P (g kg^{-1})	0.30–0.51	0.35–0.70	1.1–1.2

more than 130 years but differed in their stand density and basal area due to different management history. Amounts of annual rainfall and minimum and maximum temperatures varied among the sites with highest rainfall and lowest temperature values on the Solling site. Forest sites vary in their soil properties due to differences in parent materials (Solling – loess overlying Triassic red sandstone, Göttinger Wald – Triassic limestone, Zierenberg – basalt). They show differences in chemical characteristics and proton buffering, and in biological processes and climatic conditions (altitude, annual precipitation) (see Chaps. 1–7, this volume, for more information on the three beech sites).

Litterfall was collected periodically by using litter traps, the number of which varied on different sites, but they were commonly 12–16 on each site, which were usually combined into four replicates for chemical analysis. The period of collection was mostly after long intervals of one or more months during winter and the initial periods of foliage production. During the active leaf litter period, it was collected at fortnightly or shorter periods. Litter was dried, sorted into various components, weighed, ground and analysed for different elements. Leaves and other components were always analysed. The other components included floral or leaf buds, twigs, flower buds, seeds, seed capsules and small bark. The seeds were analysed separately, while all other components were merged together and analysed as residue. Any non-plant component was discarded.

The period of measurements of litterfall on the various sites were periodically prior to 1991 starting in the 1960s on the Solling site, in the 1980s on the Göttinger Wald site and in 1990 on the Zierenberg site. However, since 1991, measurements were made regularly on all three sites, so that this period provided the possibility of making a comparison among the three sites. Most years, the amount of leaf litter on all the three sites showed a narrow range with values lying between 2.5 and 3.5 Mg ha^{-1}.

Fully expanded leaves were collected periodically once a year from the upper third of the tree canopy of the beech stands. They were dried, ground and analysed for element contents. Since 1995, concentrations of elements have been followed on all three sites allowing a comparison among the sites. As the weight of leaves was not known, it was not possible to calculate the amount of nutrients in the green foliage. However, an attempt was made by using Ca as an indicator of constancy in the values to calculate the weights of leaves prior to their shedding.

For assessing the total litterfall on the forest site together with the above ground litter components, the belowground component of fine root litter should be considered. Estimated values of fine root litter production is provided by Murach et al. (Chap. 9, this volume) for the Göttinger Wald and Solling sites. A value of 2,600 kg ha^{-1} per year for the Solling site was assessed where data for 50 cm soil depth were available, and 1,200 kg ha^{-1} per year for the Göttinger Wald site for 20 cm soil depth. In terms of nutrients, fine root litter contained (kg ha^{-1} per year) for the two sites Göttinger Wald and Solling, respectively: N (26 and 39), P (0.9, 1.9), K (2.4, 9.0), Ca (3.2, 2.1) and Mg (0.7, 1.4). These are significant and comparable values with those of aboveground litterfall which should be considered appropriately when nutrient budgets are developed. The rest of the chapter will only consider the aboveground components of the litterfall.

11.3 Components and Annual Patterns of Aboveground Litterfall

Total aboveground litterfall under beech stands has a number of components which include leaf, as the main component, and other components of twigs, discarded buds, small pieces of woody components and pieces of bark. Periodically, a significant proportion of litterfall may contain inflorescence and fruit components during mast years.

The three sites varied in the annual total aboveground litterfall. The highest values were observed on the Zierenberg site (Table 11.2). As a mean for the measurement period, the amount of litter fall followed the decreasing order of sites – Zierenberg site (5,763 kg ha^{-1} per year) < Göttinger Wald (5,032 kg ha^{-1} per year) < Solling (4,894 kg ha^{-1} per year). These mean values of total litterfall

Table 11.2 Annual weights and N contents of total litterfall and leaf litter at the three experimental sites Solling (*SO*), Göttinger Wald (*GW*) and Zierenberg (*ZB*). Mean, minimum and maximum values (kg ha^{-1}) are given for different periods and single years respectively

Site	Mean	Period	SD	Min	Year	Max	Year[a]
		Aboveground litterfall – weights					
SO	4,894	1991–2003	1,287	3,569	1993	7,657	2000
GW	5,032	1990–2003	1,572	3,519	1993	8,517	2000
ZB	5,763	1990–2002	2,065	3,817	1991	9,363	2000
		Leaf litter – weights					
SO	2,772	1991–2003	253	2,423	1992	3,160	2001
GW	2,902	1990–2003	279	2,506	1992	3,450	2000
ZB	3,080	1990–2002	436	2,254	1990	3,882	1994
		N in aboveground litterfall					
SO	67.5	1991–2003	32.5	39.0	1993	128	2000
GW	63.3	1990–2003	31.5	29.5	1993	139	1995
ZB	74.9	1990–2002	33.9	43.6	1993	131	2000
		N in leaf litter					
SO	34.1	1991–2003	4.0	27.8	2001	40.2	1991
GW	32.7	1990–2003	4.6	23.5	1993	40.4	2000
ZB	39.6	1990–2002	5.4	31.1	1990	48.2	1991
		C in the aboveground total litter					
SO	2,581	1991–2003	705	1,871	1993	4,132	2000
GW	2,586	1990–2003	835	1,775	1993	4,450	2000
ZB[b]	2,651	1990–2002	950	1,775	1997	4,207	2003
		C in the leaf litter					
SO	1,456	1991–2003	132	1,262	1992	1,662	2001
GW	1,482	1990–2003	154	1,243	1992	1,783	2000
ZB	1,431	1990–2002	191	1,048	1990	1,773	1994

[a]Years of heavy mast were: 1992, 1995, 1998(?), 2000, 2002

[b]Values for ZB site were calculated by using average values of 46% C content. These values may be underestimated

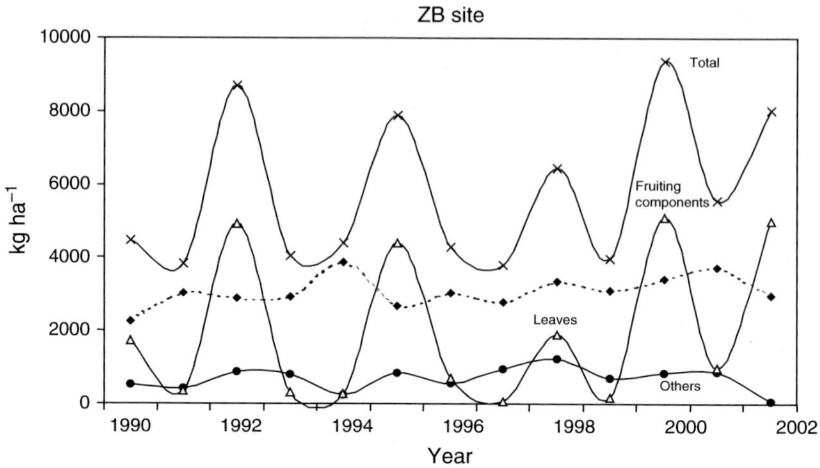

Fig. 11.1 Different components of litterfall measured at Zierenberg site from 1990 to 2002

depended on the frequency and the amount of mast produced on each site. For example, in the heavy mast year of 2000, the total amount of litterfall was 1,700 kg ha^{-1} more on the Zierenberg site than on the Solling site (9,363 kg ha^{-1} on the Zierenberg site and 7,657 kg ha^{-1} on the Solling site). Temporal patterns of various components of total litterfall are shown in Fig. 11.1 as an example for the Zierenberg site. In the common years, the leaf litter component of the litterfall is the major fraction (more than 70%) of aboveground litterfall. Leaf litter values of 60–70% of the total litter have been commonly reported in the literature, which amount to 1.8–4.7 Mg ha^{-1} (Santa-Regina and Tarazona 1999; Pedersen and Bille-Hansen 1999). Meentemeyer et al. (1982) calculated that leaf litter contributed 70% of the total litterfall in forests around the world. Lebret et al. (2001) reported that leaves represented 90% of the total litterfall in the young stand and 70% in the oldest stand, thus age and basal area provided a good prediction of leaf litter in the beech stands. But, as shown in Fig. 11.1, during mast years the fruiting components of litterfall were high, ranging for a year from 1.7 to 5.1 Mg ha^{-1}; the high value was even more than the total leaf litter in some cases. In the mast years, amount of leaf litter constituted less than 40% of the total litterfall. On average over the measuring period of 12 years, leaf litter formed about 53–57% of the total above-ground litterfall.

The mean annual weights of leaf litter measured for more than 10 years ranged from 2,254 to 3,882 kg ha^{-1} with similar overall means for different sites where a difference of about 150–300 kg ha^{-1} among the various sites (Table 11.2) was observed. This small difference in the leaf litter on different sites during the measurement period of 1990–2002 indicates that these stands had a stabilised canopy with very small variation in the amount of foliage. As the mass of leaves would depend on the number and size of leaves, and the size of leaves would be determined by the current year weather conditions at the time of leaf expansion (intensity and pattern of rainfall, sun light conditions and period of sunshine), the similarity in the leaf fall on the three sites indicates that these and other factors determining foliage mass were similar on the three sites. In a recent study of 36 old growth stands of *Fagus sylvatica* across a broad gradient of soil fertility covering nine mesozoic and kaenozoic parent material types (three limestones, two sandstones, two clay stones, one sand and one loess substrate), Meier et al. (2005) reported similar annual leaf litter masses irrespective of soil fertility or acidity of the sites.

When the minimum and maximum values of annual leaf litter are considered, the inter-annual variation, given as the difference in percentage of the mean, on each of the three sites is quite high, 26% on Solling, 33% on Göttinger Wald and 53% on Zierenberg. This is, however, not strictly related to the mast years (Table 11.2). As is evident from Fig. 11.1, the amount of mast did not affect the leaf litter in any significant way though a small trend was observed in the year following the mast years when low amounts of leaf litter were recorded. However, in some of the mast years or the year following the mast year, the highest value of annual leaf litter was recorded. It is expected that any diversion of resources for the production of mast will have only a limited effect on the weight of leaf litter. There are other retranslocation processes which may affect the weight occurring at or prior to the shedding of leaf, suggesting that the mast production and the leaf weights are not related. Due to lack of additional data it was not possible to further explain and evaluate the small differences in the leaf litter at the three beech stands (Fig. 11.2). Despite the small differences among the three sites in the annual amount of leaf litter, the three sites differed in their tree density and BA values (Table 11.1), and these were also reflected in their annual productivity values. Average annual volume increments (m^3 ha^{-1} per year) on the three sites were reported in Chap. 8 (this volume) as 7.7 (Solling), 14.3 (Göttinger Wald) and 9.3 (Zierenberg). These differences in productivity were not reflected by the leaf litter values indicating that there are other factors which determine the productivities.

No long-term trend was observed in the amount of leaf litter during the period of measurement in this study (Fig. 11.2). However, at Göttinger Wald where longer term measurements extending over the last 20 years were taken, a slight increase in leaf litter over time was recorded, though it was only a weak trend (Schmidt 2006). It is possible that high amount of atmospheric N inputs has prolonged the growing period, causing higher plant productivity including an overall increase in the frequency of mast production.

11.4 Amount of Nutrients in Litterfall

The amount of N transfer through litterfall varied among the three sites but more significantly from year to year, which was related to the periodical occurrence of mast years (Table 11.2). In the mast year of 2000, 128, 139 and 131 kg N ha^{-1} were transferred through litterfall at the Solling, Göttinger Wald and Zierenberg sites,

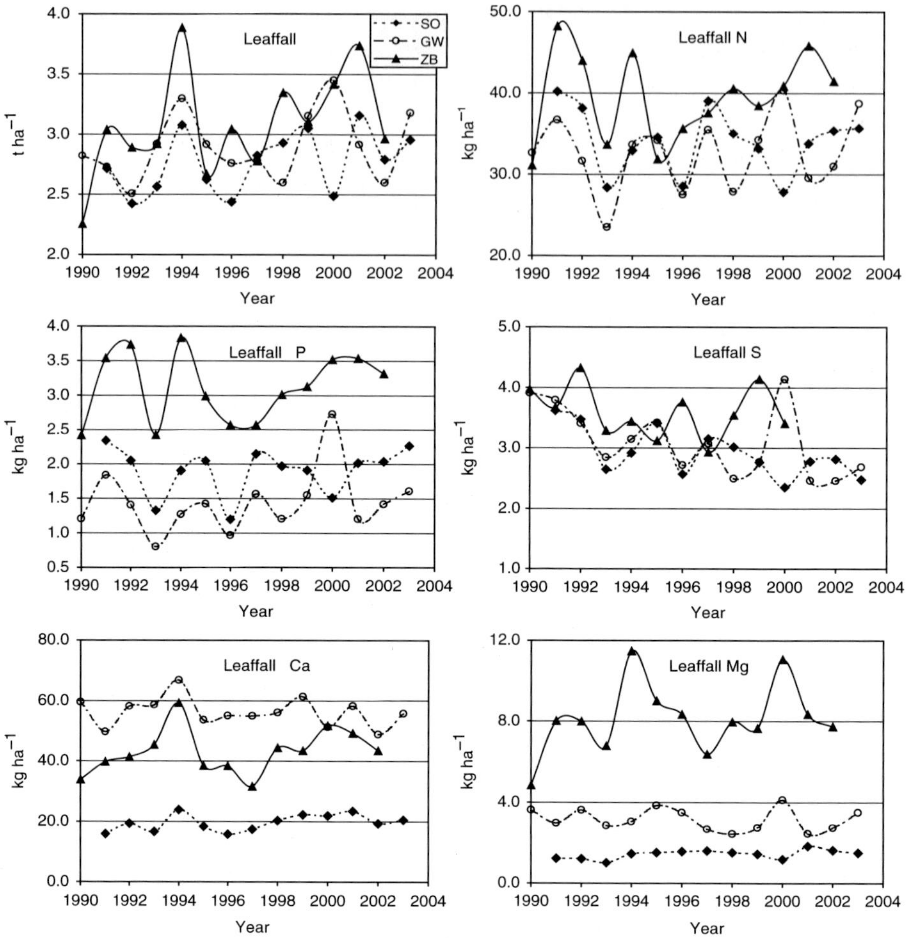

Fig. 11.2 Annual amount of leaf fall and a number of nutrients on the Zierenberg (*ZB*), Göttinger Wald (*GW*) and Solling (*SO*) sites from 1990 to 2003

respectively. As a 10-year average, the amount of N in the litterfall ranged between 63 and 75 kg N ha^{-1}. For the nitrogen budgets, it is customary to take the leaf litter data, usually measured on a short-term basis. Nitrogen content in leaf litter ranged from 24 to 48 kg ha^{-1}annually depending mostly on the site and N content of the leaves shed, but as long-term means for the three sites, the values of leaf litter were 34, 33 and 40 kg ha^{-1}for Solling, Göttinger Wald and Zierenberg sites, respectively (Table 11.2). For the 36 old growth beech forests Meier et al. (2005) reported an annual N transfer through leaf litter of 25–38 kg ha^{-1}, which were not related to any of the soil parameters measured (soil N pool, available P, exchangeable Ca, K and Mg, soil acidity or exchangeable Al). A similar mean value of 45 kg N ha^{-1} (range 31–69 kg ha^{-1}) was reported for 13 IBP sites by Cole and Rapp (1981). When annual leaf litter data are used to define N status, disregarding the other litter

Table 11.3 Mean and SD values of nutrients in the litterfall for the mast years (mast years were based on the information in literature and the relative amount of fruiting components to leaves) 1992, 1995, 2000, 2002 for all sites, plus, for the Zierenberg site, 1998, and for the Göttinger Wald site, 1990, and normal years on the three beech sites. All values are given in kg ha^{-1}

	N	P	S	Ca	K	Mg
Solling						
Mast years ($n = 4$)	111.1	8.1	8.0	33.5	25.9	4.9
SD	21.8	2.1	1.8	4.8	7.4	1.0
Normal years ($n = 9$)	48.1	2.7	4.0	24.6	10.8	2.0
SD	5.7	0.6	0.5	3.6	2.6	0.4
Overall mean	67.5	4.3	5.2	27.4	15.5	2.9
SD	37.0	3.2	2.5	5.7	9.0	1.6
Göttinger Wald						
Mast years ($n = 5$)	97.9	6.0	7.9	98.9	26.6	7.1
SD	28.6	1.9	2.1	17.7	5.1	2.0
Normal years ($n = 9$)	44.0	1.9	3.8	70.3	13.6	3.9
SD	6.4	0.4	0.5	9.7	3.1	0.7
Overall mean	63.3	3.4	5.3	82.5	18.9	4.9
SD	31.5	2.3	2.3	17.8	7.0	2.1
Zierenberg						
Mast years ($n = 5$)	112.7	11.2	8.5	67.3	nd	16.2
SD	21.6	2.7	1.5	6.8	nd	3.4
Normal years ($n = 8$)	51.3	4.1	4.4	50.0	nd	9.0
SD	7.2	0.8	0.6	8.1	nd	1.6
Overall mean	74.9	6.9	5.9	56.6	nd	11.8
SD	33.9	3.9	2.4	11.4	nd	4.3

components, they underestimate the amount of N taken up, transferred through litterfall and added to soils. However, to assess the real amount of N transfer on a site, long-term measurements of litterfall are needed. In order to understand the functioning of ecosystems, it is important that events such as production of mast should be appropriately included in the measurement period. For example, the amount of N taken up for mast production would increase many fold over that of a normal year. This poses the question of physiological changes required in the tree to cope with a very high (Table 11.3) requirement of N and other elements for the production of mast.

Amounts of P and other elements in the litterfall are given for the three sites in Table 11.3 where the years with clear mast production are separated from those of normal years. The long-term mean values of P in the litterfall were 4.3, 3.4 and 6.9 kg ha^{-1} on the Solling, GW and Zierenberg sites, respectively, indicating clearly the high P levels on the site with basaltic soil parent material, and the low values on soil developed from calcareous material. P inputs are similar to N (in pattern as well as in relative amounts in leaf litter and in fruiting components). Leaves contribute 3.1 kg of P (range 2.4–3.7 kg) annually to the litter on all three sites.

The fruiting components have higher P contents than the leaf litter (2.7–3.2 times). P contribution of the twigs and buds, etc. to the total litter is small, less than

1 kg ha^{-1} (mean 0.6 kg ha^{-1}), whereas total P input can vary between 3.3 and 12.9 kg depending upon the amount of mast produced. It is very hard to assign an annual P transfer value from litter to the soil because the final fate and the decomposition constants for fruiting bodies are not precisely known. Fruit shell will decompose much more slowly than the fruits with high nutrient levels.

Differences among the three sites were small for the overall mean amount of S in the litterfall, (5.2–5.9 kg ha^{-1}) and the mean difference between mast years and non-mast years was similar (about 4 kg S) on all sites. Litterfall at Göttinger Wald site is richer in Ca whereas that at Zierenberg showed much higher values of Mg in the total litterfall. However, the amount of Ca in the fruiting components was relatively low, contrasting with that of P, and therefore fruiting components of litter had a relatively small effect on the long-term total transfer of Ca through litterfall. Ca levels in the leaves are much higher than in the fruit components. The leaves therefore determine the Ca inputs to the soil through litterfall. This is probably an indication of the lower mobility of Ca from the cell walls. Zierenberg site has a higher amount of Mg than the Solling and Göttinger Wald sites, Solling being the lowest of the three. Differences in Mg amounts on the three sites are also reflected in the mast components. Data on K was only available for the Solling and Göttinger Wald sites with Göttinger Wald showing slightly higher values than Solling.

The data on the amount of nutrient show the importance of long-term measurements of litterfall to include the mast years in assessing the overall mean of nutrients involved in the nutrient cycling processes.

11.5 Nutrient Concentrations in the Total Litterfall and Leaf Litter Components of the Different Sites

Annual weighted mean values of concentration of nutrients in the litterfall and leaf litter are given in Table 11.4 and the changes in time are shown in Fig. 11.3. Differences between the nutrient concentrations of litterfall and leaf litter point mostly to the composition of fruiting components. The mean values of nitrogen, sulphur, potassium and magnesium did not differ much in the total litterfall and the leaf components at the individual sites, whereas Ca was low but P was high in the fruiting components. Therefore, N/P ratios were relatively lower in the total litter fall than in the leaf litter (Table 11.4).

Of the three sites, leaf litter at the Göttinger Wald site showed continuously low values of N and P but higher values of Ca (Ca-rich soil at Göttinger Wald site) (Fig. 11.3), though there was no evident continuous trend in the concentration of these elements in the leaf litter during the course of these measurements. However, concentrations of S decreased in all cases with time which was the reflection of a decrease in atmospheric S inputs in these systems (Chap. 15, this volume). This decrease in S concentrations increased the N/S ratios on all sites, especially on the Göttinger Wald sites where N/S ratio of 8.3 in 1990 increased to 12.0 in 2001. Usually, a constant N to S ratio of 16 is expected in the plant tissue and values

Table 11.4 Weighted mean and standard deviation (SD) values of concentration (g kg^{-1}) and ratios of different elements in the leaf litter and total litterfall on the three beech sites

	Solling		Göttinger Wald		Zierenberg	
	Mean	SD	Mean	SD	mean	SD
			Leaf litter			
C	525.2	4.07	510.3	6.18	465.0	9.59
N	12.4	1.63	11.3	1.40	12.9	1.44
P	0.79	0.12	0.50	0.13	1.02	0.13
S	1.06	0.19	1.07	0.20	1.14	0.11
Ca	7.10	0.79	19.5	1.98	14.0	1.22
K	3.13	0.68	4.03	0.63	nd	nd
Mg	0.52	0.07	1.09	0.19	2.63	0.38
C/N	43.2	5.38	46.0	6.36	36.4	3.65
N/P	18.2	2.14	23.6	3.67	12.8	1.13
N/S	11.7	1.06	10.7	1.74	11.0	1.15
			Total litterfall			
C	526.5	6.09	512.6	5.59	nd	nd
N	13.2	2.85	12.1	2.31	12.7	1.74
P	0.82	0.31	0.62	0.23	1.12	0.27
S	1.04	0.18	1.03	0.17	1.02	0.09
Ca	5.69	0.80	16.9	2.37	10.4	2.03
K	3.01	0.89	3.71	0.59	nd	0.00
Mg	0.57	0.13	0.95	0.15	2.06	0.29
C/N	41.3	7.82	43.8	7.59	nd	nd
N/P	17.1	3.74	20.7	4.01	11.7	1.51
N/S	12.7	1.32	11.8	1.61	12.1	1.52

lower than that indicate excessive amounts of sulphate present in the senescent leaf. An increase in that ratio indicates that the amount of free sulphate is decreasing slowly over the period as the atmospheric depositions of S are decreasing.

These differences in the element composition of the leaf litter reflected in general the differences in composition of the three soils suggesting the composition of the cell wall components (Ca and P) and the nutrient withdrawal patterns in association with plant sinks are the major factors affecting the element composition on the three beech sites. Khanna and Ulrich (1991) and Meier et al. (2005) have provided evidence of the role of soil composition on the levels of Ca and other elements in the leaf litter.

11.6 Foliar Chemistry

Mean values of N concentration were quite high on all sites (22.3–24.6 mg g^{-1}) with very little differences among them. This is related to high atmospheric N inputs. Most of the N concentrations lie between high and very high diagnostic

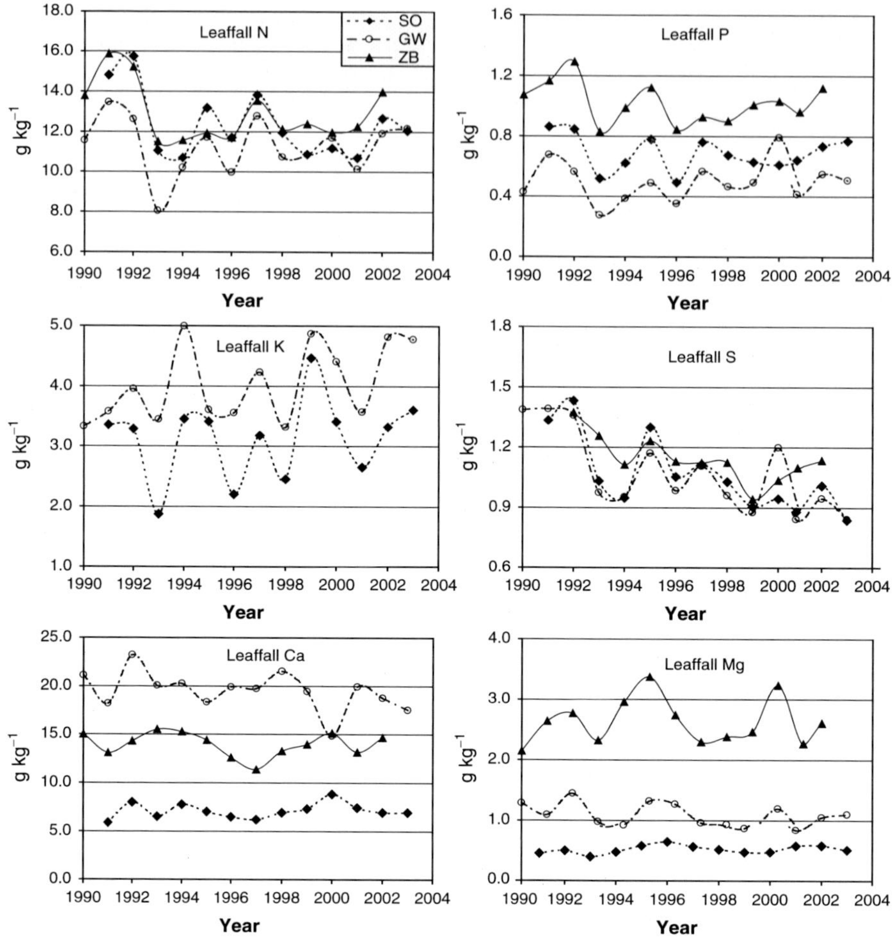

Fig. 11.3 Mean annual concentrations of nutrients in the leaf litter on the Zierenberg (*ZB*), Göttinger Wald (*GW*) and Solling (*SO*) sites during the measurement period

values (Hüttl 1992). However, the concentrations of Ca, Mg, P and Mn in the foliage varied on different sites depending upon the chemical soil characteristics (Table 11.5).

Concentrations of P in the green foliage of the Göttinger Wald and Solling sites were <1.3 mg g^{-1}, ranging from 0.98 to 1.18 mg P g^{-1} in Göttinger Wald and from 1.14 to 1.37 mg P g^{-1} for Solling. Foliar concentrations of P at the Zierenberg site were high, ranging from 1.14 to 1.69 mg g^{-1}. Different values are given in literature to define the critical values of P in beech stands. For example Krauß and Heinsdorf (2005) suggested that the values of 1.14–1.52 mg P g^{-1} are in the optimum range and values less than 1.14 should be considered low. Hüttl (1992) gives much lower values as critical values (<1.3 mg g^{-1} will be considered low). Considering these

Table 11.5 Annual mean and SD values of concentrations of a number of elements in the live foliage of beech from three sites (data for the period 1995–2002)

	Solling		Göttinger Wald		Zierenberg	
	Mean	SD	Mean	SD	Mean	SD
Nutrient and other elements (mg g^{-1})						
C	526	6.20	514	8.04	469	16.0
N	24.6	1.14	24.14	1.76	22.30	1.22
P	1.26	0.07	1.06	0.08	1.33	0.18
S	1.52	0.14	1.58	0.11	1.44	0.13
K	5.01	0.58	7.42	1.06	6.05	0.62
Ca	3.65	0.44	10.80	0.79	8.75	0.71
Mg	0.71	0.10	0.98	0.22	2.84	0.39
Mn	1.88	0.24	0.55	0.10	nd	nd
Fe	0.09	0.04	0.09	0.03	nd	nd
Al	0.06	0.04	0.04	0.04	nd	nd
Heavy metals (mg kg^{-1})						
Zn	27.02	4.23	26.19	4.87	nd	nd
Cu	7.18	0.57	6.84	0.74	nd	nd
Cr	0.31	0.10	0.31	0.07	nd	nd
Co	0.03	0.04	0.04	0.04	nd	nd
Ni	0.60	0.16	0.73	0.24	nd	nd
Cd	0.04	0.03	0.05	0.04	nd	nd
Pb	0.52	0.27	0.76	0.35	nd	nd
Element ratios						
C/N	21.42	0.97	21.40	1.59	21.08	1.32
N/P	19.50	0.86	22.92	1.51	16.97	1.92
N/Ca	6.82	0.83	2.24	0.22	2.56	0.23
P/Ca	0.35	0.03	0.10	0.01	0.15	0.02
Mg/Ca	5.24	0.79	11.52	2.61	3.11	0.27
N/S	16.29	1.29	15.31	0.67	15.53	0.93

criteria, one can safely say that the foliar P indicates adequate to high levels at the Zierenberg site and probably at the borderline to deficient range for the Göttinger Wald site. For the Solling site, the foliar levels cannot be assigned to a definite group of low or adequate levels. As the foliage data represented a short period of measurements (1995–2002), it is very hard to assess any long-term trends in these values. However, it seems certain that at present the productivity values are not related to foliar P values. Beech is considered to have a high level of physiological plasticity, and whether it can also adjust to low content of P is a possibility which has not received sufficient attention (Meier et al. 2005). Due to the short period of measurements it is hard to study the retranslocation patterns of P on each site and if the patterns have undergone any changes.

If the critical values (Hüttl 1992) for K, Ca and Mg in the beech leaves are compared with the measured concentrations (Table 11.5), all the three elements are in the very low range on the Solling site whereas their values for the other two sites lie in the adequate to high levels. Again, the low values of these elements on the Solling site are associated with the highly acid soil.

No change in the concentration of most of the elements was measured in the foliage samples during 1995–2002 on all sites except that the concentration of S on the Zierenberg site, and that of K on the Solling and Zierenberg sites, has decreased continuously during the last 10 years (Fig. 11.4). However, when long-term measurements were considered on the Solling site, decreasing trends in P and K concentrations with time were observed (Fig. 11.5). Prior to 1975, the values of K concentrations in the foliage ranged from 8 to 12 mg g^{-1} which had dropped to

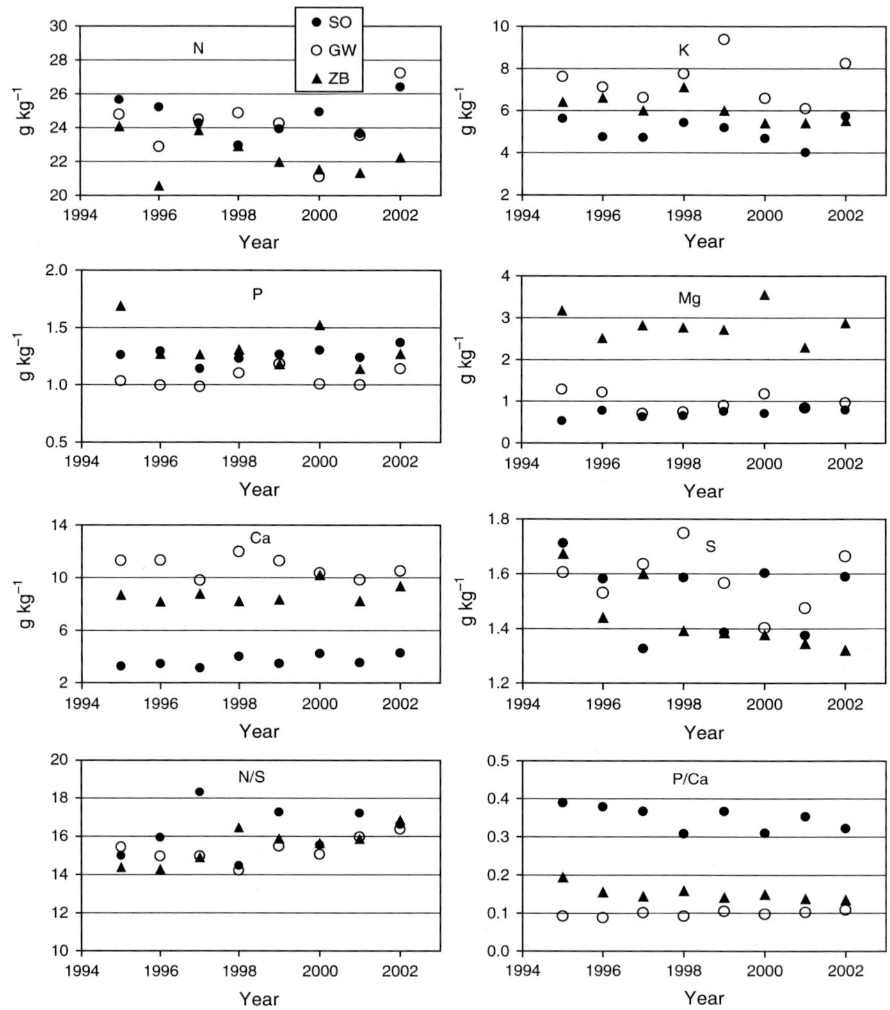

Fig. 11.4 Annual mean values of nutrient concentrations and nutrient ratios in the foliage of the Zierenberg (*ZB*), Göttinger Wald (*GW*) and Solling (*SO*) sites for 1995–2002

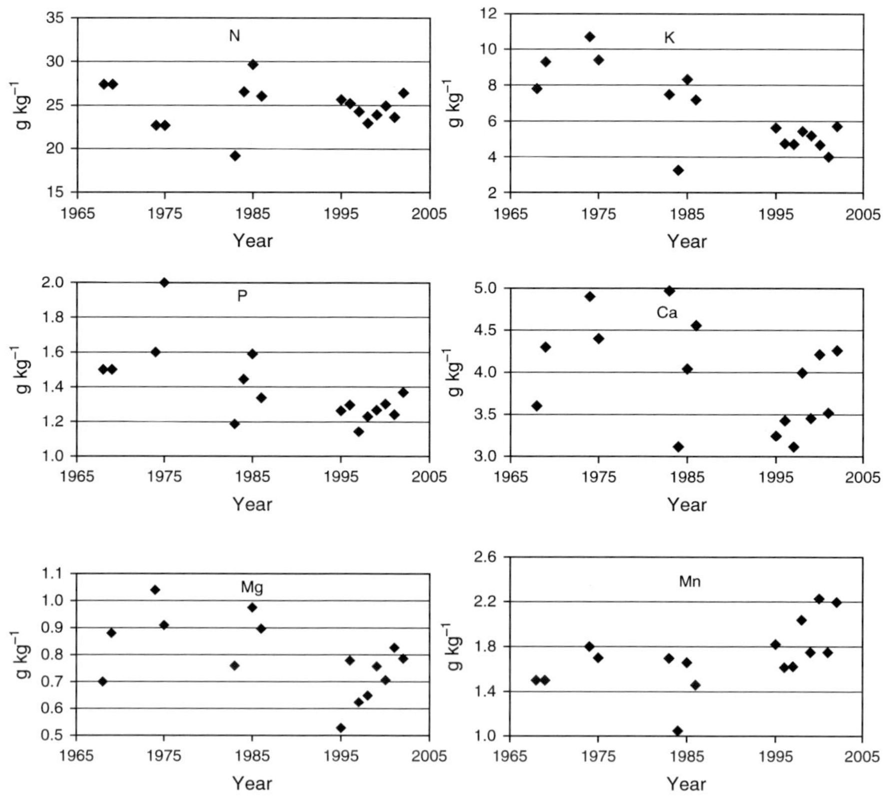

Fig. 11.5 Long-term trends in the foliar concentration of some nutrients at the Solling site

4–6 mg g^{-1} after 1995. For the same periods, the values of P concentrations ranged from 1.5–2 mg g^{-1} (prior to 1995) to 1.0–1.5 mg g^{-1} (after 1995).

In a study on changes in foliar nutrient levels in northeastern France, Duquesnay et al. (2000) sampled 118 beech stands in 1996–1997 and compared the analysed values with those obtained in 1969–1971. Between the two sampling periods, the concentration of foliar N increased (+12%), whereas that of others decreased by 23% (P), 38% (Mg) and 16% (K). Ratios of N/P and N/Mg increased by 42 and 77%. This decrease in concentrations of P and cations was particularly marked for trees growing on acidic soils. They assigned this decrease in P concentration in the foliage of beech trees to increasing atmospheric CO_2 concentrations and acidification of forest soils. In a parallel study where the foliage concentrations of beech stands were monitored continuously for 5 years, a decrease in P levels was observed, which resulted in an increase in N/P ratios. Similarly, Flückiger and Braun (1998, 1999) reported an increase in nitrogen concentration and a continuous decrease in phosphorus concentration measured every 3–4 years in the beech foliage. The values of P decreased from 1.25 mg P g^{-1} in 1984 to 1.08 mg P g^{-1} in 1995 in 51 beech stands in northwestern Switzerland, with an accompanying increase in the N to P ratios from 18.6 to 25.0.

In a study of 49 stands European beech (60-or more years-old) in the Werdenfelser Land region, Ewald (2000) analysed foliage and soil samples from eight sites for nutrient composition. High crown transparency and slow tree growth were related to low availability of P, Mn and N in soils. Sites could be divided into two groups. Moderate P levels were found in trees from well-developed mineral soils, and extremely low P levels (0.6–1.0 mg g^{-1}) in immature carbonate soils where total P levels in soils were low and Ca saturation very high. Very high N/P ratios of 30 or more were associated with low P level. P was a critical element in the nutrition of beech in the calcareous Alps, which may be partly responsible for the poor vitality of this species in mixed mountain forests. In their study of 50 European beech stands, Leuschner et al. (2006) observed that N/P values of the surface organic layers were higher on acid soils under beech stands than on the calcareous soils, indicating the possibility of P limitation on acid soils, probably in association with high atmospheric N inputs. Composition of surface organic layer for N and P is expected to relate strongly with their litterfall values.

Of 17 beech forests, which are part of the German contribution to the Forest Focus initiative of the EU (Level II sites), those with inadequate levels of nutrients were seven sites for Ca, four for Mg, three for P and two for K. None of the sites showed deficiency of N in the foliage (Chap. 23, Table 23.3). On a wide ranging dataset from beech stands (n = 75) in Lower Saxony, Brandenburg and Bavaria, Wolff and Riek (1997) reported that about 54% of the sites would show P deficiency using the commonly used critical values for beech foliage.

On the three sites under investigation N/P ratios of the live foliage was always >20 on the Göttinger Wald site and <18 on the Zierenberg site though the ratios did not change in time during 1991–2002 reflecting the P status of the sites. However, a decrease in P/Ca ratios were observed during this period on the Solling and Zierenberg sites (Fig. 11.4). As the value of Ca concentration was not expected to vary much on each site, a decrease in P/Ca ratio indicated a decrease in P concentration on the Solling and Zierenberg sites. Similarly, K/Ca ratio in the foliage decreased indicating a decrease in K concentrations on the Solling site (data not shown). However, as discussed above, only longer term measurements indicate a decrease in P and K foliar levels on the Solling site. Usually, a constant N/S ratio of 16 is expected in the plant tissue when no protein-free S and N are stored. Foliage samples prior to 1998 showed values of <16 on all the three sites, but after 1998 the ratio has slowly increased towards 16 (Fig. 11.4). A value of <16 would indicate a high amount of free sulphate in the tissue (excessive amount of sulphate) which would have occurred due to high amounts of atmospheric inputs of S.

11.7 Nutrients and Heavy Metals in Green Foliage and Litterfall in Relation to Atmospheric Inputs

The atmospheric inputs of nutrients and heavy metals have changed during the last 30 years (low acidity, low S, low Ca and other cations, low content of heavy metals) (Chap. 15). For example, at the Solling site, atmospheric inputs of S were around

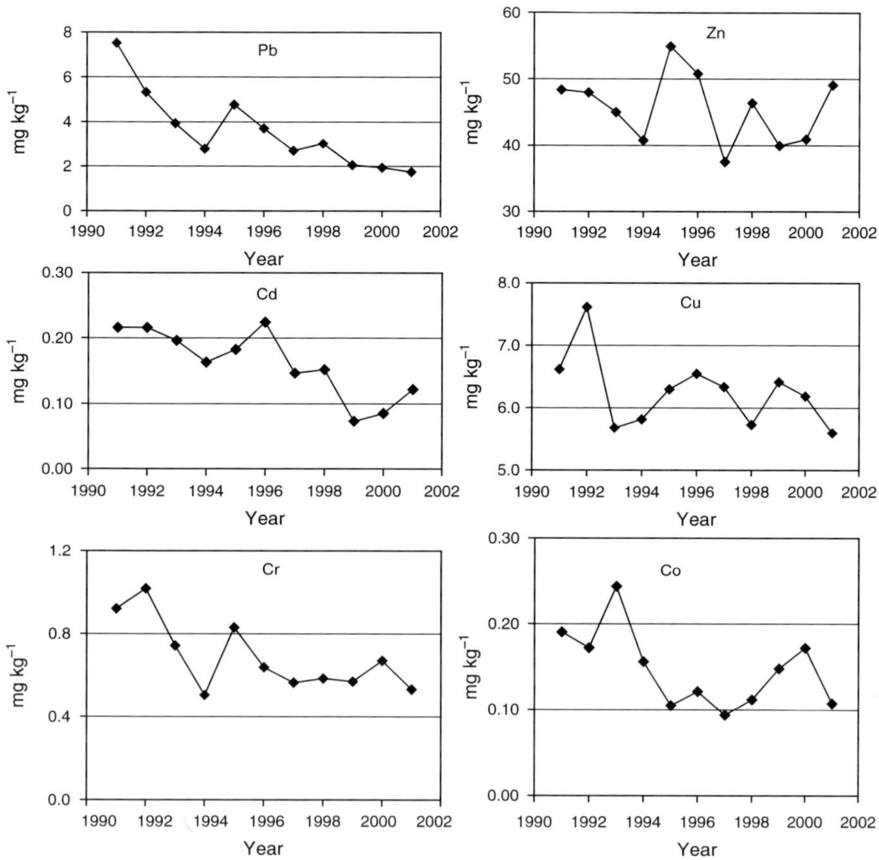

Fig. 11.6 Mean values of some heavy metals in the leaf litter at the Solling site from 1990 to 2001

60 kg ha^{-1} in 1968 which have decreased to 10 kg ha^{-1} in 2002. However, atmospheric inputs of N, which were about 20 kg ha^{-1} in the early 1970s had increased to 28 kg ha^{-1} in the 1980s, but have since decreased to 20 kg ha^{-1} and have remained constant for the last 13 years (Chap. 15). During the last 20 years, atmospheric inputs of heavy metals have also decreased (low Pb, Cd and other heavy metals). This decrease in atmospheric depositions is reflected in the concentration of S (Fig. 11.4), and a number of heavy metals in the leaf litter (Fig. 11.6). For example, the concentration of Pb in the leaf litter decreased from 7.5 mg kg^{-1} in 1991 to 1.7 mg kg^{-1} in 2001.

There is very little known of the consequences of changes in heavy metal concentrations of the litter on the decomposition and mineralisation processes. Some of the factors related to litter decomposition processes would be the content and solubilities of heavy metals, the interactions of heavy metals with other elements, the formation of stable complexes with organic matter, and the toxicity effects on soil fauna and microbial populations.

Dynamics of heavy metals in decomposing litter are particularly poorly known, with only a few studies on decomposition processes involving heavy metals as pollutants (Coughtrey et al. 1979; Killham and Wainwright 1981; Bengtsson et al. 1988; Grodzinski et al. 1990; Laskowski et al. 1994). During litter decomposition, the concentrations of heavy metals increase (Ruhling and Tyler 1973; Berg et al. 1991). Staaf (1980) and McBrayer and Cromack (1980) found significant accumulations of Fe, Zn, Pb, Cu and Cd in the litter of beech, oak, Scots pine and oak–hornbeam forests.

11.8 Autumn Withdrawal of Nutrients at the Time of Leaf Senescence

The amount of nutrients measured in the leaf litter is the net amount present after their withdrawal into the old tissue and any subsequent losses that may have occurred due to leaching during the period when the leaf is shed and when it is collected and measured. The leaching component could be quite significant for a number of elements such as K, Mg, P and N. Chapin and Moilanen (1991) reported that leaching by autumn rains accounted for 25% of the N disappearance and 55% of the P disappearance from senescing leaves of birch in a high-fertility field. From our presently available data, it is not possible to assess the amount or the fraction of nutrients which will be leached. This creates some uncertainty in the interpretation of the results by describing the whole amount as the withdrawal fraction, especially for K which is most mobile element and is easily lost.

Retranslocation of nutrients from senescing tissues, often referred to as 'resorption' (Killingbeck 1986), is an important plant strategy to economically use the accumulated nutrients. For example, Aerts (1996) suggested that 40–65% of nitrogen and phosphorus from leaves may be resorbed by plants before leaf abscission. Resorption of nutrients thus acts as an important feed-back mechanism in the nutrient cycling and litter decomposition processes in forests (Staaf 1982; Killingbeck 1996; Cherbuy et al. 2001; Cote et al. 2002). A higher rate of nutrient resorption would cause a low amount of nutrient uptake for the next growing season, but at the same time would produce litter of low quality affecting its decomposition.

Another unknown factor in the calculation of the nutrient withdrawal amount is the concentration of nutrients at the time of leaf senescence. Usually, the leaf samples are collected from a specific part of the canopy to assess the nutritional levels through critical amounts or balance of elements in the fully grown leaves. Some nutrients are constantly transferred in and out of leaves, probably affecting the value for the nutrients used as the amount in the live foliage at the time of leaf senescence (Nambiar and Fife 1991).

Another error involved is the assessment of foliage biomass from the litter biomass measured on the surface organic layer, even of deciduous foliage. It is well known that some components of the leaf are retranslocated to the living tissue for future use. This involves nutrients and also organic components. There are additional losses of organic components through decomposition and respiration of C during leaf shedding and prior to leaf shedding through consumption by insects. This has been estimated to be about 15% as a mean value for beech forests, though this value can be quite high in some cases.

By assuming no losses of organic components from the leaf litter following leaf senescence, i.e. foliage mass equals the measured leaf litter weights, one can calculate the minimum amount of nutrients which may have been withdrawn at leaf senescence. The values of nutrient withdrawal will thus be approximate as it is hard to define the amount present in the foliage sample prior to its becoming leaf litter. The time of sampling is crucial in such studies which have not been actually taken into consideration in most studies.

The results obtained from this study are presented in Table 11.6 where mean values of nutrient withdrawal for the three sites during the period of study are given. It also gives the fraction of nutrients removed or lost from the initial amount in the foliage. The results are approximate values due to deficiency of related data on different processes during leaf senescence as described above:

(a) All nutrients, except Ca and in some individual years Mg, were withdrawn in significant amounts.
(b) The amount of N withdrawn was high (31–37 kg ha^{-1}) despite high atmospheric N inputs in these systems. As a mean, the fraction of N withdrawn was 51% on the Solling, 53% on the Göttinger Wald and 44% on the Zierenberg sites.
(c) The amount of P withdrawn was similar on the Solling and Göttinger Wald sites (1.5 kg ha^{-1}) but was low (1.1 kg ha^{-1}) on the Zierenberg site. As the P

Table 11.6 Mean and SD values of amount of nutrients withdrawn at leaf senescence (kg ha^{-1}) and the amount withdrawn as a percentage of the amount in the live foliage. Live foliage mass was assumed to be equal to the measured leaf litter

Site		N	P	S	Ca	K	Mg
		kg ha^{-1}					
Solling	Mean	34.6	1.6	1.3	−10.0	4.9	0.4
	SD	4.19	0.35	0.38	1.87	2.16	0.35
Göttinger Wald	Mean	36.6	1.5	1.6	−23.8	9.7	−0.2
	SD	3.47	0.46	0.50	4.53	2.92	0.29
Zierenberg	Mean	30.6	1.1	1.0	−15.1	nd	0.5
	SD	3.86	0.47	0.28	3.56	nd	0.81
		Percentage of amount in live foliage					
Solling	Mean	51.1	47.4	31.5	−102	36.4	19.7
	SD	4.72	9.52	8.57	15.51	14.48	16.06
Göttinger Wald	Mean	53.2	51.3	34.7	−77.3	45.3	−9.8
	SD	5.26	15.06	11.28	21.66	9.06	15.09
Zierenberg	Mean	43.9	25.1	23.0	−55.3	nd	5.9
	SD	3.82	9.17	6.38	12.14	nd	9.54

concentration in the foliage of the three sites differed significantly, the proportion of P withdrawn was low on the Zierenberg site (25%) but high (47–51%) on the other two sites. The high values lay in the same magnitude as the amount present in the leaf litter, indicating that the tree has the capacity to develop half of its foliage with similar element concentrations without depending upon any uptake during the next growing season.

(d) Values for Ca withdrawal were negative, pointing to the error involved in assessing the mass of the foliage or the concentration of Ca in the leaves prior to senescence.
(e) The Göttinger Wald and Zierenberg sites have Mg-rich soils showing very little withdrawal of Mg or the withdrawal is of the same proportions as the organic substances removed or lost during the leaf litter. The fraction of Mg withdrawn at the Solling site was high (20%).
(f) Nutrient withdrawal in autumn during leaf senescence was related to the nutrient status of the three sites either in relation to nutrient inputs (N and S) or chemical soil properties influencing the concentration of nutrients in the live foliage.

An attempt was made to determine the weight of the live foliage by assuming the conservation of Ca in the leaf litter. This method was tested by using the total amount of Ca in the leaf litter collected a number of times during autumn. For leaf analysis, only one sample was taken. This produced two to three times higher values of foliage mass than measured at litterfall suggesting that these values were far too high to be plausible. It seems that Ca content in the leaves goes on increasing until they are shed, and conservation of Ca can only be valid and useful if the live leaves prior to leaf senescent are taken. In the present scheme of sampling, live leaves were collected a few months in advance of when the leaves were shed.

There are two common hypotheses pertaining to the fraction and amount of nutrients re-absorbed from leaves.

(a) The first one considers that the amount of nutrient resorption increases as an inverse function of soil nutrient availability (Grime 1979; Chapin 1980; Berendse and Aerts 1987). Nutrient resorption from senescing leaves is therefore expected to be more highly developed on sites of low fertility. For example, Cote et al. (2002) reported 56–71% N re-absorption efficiency for eight hardwood species on two sites of contrasting fertility. However, Staaf (1982) did not find any relationship between translocation efficiency of N and P and soil factors for beech trees on 24 sites. The retranslocation for N and P was always more than 55% of the amount present in the leaf prior to its senescence. For elements N, P, K, and S, the amount withdrawn at leaf senescence was related to the concentration of respective element in the leaf prior to its senescence. This was not the case with Ca and Mg where site factors such as soil pH and site index were also important. From a literature review, Aerts (1996) concluded that trees and shrubs showed poor intra-specific responses to soil nutrient availability with 63% of experiments reporting no response by 60 plant species.

(b) The second hypothesis considers that resorption is influenced by sink strength within the plant (Nambiar and Fife 1991) so that resorption efficiency is positively correlated with plant growth rates, because the production of new tissues constitutes a strong nutrient sink. Thus, one can consider the significance of nutrient withdrawal in terms of growth phases as suggested by Oron and Schulze (1989) which act as strong sinks for an element and can thus lead to nutritional disharmony when the demand of an element for the new growth is higher than its supply either through resorption or through uptake. However, Chapin and Moilanen (1991) noted that sink strength was unlikely to explain the strong resorption shown by many deciduous trees in autumn and concluded that nutrient resorption efficiency was influenced more strongly by carbohydrate flux from leaves (i.e. source-sink interactions) than by factors governing hydrolysis of nutrient-containing fractions in leaves. In our study, relationships between nutrient concentrations and nutrient withdrawal for some elements depended on nutrient status of the sites.

11.9 Conclusions

1. Leaf litter constituted about 70% of the total litterfall in normal years and about 40% in mast years with absolute values showing only small differences among the three sites and during 12 years of measurements. This was despite the major differences observed in the productivity on the three beech sites.
2. There was no clear evidence of any change in the amount of nutrients in litterfall in time over the measuring period. However, the amount of nutrients differed among the sites depending upon the soil bedrock material. Litterfall was high in P and Mg at the basaltic sites and in Ca in the calcareous Göttinger Wald soil.
3. Frequency and the amount of fructification were the most important factors affecting the amount of nutrients transferred in litterfall to the surface organic layer, due to very high amount of nutrient (especially P) contained in the fruit components (except Ca) and doubling of the frequency of masting. However, due to the periodicity and irregular nature of mast production, long-term measurements of litterfall are required to make an useful assessment of nutrient transfer that would include mast production.
4. Foliar concentration indicated that, different from the Zierenberg site, the Solling and Göttinger Wald sites were low to very low in P, and Solling also in K, Ca, Mg.
5. Following the Clean Air policies, atmospheric inputs of S and heavy metals have decreased during the course of the measurements, which reduced the amount of these elements in the litterfall, especially those of some heavy metals. For example, concentration of Pb in the leaf litter decreased from 7.5 mg kg^{-1} in 1991 to 1.7 mg kg^{-1} in 2001.

6. Nutrient withdrawal during leaf senescence in autumn was related to the nutrient status of the three sites either in relation to nutrient inputs (N and S) or chemical soil properties influencing the concentration of nutrients in the live foliage. A significant amount of N (44–53%) was withdrawn on the three sites whereas that of P varied according to the P status of the sites, low withdrawal on the Zierenberg site (25%), but high (47–51%) on the Solling and Göttinger Wald sites. Similarly, Mg was withdrawn mostly on the Solling site. The amount of nutrient withdrawn will provide the nutrients for the growth of at least half the amount of canopy without any further uptake.

References

Aerts R (1996) Nutrient resorption from senescing leaves of perennials: are there general patterns? J Ecol 80:131–140

Bengtsson G, Berden M, Rundgren S (1988) Influence of soil animals and metals on decomposition processes: a microcosm experiment. J Environ Qual 17:113–119

Berendse F, Aerts R (1987) Nitrogen-use efficiency: a biologically-meaningful definition? Funct Ecol 1:293–296

Berg B, Ekbohm G, Soderstrom B, Staaf H (1991) Reduction of decomposition rates of Scots pine needle litter due to heavy-metal pollution. Water Air Soil Pollut 59:165–177

Chapin FS (1980) The mineral nutrition of wild plants. Annu Rev Ecol Syst 11:233–260

Chapin FS, Moilanen L (1991) Nutritional controls over nitrogen and phosphorus resorption from Alaskan birch leaves. Ecology 72:709–715

Cherbuy B, Joffe R, Gillon D, Rambal S (2001) Internal remobilisation of carbohydrates, lipids, nitrogen and phosphorus in Mediterranean evergreen oak *Quercus ilex*. Tree Physiol 21:9–17

Cole DW, Rapp M (1981) Element cycling in forest ecosystems. In: Reichle DE (ed) Dynamic properties of forest ecosystems. Cambridge University Press, Cambridge, pp 314–409

Cote B, Fyles JW, Djalivand H (2002) Increasing N and P resorption efficiency and proficiency in northern deciduous hardwoods with decreasing foliar N and P concentrations. For Sci 59: 275–281

Coughtrey PJ, Jones CH, Martin MH, Shales SW (1979) Litter accumulation in woodlands contaminated by Pb, Zn, Cd and Cu. Oecologia 39:51–60

Duquesnay A, Dupouey JL, Clement A, Ulrich E, Tacon Le F (2000) Spatial and temporal variability of foliar mineral concentration in beech (*Fagus sylvatica*) stands in northeastern France. Tree Physiol 20:13–22

Ewald J (2000) Ist Phosphormangel für die geringe Vitalität von Buchen (*Fagus sylvatica* L.) in den Bayerischen Alpen verantwortlich? Forstwissenschaftliches Centralblatt 119:276–296

Flückiger W, Braun S (1998) Nitrogen deposition in Swiss forests and its possible relevance for leaf nutrient status, parasite attacks and soil acidification. Environ Pollut 102:69–76

Flückiger W, Braun S (1999) Nitrogen and its effect on growth, nutrient status and parasite attacks in beech and Norway spruce. Water Air Soil Pollut 116:99–110

Grime JP (1979) Plant strategies and vegetation processes. Wiley, Chichester

Grodzinski W, Greszta J, Laskowski R, Maryanski M, Rozen A (1990) Effect of the chemical composition of industrial dusts on forest floor organic matter accumulation. Water Air Soil Pollut 53:169–178

Gruber F (2006) Über die Vitalität der Rotbuche (*Fagus sylvatica* L.). Schriftreihe Dendrologie, Baumpflege und Waldbotanik. Shaker Verlag, Aachen, Germany

Hüttl R (1992) Die Nährelementversorgung geschädigter Wälder in Europa und Nordamerika. Freiburger Bodenkundl. Abhandlungen 28. Habil. Arbeit

Khanna PK, Ulrich B (1991) Ecochemistry of temperate deciduous forests. In: Röhrig E, Ulrich B (eds) Ecosystems of the World 7 – temperate deciduous forests. Elsevier, Amsterdam, pp 121–163

Killham K, Wainwright M (1981) Deciduous leaf litter and cellulose decomposition in soil exposed to heavy atmospheric pollution. Environ Pollut 26:70–85

Killingbeck KT (1986) The terminological jungle revisited: making a case for use of the term resorption. Oikos 46:263–264

Killingbeck K (1996) Nutrients in senescent leaves: keys to the search for potential resorption and resorption proficiency. Ecology 77:1716–1727

Kittredge J (1948) Forest Influences. McGraw Hill Book Company, New York

Krauß HH, Heinsdorf D (2005) Ernährungsstufen für wichtige Wirtschaftsbaumarten. Beitr Forstwirtsch Landschökol 39:172–179

Laskowski R, Maryanski M, Niklinska M (1994) Effect of heavy metals and mineral nutrients on forest litter respiration rate. Environ Pollut 84:97–102

Lebret M, Nys C, Forgeard F (2001) Litter production in an Atlantic beech (*Fagus sylvatica* L.): time sequence. Ann For Sci 58:755–768

Leuschner C, Meier IC, Hertel D (2006) On the niche breadth of *Fagus sylvatica*: soil nutrient status in 50 Central European beech stands on a broad range of bedrock types. Ann For Sci 63:355–368

McBrayer JF, Cromack K (1980) Effect of snow-pack on oak-litter release in a Minnesota forest. Pedobiologia 20:47–54

Meentemeyer V, Box EO, Thompson R (1982) World patterns and amounts of terrestrial plant litter production. BioScience 32:125–128

Meier IC, Leuschner C, Hertel D (2005) Nutrient return with leaf litter fall in *Fagus sylvatica* forests across a soil fertility gradient. Plant Ecol 177:99–112

Nambiar EKS, Fife DN (1991) Nutrient retranslocation in temperate conifers. Tree Physiol 9:185–207

Oren R, Schulze E-D (1989) Nutritional disharmony and forest decline: a conceptual model. In: Schulze E-D, Lange OL, Oren R (eds) Forest decline and air pollution, Ecological Studies 77. Springer, Berlin, pp 425–443

Paar U, Kirchhoff A, Westphal J, Eichhorn J (2000) Fruktifikation der Buche in Hessen. AFZ-Der Wald 55:1362–1363

Pedersen LB, Bille-Hansen J (1999) A comparison of litterfall and element fluxes in even aged Norway spruce, sitka spruce and beech stands in Denmark. For Ecol Manage 114:55–70

Piovesan G, Adams JM (2001) Masting behaviour in beech – linking reproduction and climatic variation. Can J Bot/Rev Can Bot 79:1039–1047

Ruhling A, Tyler G (1973) Heavy metal pollution and decomposition of spruce needle litter. Oikos 24:402–416

Santa-Regina I, Tarazona T (1999) Organic matter dynamics in beech and pine stands of mountainous Mediterranean climate area. Ann For Sci 56:667–677

Schmidt W (2006) Zeitliche Veränderung der Fruktifikation bei der Rotbuche (*Fagus sylvatica* L.) in einem Kalkbuchenwald (1981–2004). Allg Forst Jagdzeitung 177:9–19

Staaf H (1980) Release of plant nutrients from decomposing leaf litter in a South Swedish beech forest. Holarctic Ecol 3:129–136

Staaf H (1982) Plant nutrient changes in beech leaves during senescence as influenced by site characteristics. Oecol Plant 3:161–170

Stachurski A, Zimka JK (1975) Methods of studying forest ecosystems: leaf area, leaf production and withdrawal from leaves of trees. Ekol Polska 23:637–648

Viro PJ (1955) Investigations on forest litter. Communicationes Instituti Forestalis Fenniae 45 (6):1–65

Wolff B, Riek W (1997) Deutscher Waldbodenbericht 1996 Bd 1. BMELF, Bonn, Germany

Chapter 12
The Role of Soil Fauna for Decomposition of Plant Residues

M. Schaefer, S. Migge-Kleian, and S. Scheu

12.1 Introduction

The organic remains of plants and animals on the soil surface or in the soil are termed litter. This includes leaves, twigs, fruits, fructescences, bracts, scales of buds, bark flakes and even whole trees, dead roots, carrion and other resource types. Dead materials still attached to the living plants are distinguished from litter as 'standing dead'. Comprehensive reviews of plant litter decomposition are given by Swift et al. (1979), Cadisch and Giller (1997), Lavelle and Spain (2001) and Laskowski et al. (2006).

A special – mainly fungal – flora associated with living or senescent plant tissue initiates decomposition before plant litter originates and enters the soil subsystem. On the soil, leaching of low molecular weight compounds takes place; this material is fairly rapidly mineralised by bacteria and "sugar fungi". The remaining organic matter is degraded by further saprotrophic organisms – fungi, bacteria and saprophagous animals. Fungi, in particular Basidiomycetes, are generally believed to be the most important decomposers of structural plant compounds, e.g. cellulose and especially lignin. The decomposition subsystem performs two major functions: the mineralisation of essential plant nutrients and the formation of soil organic matter (Swift et al. 1979; Lavelle and Spain 2001).

Decomposition may be viewed as a cascade process whereby primary dead organic material experiences a succession of physical and chemical transformations in the soil leading to mineralisation of part of the resource and storage of resistant compounds as 'humus' (Wolters 1991a; Lavelle et al. 1997; Wardle and Lavelle 1997). Mechanisms within the process of decomposition are leaching of litter material, catabolism by the soil biota (evolution of carbon dioxide and the excretion of nitrogenous compounds), comminution by animals (transformation of primary resources to 'faecal material' which may be transferred to the soil via 'bioturbation') and humification (Swift et al. 1979).

An important analytical tool is the use of decomposition constant, k, describing the decay rates of individual litter substrates. Often, a negative exponential model is

R. Brumme and P.K. Khanna (eds.), *Functioning and Management of European Beech Ecosystems*, Ecological Studies 208,
DOI: 10.1007/978-3-642-00340-0_12,

a useful descriptor to define the rate of decomposition (Heal et al. 1997). However, the loss rate declines with time, because more readily decomposed substrates are successively lost, leaving the more resistant fractions, and possibly because resistant polymeric compounds are formed.

Decomposition of litter is controlled by a number of variables, including the intrinsic factors described by the physical and chemical properties ('resource quality') of the litter, and the extrinsic factors such as climate, soil chemistry, macro-, meso- and microfaunal responses, and activity of micro-organisms (Swift et al. 1979; Heal et al. 1997; Lavelle and Spain 2001) (Fig. 12.1). Though the decay process is very complex, the two most important controllers of litter decomposition rates are prevailing climatic environment and substrate quality, i.e. the susceptibility of the substrate to attack by specialised decomposers.

One of the aims of our analysis is the comparison between the decomposition subsystems in the three beech forests near Göttingen with a gradient from mull to moder soil (Göttinger Wald, Kleiner Gudenberg (Zierenberg) and Solling; see Chap. 7). Faunal influences will differ in the three forests because of their different soil animal communities (Fig. 12.7, Tables 7.1 and 7.2 in Chap. 7). Litterfall data for the three experimental sites are given by Khanna et al. (Chap. 11, this volume).

We intend to take an innovative approach by basing tentative generalisations upon a comparison of forests within one region and the synthesis of different descriptive and experimental analyses within one forest. The partly long-term studies can be sorted into three approaches (Fig. 12.2): (1) a resource-based approach studying the process of decomposition of plant material with its inherent intrinsic and extrinsic controlling factors, (2) a food web-based approach focusing

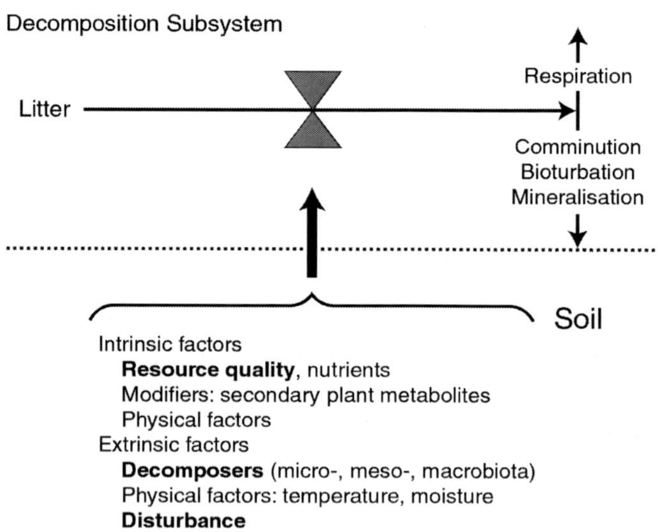

Fig. 12.1 Decomposition as a process – influence of intrinsic and extrinsic factors

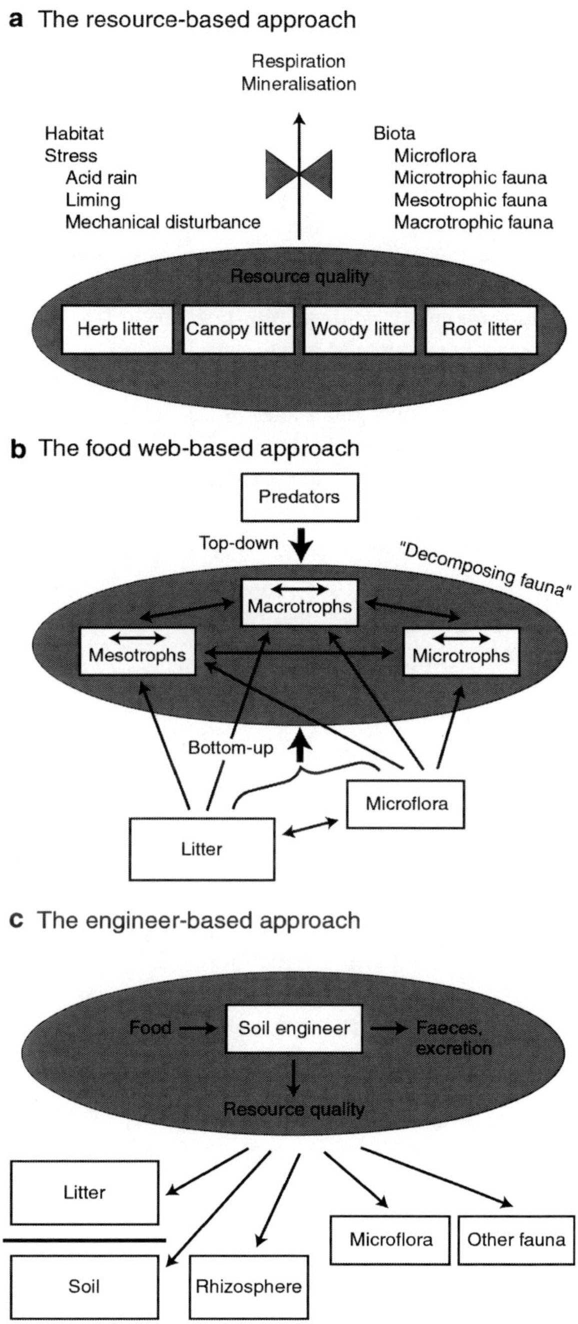

Fig. 12.2 The resource-based approach (**a**), the food web-based approach (**b**), and the engineer-based approach (**c**) in decomposition studies

on interactions in the soil food web, and (3) an engineer-based approach analysing the central role of earthworms in the soil–litter system.

Important insights will be:

- Beech leaf litter is less rapidly decomposed in acid sites as compared to litter in base-rich sites.
- Rate of litter decomposition increases in the sequence: coarse roots – wood – leaves.
- Rate of leaf litter decomposition increases in the sequence: canopy – herb material.
- Anthropogenic acidification leads to a decrease of the rate of mass loss in the litter layer. Soil fauna may partially buffer the effects of acid rain.
- Macro- and mesofauna tend to accelerate decomposition rate, but inhibitory effects may occur in a later phase of decay.
- There are key faunal interactions in the soil food web, e.g. between earthworms and saprophagous macroarthropods.
- Soil food webs are mainly donor-controlled. Resources and microflora are a bottom-up force.
- Bioturbation is a key factor in decomposition pathways.
- Earthworms are soil engineers.

12.2 Influence of Soil Fauna on Decomposition of Canopy Leaf Litter

Leaf litter of beech (*Fagus sylvatica*) trees decomposes slowly because, among other factors, the leaf is tough and contains polyphenols as secondary plant substances. However, the rate of mass loss of beech leaf litter is higher in forests with high faunal activity in the mineral soil. Litterbag studies demonstrated that beech leaf litter mass decreased by about 20% annually in the base-rich sites (Göttinger Wald, Zierenberg) in comparison to the base-poor forest, Solling with a mass loss of 10–15% (Schauermann, unpublished results). The decomposition rate *k* as the ratio of the annual input of canopy leaf litter to the mean annual standing crop of canopy leaf litter is about 0.6–0.8 for Göttinger Wald and 0.25 for Solling (Schaefer and Schauermann 1990). These results are not surprising and concur with theoretical expectations. Leaves from different origin may decompose at a different rate. Nicolai (1988) found that decay of beech leaves from base-poor sites was slower as compared to leaves from forests on base-rich soil. However, Herlitzius and Herlitzius (1977) demonstrated that, in the beech forests of Solling and Göttinger Wald, native shade leaves and leaves introduced from the other forest were decomposed at an equal rate.

Litter of canopy species of the beech forest Göttinger Wald decomposed with a different rate depending on species. Fifty percent of a mixture of ash (*Fraxinus excelsior*) and maple (*Acer* spp.) litter disappeared within 280 days from 45-μm mesh bags (a treatment that excluded the fauna) and in 60 days from 3.5-mm mesh bags (a treatment that permitted access to both meso- and macrofauna) (Fig. 12.3) (Wise and Schaefer 1994). Freshly fallen leaves of beech and year-old beech leaf litter decomposed much more slowly than litter of herbs (see below) or ash–maple mixture. Over 50% of both categories of beech leaf litter remained after 1.4 years, even in the 3.5-mm mesh bags. The fauna had no significant impact on the rate of disappearance of fresh beech leaves, and only the macrofauna accelerated the disappearance of aged beech leaf litter.

12.3 Influence of Soil Fauna on Decomposition of Herb Litter

Generally, herb litter is a resource of high quality and is readily utilised by microfloral and/or animal populations. In experiments with litterbags in the beech forest Göttinger Wald, herb litter was decomposed rapidly. For the two herbaceous species, *Anemone nemorosa* and *Mercurialis perennis*, half the litter disappeared within 48 days from 45-μm mesh litterbags (Wise and Schaefer 1994) (Fig. 12.4).

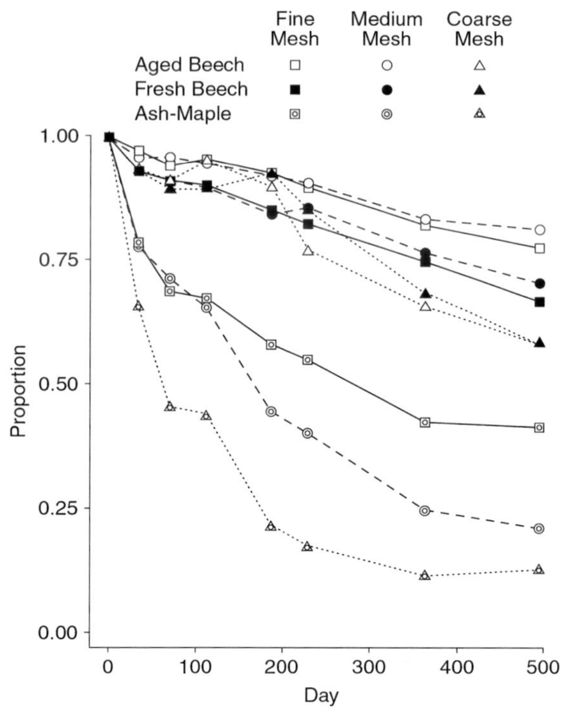

Fig. 12.3 Change in amount of canopy litter (beech forest Göttinger Wald), expressed as proportion of initial dry mass, in litter bags of fine (45 μm), medium (1 mm) and coarse mesh (3.5 mm). (after Wise and Schaefer 1994, values for SE are given in this paper)

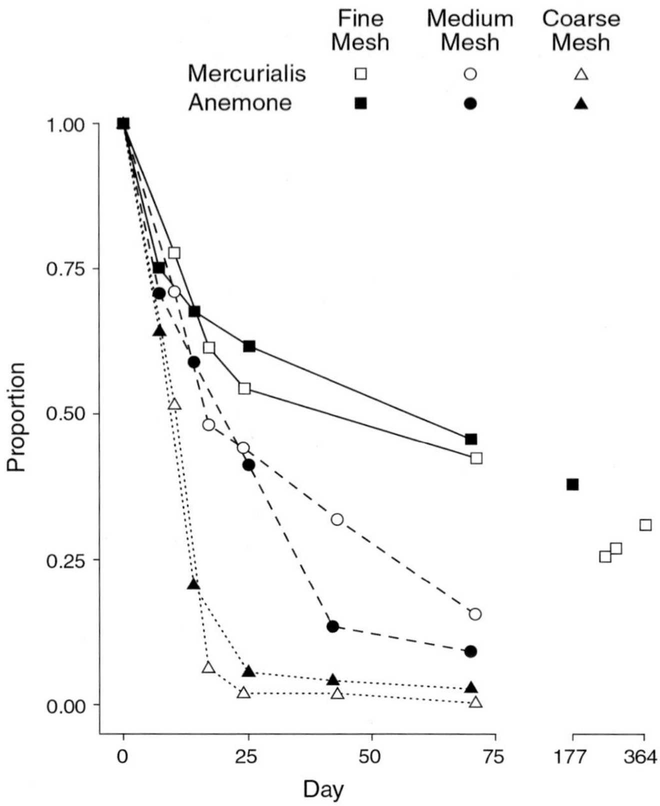

Fig. 12.4 Change in amount of herbaceous litter (beech forest Göttinger Wald), expressed as proportion of initial mass, in bags of fine, medium and coarse mesh. For further explanation see Fig. 12.3. (after Wise and Schaefer 1994, values for SE are given in this paper)

Both meso- and macrofauna accelerated decomposition of herb litter. Within 10 days, 50% of herb litter had disappeared from the 3.5-mm bags. There were no experiments with *Allium ursinum*, a very dominant herb in many places. However, it can be observed in the field that decay of aboveground material of *Allium* is extremely rapid. For *M. perennis* litter in the beech forest Zierenberg, decomposition rates were similarly high (Pawellek et al. 1993). The authors differentiated between the qualities leaf, stem and rhizome litter. After 1 month, the loss of ash-free dry mass was 50–90% for leaf, 30–60% for stem and 40–50% for rhizome litter. Macrofauna influence was pronounced; the presence of mesofauna led to an increase of decomposition only for leaf litter. For the stinging nettle *Urtica dioica* in the Zierenberg site, leaves were decomposed most rapidly, while coarse roots and stems were decomposed slowly (Göbel et al. 1993; Judas et al. 1995). The macrofauna accelerated the mass loss of leaves and fine roots, the mesofauna contributed to a more rapid decay of leaves. In microcosm experiments, earthworms rapidly removed nettle litter from the soil surface; after 4–6 weeks it disappeared entirely (Scheu 1997).

Litter of herbaceaous species decomposes much more rapidly than canopy litter (Wise and Schaefer 1994). One reason is that the fauna readily consumes herb litter (Schaefer 1991a, b). There are indications that herbaceous litter plays an important role in maintaining high densities of the macrofauna in beech forests with mull-structured soils.

12.4 Decomposition of Roots and Woody Material

The decomposition of woody materials is slow; important influencing factors are: the size of the woody debris, its N and lignin content and the composition of the microbial community.

Carbon loss in beech and ash roots and beech twigs of 0–3, 3–10 and 10–40 mm diameter was studied during 36 months of exposure in litter bags of 1- and 4-mm mesh (Scheu and Schauermann 1994). Beech and ash roots (3 size classes) were placed in a soil depth of about 5 cm in the beech stand of Göttinger Wald; beech twigs (3 size classes) and beech roots (3–10 mm) were placed on the soil surface of the beech stand. From beech twigs, carbon loss was in the order: large twigs > medium twigs > small twigs (Fig. 12.5). Ash roots generally lost more C than beech roots (Fig. 12.6). Loss in C of ash roots was similar for each of the size classes, whereas for beech roots, C loss was in the order: large roots > medium roots > small roots. Beech roots (3–10 mm) placed on the soil surface lost considerably less C than beech twigs (3–10 mm). Generally, mesh size affected C-loss only slightly, which is attributed to the comparatively short exposure time. Presumably, soil animals will contribute to wood decomposition during later stages. However, the fungal flora was important as the degree of colonisation of wood materials by white rot fungi significantly affected C mineralisation. During exposure of the beech twigs (and partly the roots) N accumulated (N storage); afterwards a net loss of N (N mineralisation) occurred (cf. Fig. 12.5).

12.5 Faunal Control of Litter Decay

Two principal approaches allow the analysis of faunal effects on litter decomposition: (1) the measurement of mass loss of organic matter after the exposure of litter in the field using litterbags of different mesh sizes; (2) the measurement of litter mass loss or carbon dioxide production or nutrient mineralisation in microcosms under defined conditions in the laboratory. As demonstrated above, faunal influence in general on decomposition measured as mass loss in litterbags may be high. In microcosm studies in the laboratory, the fauna may lead to an increase of CO_2 production or nutrient mineralisation thus indicating acceleration of decomposition. However, in many cases, both stimulatory and inhibitory effects can be observed depending on substrate quality, milieu factors, composition of the microflora and the type of the fauna.

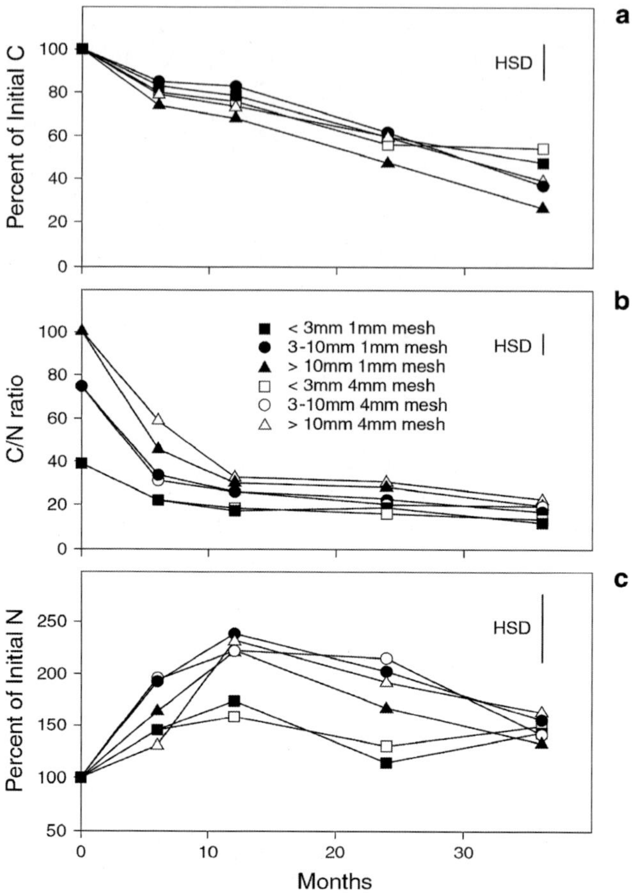

Fig. 12.5 Carbon loss (**a**), C/N ratio (**b**) and changes in the amount of N (**c**) of beech twigs of diameters of <3 mm, 3–10 mm and >10 mm in 1-mm and 4-mm mesh bags during 36 months of exposure on the soil surface of the beechwood Göttinger Wald. HSD = Tukey's Honest significant difference. (after Scheu and Schauermann 1994)

In the context of the resource-based approach, two further questions were studied: the effect of specific life forms of the fauna on decomposition of organic substrates and the influence of stress factors (acid rain and liming).

12.5.1 Microfauna, Mesofauna, and Macrofauna

Soil animal populations can either directly participate in the degradation of dead organic material or indirectly control the mineralising microflora, presumably a much more important contribution. For the decomposing soil fauna, three trophic

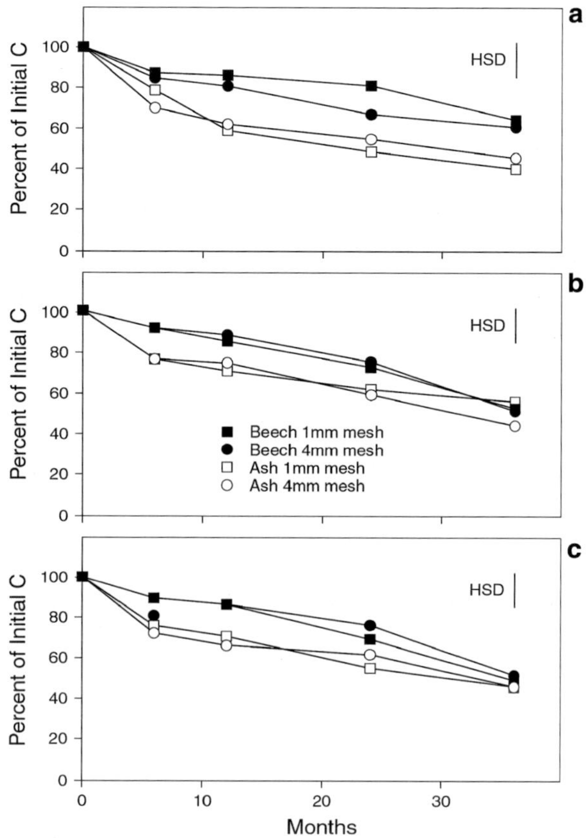

Fig. 12.6 Carbon loss of roots of beech and ash of diameters of <3 mm (**a**), 3–10 mm (**b**) and >10 mm (**c**) in 1-mm and 4-mm mesh bags during 36 months of exposure in the soil of the beechwood Göttinger Wald (after Scheu and Schauermann 1994)

systems can be distinguished on three different scales: the microtrophic, the mesotrophic and the macrotrophic system (Wolters 1991a). The microtrophs inhabit the water film around soil particles and mainly comprise Protozoa and Nematoda. To a large extent, they are bacterial feeders (bacteriophages). The mesotrophic groups, Collembola, oribatid mites and Enchytraeidae, are predominantly fungal feeders (mycetophages), but may also ingest decomposed plant material and mineral particles. The nutrition of the macrotrophs (earthworms, gastropods and saprophagous macroarthropods) is based on substrate use as well as on the utilisation of associated microflora and microfauna. In a similar approach, Wardle and Lavelle (1997) define interactions at three levels of resolution as a 'micro-food-web', a 'litter-transforming system' and 'ecosystem engineers'.

A fourth category are predators. Data for the influence of predacious microfauna, mesofauna and macrofauna on decomposition are scarce. It is well established that the influence of predators may lead to a retardation of decomposition. Lawrence and Wise (2000) reported that reducing spider predation in the leaf litter of a deciduous forest increased the rate of disappearance of a test straw litter from litterbags by 25% over a 6-week period. However, spiders may also enhance the decomposition of canopy litter under certain environmental conditions (Lawrence and Wise 2004), indicating the specificity and peculiarity of faunal effects.

Schaefer (1990a) calculated the relative contributions of microtrophic, mesotrophic and macrotrophic components to carbon flow in the beech forest of Göttinger Wald. Consumption of organic material by macrotrophs (mainly earthworms) amounted to about 9,000 kJ m^{-2} per year, of mesotrophs (mainly enchytraeids and collembolans) to about 1,000 kJ m^{-2} per year, and of microtrophs (mainly amoebae) to about 3,000 kJ m^{-2} per year. Animal production values were 200, 100 and 600 kJ m^{-2} per year, respectively. Total annual consumption by the soil fauna almost equals annual litterfall (Schaefer 1996), a fact that demonstrates the importance of the macrofauna for the carbon budget of the beech ecosystem. However, precise and detailed estimates of faunal effects for parts of the system and for the whole ecosystem do not exist. For the three forests, some groups of the mesofauna and macrofauna had been studied in more detail. Often their effect on decomposition is stimulatory; however, idiosyncratic effects depending on resource type and other system properties also occur and are documented in the following paragraphs.

12.5.2 Carbon and Nitrogen Mineralisation

Mesofaunal influence on decomposition is mainly exerted by feeding on fungal hyphae which normally has a stimulatory effect on decay, but may also cause a retardation of the decay process. Schulz and Scheu (1994) studied the effect of oribatid mites on decomposition of ^{14}C-labelled holocellulose mixed with brown-rotted or white-rotted beech wood. The effect of oribatid mites on $^{14}CO_2$–C release from holocellulose depended on the mite species. Generally, phytosaprophagous mite species (*Steganacarus magnus* and *Phthiracarus piger*) caused an increase in cumulative $^{14}CO_2$–C release from holocellulose; however, for the latter species it depended on feeding intensity (number of specimens per chamber). Presumably, the effects of mites were caused by inoculation with microbial propagules and increased mineralisation of nutrients (Scheu and Schulz 1994).

In microcosms with ^{14}C-labelled fresh beech litter, increasing the density of the collembolan *Isotoma tigrina* had an increasing effect on C mineralisation. The effect was positive in mull-structured soil but negative under moder conditions (Wolters, unpublished results, in Schaefer 1996).

The main macrofaunal influences on decomposition concern feeding on litter by primary consumers, comminution and fragmentation of litter by the production of faeces, transfer of organic material between the strata and carbon and nutrient mineralisation. These activities are accompanied by a diverse array of effects on the state of the litter resource and on soil microflora.

The effect of the earthworm *Aporrectodes caliginosa* depended on the recalcitrance of the substrate. In microcosm systems with old beech litter and ^{14}C-labelled beech litter, the worms stimulated $^{14}CO_2$–C release from 188 to 243 nCi (22.4% $\pm$ 11.0 SD) in systems with a resistant fraction of litter and depressed $^{14}CO_2$–C release for an easily degradable fraction of litter from 147 to 128 nCi (13% ± 8.1 SD) (Scheu, unpublished results, in Schaefer 1996).

Scheu (1993a) tested the effect of the endogeic earthworm species *Octolasion lacteum* on mineralisation of ^{14}C-labelled holocellulose and ^{14}C-labelled lignin in laboratory experiments. Earthworm processing caused a two-phase alteration in holocellulose and lignin mineralisation: mineralisation rates were increased for 6–15 weeks (holocellulose) or about 10 weeks (lignin), but decreased later in the experiment in both cases. In another experiment, Scheu (1993b) analysed the influence of the earthworms *O. lacteum* and *Lumbricus castaneus* on decomposition of ^{14}C-labelled (lignin) lignocellulose mixed with brown-rotted or white-rotted beech leaf litter. $^{14}CO_2$–C release from lignin mixed in brown-rotted beech leaf litter exceeded that of lignin mixed in white-rotted beech leaf litter due to excessive nutrient supply in the latter treatment. Earthworm activity increased the cumulative N mineralisation and $^{14}CO_2$–C release from lignin, but did not affect total CO_2 production by the systems during the experiment. The presence of *Glomeris marginata* enhanced cumulative N mineralisation and total CO_2 production by the systems but did not affect overall $^{14}CO_2$–C release from lignin.

Thus, it appears that there is a tendency for an idiosyncratic response of the system. At least in the laboratory, a two-phase sequence in macrofauna-mediated decomposition can often be observed. In the above-mentioned experiments with earthworms, holocellulose and lignin mineralisation was increased for the first weeks and decreased afterwards (Scheu 1993a). In fresh faeces, mineralisation rates are high, due to additional nutrients being available, and decrease with age. Microbial carbon turnover in soils freshly worked by the earthworm *A. caliginosa* were significantly different from the microbial C turnover in aged casts (Wolters and Joergensen 1992). Generally, aged faecal pellets tend to be resistant to decomposition, as is also the case for the millipede *Glomeris* (Scheu and Wolters 1991b).

The soil fauna affects nitrogen mineralisation by modifying the soil microflora directly (by grazing) or indirectly (by modifying the microbial environment) (Wardle and Lavelle 1997; Wardle 2002; Scheu and Setälä 2002; Paul 2007). In general, animals enhance the microflora-mediated net mineralisation rate by about 30% (Verhoef and Brussaard 1990). However, the role of the fauna in nitrogen dynamics is difficult to quantify for the whole system.

The additional mineralisation of nitrogen caused by the burrowing activity of the substrate feeding earthworm *A. caliginosa* was measured by Scheu (1987b) in microcosm experiments. At 15°C, the additional N mineralisation in comparison to controls without earthworms was 250 μg N g^{-1} fresh body wt per day and was calculated to be 4.2 kg ha^{-1} per year, which compared to about 160 kg N ha^{-1} per year net mineralisation in the whole beech forest Göttinger Wald. The whole earthworm fauna may contribute about 23 kg N ha^{-1} per year, which corresponds to about 15% of annual N net mineralisation (154 N ha^{-1} per year) or 25% of annual nitrogen uptake of the trees (93 N ha^{-1} per year) (Scheu 1994). It should be emphasised that the main effect of earthworms is not excretion of N but stimulation of microbially-mediated ammonification and nitrification.

Nitrogen mobilisation during the gut passage significantly exceeds the amount of nitrogen excreted especially in endogeic earthworm species (Scheu 1993c). The presence of *O. lacteum* in microcosms caused an increase in cumulative N mineralisation by factors of 2.7 and CO_2 production by a factor of 2.5. Obviously, these earthworms mobilise a physically protected nitrogen and carbon pool by breaking up soil aggregates.

12.5.3 Acid Rain and Liming

Human impact on decay of organic material is manifold. Man influences the processes in the soil system by immission inputs of acidity, N and other elements, by logging and by liming. It is well established that addition of protons depresses decomposition of organic matter (Wolters and Schaefer 1994), and that logging and liming enhance decomposition. Berg et al. (1998) analysed the effects of high deposition of inorganic nitrogen on long-term decomposition of organic matter in a pine forest soil. They found that in the long-term enhanced atmospheric-N depositions aggravate C-limitation for microbial degradation.

The input of protons ('acid rain') leads to acidification of freshwater and terrestrial habitats. Chemical changes in the soil and water can be measured. The effect of acid rain on the fauna is less clear-cut (Schaefer 1990b; Wolters 1991b; Wolters and Schaefer 1994). In an analysis of a 30-year time series in the moder-structured beech forest Solling, no effect of a further decrease in soil pH of already acid soil on the fauna was observed (Schauermann, unpublished results; Schaefer and Scheu 1996). However, small-scale acidification experiments in the laboratory and in the field demonstrated changes in the soil faunal community and/or decomposition rate. For instance, in a laboratory system with 4-month-old ^{14}C-labelled beech leaf litter, acid rain caused a strong decrease in C mineralisation of intact litter situated on the surface of a carbonate-rich mineral soil (to 0.43 of control) (Scheu and Wolters 1991a). A large portion of the protons is transferred to the ground by stemflow. In the mull-structured forest at the Göttinger Wald site, the density of endogeic earthworms is drastically reduced near tree trunks (Scheu and Poser 1996), which certainly has effects on decay processes.

Anthropogenic acidification of forests may ultimately result in more moder-like conditions with decreased macrofauna abundance and biomass. This trend can be monitored by comparing forest soils of different acidity (Fig. 12.7) (Schaefer and Scheu 1996). However, the comparison of a spatial series allows conclusions on temporal trends only with a certain degree of certainty.

Liming may compensate for acidification and lead to an increase of faunal biomass (Ammer and Makeschin 1994; Rundgren 1994) and of litter mass loss (Bauhus and Barthel 1995). Application of lime to counteract soil acidification leads to long-term changes of the soil animal community. Heiligenstadt and Schauermann (1993) observed an increase in earthworm biomass and a decrease in mesofauna numbers in a moder-structured spruce forest (Solling). Earthworms and dipterous larvae of the families Sciaridae, Scatopsidae and Tipulidae were most

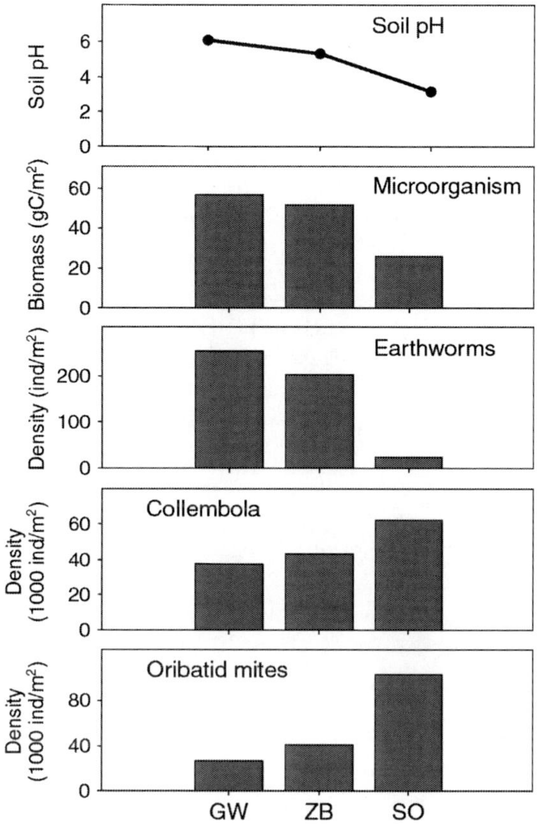

Fig. 12.7 pH, microbial biomass and density of some soil fauna groups in three different beech forests (*GW* Göttinger Wald, *ZB* Kleiner Gudenberg (Zierenberg), *SO* Solling). Values refer to litter plus Ah soil layer up to a depth of 6 cm (Göttinger Wald, Zierenberg) or 3 cm (Solling). (from Schaefer and Scheu 1996)

abundant on limed plots of the Solling site (Theenhaus and Schaefer 1995). The humus form moder developed some characteristics of mull ('mull-like moder') (Heiligenstadt and Schauermann 1993; Theenhaus and Schaefer 1995). Earthworms (*Lumbricus terrestris*) were artificially introduced into the moder soil of a spruce stand near the beech forest at Solling and were able to build up permanent populations (Judas et al. 1991, 1997). The population started in 1986 with 10,000 individuals on an area of 400 m^2 (=25 individuals m^{-2}) and exhibited a density in 1990 of 12 adults and 18 juveniles per m^2.

12.6 The Decomposer Web

Soil organisms are interconnected by a complex web of trophic relationships (Scheu and Falca 2000; Scheu and Setälä 2002; Wardle 2002; De Ruiter et al. 2005; Brussaard et al. 2007). Important trophic groups are litter feeders (saprophages), fungal and bacterial feeders (microphytophages) and predators and parasites (zoophages) (Schaefer 1996). This decomposer web is not easy to describe. It is ‘opaque’ and contains many generalists. Specific soils are inhabited by specific soil communities. Interactions can be of different types (Scheu and Setälä 2002) and comprise predation, inhibition, facilitation, stimulation and mutualism.

12.6.1 Important Interactions Between Decomposer Groups

In terms of biomass, Protozoa, Nematoda, Lumbricidae and saprophagous macroarthropods are among the most dominant soil animal groups in the beech forests at Göttinger Wald and Zierenberg. It can be presumed that interactions between populations of these groups are pronounced. However, not much is known about such specific interrelationships. We discuss some evidence of closer links between nematodes and Protozoa, between earthworms and Protozoa and between earthworms and macroarthropods.

Nematodes and *Protozoa* can interact by direct feeding or by stimulating the development of bacterial populations. Alphei et al. (1996) studied the interaction between nematodes, Protozoa and earthworms in the rhizosphere of the grass *Hordelymus europaeus* in microcosm experiments. Nematodes (*Pellioditis pellio*) strongly increased the number of flagellates in the rhizosphere and non-rhizosphere compartment, while the number of ciliates remained unaffected. Whereas nematodes decreased the number of amoebae in non-rhizosphere soil, they increased their number in rhizosphere soil. The positive effect of nematodes on soil amoebae was more pronounced in the presence of earthworms (*A. caliginosa*). It is possible that probiotic effects of nematodes on Protozoa in earthworm treatments may be an outcome of the tunnelling of nematodes in faecal material produced by earthworms, facilitating access to microorganisms in these aggregates.

Protozoa can be food for *earthworms* and earthworms may modify the substrates and thus protozoan populations. A multiple choice feeding experiment demonstrated that the distribution of *A. caliginosa* was correlated with number of amoebae in soil (Bonkowski and Schaefer 1997). *A. caliginosa* is actively searching for places with high protozoan densities; thus Protozoa may play a significant role in earthworm nutrition. In this experiment, weight gain of young *A. caliginosa* in soil with amoebae was twice to that in soil without Protozoa. The earthworms digested active Protozoa; however, most of the encysted Protozoa survived the gut passage, an indication that earthworms may be important in distributing Protozoa in the soil.

Interactions between *earthworms* and *macroarthropods* might be a key process in the soil-litter system. The faeces of litter-feeding arthropods can form food for all those secondary decomposers who ingest only fragmented litter. This metabiotic relationship may be an important functional sequence of faunal activity in the soil-litter system. In microcosm systems, the earthworm *O. lacteum* did not feed on intact beech leaf litter, but removed the faecal material of *G. marginata* from the soil surface (Bonkowski et al. 1998). Thus, the combined action of faeces-producing saprophagous macroarthropods and endogeic earthworms is assumed to be of crucial importance for the formation and maintenance of the mull soil in the beech forests of Zierenberg and Göttinger Wald. The intimate mixing of ingested diplopod faeces with mineral soil by endogeic earthworms lead to a decrease in C mineralisation and consequently to the stabilisation of organic material in the mineral soil (Scheu and Wolters 1991b).

A mainstream metabiotic interaction is the continuum from *primary to secondary decomposers*. Scheu and Falca (2000) studied the structure of the soil food web of the beech forests at Göttinger Wald and Solling using variations in tissue ^{15}N concentrations of animal species or taxa. The delta^{15}N values indicated that most of the studied saprophagous taxa (Lumbricidae, Diplopoda, Isopoda) or microphytophagous taxa (Collembola, Oribatida) consisted of primary and secondary decomposers. The high variations of delta^{15}N values of species within these groups demonstrated that the species formed a continuum from primary to secondary decomposers ('microbi-saprophages'). It can be hypothesised that the facilitation of secondary decomposers by primary decomposers is an organising principle of the soil system. Interestingly, in a wood-web analysis with stable isotopes, the cluster of predatory groups is distinctly separated from the cluster of microbi-saprophagous taxa (Scheu and Falca 2000).

12.6.2 Bottom-up and Top-down Effects

The populations of the primary consumers in the beech forests are more or less persistent in time (Schaefer 1996), whereas litter does not accumulate. Therefore, theoretically (Hairston et al. 1960; Halaj and Wise 2001; Scheu and Setälä 2002; Wardle 2002), this observation leads to the hypothesis that these populations are regulated by the availability of food resources. If litter-utilising populations were top-down controlled by predators, decomposition processes would proceed more erratically with varying depths of the litter layer.

A long-term experiment was set up in the Göttinger Wald site where *litter inputs* were manipulated. In 1981 and 1982, litter fall was excluded on plots of 10×10 m in the beech forest; and on other plots, litter inputs were increased to fivefold of the normal value for these 2 years. Earthworms and dipterans were studied in detail. In the following years, earthworms (Judas 1990) and dipterans (Hövemeyer 1992) decreased on the litter-exclusion plots. Contrary to expectation, on the plots with higher litter inputs, earthworms declined in number (Judas 1990), indicating unfavourable conditions in the thick litter layer for most of the earthworm populations. However, among dipterans, microhumiphagous species and 'surface scrapers' strongly increased in numbers on the plots with high litter input (Hövemeyer 1992).

Experimental *fertilisation* was a second approach where a 15-month fertilisation experiment on plots of 1 m^2 in the mull-structured forest at Göttinger Wald was conducted. It supported the bottom-up hypothesis for earthworms (Scheu and Schaefer 1998). Addition of nitrogen (as NH_4NO_3), phosphorus (as NaH_2PO_4) and carbon (as glucose) in different combinations led to an increase in microbial

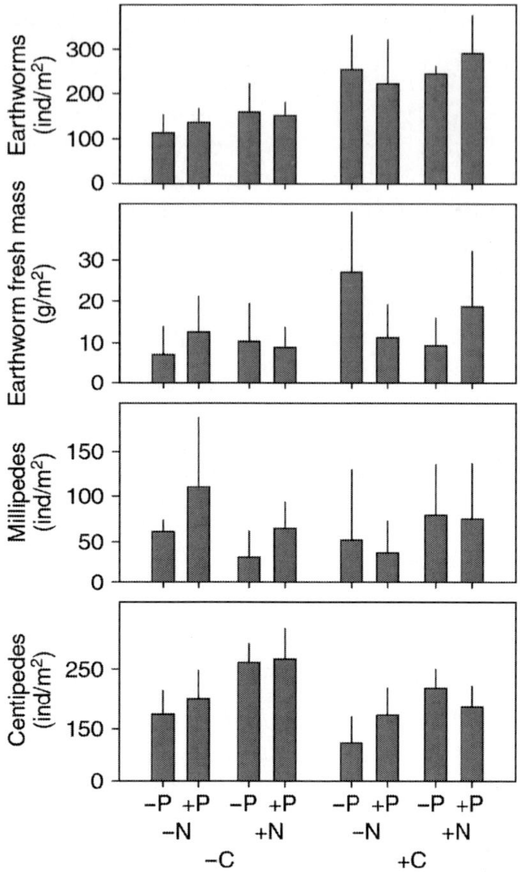

Fig. 12.8 Response of earthworms (excluding *Lumbricus terrestris*), millipedes and centipedes to addition of carbon (C), nitrogen (N), and phosphorus (P) (means + SD) (after Scheu and Schaefer 1998)

biomass and changes in the animal community. The microorganisms in the litter layer were primarily limited by nitrogen, whereas soil microorganisms were primarily limited by phosphorus. None of the animal groups studied (Lumbricidae, Diplopoda, Isopoda and Chilopoda) exhibited a response that was parallel to that of microorganisms. The additional availability of carbon alone (or in combination with phosphorus) resulted in an increase of earthworm numbers and biomass (Fig. 12.8). Increased earthworm densities were always associated with an increase in soil moisture.

Small microbi-saprophagous soil invertebrates like collembolans or oribatid mites are not directly linked to litter input because they predominantly feed on fungi, bacteria or faeces. This may weaken bottom-up forces. In the above-mentioned fertilisation experiment, soil meso- and microfauna responded differentially to the addition of C and nutrients. Collembolans and oribatid and uropodid mites tended to decline, whereas nematodes and protozoans increased in density (Maraun et al. 2001). Obviously, the responses were indirect effects of increased number and biomass of earthworms in additional carbon and carbon + phosphorus treatments (cf. Fig. 12.8). Earthworms detrimentally affected soil mesofauna, whereas protozoans and nematodes benefited from higher soil moisture combined with increased microbial biomass.

Thus, the effects of additional litter input or of fertilisation are not clear-cut and depend on the system. In general, there is a differential response of soil animal groups to an increased supply of food resources.

Important *predatory groups* in the beech forest at Göttinger Wald are chilopods, spiders, carabid and staphylinid beetles and gamasid mites. Chilopods consume a high proportion of collembolan populations and of other saprophagous groups (Poser 1988, 1991). However, Judas (1989) found no evidence for high predation rates on earthworms, a group with two-thirds of total faunal biomass. It can be presumed that top-down forces are less pronounced than bottom-up forces in the soil food web of this forest. On the other hand, recent analyses found cascading effects of predators that may also affect the rate of decomposition at the base of detrital food webs (Scheu and Setälä 2002; Wardle 2002; Moya-Larano and Wise 2007). It emerges that the relative contributions of bottom-up and top-down forces in soil food webs are not predictable, and are one of the key issues of future studies in soil biology (Halaj and Wise 2002).

12.7 Earthworms as Ecosystem Engineers

Ecosystem engineering is the physical modification, maintenance or creation of habitats, and the organisms involved. The physical ecosystem engineers are species that directly or indirectly control the availability of resources to other organisms by causing physical state changes in biotic or abiotic materials (Lawton 2000).

Effects of engineers on soil organic matter dynamics occur at four different scales of space and time (Lavelle 1997; Lavelle et al. 1997): (1) the transit of the material through intestine (hours); (2) the enhancement of microbial activity in, and the release of labile nutrients from, fresh faecal material (days); (3) the sequestration and protection of undegraded organic matter in ageing casts and construction (weeks); and (4) the distribution and turnover of soil organic matter (bioturbation) in whole profiles (years to decades).

Macrofaunal species are assumed to be the major agents responsible for the formation of mull humus. In particular, anecic earthworms are known to feed on litter materials and incorporate them into the mineral soil (Lee 1985; Edwards and Bohlen 1996). Hence, lumbricids are candidates for being ecosystem or soil engineers (Lavelle et al. 1997; Edwards 2000, 2004; Parkinson et al. 2004). Their possible influence on the soil and litter environment is manifold: they may select high quality food as hot spots, they modify the food substrate during the gut passage, they produce mucus, they transform the soil matrix into macro-aggregates, they create burrows, they produce casts and middens, they feed in the rhizosphere. Thus, they influence soil aggregate stability, water flow, microbial activity (see Chap. 6), nutrient mineralisation (see Chap. 13) and plant growth (see Chap. 8).

12.7.1 Food Ingestion, Gut Passage, Production of Casts and Mucus

The results presented here mainly relate to earthworms of the forest at Göttinger Wald. *L. terrestris* (the only anecic species) was the only earthworm with high amounts of particulate plant remains in the gut (median 38% dry wt) (Judas 1992). The worms preferred non-beech litter: only 25% of all particles in *L. terrestris* crops and gizzards were beech leaf litter; 41% were non-beech tree leaf litter and herbaceous litter. For *L. terrestris*, Wolter and Scheu (1999) found that the number of bacteria changed in a very uniform pattern during the gut passage irrespective of food materials (leaf litter of *F. sylvatica* or *M. perennis*) and season. Bacterial numbers increased from the crop/gizzard to the foregut and then declined or remained constant in the hindgut and decreased further in the faeces. When fed with soil or beech litter, the length of fungal hyphae was generally at a maximum in the crop/gizzard and tended to decline during the gut passage to the faeces. Bacterial numbers and hyphal lengths in the crop/gizzard generally exceeded those found in the soil, indicating the selective feeding nature of *L. terrestris* in microsites rich in microorganisms. However, the authors concluded that microorganisms were not a major component of the diet of *L. terrestris* and other earthworms.

Microbial biomass in soil is only slightly affected by the gut transit in earthworms (Scheu 1993d; Tiunov and Scheu 2000). However, during gut passage of soil, microbial nutrient status changes. This effect depends on the nutrient deficiencies of the microflora and is assumed to vary among soil types and ecosystems. For instance, in the beech forest at Zierenberg, the gut passage through *O. lacteum* eliminated microbial P limitation in limestone soil, and increased microbial N supply in soil from the basalt and intermediate site (Scheu 1993d).

Microbial activity in casts is a two-phase process: in most cases, stimulation is followed by inhibition (Scheu 1987c, 1993a; Tiunov and Scheu 2000). Stimulation in mineralisation of litter processed by earthworms into casts is most likely caused by increasing the ability of the microflora to attack soluble and easily degradable components (e.g. starch, proteins). Long-term alterations in litter decomposition are presumably caused by modifications in cellulose and lignin mineralisation (Scheu 1993a).

12.7.2 Bioturbation

Earthworms mix soil layers by feeding on substrates and depositing faeces (Lavelle et al. 1997). Scheu (1987a) measured the cast production of the earthworm *A. caliginosa* in microcosm experiments in relation to temperature and moisture. For the forest at Göttinger Wald, the *A. caliginosa* population produced 3 kg faeces per m^2 per year (which corresponds to a soil layer of 4.3 mm). All endogeic species produce faeces with a mass of 6 kg m^{-2} per year, corresponding to a soil layer of 9 mm. If the anecic species *L. terrestris* is included, the total amount of egesta produced by the lumbricids should exceed 10 kg m^{-2} per year, a value equal to a soil layer of 14 mm. The actual transport might be high, because comminuted litter on the surface of the mineral soil layer (mainly faeces of macroarthropods and gastropods) is consumed by endogeic earthworms (Bonkowski et al. 1998) and thus transferred into the soil.

Generally, bioturbation stimulates the development of the microflora; however, effects on the fauna may be of positive (facilitation) or negative (inhibition) (Maraun et al. 2001) nature.

12.7.3 Earthworm Burrows and Middens

The *burrows* of earthworms consist of faeces, litter material and mucus. They are hot spots of biological activity caused by the stimulative effects of earthworms on microbial activity and nutrient dynamics. Mucus is considered to affect soil microflora and soil structure. Scheu (1987c) observed an increase of microbial activity in burrow walls of *A. caliginosa* by 22% for 2 weeks. Tiunov and Scheu (1999) found that the microbial community in the burrow walls of *L. terrestris* contained a larger fraction of metabolically-active microorganisms, adapted to continuous resource addition by earthworm faeces and mucus. Burrow walls were strongly enriched with mineral nitrogen and phosphorus. The number and biomass of naked amoebae and flagellates were significantly greater in burrow walls than in the control soil, whereas total nematode density was not affected (Tiunov et al. 2001). The walls are very stable microhabitats inhabited by a specific community of microphytophagous soil animals grazing on the wall microflora, strongly affecting nutrient dynamics.

Anecic earthworms produce high numbers of casts on the soil surface, which are intensively mixed with decomposing litter material. These conspicuous structures are termed *middens*. In middens of the earthworm species *L. terrestris* in the forest at Göttinger Wald, C and N content, C/N ratio, microbial biomass, basal respiration and specific respiration were significantly higher than in non-midden soil (Maraun et al. 1999). The abundance of Gamasina, Uropodina and (less pronounced) Collembola and the biomass of Nematoda was increased. However, most oribatid taxa preferred the non-midden soil. Obviously, the increased biomass of bacterial feeding nematodes and euryoecious Collembola was caused by an increased microbial biomass in the middens, and predacious Gamasina and Uropodina benefited from this increased prey density. Earthworm middens are beneficial for most soil animals, in contrast to the actual mixing of soil layers by bioturbation. In the Göttinger Wald stand, up to 35 middens may occur per square metre. Thus, this relationship between trophic levels in a microhabitat with microbial biomass as a key resource is a further case of an important bottom-up process in the soil food web.

12.7.4 Earthworms in the Rhizosphere

Earthworms can be active in the rhizosphere and influence plant growth. In the microcosm experiment by Alphei et al. (1996), earthworms (*A. caliginosa*) reduced the root biomass (and the total biomass) of *H. europaeus*; however, shoot/root ratio drastically increased from 3.4 to 5.6. Thus, *H. europaeus* could sustain a high shoot biomass with a strongly reduced root system. The effects of earthworms were caused by an increase in nutrient supply to the plants. High amounts of extractable mineral nitrogen and increased concentrations of nitrogen in plant tissue in the earthworm treatments are evidence for this interpretation.

12.8 Conclusions

As expected, decomposition of litter material is strongly influenced by litter quality. Rates of litter disappearance increase in the order: beech roots – beech twigs – beech leaf litter – canopy litter of other tree species – herbs (*Mercurialis*, *Anemone*). Soil fauna, especially the macrofauna, is an important controlling factor of litter decomposition. In many cases, decomposition rate is enhanced; however, under certain conditions, faunal influence may result in a decrease of litter decay rate, for instance by the production of stable faeces particles.

The decomposition rate of beech leaf litter decreases from the base-rich forest to the acid forest site. This gradient from mull to moder of the litter layer is accompanied by a decrease in macrofauna abundance and faunal biomass (Schaefer and Schauermann 1990) (see Chap. 7). Soil engineering is responsible for mull-structured soils. Comminution of organic substrates and the modification of the soil environment by the fauna appears to be an important causal factor for rapid decay in these mull soils.

The functional interactions between faunal groups in soil and litter and their combined action on microfloral activity and carbon and nutrient dynamics are scarcely understood (Wolters 1991a; Schaefer 1995; Scheu 1995; Scheu and Setälä 2002; Bardgett et al. 2005; Brussaard et al. 2007). The decomposition web is a network of faunal–faunal and faunal–microfloral interactions, with a continuum from primary to secondary decomposers, with more pronounced bottom-up control and less developed top-down control, and with partly idiosyncratic responses to environmental influences. Thus, it is difficult to relate single processes in the soil web to total material dynamics.

Differences in litter decay arise due to differences in resource quality and activity of the biota which may contribute to the resilience of the decomposition subsystem with its large pools of soil carbon which is variably distributed on the surface organic layer.

Acknowledgement We thank David H. Wise (Chicago, Illinois) for reading and commenting on the manuscript.

References

Alphei J, Bonkowski M, Scheu S (1996) Protozoa, Nematoda und Lumbricidae in the rhizosphere of *Hordelymus europaeus* (Poaceae): faunal interactions, response of microorganisms and effects on plant growth. Oecologia 106:111–126

Ammer S, Makeschin F (1994) Effects of simulated acid precipitation and liming on earthworm fauna (Lumbricidae, Oligochaeta) and humus type in a mature stand of Norway spruce (Hogwald Experiment). Forstwissenschaftliches Centralblatt 113:70–85

Bardgett R, Usher M, Hopkins D (2005) Biological diversity and function in soils. Cambridge University Press, Cambridge

Bauhus J, Barthel R (1995) Mechanismen for carbon and nutrient release and retention in beech forest gaps. 2. The role of soil microbial biomass. Plant Soil 169:585–592

Berg MP, Kniese JP, Zoomer R, Verhoef HA (1998) Long-term decomposition of successive organic strata in a nitrogen saturated Scots pine forest soil. For Ecol Manage 107:159–172

Bonkowski M, Schaefer M (1997) Interactions between earthworms and soil protozoa: a trophic component in the soil food web. Soil Biol Biochem 29:499–502

Bonkowski M, Scheu S, Schaefer M (1998) Effects of endogeic earthworms (*Octolasion lacteum*), diplopods (*Glomeris marginata*) and plants (*Hordelymus europaeus*) on the decomposition of litter in a beechwood on a basalt hill. Appl Soil Ecol 9:161–166

Brussaard L, Pulleman MM, Ouedraogo E, Mando A, Six J (2007) Soil fauna and soil function in the fabric of the food web. Pedobiologia 50:447–462

Cadisch G, Giller KE (eds) (1997) Driven by nature: plant litter quality and decomposition. CABI, Wallingford

De Ruiter PC, Neutel A-M, Moore J (2005) The balance between productivity and food web structure in soil ecosystems. In: Bardgett R, Usher M, Hopkins D (eds) Biological diversity and function in soils. Cambridge University Press, Cambridge, pp 139–153

Edwards CA (2000) Soil invertebrate controls and microbial interactions in nutrient and organic matter dynamics in natural and agroecosystems. In: Coleman DC, Hendrix PF (eds) Invertebrates as webmasters in ecosystems. CABI, Wallingford, New York, pp 141–159

Edwards CA (2004) Earthworm ecology, 2nd edn. CRC, Boca Raton

Edwards CA, Bohlen PJ (1996) Biology and ecology of earthworms. Chapman and Hall, London

Göbel M, Judas M, Schauermann (1993) Die Zersetzung von Krautarten in einem heterogenen Buchenwald. I. Einfluß der Fauna auf den Abbau von Brennesselstreu. Berichte des Forschungszentrums Waldökosysteme B32:9–68

Hairston NG, Smith FE, Slobodkin LB (1960) Community structure, population control, and competition. Am Nat 879:421–425

Halaj J, Wise DH (2001) Terrestrial trophic cascades: how much do they trickle? Am Nat 157:262–281

Halaj J, Wise DH (2002) Impact of a detrital subsidy on trophic cascades in a terrestrial grazing food web. Ecology 83:3141–3151

Heal OW, Anderson JM, Swift MJ (1997) Plant litter quality and decomposition: an historical overview. In: Cadisch G, Giller KE (eds) Driven by nature: plant litter quality and decomposition. CABI, Wallingford, pp 3–30

Heiligenstadt B, Schauermann J (1993) Auswirkungen von Kalkungsmaßnahmen auf Bodentiere in einem Moderbuchenwald. Berichte des Forschungszentrums Waldökosysteme B32:5–6

Herlitzius R, Herlitzius H (1977) Streuabbau in Laubwäldern. Oecologia 30:147–177

Hövemeyer K (1992) Response of Diptera populations to experimentally modified leaf litter input in a beech forest on limestone. Pedobiologia 36:35–49

Judas M (1989) Predator pressure on earthworms: field experiments in a beechwood. Pedobiologia 33:339–354

Judas M (1990) The development of earthworm populations following manipulation of the canopy leaf litter in a beechwood on limestone. Pedobiologia 34:247–255

Judas M (1992) Gut content analysis of earthworms (Lumbricidae) in a beechwood. Soil Biol Biochem 24:1413–1417

Judas M, Schauermann J, Meiwes KJ (1991) Die erfolgreiche Ansiedlung des tiefgrabenden Regenwurms *L. terrestris* in einem gekalkten Siebenstern-Fichtenforst auf Buntsandstein. Berichte des Forschungszentrums Waldökosysteme Göttingen B22:377–379

Judas M, Göbel M, Schauermann J (1995) Mass loss of *Urtica dioica* L. roots in a beech forest as affected by soil fauna and soil type. Pedobiologia 39:528–537

Judas M, Schauermann J, Meiwes KJ (1997) The inoculation of *Lumbricus terrestris* L in an acid spruce forest after liming and its influence on soil properties. Soil Biol Biochem 29:677–679

Laskowski R, Berg B, Caswell H (2006) Litter decomposition – A guide to carbon and nutrient turnover. Academic Press, San Diego

Lavelle P (1997) Faunal activities and soil processes: adaptive strategies that determine ecosystem function. Adv Ecol Res 27:93–127

Lavelle P, Spain AV (2001) Soil ecology. Kluwer, Dordrecht

Lavelle P, Bignell D, Lepage M, Wolters V, Roger P, Ineson P, Heal OW, Dhillion S (1997) Soil function in a changing world: the role of invertebrate ecosystem engineers. Eur J Soil Biol 33:159–193

Lawrence KL, Wise DH (2000) Spider predation on forest-floor Collembola and evidence for indirect effects on decomposition. Pedobiologia 44:33–39

Lawrence KL, Wise DH (2004) Unexpected indirect effect of spiders on the rate of litter disappearance in a deciduous forest. Pedobiologia 48:149–157

Lawton JH (2000) Community ecology in a changing world. Excellence in Ecology Series Book 11. Inter-Research Science and Ecology Institute, Oldendorf/Luhe

Lee KE (1985) Earthworms: their ecology and relationships with soils and land use. Academic Press, Sydney

Maraun M, Alphei J, Bonkowski M, Buryn R, Migge S, Peter M, Schaefer M, Scheu S. (1999) Middens of the earthworm *Lumbricus terrestris* (Lumbricidae): microhabitats for micro- and mesofauna in forest soil. Pedobiologia 43:276–287

Maraun M, Alphei J, Beste P, Bonkowski M, Buryn R, Migge S, Peter M, Schaefer M, Scheu S (2001) Indirect effects of carbon and nutrient amendments on the soil meso- and microfauna of a beechwood. Biol Fertil Soils 34:222–229

Moya–Larano J, Wise DH (2007) Direct and indirect effects of ants of a forest-floor food web. Ecology 88:1453–1465

Nicolai V (1988) Phenolic and mineral content of leaves influences decomposition in European forest ecosystems. Oecologia 75:575–579

Parkinson D, McLean MA, Scheu S (2004) Impacts of earthworms on other biota in forest soils, with some emphasis on cool temperate montane forests. In: Edwards CA (ed) Earthworm ecology, CRC, Boca Raton, pp 241–259

Paul EA (2007) Soil microbiology, ecology, and biochemistry, 3rd edn. Academic Press, San Diego

Pawellek U, Judas M, Schauermann J (1993) Die Zersetzung von Krautarten in einem heterogenen Buchenwald. II. Einfluß der Fauna auf den Abbau von Bingelkrautstreu. Berichte des Forschungszentrums Waldökosysteme B32:69–131

Poser T (1988) Chilopoden als Prädatoren in einem Laubwald. Pedobiologia 31:261–281

Poser T (1991) Die Hundertfüßer (Myriapoda, Chilopoda) eines Kalkbuchenwaldes: Populationsökologie, Nahrungsbiologie und Gemeinschaftsstruktur. Berichte des Forschungszentrums Waldökosysteme Göttingen A71:1–211

Rundgren S (1994) Earthworms and soil remediation – Liming of acidic coniferous forest soils in southern Sweden. Pedobiologia 38:519–529

Schaefer M (1990a) The soil fauna of a beech forest on limestone: trophic structure and energy budget. Oecologia 82:128–136

Schaefer M (1990b) Effect of acid deposition on soil animals and microorganisms: influence on structures and processes. Proceedings of the International Congress on Forest Decline Research, Friedrichshafen 1989. Kernforschungszentrum, Karlsruhe, pp 415–430

Schaefer M (1991a) Ecosystem processes: secondary production and decomposition. In: Röhrig E, Ulrich B (eds) Temperate deciduous forests (Ecosystems of the World). Elsevier, Amsterdam, pp 175–218

Schaefer M (1991b) The animal community: diversity and resources. In: Röhrig E, Ulrich B (eds) Temperate deciduous forests (Ecosystems of the World). Elsevier, Amsterdam, pp 51–120

Schaefer M (1995) Biological interactions in the soil community. Acta Zool Fennica 196:101–106

Schaefer M (1996) Die Bodenfauna von Wäldern: Biodiversität in einem ökologischen System. Abhandlungen der Mathematisch-Naturwissenschaftlichen Klasse. Akademie der Wissenschaften und der Literatur, Mainz 2:1–81

Schaefer M, Schauermann J (1990) The soil fauna of beech forests: comparison between a mull and moder forest. Pedobiologia 34:299–314

Schaefer M, Scheu S (1996) Animal-oriented ecosystem research – the importance of long-term studies. Verhandlungen der Deutschen Zoologischen Gesellschaft 89. 2:189–203

Scheu S (1987a) The role of substrate feeding earthworms (Lumbricidae) for bioturbation in a beechwood soil. Oecologia 72:192–196

Scheu S (1987b) The influence of earthworms (Lumbricidae) on the nitrogen dynamics in the soil litter system of a deciduous forest. Oecologia 72:197–201

Scheu S (1987c) Microbial activity and nutrient dynamics in earthworm casts (Lumbricidae). Biol Fertil Soils 5:230–234

Scheu S (1993a) Cellulose and lignin decomposition in soils from different ecosystems on limestone as affected by earthworm processing. Pedobiologia 37:167–177

Scheu S (1993b) Litter microflora – soil macrofauna interactions in lignin decomposition: a laboratory experiment with 14-C-labelled lignin. Soil Biol Biochem 25:1703–1711

Scheu S (1993c) There is an earthworm mobilizable nitrogen pool in soil. Pedobiologia 38:243–249

Scheu S (1993d) Analysis of the microbial nutrient status in soil microcompartments: Earthworm faeces from a basalt-limestone gradient. Geoderma 56:575–586

Scheu S (1994) The influence of earthworms (Lumbricidae) on nitrogen mineralization in different ecosystems on limestone. Mitteilungen des Hamburger Zoologischen Instituts und Museums 89:129–137

Scheu S (1995) Bodenfauna und Bodenmikroflora in Buchenwäldern: Wechselwirkung von Populationen, Stoffflüsse und Pflanzenwachstum. Göttingen: Habilitationsschrift

Scheu S (1997) Effect of litter (beech and stinging nettle) and earthworms (*Octolasion lacteum*) on carbon and nutrient cycling in beech forests on a basalt-limestone gradient: a laboratory experiment. Biol Fertil Soils 24:384–393

Scheu S, Falca M (2000) The soil food web of two beech forests (*Fagus sylvatica*) of contrasting humus form: stable isotope analysis of a macro- and mesofauna dominated community. Oecologia 123:285–296

Scheu S, Poser G (1996) The soil macrofauna (Diplopoda, Isopoda, Lumbricidae and Chilopoda) near tree trunks in a beechwood on limestone: indications for stemflow induced changes in community structure. Appl Soil Ecol 3:115–125

Scheu S, Schaefer M (1998) Bottom-up control of the soil macrofauna community in a beechwod on limestone: manipulation of food resources. Ecology 79:1573–1585

Scheu S, Schauermann J (1994) Decomposition of roots and twigs: effects of wood type (beech and ash), diameter, site of exposure and macrofauna exclusion. Plant Soil 163:13–24

Scheu S, Schulz E (1994) Simulation of mesofauna mediated changes in litter decomposition: laboratory experiments with 14-C-labelled holocellulose. Pedobiologia 38:353–360

Scheu S, Setälä H (2002) Multitrophic interactions in decomposer food-webs. In: Tscharntke T, Hawkins BA (eds) Multitrophic level interactions. Cambridge University Press, Cambridge, pp 223–264

Scheu S, Wolters V (1991a) Buffering of the effect of acid rain on decomposition of C-14-labelled beech leaf litter by saprophagous invertebrates. Biol Fertil Soils 11:285–289

Scheu S, Wolters V (1991b) Influence of fragmentation and bioturbation on the decomposition of 14C-labelled beech leaf litter. Soil Biol Biochem 23:1029–1034

Schulz E, Scheu S (1994) Oribatid mite mediated changes in litter decomposition: model experiments with 14-C-labelled holocellulose. Pedobiologia 38:344–352

Swift MJ, Heal OW, Anderson JM (1979) Decomposition in terrestrial ecosystems. Blackwell, Oxford

Theenhaus A, Schaefer M (1995) The effects of clear-cutting and liming on the soil macrofauna of a beech forest. For Ecol Manage 77:35–51

Tiunov AV, Scheu S (1999) Microbial respiration, biomass, biovolume and nutrient status in burrow walls of *Lumbricus terrestris* L. (Lumbricidae). Soil Biol Biochem 31:2039–2048

Tiunov AV, Scheu S (2000) Microbial biomass, biovolume and respiration in *Lumbricus terrestris* L. cast material of different age. Soil Biol Biochem 32:265–275

Tiunov AV, Bonkowski M, Alphei J, Scheu S (2001) Microflora, protozoa and nematoda in *Lumbricus terrestris* burrow walls: a laboratory experiment. Pedobiologia 45:46–60

Verhoef HA, Brussaard L (1990) Decomposition and nitrogen mineralization in natural and agroecosystems: the contribution of soil animals. Biogeochemistry 11:175–211

Wardle DA (2002) Communities and ecosystems: linking the aboveground and belowground components. Princeton University Press, Princeton

Wardle DA, Lavelle P (1997) Linkages between soil biota, plant litter quality and decomposition. In: Cadisch G, Giller KE (eds) Driven by nature. Plant litter quality and decomposition. CABI, Wallingford, pp 107–124

Wise DH, Schaefer M (1994) Decomposition of leaf litter in a mull beech forest: comparison between canopy and herbaceous species. Pedobiologia 38:269–288

Wolters V (1991a) Soil invertebrates – effects on nutrient turnover and soil structure. Zeitschrift für Pflanzenernährung und Bodenkunde 154:389–402

Wolters V (1991b) Effects of acid rain on leaf litter decomposition in a beech forest on calcareous soil. Biol Fertil Soils 11:151–156

Wolters V, Joergensen RG (1992) Microbial carbon turnover in beech forest soils worked by *Aporrectodea caliginosa* (Savigny) (Oligochaeta Lumbricidae). Soil Biol Biochem 24:171–177

Wolters V, Schaefer M (1994) Effects of acid deposition on soil organisms and decomposition processes. In: Godbold DG, Hüttermann (eds) Effects of acid rain on forest processes. Wiley, New York, pp 83–127

Wolter C, Scheu S (1999) Changes in bacterial numbers and hyphal lengths during the gut passage through *Lumbricus terrestris* (Lumbricidae, Oligochaeta). Pedobiologia 43:891–900

Chapter 13
Nitrogen and Carbon Transformations

R. Brumme, J. Priess, C.P. Wang, M. Raubuch, G. Steinmetz, and H. Meyer

13.1 Introduction

Carbon and nitrogen are quantitatively important elements in forest ecosystems. Most of the carbon and nitrogen are stored in the soil and have a long turnover time. Only a small fraction of the N-pool is cycled every year through growth and litter fall and mineralisation of litter. Industrialisation has caused high acid and nitrogen inputs to ecosystems which have affected carbon and nitrogen stocks and mineralisation processes in forest ecosystems. In addition, global change of temperature and precipitation will also have a direct influence on carbon and nitrogen cycles in the short-term and on their stocks in the long-term.

In order to detect measurable changes in the carbon and nitrogen stocks of forest soils, several decades and sometimes centuries may be needed because they have by nature high spatial variability and occur in large stocks. Therefore, it is very difficult to obtain information on fluxes of these elements by measuring changes in the bulk stocks. Direct measurements of input and output fluxes may provide useful information to assess any changes of the bulk stocks, and on understanding of human-induced changes in the mineralisation processes, as long as the values of input and output fluxes can be accurately measured.

Net carbon mineralisation is usually measured as heterotrophic respiration in trenched, root-free plots in the field or in laboratory studies. Net nitrogen mineralisation and net nitrification are often studied by the in situ buried-bag method (Eno 1960; Runge 1971; Nadelhoffer et al. 1983), or the in situ tube method (Raison et al. 1987; Rapp et al. 1979). We used the leaching method in the field and laboratory experiment where the mineralised nitrogen was leached out of undisturbed soil cores through irrigation with artificial rainfall (Stanford and Smith 1972, modified by Mochoge and Beese 1983; Brumme and Aden 1995) in order to simulate close to field conditions.

The objectives of this study were: (1) to quantify the net N-mineralisation and net C-mineralisation in soils of three beech forest ecosystems which differed in their soil acidity, (2) to quantify the contribution of different soil depths to total net N-mineralisation and net C-mineralisation, (3) to include the C/N ratios of soil

R. Brumme and P.K. Khanna (eds.), *Functioning and Management of European Beech Ecosystems*, Ecological Studies 208,
DOI: 10.1007/978-3-642-00340-0_13,

organic matter, the microbial biomass, and mineralisation products in gaining a better understanding of their interactions, (4) to discuss the relative nitrification in the context of autotrophic versus heterotrophic nitrification processes, and (5) to describe the temperature sensitivity of net N-mineralisation, net C-mineralisation, and nitrification processes.

13.2 Experimental Details

Two experiments, one field-based study and the laboratory-based, were conducted to study nitrogen and carbon transformations in the forest soils of the Solling, Göttinger Wald and Zierenberg sites. In the *field study*, eight undisturbed soil cores (15 cm diameter, 20 cm length) from each of the sites were collected and placed at the Solling site in such a way that the surface of the soil cores was in level with the surrounding soil (Raubuch 1997). The cores were kept covered at the surface to prevent natural rainfall but were open between the soil cores and the lid to enable gas exchange. Every 2 weeks, all soil cores were irrigated with 486 mL of an artificial throughfall each (equal to an irrigation of 2.1 mm per day). Ammonium in the irrigation solution was applied at a rate of 10 kg N ha^{-1} per year as $(NH_4)_2SO_4$. Ceramic suction cups were installed at the bottom of sealed soil cores to collect leached water twice a month, which was combined to a monthly sample and analysed for nitrate and ammonium. The net nitrification, net ammonification and the sum of both, defined as the net nitrogen mineralisation, were calculated by subtracting the amount of N-input from that of the N-output in the leachate. Net carbon mineralisation was measured every 2 weeks by closing the top of the soil cores to take gas samples after 0 and 30 min of closure with evacuated glass bottles, which were then analysed for CO_2 concentration in the laboratory (Loftfield et al. 1997).

In the *laboratory study*, undisturbed soil cores (15 cm diameter) from different soil layers were used to study the vertical distribution of C- and N-transformations at different temperatures and for different form of N applied: at 3–4°C (Solling: $n = 4$) (Göttinger Wald: $n = 6$), at 7–8°C (Solling: NH_4^+, $n = 8$; NO_3^-, $n = 4$) (Göttinger Wald: $n = 3$), and at 16°C (Solling: $n = 4$) (Göttinger Wald: $n = 9$) (Wang 1998). From the Solling site, soil cores of the surface organic layer, 0–10 cm and 10–20 cm soil depth were collected, and from the Göttinger Wald site mineral soil layers were used (L–10 cm, 10–20 cm, and 20–30 cm soil depth). The L layer was included in the L–10 cm layer at Göttinger Wald. The soil cores were installed in a microcosm system (Hantschel et al. 1994) and automatically irrigated six times per day (Solling: 4 mm per day; Göttinger Wald: 6 mm per day) with an artificially prepared throughfall solution over a 6-month period. Nitrogen was applied at the rate of 20 kg N ha^{-1} per year either as ammonium or nitrate. CO_2 emissions were measured continuously in the automatic set up and were used to calculate net C-mineralisation. Soil solutions were sampled from the bottom of the soil cores, where a 0.45 μm nylon membrane (Hybond-N, Amersham, England) was placed, and analysed for ammonium and nitrate to calculate the net mineralisation and nitrification rates.

13.3 Nitrogen Mineralisation in Soil Cores

The amount of nitrogen leached from soil cores was primarily related to changes in soil temperature and water content. The water content in the field leaching study remained more or less constant near field capacity because of fortnightly irrigations, no transpiration and low soil evaporation. Soil temperature was therefore the primary factor affecting nitrate production and leaching (Fig. 13.1). The high rate of nitrate leaching at the beginning of the experiment was due to the experimental setup. An increase in temperature increased the nitrate leaching resulting in higher values at the beginning of September. There was a delayed response to nitrates measured due to the time required for the leaching of the nitrates produced in the surface layers of soil columns. Transport processes required about 1 month to move

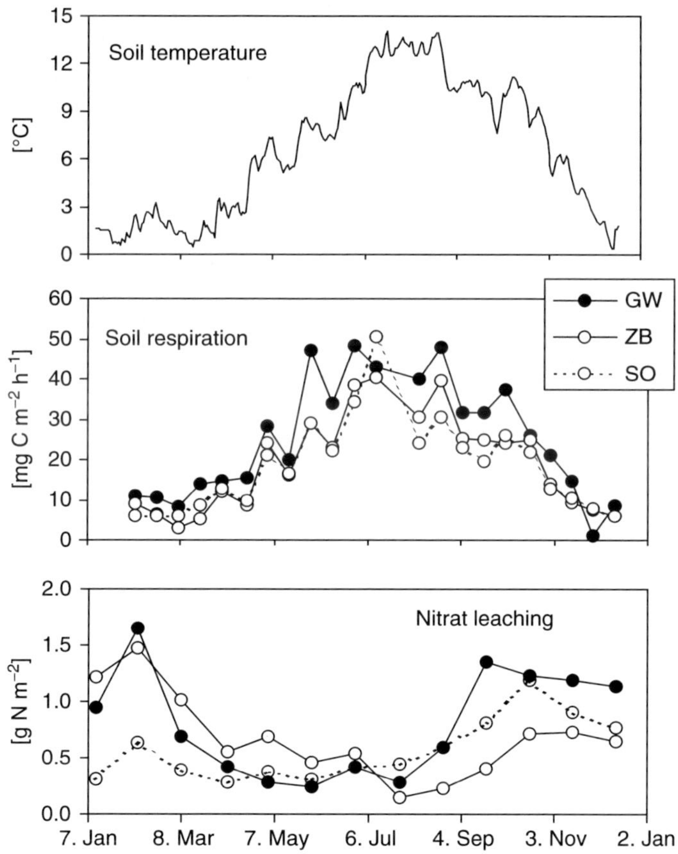

Fig. 13.1 Carbon mineralisation and nitrate leaching from soil columns in 20 cm depth of the Göttinger Wald (*GW*), Zierenberg (*ZB*) and Solling (*SO*) sites incubated at the Solling site (field experiment), soil temperature values were taken at 5 cm depth

nitrate through the soil cores of 20 cm length and explained some of the observed delay. Ammonium leaching was low from the Göttinger Wald and Zierenberg cores (respective values of 1% and 2% of the total mineral nitrogen leached) whereas it was high from Solling cores (7%).

Net N-mineralisation was calculated from the amount of nitrate and ammonium leached (Table 13.1). The high amount of nitrate leaching at the beginning of the experiment until March, which were assumed to be influenced by the experimental set up, were corrected assuming a linear decline of N-leaching between December and April. Since the soil cores were incubated at the Solling site, the net N-mineralisation was calculated for the higher soil temperatures of 7.4°C at Göttinger Wald and 8.1°C at Zierenberg, by using the Q_{10} values developed for net C-mineralisation (Chap. 17). The use of these Q_{10} values was justified on the basis that the differences in the temperature dependence between net C-mineralisation and net N-mineralisation seem to be small (Kirschbaum 1995). The estimated values of net N-mineralisation increased from 62 to 80 kg N ha^{-1} per year for Zierenberg soil and from 86 to 99 kg N ha^{-1} per year for Göttinger Wald soil (Table13.1). Thus, there is a trend of increasing net N-mineralisation rates with decreasing acidity from Solling and Zierenberg to Göttinger Wald.

Several methods have been used in the past to study the net N-mineralisation at the three beech sites. At the Solling site, plastic bags were used for in situ incubation of homogenised soils which were placed back at the same depth in the soil to which the sample belonged to (in situ buried-bag method; Runge 1971). The net mineralisation was calculated as the difference between the initial and final amount of ammonium and nitrate in the soil during 6 weeks of placement of bags in the field. Runge (1974a) calculated a mineralisation rate of 112 kg N ha^{-1} per year (Table 13.1) which was considered to be an overestimation due to high microbial activity in homogenised samples resulting from a better aeration after disturbance,

Table 13.1 Net nitrogen mineralisation rates (kg N ha^{-1} per year, L – 20 soil depth) from different studies (standard deviation in parenthesis) at the Göttinger Wald, Zierenberg, and Solling sites conducted with the in situ tube method, the in situ buried-bag method and the field leaching method

Method	Year of study	Göttinger Wald	Zierenberg	Solling	References
Buried bag	1967–1969	–	–	112	Runge (1974a)
Buried bag	1981	190	–	–	Reichardt (1982)
Tube	1991–1992	–	–	90 (17)	Bauhus and Barthel (1995)
Tube	1995–1996	–	128–167[c]	–	Mrotzek (1998)
Leaching	1991	86 (14)	62 (12)	74 (14)	this study[a]
Leaching	1991	99 (16)	80 (15)	–	this study[b]

[a]‘This study’ refer to the field leaching study where undisturbed soil cores were incubated at the Solling site over 1 year at a mean annual soil temperature of 6.2°C
[b]Net N-mineralisation for Göttinger Wald and Zierenberg were calculated for the annual soil temperatures at the sites, 7.4°C for Göttinger Wald, 8.1°C for Zierenberg
[c]Range of mean values of three subplots

a better access of organic matter by micro-organisms, and from decomposition of fine plant roots. The same method was used again by Ibrom and Runge (1989) to study the effect of liming and fertilisation from April to November 1986 at Solling. The mineralisation rate at the control site was 113 kg N ha^{-1} (L-12 cm) during 39 weeks.

Another method to study in situ mineralisation rates was developed by Raison et al. (1987). Steel or plastic tubes are hammered into the soil and the net mineralisation is calculated by the differences in the amount of ammonium and nitrate in the soil during the time of incubation, which was kept short enough to avoid any effect of disturbance and any decay of roots (in situ tube method). Bauhus and Barthel (1995) used this method to study the mineralisation rates of a beech forest gap and a control site at Solling. They found a mineralisation rate of 90 kg N ha^{-1} per year at the control site (Table 13.1).

Buried bags were also used at the Göttinger Wald site over 1 year by Reichardt (1982) who estimated a mineralisation rate of 190 kg N ha^{-1} per year (Table 13.1), which is about twice as high as our result. At Zierenberg, the in situ tube method was used by Mrotzek (1998) on three subplots. At one plot, *Mercuralis perennis* dominated the ground vegetation, another was dominated by *Urtica dioica*, and at the third plot, both plants were present. Similar mineralisation rates of about 128 kg N ha^{-1} per year were measured at the *Mericuralis* and *Urtica* plots, and 167 kg N ha^{-1} per year at the mixed plot (Table 13.1). The differences could not be explained and the expected higher mineralisation at the *Urtica* plot was not found. Compared to the leaching method, the mean net N-mineralisation rate of Mrotzek (1998) (136 kg N ha^{-1} per year) was 41% higher.

Nitrogen mineralisation has been shown to be very variable in space and time and the values obtained by different methods indicate that they are not always comparable. The comparison suggests that our field leaching method delivered the lowest net N-mineralisation followed by the in situ tube and in situ buried-bag method. There are probably three factors responsible for these differences. Adsorbed ammonium and litter production by herbs during the year is not covered in the leaching method in contrast to the buried-bag and tube method and may have lead to an underestimation of N-mineralisation in our experiment. By the use of the buried bag and the tube methods mycorrhizal fine roots were cut and additional nitrogen from mineralisation of dying mycorrhizal fine roots might increase net N-mineralisation. A monthly repeated incubation would add exudates and respiration of mycorrhizal fine root shortly after cutting and their subsequent decomposition 12 times per year in contrast to leaching method where root mineralisation is included only once. However, decomposition of root is a slow process (Chap. 12) and N-mineralisation of mycorrhizal fungi is not known and prevents a conclusive evaluation of the three methods. Another factor is related to changes in soil density, which will influence the accessibility of physically protected soil organic matter by microorganisms through the aeration of the soil. Comparing disturbed and undisturbed soil samples from Solling and Göttinger Wald indicated an increase or decrease of net N-mineralisation depending on the availability of C to drive the process and that of excess C to immobilise the mineralised amount. A positive effect between 35–140% at Solling and 4–41% at Göttinger Wald was observed in the surface 5 cm of the mineral soil (Beese 1986). In soil from 5 to 20 cm depth, mineralisation rates either increased or decreased. Any marked effect on net N-mineralisation was not found by comparing sieved and intact soil cores after adjusting for differences in soil moisture (Persson et al. 2000b). Effect of disturbance is evident in the in situ buried-bag method, but a recent study (Vor and Brumme 2002) indicated that even the in situ tube method, which is expected to create less disturbances, is affected by the experimental design. Comparing CO_2 respiration in the in situ tube method and the closed chamber technique indicated an increase in microbial activity (80–260% at different plots). A simultaneously large increase in N_2O emissions reduced the net N-mineralisation by between zero and 80%. A comparison with the in situ buried-bag method in the same study showed that net N-mineralisation was on an average higher by using buried bags (+25 – + 104%). It was concluded that soil disturbance reduced denitrification losses and/or enhanced mineralisation in the buried bags. To avoid soil disturbances in the tube method, it was recommended to increase the inner diameter of the incubation tubes to reduce edge effects.

Comparing the net N-mineralisation of the leaching (74 kg N ha^{-1} per year) and tube method (90 kg N ha^{-1} per year) at 8.1°C at Solling site with N-input by tree litterfall (68 kg N ha^{-1} per year, Chap. 11), fine-root litter (36–39 kg N ha^{-1} per year, Chap. 9) and by herb litter (0.4 kg N ha^{-1} per year, Chap. 10), in total about 106 kg N ha^{-1} per year, would indicate that nitrogen is accumulated in the soil. N-accumulation was confirmed by long-term observations of the C- and N-stocks at Solling where 21 kg N ha^{-1} per year accumulated in the surface organic layer (Chap. 4). The higher value for N-mineralisation by the buried-bag method in comparison to the N-input by litter indicates that this method may overestimate the N-mineralisation.

The net N-mineralisation derived from the field leaching method at Göttinger Wald of 99 kg N ha^{-1} per year is about 12% lower compared to litter N-input by trees, fine-roots and herbs litter (in total about 112 kg N ha^{-1} per year). This would suggest that the soil is accumulating nitrogen. The two times higher N-mineralisation analysed with the buried-bag method suggest a stimulation of mineralisation which is not evident from the input/output measurements (Chap. 16).

For the Zierenberg site the leaching method calculated 80 kg N ha^{-1} per year and the tube method 128–167 kg N ha^{-1} per year of which the value of the tube method is comparable to litter N-input by trees, fine-roots, and herbs (in total about 133 kg N ha^{-1} per year assuming a similar value for fine roots as observed at Göttinger Wald). Eichhorn and Hüttermann (1994) observed a high N-leaching and concluded that the Zierenberg forest is saturated with nitrogen and loses nitrogen by humus disintegration. N-mineralisation in excess of litter N-input may be present at Zierenberg site since the deeper soil layers have not been considered in the mineralisation study. The leaching method underestimated the N-mineralisation in part because litter from nettle plants was excluded.

13.3.1 Depthwise Distributions of Net N-Mineralisation

The depthwise distributions of nitrogen mineralisation in the soil cores at Göttinger Wald and Solling were studied in the laboratory experiment using undisturbed soil cores. At the Solling site, 77% of the total net N-mineralisation (L-20 cm) was measured at 8°C in the surface organic layer, 16% in 0–10 cm and 7% in 10–20 cm depth (Fig. 13.2). These values are in the upper range of forest soils with a FH layer within a European gradient (34–79% in LFH from L–20 cm, $n = 8$; Persson et al. 2000b). Most active in mineralisation is the F horizon as observed by a study with the in situ buried-bag method (Ibrom and Runge 1989). The net N-mineralisation in the F1 and F2 horizons was 240 mg N 100 g^{-1} which was four and six times higher than in the H and L horizons, respectively. The relative contribution of the L horizon to the total net N-mineralisation in the surface organic layer with a C/N ratio of 29 was low, but increased to 16% and 39% in the F1 and F2 horizons with a C/N ratio of 22 and 21, respectively (Fig. 13.3). A C/N ratio between 20 and 25 has been found to be necessary for a substrate to cause a net release of nitrogen (Killham 1994; Swift et al. 1979). Further narrowing of the C/N ratio to 19 did not increase but decreased the net N-mineralisation in the H horizon which might indicate that nitrogen was chemically stabilised by the formation of humic substances in this layer. A similar result was found by Bauhus (1994) and Bauhus and Barthel (1995) at an adjacent beech stand at Solling using the in situ tube method of Raison et al. (1987).

For the Göttinger Wald soil, the thin L layer was not studied separately but included in the L–10 cm layer in the laboratory study. This L–10 cm layer contributed to 59% of the net N-mineralisation in the surface 30 cm at 7°C (Fig. 13.2) and indicated the importance of the surface mineral soil for N-transformation at Göttinger Wald. This is in contrast to the acid mineral soil at Solling where only 16% of the net N-mineralisation occurred in 0–10 cm mineral soil

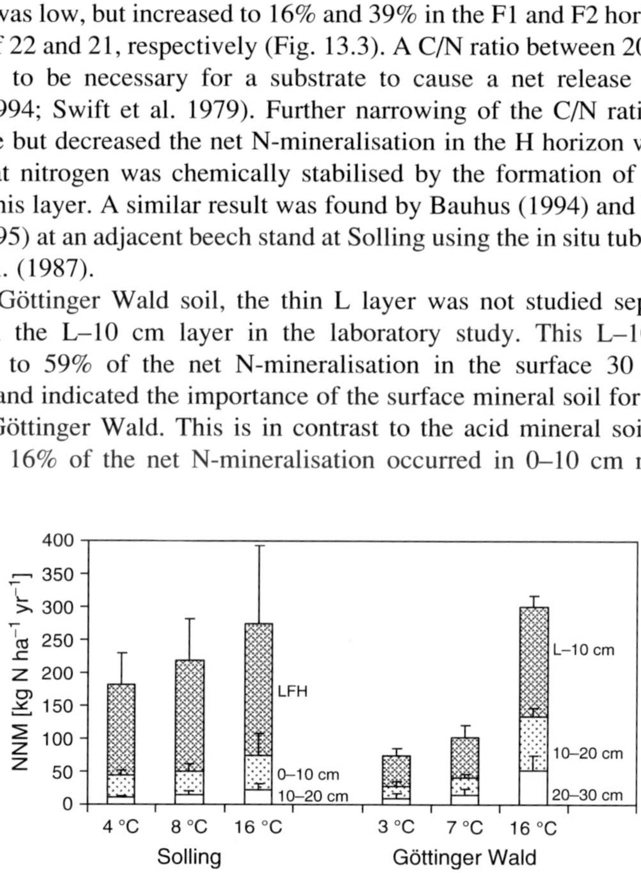

Fig. 13.2 Means and standard deviation of net N-mineralisation (*NNM*) in the soils of Solling with moder humus type (surface organic layer (LFH), 0–10 cm, 10–20 cm) and Göttinger Wald with mull humus type (L–10 cm, 10–20 cm, 20–30 cm) at three different temperatures in the laboratory experiment

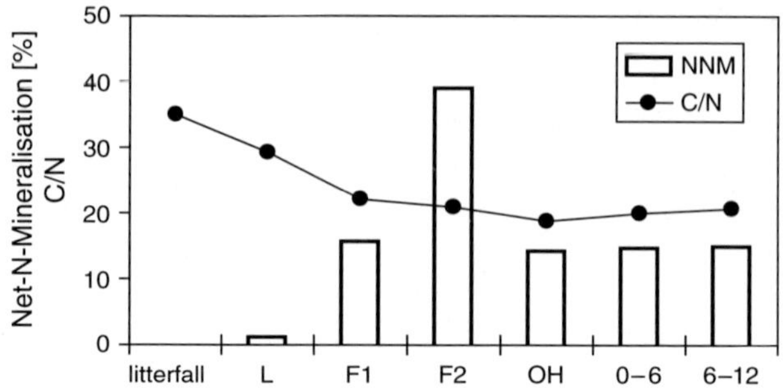

Fig. 13.3 Depthwise distribution of relative Net-N-Mineralisation (*NNM*) which was measured by the buried-bag method from April to December 1986 are shown together with the C/N ratios in the litterfall, the horizons of the moder humus and the upper 12 cm of the mineral soil at Solling (Ibrom and Runge 1989)

depth. Even the 10–20 cm and the 20–30 cm soil layers at Göttinger Wald contributed significant amounts of 26% and 15% to the net N-mineralisation, and may probably result from litter incorporation by earthworms.

13.3.2 Temperature Sensitivity of Net N-Mineralisation

The net N-mineralisation values in the laboratory study increased with incubation temperatures, and were 75 (±25) kg N (3°C), 100 (±32) kg N (7°C) and 300 (±55) kg N (16°C) ha^{-1} per year for L–30 cm soil depth from Göttinger Wald (Fig. 13.2). The rate at 7°C is comparable to the value found with the field leaching method (Table 13.1). In another incubation study with the same method, net N-mineralisation rates of 137 (±23) (8°C) and 370 (±26) (23°C) kg N ha^{-1} per year were calculated for the surface 30 cm of the Göttinger Wald soil (Brumme and Beese 1991).

For the Solling soil, net N-mineralisation rates increased from 180 (±57) kg N ha^{-1} per year (4°C) and 220 (±79) kg N ha^{-1} per year (8°C) to 273 (±162) kg N ha^{-1} per year (16°C) for the L–20 cm soil sample in the laboratory experiment (Fig. 13.2). Compared with the field leaching method, these rates are more than two times higher (Table 13.1), and are probably induced by high water content under the conditions of semi-continuous water leaching. This condition has probably induced higher leaching losses of nitrate from the N-enriched moder like humus in contrast to the mull humus type at the Göttinger Wald site. The Q_{10} values of net N-mineralisation for Solling soil were lower (1.3 at the surface organic layer, 1.5 at 0–10 cm and 1.7 at 10–20 cm depth) than those for Göttinger Wald (2.9 at the L–10 cm, 3.4 at 10–20 cm and 3.8 at 20–30 cm depth), and suggest that the enrichment of N in the surface organic layer from atmospheric deposition has reduced the temperature sensitivity at the Solling site. Similar Q_{10} values were calculated from an incubation study by Beese (1986) for Göttinger Wald soil. In that study, the net N-mineralisation for undisturbed soil cores incubated for 160 days at 8°C, 16°C and 24°C showed Q_{10} values between 2.3 and 2.8 (0–5 cm depth soil sample). A literature review by Kirschbaum (1995) suggested that net N-mineralisation appeared to have overall lower temperature sensitivities and had smaller increases in temperature sensitivity at low temperatures than the similar data for CO_2 efflux. This finding could be explained by higher C metabolism for the maintenance of a constant microbial biomass under elevated temperatures which increased the microbial demand for carbon more than for nitrogen.

13.4 Carbon Mineralisation

The net C-mineralisation values for the soils from Solling, Zierenberg and Göttinger Wald studied in the field leaching method under the environmental conditions of the Solling site are given in Table 13.2. A higher mean annual soil temperature at the Göttinger Wald and Zierenberg sites than at the Solling site required a correction for comparison purposes. This was done by using Q_{10} values and resulted in 2.1 Mg C ha^{-1} per year for Zierenberg and 2.4 Mg C ha^{-1} per year for Göttinger Wald (Table 13.2). These net C-mineralisation rates are very low when compared to leaf litter input from trees (2.6–2.7 Mg C ha^{-1} per year at the three beech sites; Chap. 11). Additional litter produced by roots (1.3 Mg C ha^{-1} per year at Solling and 0.6 Mg C ha^{-1} per year at Göttinger Wald; Chap. 9) and herbaceous plants (0.01 Mg C ha^{-1} per year at Solling, 0.22 Mg C ha^{-1} per year at Göttinger Wald and 0.42 Mg C ha^{-1} per year at Zierenberg; Chap. 10) were not covered by the calculated respiration values. This would indicate accumulation of large amounts of C in the soils.

The long-term measurements at the Solling site suggested that the C stock in the mineral soil remained constant while C stock in the surface organic layer increased by 0.35 Mg C ha^{-1} per year (Chap. 4). A simple calculation showed that any unaccounted C from the input/output balance would have been detectable in the soil. For example, at the Solling site, the total litter input from leaves and roots amounted to 3.9 Mg C ha^{-1} per year of which 0.35 Mg C was found to accumulate in the surface organic layer and 1.57 Mg were respired. The remaining 2 Mg C ha^{-1} which is accumulated annually in the mineral soil would have amounted to 70 Mg C ha^{-1} within 35 years, the long-term observation period at Solling. This value is higher than the standard deviation of the C stock of 26 Mg C ha^{-1} (0–20 cm;

Table 13.2 In situ net C-mineralisation rates (Mg ha^{-1} per year, standard deviation in parenthesis) studied on undisturbed soil cores (20 cm length) from Göttinger Wald (GW), Zierenberg (ZB), and Solling and incubated at the Solling site over 1 year at a mean annual soil temperature of 6.2°C. Net N-mineralisation for Göttinger Wald and Zierenberg were also calculated for the annual soil temperatures at the sites, 7.4°C for Göttinger Wald, 8.1°C for Zierenberg

Mean annual soil temperature	Göttinger Wald	Zierenberg	Solling
6.2°C	2.06 (0.23)	1.60 (0.23)	1.57 (0.20)
7.4°C (GW), 8.1°C (ZB)	2.4 (0.27)	2.1 (0.30)	–

Chap. 4). Therefore, it seems plausible that net C-mineralisation measured in the field leached cores only covered the decomposition of tree leaf litter and not root-associated carbon release and litter from herbs. The girdling (Högberg et al. 2001) and gap studies (Chap. 17) showed that up to half of the carbon released from the soil is from living plant roots, their mycorrhizal fungi and other root-associated microbes and that this release is driven directly by recent photosynthesis. Mineralisation measurements change the carbohydrate flux from the plant via mycorrhizae to soil organisms (Högberg and Read 2006) and thus underestimate conventional mineralisation measurements.

13.4.1 Depthwise Distributions of Net C-Mineralisation

The depthwise distributions of carbon mineralisation were studied in the laboratory experiment by the use of undisturbed soil cores. Similar to net N-mineralisation, the net C-mineralisation was highest in the surface organic layer at the Solling site (73% of the total net C-mineralisation in L–20 cm depth at 8°C) (Fig. 13.4). Within the surface organic layer when measured at 22°C, a strong decrease in respiration occurred from 59 g CO_2–C h^{-1} per g organic matter in the L layer to 12 and 4 μg CO_2–C h^{-1} g^{-1} in the F and H layers. If the respiration was related to soil organic carbon, the respiration still decreased from 118 g CO_2–C g^{-1} C_{org} h^{-1} in the L layer to 25 and 10 μg in the F and H layers. Persson et al. (2000a) reported values for the L layer in the range of 26–62 μg CO_2–C g^{-1} C h^{-1} and for the FH layer 4–31 μg CO_2–C g^{-1} C_{org} h^{-1} from 15 European forest soils. This indicates that less amount of carbon is available for microorganisms from more decomposed F and H layers.

At Göttinger Wald, the surface mineral soil (L–10 cm) contributed 72% of the total net C-mineralisation in L–30 cm depth and this was 13% higher than the similar values for net N-mineralisation in the same layer. In 10–20 cm soil depth,

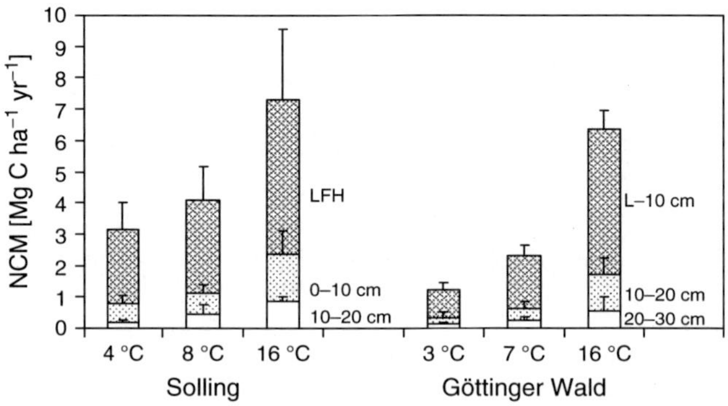

Fig. 13.4 Means and standard deviation of net C-mineralisation (*NCM*) in the soils of Solling (surface organic layer (*LFH*), 0–10 cm, 10–20 cm) and Göttinger Wald (L–10 cm, 10–20 cm, 20–30 cm) at three different temperatures in the laboratory experiment

the net C-mineralisation was 10% lower than those for net N-mineralisation. This indicated that the net C-mineralisation was high in the surface soil whereas lower depths were relatively more important for net N-mineralisation.

13.4.2 Temperature Sensitivity of Net C-Mineralisation

The seasonal patterns of net C-mineralisation were similar at Göttinger Wald, Zierenberg, and Solling soils and followed the course of the soil temperature in our field experiment conducted at the Solling site (Fig. 13.1). The net C-mineralisation in the laboratory study increased with the incubation temperature. Net C-mineralisation increased from 1.2(±0.49) Mg at 3°C, to 2.3(±0.64) Mg at 8°C, to 6.4(±1.6) Mg C ha^{-1} per year and 16°C in the surface soil (L–'30 cm) of Göttinger Wald (Fig. 13.4), which were similar as found under field conditions. The net C-mineralisation at Solling site increased from 3.3(±1.2) Mg at 4°C, to 4.1(±1.6) Mg at 8°C, to 7.3 (±3.2) Mg C ha^{-1} per year at 16°C (L–20 cm) and was about two times higher than those measured in the field study. Two times higher values under laboratory conditions than those measured under field conditions were also observed for net N-mineralisation in the Solling soil and suggest that both rates were stimulated by the irrigation which was done six times per day in this N-enriched moder-like humus layer at the Solling site. The calculated Q_{10} values from the laboratory data were 3.4 at L–10 cm, 3.6 at 10–20 cm, and 2.4 at 20–30 cm samples from Göttinger Wald, and 1.9 at the surface organic layer, 2.5 at 0–10 cm, and 2.7 at 10–20 cm samples for the Solling soil. The Q_{10} value for the Göttinger Wald soil was the same as found in the field experiment ($Q_{10} = 3.3$). The very low values in the laboratory study compared to the field study ($Q_{10} = 3.4$) at the Solling site resulted from high mineralisation rates at low temperatures.

13.5 Relationship Between N-Mineralisation and C-Mineralisation

The ratio between net carbon mineralisation and net nitrogen mineralisation ratio in the field leaching study was higher (24–26) than the C/N ratios of soil organic matter (15) at the Göttinger Wald and Zierenberg sites (Fig. 13.5). Micro-organisms mineralise organic matter for their energy metabolism (maintenance) and for construction of microbial tissue. Nitrogen is released to the soil solution in excess of what remains after microbial immobilisation. Due to low C/N ratio of micro-organisms biomass they immobilise more nitrogen than carbon and increased the C/N ratio in the mineralised products as was observed at Göttinger Wald and Zierenberg. The ratio between net C-mineralisation and net N-mineralisation (21.2) in Solling soil was similar to C/N ratios of soil organic matter (18.9) suggesting that microbial growth was less in this soil.

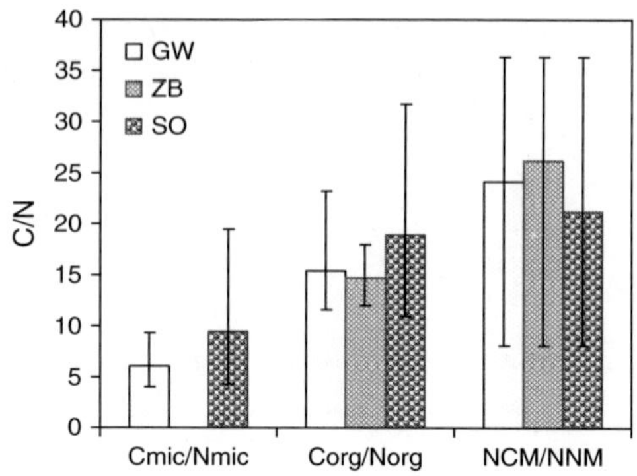

Fig. 13.5 C/N quotients end estimations of errors based on error propagation of the microbial biomass (C_{mic}/N_{mic}, CFE method), the soil organic matter (C_{org}/N_{org}, calculated by considering fixed ammonium), and mineralisation of carbon and nitrogen (*NCM*/*NNM*) in the field leaching experiment for Göttinger Wald (*GW*), Solling (*SO*) and Zierenberg (*ZB*) sites (no value for C_{mic}/N_{mic} at ZB)

A simple calculation may quantify the interaction between carbon and nitrogen pools (microbial biomass, soil organic matter: SOM) and fluxes (mineralisation, immobilisation) Equation 13.1 assuming that gaseous N losses were negligible. The following parameters in equations (13.1) and (13.2) are known from the field experiment: the fluxes of net C-mineralisation (NCM) (Table 13.2) and net N-mineralisation (NNM) (Table 13.1), and the C/N ratios of the soil organic matter (C/N_{SOM}, calculated by considering fixed ammonium of about 8% of total N in the top 30 cm of the Göttinger Wald and Solling sites, Beese 1986; Brumme 1986) and the microbial biomass (C/N_{MB}) (Chap. 6). Thus, there are two unknown parameters in equations (13.1) and (13.2) which can be calculated as follows. The amount of carbon immobilised by micro-organisms is found by following the equation (13.3) which was derived by combining equations (13.1) and (13.2). Net N-immobilisation (NNI) is subsequently derived from equation (13.2).

$$(\mathrm{NCI} + \mathrm{NCM})/(\mathrm{NNI} + \mathrm{NNM}) \approx C/N_{SOM}, \qquad (13.1)$$

$$\mathrm{NNI} = \mathrm{NCI} \,/\, C/N_{mic}, \qquad (13.2)$$

$$\mathrm{NCI} = C/N_{SOM} \,/\mathrm{NNM} - \mathrm{NCM}/(1 - (C/N_{SOM}/C/N_{mic})). \qquad (13.3)$$

The calculated microbial immobilisation reached up to 94 times microbial C and N at Göttinger Wald, and only 18 times microbial C and N at Solling per year (Fig. 13.6). Such a low microbial immobilisation at Solling suggests that the decomposition of different fractions of SOM occurred with different rates and not – as assumed in the calculation – with the same rate.

The laboratory study confirmed the finding in the field leaching method. Net C-mineralisation/net N-mineralisation values in the moder humus layer of the Solling soil were similar to the C/N ratios of soil organic matter while at Göttinger Wald the ratios of the fluxes were lower than for the organic matter (Fig. 13.7). We hypothesise that a fraction of the organic matter with a lower C/N ratio is

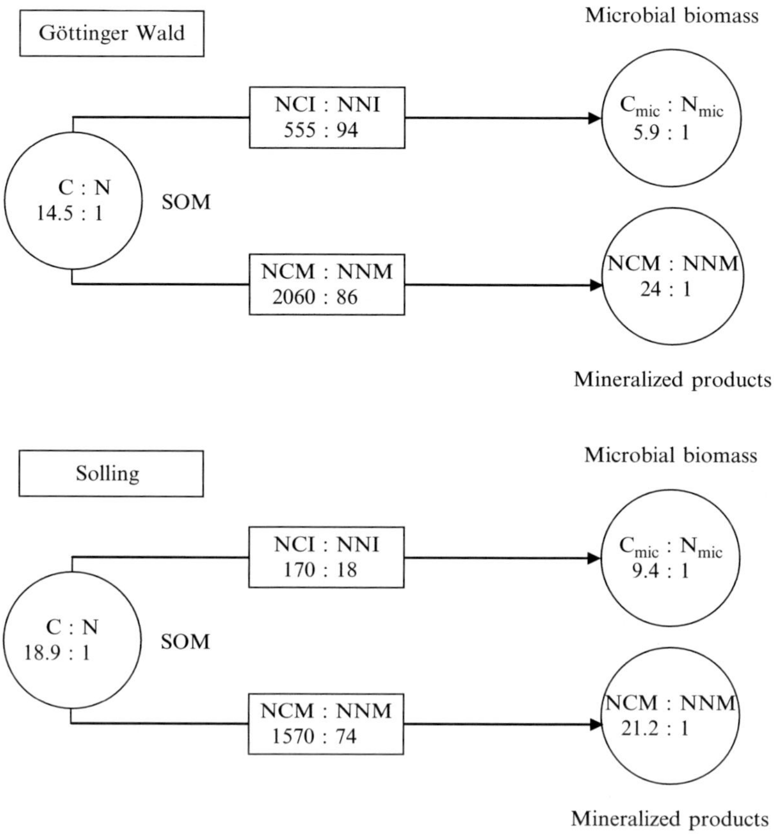

Fig. 13.6 Nitrogen and carbon transformation relationship between soil organic matter (*SOM*), microbial biomass and mineralisation products for Göttinger Wald and Solling. Data measured in the soil and field leaching experiment: net C-mineralisation (*NCM*), net N-mineralisation (*NNM*), C/N ratios of SOM and microbial biomas (C_{mic}, N_{mic}). Calculated values: net C-immobilisation (*NCI*), net N-immobilisation (*NNI*) (fluxes were given in kg ha^{-1} per year)

mineralised presumably in the F layer which has been identified as the layer with the highest net N-mineralisation. The F layer sample might have two types of pools. One with a low C/N ratio is preferentially respired and the other recalcitrant fraction which is later converted to H material.

13.6 Nitrification

Production of nitrate on oxidation of ammonium is the most important process causing leaching of nitrate from soils. Ammonium as a cation is adsorbed or fixed by clay minerals and its leaching losses are usually low. Nitrification is carried out primarily by autotrophic organisms, the dominant form of which occur in neutral or

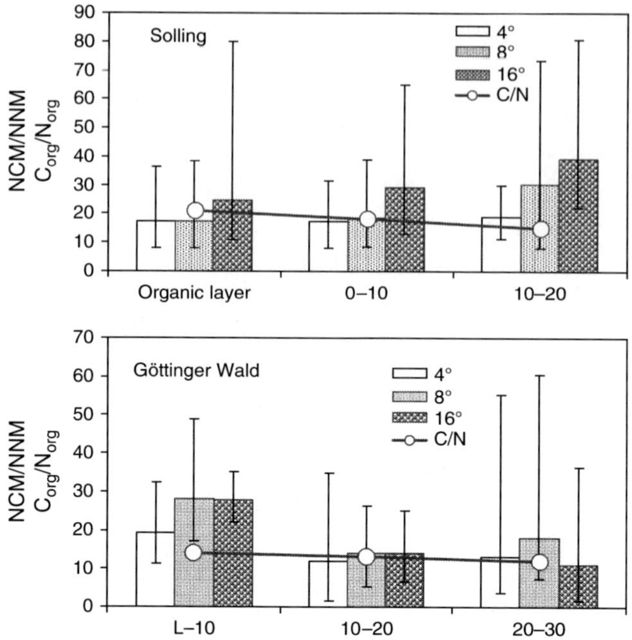

Fig. 13.7 Relationships between net carbon and net nitrogen mineralisation (*NCM/NNM*) and C/N ratios of soil organic matter (C_{org}/N_{org}, calculated by considering fixed ammonium) end estimations errors based on error propagation, in different soil depths and soil temperatures at the Solling and Göttinger Wald sites (Laboratory study)

less acid soils (gram-negative autotrophic bacteria *Nitrosomonas* and *Nitrobacter*). However, in addition to autotrophic nitrifiers, heterotrophic bacteria and certain fungi have been shown to nitrify organic N, especially in acid forest soils (Killham 1990; De Boer and Kowalchuk 2001). For heterotrophic nitrification, two pathways have been identified: an inorganic one which involves ammonium as an intermediate product in mineralisation, and an organic pathway where organic N is directly mineralised to nitrate (Killham 1990). It has been shown by the use of ^{15}N labelled ammonium that nitrifying fungi can nitrify organic nitrogen without the production of ammonium as an intermediate product (Schimel et al. 1984). But the predominance of heterotrophic fungal nitrifiers depends upon the type and quality of the decomposing substrate and soil pH (Killham 1986). In general, autotrophic nitrifiers are more acid sensitive than heterotrophic nitrifiers suggesting that an acid soil environment may be conducive for a dominantly heterotrophic nitrifying microbial population to develop.

We will consider here three issues: (1) the effect of soil pH on net nitrification rates, (2) autotrophic versus heterotrophic nitrification, and (3) the effect of temperature and moisture conditions on nitrification at the three beech forest soils of Solling, Göttinger Wald, and Zierenberg.

Table 13.3 Relative proportion of nitrate (%) in the in situ mineralised N (nitrate +ammonium) at Solling (*SO*), a *Mercurialis* and *Allium* dominated plot at Göttinger Wald (*GW*), and a *Mercurialis perennis* and *Urtica dioica* dominated plot at Zierenberg (*ZB*). (means, range in parentheses)

	GW[a]	SO[b]	ZB[c]
LF	97, 87	12 (7–25)	100, 100
H	–	51 (16–64)	–
0 - 5	99, 99	50 (0–77)	99, 100
5 - 10	92, 96	29 (0–86)	–
10 - 20	97, 100	16 (0–16)	88, 100

[a]Buried-bag method (Reichardt 1982), 14 samplings during 12 months
[b]Tube method (Bauhus 1994), 8 samplings during 12 months
[c]Tube method (Mrotzek 1998), 11 samplings during 12 months

Most investigations on nitrification are carried out under laboratory conditions where ammonium is added to the soil. Nitrification measured under such conditions provides a potential rate. An alternative approach is to use the relative net nitrification, where the proportion of mineral N present as nitrate during in situ mineralisation studies is used. Here, data from long-term in situ incubation studies using the buried-bag (Runge 1971) method and the sequential soil column method (Raison et al. 1987) (Table 13.3) will be presented.

More than 88% of mineralised N in all the soil layers at Göttinger Wald and Zierenberg consisted of nitrate (Table 13.3). At the Solling site, the fraction of nitrate was low (12–51% of total mineralised N) as reported by Bauhus (1994). The proportion of nitrate increased from the L to the H layer at this site (Runge 1974b). Only a negligible fraction of nitrate was produced in the L horizon, whereas the fraction was 15% in the F1 horizon and 27% in F2 horizon. In the H layer, 70% of total mineralised N occurred as nitrates after 5–7 weeks of buried-bag incubation (Fig. 13.8). Liming increased the nitrification when measured 4 years after 30 Mg ha^{-1} of finely ground dolomitic limestone were applied at Solling (BK plot) (Fig. 13.8). The proportion of nitrate produced increased up to about 100% during 6 weeks of incubation in the F and H layers and was attributed to high pH. Persson et al. (2000b) showed that nitrification generally increased very fast after raising the soil pH by 1.5–2 units in a laboratory study.

It is well known that low soil pH does not restrict nitrification (Killham 1994); Kriebitzsch (1978) found that nitrate was formed in most of the 84 acid forest soils studied in Germany. Only in forests where in the past litter was removed had ammonium accumulated. Robertson (1982) reported from the literature study of 125 forest soils from the temperate and tropical zones that soil pH, C/N or total nitrogen were not good predictors of nitrification across the sites. In contrast, Persson et al. (2000b) reported that relative net nitrification from boreal, temperate and Mediterranean forest soils which were incubated for 50 days under laboratory conditions, was positively correlated with pH and negatively with C/N ratio in L and FH layers. Soils from deeper depths showed no clear relationship to pH or C/N ratios, though net nitrification occurred even at low soil pH values of 3.5.

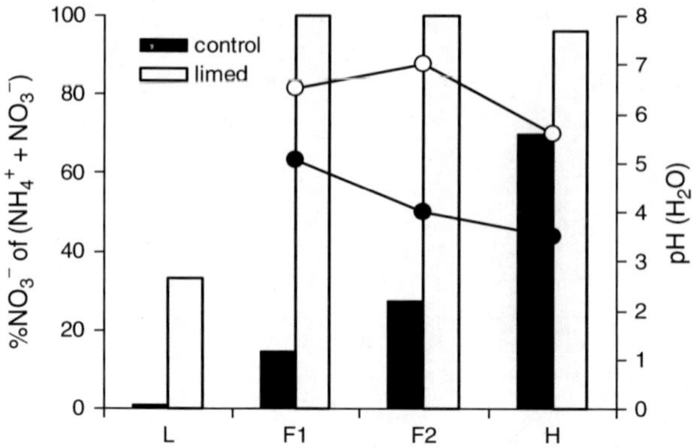

Fig. 13.8 Proportion of nitrate (%) of the total mineral N in the litter layer (*L*), fermentation layer (*F*), and humus layer (*H*) in a mineralisation study with buried bags at the Solling control (*B1*) and limed (*BK*) sites; pH(H_2O) of untreated (*filled circle*) and limed plots(*open circle*) are also given (Ibrom and Runge 1989)

13.6.1 Autotrophic Versus Heterotrophic Nitrification

Autotrophic nitrifiers were not detected in forest soil samples from Solling by using the 'most probable number' method (Lang and Beese 1985). The detection limit of most probable number method was about 200 germs g^{-1} which was far too low to explain any of the observed nitrification by autotrophic nitrifiers. In the soil from a limed plot, which had received 6 and 4 Mg ha^{-1} dolomitic limestone in 1975 and 1980, autotrophic nitrifiers had greatly increased in the surface organic layer when measured in 1983 and 1984. About 8–24 $\times$ 10^3 germs g^{-1} of nitrite and 15–53 $\times$ 10^3 germs g^{-1} of ammonium oxidiser were counted in the FH layer after liming (Lang 1986). Similar results were obtained in an acid forest soil in south Germany by Papen and von Berg (1998) where autotrophic ammonia oxidisers were not detected in the control but in a limed plot. In the less acid mineral soil of the Göttinger Wald site, similar numbers of germs as in the limed acid surface organic layer at Solling were also found (Lang 1986). The germ numbers ranged from 24 to 32 (nitrite) and from 32 to 75 $\times$ 10^3 germs g^{-1} (ammonium oxidiser) at 0–15 cm depth. This study indicated the pronounced effect of liming on the nitrifying population.

The longevity of such changes in autotropic population, though documented in literature, is not properly understood. For example, in a study by Rudebeck and Persson (1998), a comparison of short-term laboratory treatments with long-term (37–42 years) field treatments showed that, at the same soil pH, the long-term treatment had higher nitrification activity than the laboratory treatment. An explanation for the reappearance of autotrophic nitrifiers could be the presence of nitrifiers in soil pockets or micro-habitats which would then multiply quickly with an increase in the pH. It is also possible that strains of some acid-tolerant nitrifiers exist as suggested by De Boer et al. (1992). Even if autotrophic nitrifiers are not detectable with the most probable number method they may still exist in the soil pockets. Papen et al. (2002) found that autotrophic ammonia and nitrite oxidisers colonised in appreciable cell numbers the phyllosphere of spruce trees in a forest ecosystem which received high levels of atmospheric nitrogen for decades. The bacteria are predominantly located inside the spruce needles, most likely within the stomatal cavity. These authors suggested that the soil under spruce is being continuously inoculated by phyllosphere nitrifiers.

Heterotrophic nitrification is carried out by a wide range of bacteria and fungi (Remacle 1977; Focht and Verstraete 1977; Lang and Jagnow 1986; Ferguson et al. 2007) and does not appear to generate energy for growth in contrast to autotrophic nitrification (Prosser et al. 2007). NH_3 and NO_2^- oxidisers grow optimum in the pH range 7.5–8 and decreases sharply as pH falls below 7. For NH_3 oxidisers this is due to increasing ionisation of NH_3 which enters the cell by diffusion, to NH_4^+, whose uptake requires active transport (Prosser et al. 2007). At the Solling site, Remacle (1977) isolated many mould strains capable of nitrifying by using the 'continuous culture' technique. Lang and Jagnow (1986) found that 4–27% of the bacteria and 18–31% of the isolated fungi at the Solling were able to produce nitrite and nitrate. Papen and von Berg (1998) found that the fraction of chemo-organotrophic bacteria which were capable of carrying out heterotrophic nitrification, varied between 0 and 79% depending on the season and the presence of an acid surface organic layer under spruce. Nitrapyrin is used as an inhibitor for autotrophic nitrification (Gasser 1970). It reduced the nitrification rate in the surface mineral soil of Göttinger Wald by 80% whereas no reduction was observed at Solling (Lang 1986). Bauhus et al. (1996) found that nitrapyrin reduced the nitrate production in the limed organic layer at Solling by about 25% but no reduction was observed in the control plot. These results indicate that the heterotrophic nitrifiers dominated in the surface organic layer at Solling site, whereas the autotrophic nitrifies occurred in the surface mineral soil at Göttinger Wald. Lang (1986) reported that all of the isolated heterotrophic bacteria and fungi from Solling soil were able to produce small amount of nitrite and nitrate but only *Verticillum lecanti* (fungus) nitrified with rates similar to those observed under field conditions. Nevertheless, she concluded that the in vivo rates were probably too low to explain the field-observed nitrification rates because of its low fraction among 290 isolates of fungi observed.

13.6.2 Effects of Temperature and Water Stress on Nitrification

Amount of NO_3^- leached from the surface organic layer of the Solling site increased from 23% of the mineral N when incubated at 4°C to 52% at 16°C in the laboratory study. A similar increase in nitrate was observed in the surface soil of 0–10 cm depth between 4°C and 8°C (from 25 to 43% of the mineral N), and then after no further increase occurred between 8°C and 16°C. In soil samples from 10 to 20 cm depth, no temperature effect was observed in the fraction of nitrates which was lying between 62 and 70% of the total mineral N. These values are in the range

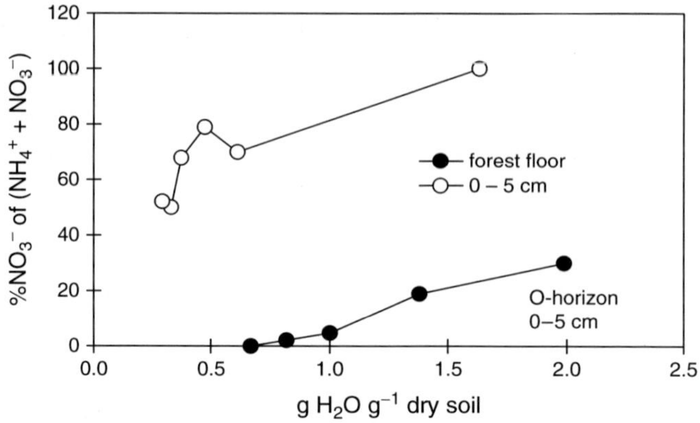

Fig. 13.9 Influence of gravimetric water content on the proportion of nitrate in the amount of N mineralised (%) in the surface organic layer (O-horizon) and 0–5 cm depth in an incubation experiment with Solling soil (Lang 1986)

of the observed values for in situ incubations (Table 13.3) indicating a temperature effect on nitrification. Low temperatures impeded the nitrification more than the ammonification and thus resulted in the built-up of exchangeable ammonium below 7°C (Frederick 1956; Campbell et al. 1971; Paul and Clark 1996). The optimum of nitrification occurs between 30°and 35°C with low values above 40°C.

Nitrification proceeds readily at –0.1 to –1 MPa moisture tension but is low in waterlogged soils because of restrictions in the diffusion of oxygen. It is also diminished in dry soils, because bacterial proliferation is retarded due to insufficiency of water (Paul and Clark 1996). Autotrophic nitrifiers are most sensitive to water stress, whereas fungi nitrifiers are generally much more tolerant (Killham 1990). Although heterotrophic nitrifiers dominated at Solling, the proportion of nitrate to mineralised nitrogen had decreased drastically from about 30% to zero with increasing dryness in the surface organic layer (Fig. 13.9). In the mineral soil, the nitrification rates were higher and the reduction was much lower than those observed for surface organic layer. The reduction of nitrate production at higher temperature and dryness indicates that nitrification may be reduced in summer months at the Solling site as reported by Runge (1974b).

13.7 Conclusions and Indications of Human Impacts

- Net N-mineralisation was studied using three different in situ methods (buried-bag method, tube method, field leaching method) at the three beech forest ecosystems. The highest rate was measured with the in situ buried-bag method fallowed by in situ tube method and the field leaching method. Of the three methods, the in situ tube method provided N-values which came close to the expected ones when the litter N-input by vegetation, input/output balances and inventories were considered. Net C-mineralisation was measured with the leaching method and was lower than the input of C by plant litter suggesting that litter from fine roots and herbs which was not included in the field leaching method might have underestimated the C (and N) mineralisation of soils.

- Proportions of net N-mineralisation and net C-mineralisation in the acid moder-humus from the Solling site were similar to those measured in the surface mineral soil of the less acid soil at the Göttinger Wald site. High soil acidity at the Solling site has reduced the earthworm and microbial activity in its mineral soil and has shifted the microbial activity to the surface organic layer.
- Lower C/N ratios of the mineralisation products than those of the soil organic matter suggest that a fraction of soil organic matter in the surface organic layer of the Solling soil with a high N concentration was preferential substrate for microbial respiration. This fraction might derive from immobilisation of atmospheric nitrogen in the surface organic layer.
- The acid soil of the Solling site contains a high fraction of heterotrophic nitrifiers causing low relative nitrification rates between 12 and 51% of the mineral N produced. However, autotrophic nitrifiers are active in the less acid soil at the Göttinger Wald site and are responsible for the transformation of more than 87% of net N-mineralisation to nitrate. Heterotrophic nitrifiers at the Solling site are sensitive to temperature and soil moisture changes resulting in low nitrification at low soil temperatures and under dry soil conditions.

References

Bauhus J (1994) Stoffumsätze in Lochhieben. Forest Ecosystems Research Centre. Göttingen University, A 113, 180 pp (in German)

Bauhus J, Barthel R (1995) Mechanisms for carbon and nutrient release and retention in beech forest gaps: II. The role of soil microbial biomass. Plant Soil 168–169:585–592

Bauhus J, Meyer AC, Brumme R (1996) Effect of the inhibitors nitrapyrin and sodium chlorate on nitrification and N_2O formation in an acid forest soil. Biol Fertil Soils 22:318–325

Beese F (1986) Parameter des Stickstoffumsatzes in Ökosystemen mit Böden unterschiedlicher Acidität. Göttinger Bodenkundliche Berichte 90

Brumme R (1986) Modelluntersuchungen zum Stofftransport und Stoffumsatz in einer Terra fusca-Rendzina auf Muschelkalk. Berichte des Forschungszentrums Waldökosysteme der Universität Göttingen, Reihe A, Bd. 24

Brumme R, Aden G (1995) The use of ^{15}N to study the N-turnover in soils. In: Alef K, Nannipiere P (eds) Methods in applied soil microbiology and biochemistry. Academic, London, pp 246–257

Brumme R, Beese F (1991) Simultane Bestimmung von N-Transformationsraten in Bodensäulen unter Verwendung von ^{15}N: Stickstoffmodell für eine Terra fusca-Rendzina. Z Pflanzenernähr Bodenk 154:205–210

Campbell CA, Biederbeck VO, Warder FG (1971) Influence of simulated fall and spring conditions on the soil system: II. Effect on soil nitrogen. Soil Sci Soc Am Proc 35:480–482

De Boer W, Kowalchuk GA (2001) Nitrification in acid soils: micro-organisms and mechanisms. Soil Biol Biochem 33:853–866

De Boer W, Tietema A, Klein Gunnewiek PJA, Laanbroek HJ (1992) The chemolithotrophic ammonium-oxidizing community in a nitrogen-saturated acid forest soil in relation to pH-dependent nitrifying activity. Soil Biol Biochem 24(3):229–234

Eno CF (1960) Nitrate production in the field by incubating the soil in polyethylene bags. Soil Sci Soc Am Proc 24:227–27

Eichhorn J, Hüttermann A (1994) Humus disintegration and nitrogen mineralization. In Godbold DL, Hüttermann A (eds) Effects of acid rain on forest processes. Wiley, New York, pp 129–162

Ferguson SJ, Richardson DJ, van Spanning RJM (2007) Biochemistry and molecular biology of nitrification. In: Bothe H, Ferguson SJ, Newton WE (eds) Biology of the nitrogen cycle. Elsevier, Amsterdam, pp 209–222

Frederick LR (1956) The formation of nitrate from ammonium nitrogen in soils: I. Effect of temperature. Soil Science Society Proceedings, Departments of Agronomy, and Botany and Plant Pathology pp 496–500

Focht DD, Verstraete W (1977) Nitrification and denitrification. Adv Microb Ecol 1:135–214

Gasser JKR (1970) Nitrification inhibitors – their occurence, production and effects of their use on cropyield and consumption. Soils Fertil 33:547–554

Hantschel RE, Flessa H, Beese F (1994) An automated microcosm system for studying ecological processes. Soil Sci Soc Am J 58:401–404

Högberg P, Read DJ (2006) Towards a more plant physiological perspective on soil ecology. Trends Ecol 21:548–554

Högberg P, Nordgren A, Buchmann N, Taylor A, Ekblad A, Högberg MN, Nyberg G, Ottosson-Löfvenius M, Read DJ (2001) Large-scale forest girdling shows that current photosynthesis drives soil respiration. Nature 411:789–792

Ibrom A, Runge M (1989) Die Stickstoff-Mineralisation im Boden eines Sauerhumus-Buchenwaldes unter dem Einfluß von Kalkung oder Stickstoffdüngung. In: Ulrich B (Hrsg) Ergebnisse neuerer, ökosystemarer Untersuchungen an Buchenbeständen im Solling. Berichte des Forschungszentrums Waldökosysteme A49:129–140

Killham K (1986) Heterotrophic nitrification. In: Prosser JI (ed) Nitrification. IRL, Oxford, Washington DC, pp 117–126

Killham K (1990) Nitrification in coniferous forest soils. Plant Soil 128:31–44

Killham K (1994) Soil ecology. Cambridge University Press, New York

Kirschbaum MUF (1995) The temperature dependence of soil organic matter decomposition, and the effect of global warming on soil organic C storage. Soil Biol Biochem 27:753–760

Kriebitzsch WU (1978) Stickstoffnachlieferung in sauren Waldböden Nordwestdeutschlands. Scripta Geobotanica, Band 14. Verlag Erich Goltze KG, Göttingen

Lang E (1986) Heterotrophe und autotrophe Nitrifikation untersucht an Bodenproben von drei Buchenstandorten. Göttinger Bodenkundliche Berichte 89:1–199

Lang E, Beese F (1985) Die Reaktion der mikrobiellen Bodenpopulation eines Buchenwaldes auf Kalkungsmaßnahmen. AFZ 43:1166–1169

Lang E, Jagnow G (1986) Fungi of a forest soil nitrifying at low pH values. FEMS Micorbiol Ecol 38:257–265

Loftfield N, Flessa H, Augustin J, Beese F (1997) Automated gas chromatographic system for rapid analysis of the atmospheric trace gases methane, carbon dioxide, and nitrous oxide. J Environ Qual 26:560–564

Mochoge B, Beese F (1983) The behaviour of nitrogen fertilizers in neutral and acid loess soils: II. Distribution and balances of 15 N-tagged nitrogen. Z Pflanzenernaehr Bodenk 146(1):504–515

Mrotzek R (1998) Wuchsdynamik und Mineralstoffhaushalt der Krautschicht in einem Buchenwald auf Basalt. Berichte des Forschungszentrums Waldökosysteme A152

Nadelhoffer KJ, Aber JD, Melillo JM (1983) Leaf litter production and soil organic matter dynamics along a nitrogen availability gradient in southern Wisconsin (USA). Can J For Res 13:12–21

Papen H, von Berg R (1998) A most probable number method (MPN) for the estimation of cell numbers of heterotrophic nitrifying bacteria in soil. Plant Soil 199:123–130

Papen H, Gessler A, Zumbusch E, Rennenberg H (2002) Chemolithoautotrophic nitrifiers in the phyllosphere of a spruce ecosystem receiving high atmospheric nitrogen input. Curr Microb 44:56–60

Paul EA, Clark FE (1996) Soil microbiology and biochemistry. Academic, San Diego

Persson T, Karlsson PS, Seyferth U, Sjöberg RM, Rudebeck A (2000a) Carbon mineralisation in European forest soils. In: Schulze ED (ed) Carbon and nitrogen cycling in European forest ecosystems. Springer, Berlin, pp 257–275

Persson T, Rudebeck A, Jussy JH, Colin-Belgrand M, Priemé A, Dambrine E, Karlsson PS, Sjöberg RM (2000b) Soil nitrogen turnover – mineralisation, nitrification and denitrification in European forest soils. In: Schulze ED (ed) Carbon and nitrogen cycling in European forest ecosystems. Springer, Berlin, pp 297–340

Prosser JI (2007) The Ecology of nitrifying bacteria. In: Bothe H, Ferguson SJ, Newton WE (eds) Biology of the nitrogen cycle. Elsevier, Amsterdam, pp 223–245

Rapp M, LeClerc MC, Loissant P (1979) Nitrogen economy in a *Pinus pinea* L stand. For Ecol Manage 2:221–231

Raison JR, Connell MK, Khanna PK (1987) Methodology for studying fluxes of mineral N in situ. Soil Biol Biochem 19:521–530

Raubuch M (1997) Field microcosmos, a new method for investigations of soil processes: calculation of mineralization and denitrification by using chloride as tracer. Verhandlung der Gesellschaft für Ökologie 27:323–328

Reichardt C (1982) Stickstoff-Nettomineralisation im Boden eines Kalkbuchenwaldes. Mitteilgn Dtsch Bodenkundl Gesellsch 34:33–38

Remacle J (1977) The role of heterotrophic nitrification in acid forest soils – Preliminary results. Ecol Bull 25:560–561

Robertson GP (1982) Nitrification in forested ecosystems. Phil Trans R Soc Lond B296:445–457

Rudebeck A, Persson T (1998) Nitrification in organic and mineral soil layers in coniferous forests in response to acidity. Environ Pollut 102:377–383

Runge M (1971) Investigation of the content and the production of mineral nitrogen in soils. In: Ellenberg H (ed) Integrated experimental ecology. Ecol Stud 2:191–202

Runge M (1974a) Die Stickstoff-Mineralisation im Boden eines Sauerhumus-Buchenwaldes. Teil I. – Mineralstickstoff-Gehalt und Netto-Mineralisation. Ecol Plant 9(3):201–217

Runge M (1974b) Die Stickstoff-Mineralisation im Boden eines Sauerhumus-Buchenwaldes. Teil II: Die Nitratproduktion. Ecol Plant 9(3):219–230

Schimel JP, Firestone MK, Killham KS (1984) Identification of heterotrophic nitrification in a Sierran forest soil. Appl Environ Microbiol 48(4):802–806

Stanford G, Smith SJ (1972) Nitrogen mineralization potentials of soils. Soil Sci Soc Am Proc 36:465–472

Swift MJ, Heal OW, Anderson JM (1979) Studies in ecology, Volume 5. Decomposition in Terrestrial Ecosystems. Blackwell, Oxford

Vor T, Brumme R (2002) N_2O losses result in underestimation of in situ determinations of net N mineralization. Soil Biol Biochem 34:541–544

Wang CP (1998) Soil and temperature influence on microbiological nitrogen transformations. Berichte des Forschungszentrums Waldökosysteme Univ Göttingen A156

Chapter 14
Fate, Transport, and Retention of Applied ^{15}N Labelled Nitrogen in Forest Soils

R. Brumme, C.P. Wang, J. Priess, M. Raubuch, and G. Steinmetz

14.1 Introduction

Forest soils in central Europe have received high atmospheric inputs of protons, sulphate and nitrogen (Chap. 15) since the beginning of industrialisation. As a result, chemical and biological processes in many German forest soils have changed (Godbold and Hüttermann 1994) and soil characteristics like pH, base saturation and C/N ratios have altered (Chap. 21). The mechanisms of the fate and the retention of this high atmospheric input of nitrogen still remain unclear. There are reports indicating that, despite high N-deposition since the beginning of the nineteenth century, most forests continue to retain nitrogen (Ulrich 1994; Johnson et al. 1997; Dise et al. 1998; Borken and Matzner 2004) even when the low C/N ratios in the surface organic layer have become low (Brumme and Khanna 2008). This raises the question of the retention mechanisms of nitrogen in forest soils.

The use of labelled nitrogen makes it easy to follow internal nitrogen cycle of N-inputs in forest ecosystems (Nadelhoffer et al. 2004; Tietema et al. 1998; Feng et al. 2008) and soils (Mochoge and Beese 1983a,b). Often, labelled nitrogen is applied in a single dose which provides information on the nitrogen transformation rates under the current environmental conditions of addition. We applied constant rates of ^{15}N labelled ammonium and nitrate as an artificial crown drip for long periods on undisturbed soil columns (Brumme and Aden 1995) under field and laboratory conditions in order to simulate continuous N-depositions so as to follow N-transformations under close to field conditions.

The aims of this study were to estimate: (1) the leaching losses of deposited ammonium and nitrate from soil of different acidity by adding ^{15}N labelled ammonium and nitrate on undisturbed soil cores from three beech forest ecosystems at the Göttinger Wald, Zierenberg and Solling sites, (2) to compare the transport of added ammonium and nitrate, which are biologically transformed, with chloride as a marker which is not biologically transformed, (3) to determine the retention processes of added ammonium and nitrate involved in forest soils, and (4) to study the temperature effect on N-transformation of added labelled nitrogen.

R. Brumme and P.K. Khanna (eds.), *Functioning and Management of European Beech Ecosystems*, Ecological Studies 208,
DOI: 10.1007/978-3-642-00340-0_14, © Springer-Verlag Berlin Heidelberg 2009

14.2 Experimental Details

The results presented in this chapter were derived from the same field and laboratory study as already described in the Chap. 13 "Nitrogen and carbon transformations." Undisturbed soil cores of the L–20 cm depth (15 cm diameter) from the Göttinger Wald, Zierenberg and Solling (n =8) sites were placed at the Solling site in such a way that the surface of the soil cores was in level with the surrounding soil (*field study*). The cores were kept covered at the surface to prevent natural rainfall but were open between the soil cores and the lid to enable gas exchange. Every 2 weeks, the soil cores were irrigated with 486 mL of an artificial throughfall (equal to an irrigation of 2.1 mm per day) which contained ^{15}N labelled ammonium sulphate (five of eight soil cores) (8.4 atom % $^{15}N_{ex}$, where ex refers to the ^{15}N in excess of the natural abundance) to achieve a deposition rate of 11 kg N ha^{-1} per year. The columns were closed at the bottom and ceramic suction cups enabled sampling and quantifying of leached water twice a month over a 12-month period, which were combined to a monthly sample and analysed for nitrate, ammonium and the labelled fraction.

In the *laboratory study*, undisturbed soil cores (15 cm diameter) from different soil layers were used to study the vertical distribution of C- and N-transformations at different temperatures and for different form of N-applied: at 3–4°C (Solling: $n = 4$ and Göttinger Wald: $n = 6$), at 7–8°C (Solling: $^{15}NH_4^+$, $n = 8$; $^{15}NO_3^-$, $n = 4$ and Göttinger Wald: $n = 3$), and at 16°C (Solling: $n = 4$ Göttinger Wald: $n = 9$) (Wang 1998). From the Solling site, soil cores of the surface organic layer, 0–10 cm, and 10–20 cm soil depth were collected, and from the Göttinger Wald site mineral soil layers were used (L–10 cm, 10–20 cm, and 20–30 cm soil depth). The L layer was included in the L–10 cm layer at Göttinger Wald. The soil cores were installed in a microcosm system (Hantschel et al. 1994) and automatically irrigated six times per day (Solling: 4 mm per day; Göttinger Wald: 6 mm per day) with an artificially prepared throughfall solution over a 4-month period. ^{15}N labelled nitrogen was applied at the rate of 20 kg N ha^{-1} per year either as ammonium or nitrate but in three different steps. During the first 3 weeks, nitrogen was applied as non-labelled NH_4NO_3 (preliminary period) to achieve a steady state condition in the water flow and in the microbial transformation processes. For the next 60 days, ^{15}N labelled NH_4NO_3 with either ammonium or nitrate ($^{15}N_{ex}$ 49.4 atom%) were applied in separate experiments together with Cl to study the anion transport processes in the soil layers. In the third period, non-labelled NH_4NO_3 was applied to study the remobilisation of ^{15}N from the soils over 4 weeks. Soil solutions were sampled from the bottom of the soil cores where a 0.45 μm nylon membrane (Hybond-N, Amersham, England) was placed, and analysed for ammonium and nitrate and the ^{15}N content to calculate the leaching losses of labelled ammonium and nitrate. Exchangeable ammonium and soluble nitrate were extracted from soil samples with 0.5 M K_2SO_4 at the end of the experiment. The ^{15}N content of ammonium and nitrate was analysed after using the diffusion technique (Brumme and Aden 1995). ^{15}N was measured with an element analyser coupled to an isotope ratio mass spectrometer (EA-IRMS) (Finnigan, Germany).

14.3 Transport and Leaching Losses of Added ^{15}N Labelled Nitrogen

The fate of ammonium and nitrate in plant-free soils depends on several processes such as adsorption and desorption, microbial uptake and remineralisation, fixation, denitrification and the physical transport described by convection–dispersion transport equations (Parker and van Genuchten 1984). In order to study the physical transport processes of anions, we used chloride simultaneous with ^{15}N labelled ammonium and nitrate in undisturbed soil columns at constant water content and steady-state water flow in the laboratory study according to Beese (1982) and Mochoge and Beese (1983a). Chloride does not enter biological and chemical reactions in soils and any difference in the fate of chloride and nitrate is attributed to microbial uptake of N and denitrification. However, for ammonium, there are additional processes such as nitrification, adsorption and fixation which should be considered. Transport studies during steady-state water flow showed that the breakthrough of applied chloride after pulse application culminated after the exchange of 0.5–0.9 pore water volume (which equals the volume of water in soil cores). The C/Co value of 1 was achieved for Cl after 2 pore water volumes had been replaced in soil columns independent of the soils used (Beese and van der Ploeg 1979). This is a typical response for anion transport in soils and is explained by a reduction of mobile soil water by anion exclusion and thereby increased the pore water velocity (i.e. irrigation intensity (cm per day) divided by volumetric water content (cm^3 cm^{-3})) (Bresler 1973; Beese 1982). Figure 14.1 showed a typical breakthrough of chloride for the Solling soil. After an exchange of 2–3 pore water volumes, the relative concentration C/Co of chloride approached the value of 1, and after that Cl-inputs to the surface of the columns were stopped. That caused the concentration of chloride in the leachates from cores with surface organic layer, 0–10 cm soil and 10–20 cm soils to decline to about zero after additional 2 pore water volume were applied. Labelled nitrate applied to the soil surface showed a small retardation in its transport compared to chloride and approximately 4 pore water volume were required for C/Co to approach the values of >0.8. At day 60 of the leaching experiment, when labelled nitrogen was replaced by unlabelled nitrogen (indicated by crosses in Fig. 14.1), the leaching losses reached between 95 and 99% of Co and indicated that there is nearly no retention of nitrate by microbial uptake or denitrification at the Solling site under these experimental conditions.

However, the leaching of applied ^{15}N labelled ammonium was slow in the Solling soil (not shown). Applied ammonium was partly nitrified and leached as nitrate after nitrification with a large delay in all soil layers (Fig. 14.1). Delayed nitrification resulted in retarded leaching of added ^{15}N labelled ammonium as indicated by a lower values (0.17) of relative concentration C/Co in the mineral soil layers and in the surface organic layer (0.33) than those observed when ^{15}N labelled nitrate (>0.95) was applied after 60 days. Another factor responsible for this retardation in the leaching of added ammonium in acid soils is its adsorption by

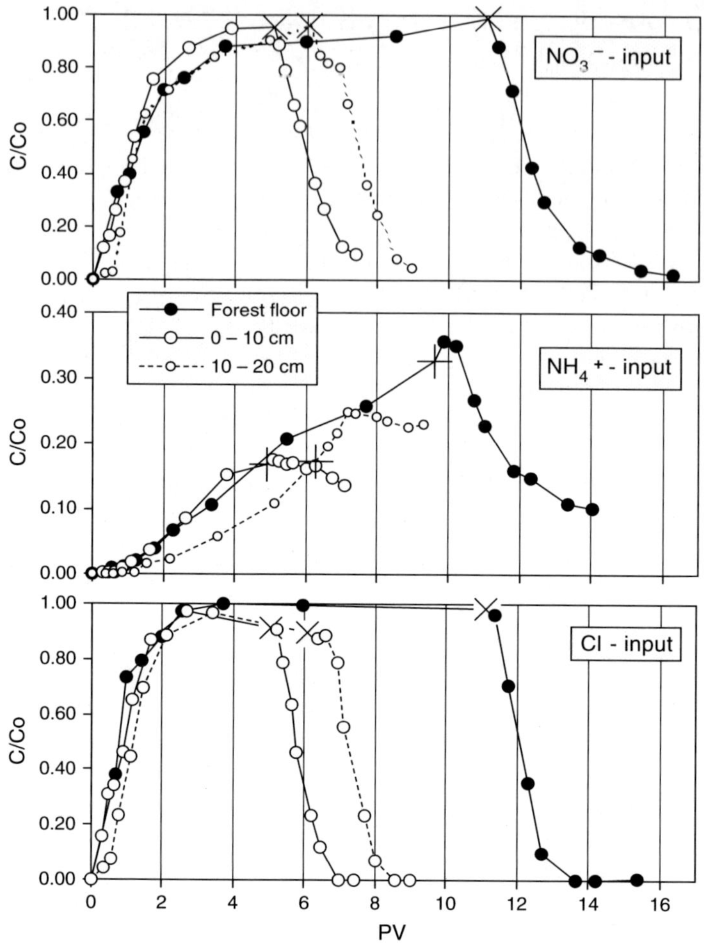

Fig. 14.1 Concentrations (C) of ^{15}N labelled nitrate and chloride in the leachates relative to the input concentration (Co) of ^{15}N labelled nitrate, ^{15}N labelled ammonium, and chloride (chloride is shown for the ^{15}N labelled nitrate treatment), applied to undisturbed soil cores of surface organic layer (forest floor), 0–10 and 10–20 cm depths from Solling at 8°C (laboratory study). Changes from labelled to non-labelled nitrogen and from chloride to sulphate application at day 60 are indicated by (×). (*PV* – Pore water volume in soil cores replaced)

negatively charged clay minerals and humus particles. The proportion of adsorbed ^{15}N labelled ammonium was higher in the mineral soil layers than in the surface organic layer at the end of the experiment (Table 14.1) despite similar nitrification rates (Chap. 13). The high adsorption of ammonium in the mineral soils caused a slow percolation of labelled nitrate after nitrification or even an increase after 60 days as observed in columns of 10–20 cm depth. Fast percolation of labelled nitrate in the surface organic layer is an indication of the lack of adsorption of ammonium in this layer.

Table 14.1 ^{15}N-balance after 60 days of irrigation (4 mm per day) with ^{15}N labelled ammonium at a rate of 20 kg N ha^{-1} per year and 30 days with non-labelled ammonium, on soils from three different soil depths from Solling at 8°C in the "laboratory study," (all values are given as % of ^{15}N applied) (n = 8)

Depth	Leachate			Soil				^{15}N-
	NO_3^-	NH_4^+	$\sum$	NO_3^-	NH_4^+	N_{org}[a]	$\sum$	Recovered
Organic layer	29 (13)	18 (4)	47 (20)	2	3	54	59 (17)	106 (29)
0–10	25 (18)	8 (3)	33 (20)	5	24	34	63 (17)	96 (38)
10–20	21 (8)	1 (1)	22 (9)	4	30	34	68 (9)	90 (18)

[a]Assuming a negligible contribution of N-fixation

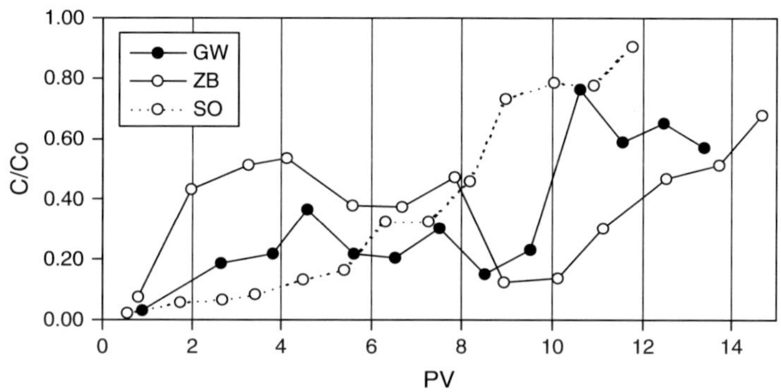

Fig. 14.2 Concentration (C) of ^{15}N labelled nitrate in the leachate relative to the concentration of applied ^{15}N labelled ammonium in the input (Co) from undisturbed soil columns (L–20 cm depth) of the Göttinger Wald (GW), Zierenberg (*ZB*), and Solling (*SO*) site (n = 8) in the field study, related to the water volume in the soil cores (pore water volume, *PV*) which was replaced by irrigation

In the field study, ^{15}N labelled nitrate and chloride were not added. However, labelled nitrate was produced through nitrification of applied ^{15}N labelled ammonium to the soil. Leaching of labelled ammonium was low (not shown). The behaviour of ^{15}N labelled nitrate in the leachates was different among the three beech forest soils of the Göttinger Wald, Zierenberg, and Solling sites. The concentration of ^{15}N labelled nitrate (C) in the leachate relative to the concentration of added ^{15}N labelled ammonium (Co) increased very quickly at the less acid soil from the Zierenberg site which was followed by those of the Göttinger Wald and Solling sites (Fig. 14.2). The leaching of ^{15}N labelled nitrate from the acid Solling soil was highly retarded as was also observed in the laboratory study involving separate soil layers.

A delay of leached nitrate from added ^{15}N labelled ammonium was also observed in a previous study with undisturbed soil cores from Solling site (Mochoge and Beese 1983a). In their study, leaching of nitrate was delayed by about 1 pore water volume after a pulse application of ^{15}N labelled ammonum sulphate with continuously applied artificially prepared throughfall solution at rates of 5 mm per day as compared to the case when ^{15}N labelled nitrate was added. The method and the amount of ammonium application will affect the rate of its transport through soil columns. Continuous application compared to pulse application of labelled ammonium may lead to differences in concentration and duration of labelled nitrogen in the soil. A pulse application may increase ammonium fixation in clay minerals if high doses were used, and may also cause some loss of ammonium by leaching. Continuously applied ammonium with rates in low concentrations similar to those in situ conditions in forest ecosystems may increase the biological retention. However, retention of ammonium by cation exchange in acid soils is of a transitory nature as indicated by the steady increase of leached ^{15}N labelled nitrate at the Solling site (Fig. 14.2). At leaching of 9 pore water volumes, a relatively constant nitrate level of C/Co = 0.8 was observed.

14.4 Retention of Applied Ammonium and ^{15}N Balance of Forest Soils

A number of factors affect the retention of ammonium by soils. Adsorption by cation exchange in the mineral soils constituted 24% (0–10 cm depth) and 30% (10–20 cm depth) of the applied ammonium in the laboratory study (Table 14.1). The high retention was observed even 30 days after inputs were changed from labelled to non-labelled ammonium. This indicated that the nitrification by heterotrophic organisms was low (see also Chap. 13) and has promoted the adsorption of ammonium in Solling soil. This would explain the retardation in leaching of N as observed in both the field (Fig. 14.2) and the laboratory studies (Fig. 14.1).

The surface organic layer retained only 3% of the applied ammonium on exchange complex and 18% of the applied ammonium was leached as ammonium (Table 14.1). In contrast to this acid soil, a continuous application of labelled ammonium sulphate to the less acid soil from Göttinger Wald resulted in a negligible adsorption of ammonium of 0.3% (Brumme and Beese 1991, Table 14.2). The lower adsorption at Göttinger Wald was related to high nitrification (see also Chap. 13). These results are in accordance with a study on undisturbed soil cores from different European countries. Application of labelled ammonium chloride to six forest soils showed the highest recovery as ammonium from soils with very low nitrification (Coûteaux and Sallih 1994).

Retention of ammonium by cation exchange is of a transitory nature in acid soils. More important for a long-term retention is the transformation of applied nitrogen into stable or meta-stable microbial-mediated organic compounds. Assuming that fixation of ammonium by clay was absent in the surface organic layer, about 54% of the added ammonium was transformed to microbial biomass or soil organic matter

Table 14.2 Balances of ^{15}N-ammonium of the Göttinger Wald soil 160 days after continuous application of 1 kg N ha^{-1} per day in form of $(NH_4)_2SO_4$ to undisturbed soil columns (0.25 m^2, 100 cm length, $n = 1$) irrigated with a rate of 3 mm per day at 8 and 23°C (% of ^{15}N applied) (Brumme and Beese 1991) (ex, exchangeable; fix, fixed; org, organic)

	Leachate			Soil					^{15}N-
	NO_3^-	NH_4^+	$\sum$	NO_3^-	NH_4^+(ex)	NH_4^+(fix)	N (org)	$\sum$	Recovery
8°C	35	0	35	27	0.3	0	25	52	87
23°C	32	0	32	27	0.3	0	29	56	88

Table 14.3 ^{15}N-balances after 1 year of irrigation (2.1 mm per day) with ^{15}N labelled ammonium sulphate at a rate of 11 kg N ha^{-1} per year in the Göttinger Wald (*GW*), Zierenberg (*ZB*) and Solling (*SO*) soil of the field study (% of ^{15}N applied, standard deviation) ($n = 5$)

	Leachate			Soil			^{15}N-recovered
	NO_3^-	NH_4^+	$\sum$	Organic layer	Mineral soil	$\sum$	
GW	38 (6.3)	0.3 (0.2)	38 (6.4)	4.0 (2.1)	41 (2.1)	45 (2.2)	84 (4.6)
ZB	43 (9.8)	0.5 (0.5)	43 (10)	7.4 (2.9)	26 (2.8)	34 (4.1)	77 (8.5)
SO	31 (8.4)	3.7 (3.3)	35 (11)	20 (3.9)	26 (5.6)	46 (4.8)	81 (7.5)

at the Solling site (Table 14.1). In the mineral soil depths, about 34% of added ammonium was retained by either fixation or transformation into organic compounds. The fixed fraction was not studied in our laboratory experiment, but the study of Mochoge and Beese (1983b) indicated that labelled ammonium could be fixed in clay minerals of the Solling soil to the extent of 17 and 15% at 4° and 23°C, respectively. Mochoge and Beese (1983b) applied 80 kg N ha^{-1} within 2 days which led to a very high ammonium concentration in the soil solution which may have increased the fixation of ammonium. Fixation of ammonium has been generally found to increase with an increase in the amount of ammonium added (Nommik and Vahtras 1982). The soil cores in our laboratory study received a low amount (20 kg N ha^{-1} distributed over 1 year), so the fixation of ammonium might be of minor importance and most of the 34% of the added amount found in the mineral soil depths was assumed to be bound to organic compounds.

The fraction of applied ammonium recovered in the surface organic layer at the end of the field study increased with the amount of organic matter from the less acid Göttinger Wald and Zierenberg sites to the acid Solling site (Table 14.3). The retention of applied ammonium per g Corg increased with an increase in microbial biomass per g Corg (Fig. 14.3) indicating a preferential microbial uptake of applied ammonium. The preferential use of ammonium than nitrate by microorganisms was shown in a ^{15}N-pool-dilution experiment at Solling site. Corre et al. (2003) incubated soil cores of the surface organic layer and the upper soil layer (0–5 cm) with ^{15}N labelled $(NH_4)_2SO_4$ and ^{15}N labelled KNO_3 and found a two times higher uptake of ammonium than nitrate.

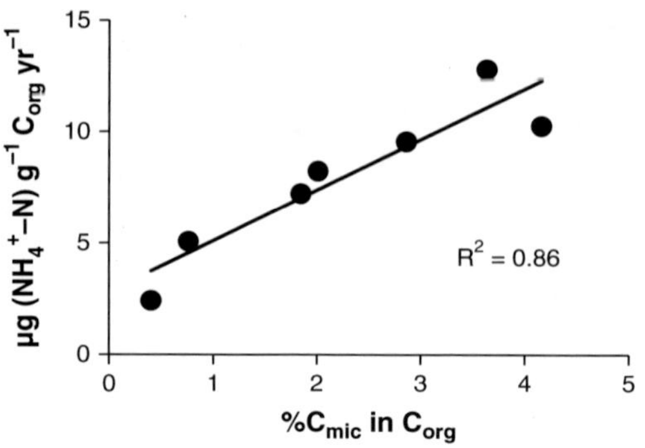

Fig. 14.3 Retention of applied ammonium in relation to microbial biomass-C of the L, F, and H layers of Solling, and the L layers and 0–5 cm layers of Göttinger Wald and Zierenberg after 12 months of application ^{15}N–NH_4^+ in undisturbed soil cores in the field study. Only the surface layers were used in this figure because of a pore water exchange by each irrigation event higher than one (1.6–1.9) to ensure ammonium infiltration into the layers

Table 14.4 ^{15}N-balance after 60 days of irrigation (4 mm per day) with ^{15}N labelled nitrate at a rate of 20 kg N ha^{-1} per year and 30 days with nonlabelled nitrate, on the top of three different soil depths from Solling at 8°C in the laboratory study, (% of ^{15}N applied) (n = 4)

Depth	Leachate		Soil	^{15}N-
	NO_3^-	NH_4^+	N_{total}	Recovered
Organic layer	100 (3)	3 (1)	3 (2)	106 (5)
0–10	94 (1)	2 (0)	2 (2)	98 (4)
10–20	93 (2)	0 (0)	0 (0)	93 (3)

14.5 Retention of Added Nitrate and ^{15}N Balances of Forest Soils

This study with labelled nitrate in the laboratory showed that the acid Solling soil has no ability to retain significant amount of applied nitrate. About 3% of the applied nitrate was found in the surface organic layer (Table 14.4) as compared to 54% of ammonium (Table 14.1). Among different horizons, L retained the most of all layers, 15- and 60-fold as much of added nitrate per gram matter as the F and H layers of the Solling site.

The soil from 0 to 10 cm depth retained about 2% of the applied nitrate whereas that from the 10 to 20 cm layer did not retain any ^{15}N. A similar fraction of labelled nitrate was recovered after a pulse application of 80 kg N ha^{-1} ^{15}N labelled calcium nitrate to the Solling soil (Mochoge and Beese 1983b). About 6 and 15% of the applied nitrate was found in the organic nitrogen pool in 0–30 soil depth at 4 and 23°C, respectively, and suggested a positive temperature effect on nitrate immobilisation. Generally, micro-organisms prefer ammonium as a nitrogen source and their ability to use nitrate is restricted (Rosswall 1981). A short-time experiment with ^{15}N labelled nitrate has shown that biotic immobilisation of nitrate was about half that of ammonium in the surface organic layer and 0–5 cm depth of the Solling soil (Corre et al. 2003). Similarly, Puri and Ashman (1999) reported from a temperate woodland soil that ammonium had twice the immobilisation rate of nitrate. These fractions are much higher than found in our laboratory study where nitrate immobilisation in soil organic matter was <6% of ammonium immobilisation. These differences might be related to methodical differences of labelled nitrate application to soils. Whereas a semicontinuous application of small ^{15}N rates with the irrigation simulates field conditions in our field and laboratory study, in other studies the labelled nitrate was injected into soils or added to soil slurries in the ^{15}N-pool dilution technique.

14.6 Temperature Effects on Transformation of Ammonium Applied to the Surface Organic Layer and to Mineral Soil Layers of the Solling Site

Temperature affected the transformation of ammonium when applied continuously to undisturbed soil cores from the Solling site for 2 months in the laboratory study. Leaching of labelled nitrate increased from 10 to 37% of the applied ammonium in the 0–10 cm depth and from 6 to 25% in the 10–20 cm depth at 4 and 16°C, respectively (Fig. 14.4d). Ammonium leaching was low (6–8% and 1–2% of the amount applied in the 0–10 cm and 10–20 cm depths, respectively) and showed no difference in relation to temperature (not shown). A similar increase in nitrate leaching with temperature was observed in the surface organic layer (19 and 39%

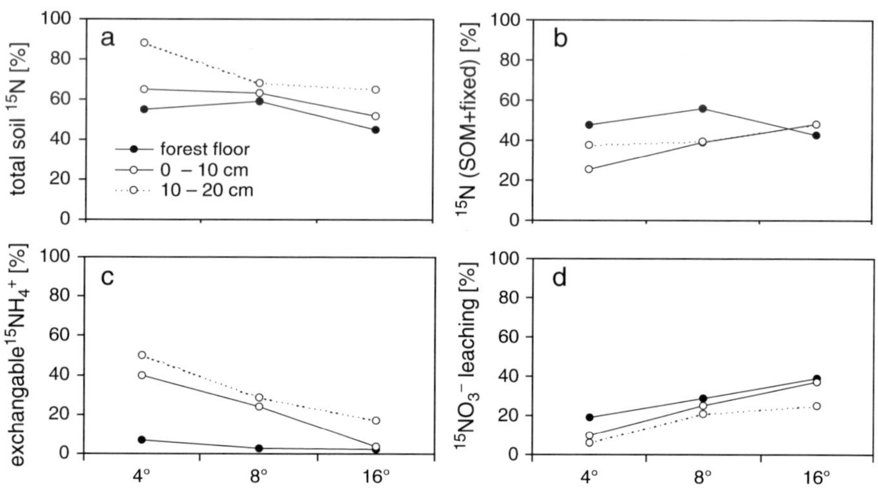

Fig. 14.4 Proportion of applied ammonium that remained as total nitrogen (**a**), organically bound plus fixed ammonium (**b**), exchangeable ammonium (**c**), or leached as nitrate (**d**) from the forest floor, 0–10 cm and 10–20 cm soil depth at Solling at 4, 8 and 16°C in the "laboratory study"

of the amount applied at 4 and 16°C; Fig. 14.4d) but a reduction in ammonium leaching (24 and 19% at 4 and 16°C, respectively, not shown). Increased leaching losses were attributed to adsorbed labelled ammonium which was reduced with increasing temperature. Figure 14.4c showed that about 40 and 50% of the applied ammonium were adsorbed in 0–10 cm and 10–20 cm soil depths at 4°C and were reduced to 4 and 17% at 16°C. The later desorption of ammonium caused leaching to increase (by about one-third in both the mineral soil depths). The moderate increased of retention by organic matter (Fig. 14.4b) could not prevent a general decrease in total retention at higher temperatures (Fig. 14.4a).

14.7 Conclusions

- Applied ^{15}N labelled ammonium was retained longer in an acid soil with moder humus but was quickly nitrified and leached as nitrate in base rich soils with mull humus type of surface organic layer. Longer retention of applied ammonium in acid soils was associated with low nitrification (heterotrophic type) rate. However, the surface organic layer in moder humus adsorbed only a small amount of the applied ammonium whereas most was adsorbed in the mineral soil layers. Retention of ammonium by adsorption depended on soil temperature which was reduced at higher temperature by increasing nitrification, and was increased with decreasing temperature by lowering nitrification. This mechanism potentially delayed leaching losses of applied ammonium as nitrates in acid soils especially at low temperatures in the nongrowing season and probably increased the uptake of N by microbes and plants.
- Retention of ^{15}N labelled ammonium in organic compounds (microbial biomass, soil organic matter) was high when microbial biomass was high. The dominant effect of microbial immobilisation, which was independent of soil layers and humus forms, suggested the importance of the fate of microbial detritus for the long-term retention of deposited ammonium in forest soils. Retention of ^{15}N labelled nitrate in organic compounds was very low in the Solling soil in contrast to other studies with different application procedures of labelled nitrate.

References

Aber J, McDowell W, Nadelhoffer K, Magill A, Berntson G, Kamakea M, McNulty S, Currie W, Rustad L, Fernandez I (1998) Nitrogen saturation in temperate forest ecosystems. Bio Science 48:921–934

Beese F (1982) Gesetzmäßigkeiten beim Transport gelöster Stoffe im Boden. Beiträge zur Hydrologie 4:267–300

Beese F, van der Ploeg RR (1979) Simulation des Anionen-Transports in ungestörten Bodensäulen unter stationären Fließbedingungen. Z Pflanzenernaehr Bodenkd 142:69–85

Berg B (1988) Dynamics of nitrogen (^{15}N) in decomposing Scots pine *(Pinus sylvestris)* needle litter. Long-term decomposition in a Scots pine forests. VI. Can J Bot 66:1539–1546

Borken W, Matzner E (2004) Nitrate leaching in forest soils: an analysis of long-term monitoring sites in Germany. J Plant Nutr Soil Sci 167:277–283

Bresler E (1973) Anion exclusion and coupling effects in nonsteady transport through unsatureated soil: 1. Theory. Soil Sci Soc Am Proc 37:663–669

Brumme R, Aden G (1995) The use of ^{15}N to study the nitrogen turnover in soils. In: Alef K, Nannipieri P (eds) Methods in applied soil microbiology and biochemistry. Academic, London, pp 246–257

Brumme R, Beese F (1991) Simultane Bestimmung von N-Transformationsraten in Bodensäulen unter Verwendung von 15-N: Stickstoffmodell für eine Terra fusca Rendzina. Z Pflanzenernähr Bodenk 154:205–210

Brumme R, Khanna PK (2008) Ecological and site historical aspects of N dynamics and current N status in temperate forests. Glob Chang Biol 14:125–141

Corre MD, Beese FO, Brumme R (2003) Soil nitrogen cycle in high nitrogen deposition forest: changes under nitrogen saturation and liming. Ecol Appl 13(2):287–298

Coûteaux MM, Sallih Z (1994) Fate of inorganic ^{15}N in the profile of different coniferous forest soils. Biol Fertil Soils 17:101–107

Dise NB, Matzner E, Forsius M (1998) Evaluation of organic horizon C:N ratio as an indicator of nitrate leaching in conifer forests across Europe. Environ Pollut 102:453–456

Feng Z, Brumme R, Xu Y-J, Lamersdorf N (2008) Tracing the fate of mineral N compounds under high ambient N deposition in a Norway spruce forest at Solling/Germany. For Ecol Manage 255:2061–2084

Godbold DL, Hüttermann A (1994) Effects of acid rain on forest processes. Wiley, New York

Hantschel RE, Flessa H, Beese F (1994) An automated microcosm system for studying ecological processes. Soil Sci Soc Am J 58:401–404

Johnson DW, Susfalk RB, Dahlgren RA (1997) Nutrient fluxes in forests of the easten Sierra Nevada mountains, United States of America. Glob Biogeochem Cycles 11(4):673–681

Mochoge B, Beese F (1983a) The behaviour of nitrogen fertilizers in neutral and acid loess soils: I. Transport and transformations of nitrogen. Z Pflanzenernährung Bodenk 146(1):89–100

Mochoge B, Beese F (1983b) The behaviour of nitrogen fertilizers in neutral and acid loess soils: II. Distribution and balances of 15 N-tagged nitrogen. Z Pflanzenernaehr Bodenk 146(1):504–515

Nadelhoffer KJ, Colman BP, Currie WS, Magill A, Aber JD (2004) Decadal-scale fates of ^{15}N tracers added to oak and pine stands under ambient and elevated N inputs at the Harvard forest (USA). For Ecol Manage 196:89–107

Nommik H, Vahtras K (1982) Retention and fixation of ammonium and ammonia in soils. In: Stevenson FJ (ed) Nitrogen in agricultural soils, agronomy monograph no 22. ASA-CSSA-SSSA, Medison, USA, pp 123–167

Parker JC, van Genuchten MTH (1984) Determining transport parameters from laboratory and field tracer experiments. Bulletin 84–3, USA Virginia Polytechnic Institute and State University,

Puri G, Ashman MR (1999) Microbial immobilization of ^{15}N-labelled ammonium and nitrate in a temperate woodland soil. Soil Biol Biochem 31:929–931

Rosswall T (1981) The biogeochemical nitrogen cycle. In: Likens GE (ed) Some perspectives of the major biogeochemical cycles. Wiley, New York, pp 25–49

Tietema A, Emmett BA, Gundersen P, Kjønaas OJ, Koopmans CJ (1998) The fate of ^{15}N-labelled nitrogen deposition in coniferous forest ecosystems. For Ecol Manage 101:19–28

Ulrich B (1994) Nutrient and acid-base budget of central European forest ecosystems. In: Godbold DL, Hüttermann A (eds) Effects of acid rain on forest processes. Wiley, New York, pp 1–50

Wang CP (1998) Soil and temperature influence on microbiological nitrogen transformations – a microcosm study by dynamic ^{15}N dilution technique. Berichte des Forschungszentrums Waldökosysteme, Universität Göttingen, Reihe A, Bd:156

Chapter 15
Atmospheric Deposition and Canopy Interactions

H. Meesenburg, J. Eichhorn, and K.J. Meiwes

15.1 Introduction

Element inputs by atmospheric deposition form a major contribution to a number of element fluxes of forest ecosystems. During the last few decades, inputs from the atmosphere have significantly altered the geochemical cycles of forest ecosystems especially in heavily polluted areas of Central Europe where forests have remained major sinks for air pollution.

The deposition of acids such as sulphuric and nitric acids was a major environmental concern during recent decades (Galloway 1995). Acid deposition has caused the acidification of soils and freshwaters in large areas of North America and Europe (Johnson et al. 1991). The effects of soil acidification are the leaching of base cations from the soil (Matzner and Murach 1995), the release of aluminium (Al) ions and heavy metals into soil solution (Tyler 1983), reduced decomposition of soil organic matter, and reduced growth of fine roots (Godbold et al. 2003). The release of acid soil solutions to the hydrosphere is detrimental to aquatic ecosystems.

During the last two decades, deposition of acids has decreased substantially in Central Europe due to improved emission controls and the closing down of industry in eastern Germany after the reunion (Meesenburg et al. 1995). However, the deposition of nitrogen (N) compounds has decreased only slightly and has become an increasingly important fraction of the total deposition of acids (Wright et al. 1995). Despite reduced deposition of free acidity, the acid load to soils is still high because of high deposition of ammonium (NH_4^+). The fate of elevated N-input on forest ecosystem remains partly unknown, but besides the impact on the acid/base balance of ecosystems there are some indications of increased tree growth, reduced root/shoot ratio, nutrient imbalances, reduced frost hardiness and elevated foliage consumption by insects (Binkley and Högberg 1997; Aber et al. 1998; Meiwes et al. 1999). In N-saturated ecosystems, soils have increased susceptibility for losses of nitrate (NO_3^-) to the hydrosphere and of trace gases to the atmosphere (Aber 2002).

The transfer of elements from the atmosphere to forests takes several pathways. According to Ulrich (1994), total deposition can be divided into wet deposition and

R. Brumme and P.K. Khanna (eds.), *Functioning and Management of European Beech Ecosystems*, Ecological Studies 208,
DOI: 10.1007/978-3-642-00340-0_15,

interception deposition. The latter is composed of particulate interception and gaseous interception and is also referred as dry deposition. After deposition to the canopy of forest stands, the deposited elements may either be taken up by the canopy or washed from the surfaces by subsequent rainfall (Harrison et al. 2000).

Precipitation beneath the canopy of forest ecosystems contains both wet deposition and interception deposition. As the canopy acts as sink or source for solutes in precipitation passing through the canopy, stand precipitation cannot be used as a measure of total deposition (Horn et al. 1989). Total deposition can be calculated using wet deposition and independent estimates of dry deposition. Frequently used methods for the estimation of dry deposition are: canopy budget models (Draaijers et al. 1996), inferential modelling (van Leeuwen et al. 1996; Gauger et al. 2002) and gradient measurements of air pollutants (Sutton et al. 1995). The use of canopy budget models for forest ecosystems was extensively discussed by Draaijers (1999). Results of canopy budget models are very uncertain for the estimation of N-deposition, because dry deposition can form a high proportion of total deposition of N (Lindberg et al. 1986), and N is involved intensively in interaction processes with the foliage (Horn et al. 1989).

For the investigation of effects of atmospheric deposition on nutrient cycles of forest ecosystems and of temporal trends of element fluxes on ecosystem processes, long-term monitoring sites are of overwhelming importance. For this study, three mature beech forests located at the northern part of the central German mountain range have been compared. The three sites (Solling, Göttinger Wald and Zierenberg) are included in the Level II European Forest Intensive Monitoring programme (de Vries et al. 2001). At the Solling site, deposition measurements started in 1968. The Solling beech forest site together with the Solling spruce forest site has – to our knowledge – the longest continuous record of throughfall measurements globally. Deposition measurements at the Göttinger Wald site started in 1981 and at Zierenberg in 1989.

In this chapter, these long-term data sets will be used: (1) to characterise the chemical composition of open field deposition, throughfall and stemflow and to analyse relationships among major input components to relate them to different sources, (2) to describe the annual deposition fluxes and their temporal changes on the three sites, and (3) to analyse the interactions of precipitation inputs with the canopy of the stands.

Data collection and evaluation procedures. Element fluxes have been measured in open field deposition, throughfall and stemflow. Stand precipitation is the sum of throughfall and stemflow. Open field deposition and throughfall were monitored with samplers, which remained continuously open to the atmosphere (bulk samplers). Samples obtained with such samplers are composed of rainwater or snow and gravitational sedimented particles. In remote areas, there is little difference between element fluxes of bulk precipitation and wet-only precipitation (Ibrom 1993; Gauger et al. 2002). Open field deposition is being sampled at clearings located close to the monitoring sites. Here, we use the term open field deposition instead of bulk deposition, because the term bulk deposition is generally used for sampling of precipitation with bulk samplers without any regard to their location (open field or under the canopy). At the Solling and Göttinger Wald sites, samplers with 50-cm^2 surface area were used in summer months (May–October) until 1990. From 1990 onwards, funnel-flask samplers with 87.5-cm^2 surface area have been used for summer sampling. In winter months (November–April), buckets with surface area of 570 (until 1990) or 500 cm^2 (from 1990 onwards) have been used (Meiwes et al. 1984). At the Zierenberg site, funnel-flask samplers with 100-cm^2 surface area have been used in summer months and buckets with 500-cm^2 surface area in winter months (Brechtel and Hammes 1984; Eichhorn 1995). At the Solling and Göttinger Wald sites, six samplers have been used for open field deposition and 15 samplers for throughfall (Meesenburg et al. 1997). At the Zierenberg site, ten replicates have been used for open field deposition and 20 for throughfall. Coarse particles (e.g. litter) have been prevented from falling into the samplers by using a polyethylene mesh at the Solling and Göttinger Wald sites and a ceramic sieve at Zierenberg site. After recording the volumes, three composite samples were formed for each of open field deposition and throughfall for the Solling and Göttinger Wald sites, and four composite samples for the Zierenberg site for laboratory analysis.

Sampling devices and sampling procedures of the three sites were tested in comparison with 18 other methods for open field deposition and throughfall used within the framework of the ICP forest level II programme. The performance of the deposition monitoring at the study sites was found to be acceptable (Draaijers et al. 2001).

Stemflow has been sampled by fixing polyurethane spirals around the stems, which were coated with paraffine. Three to five replicates were installed, which were analysed separately (Solling and Göttinger Wald sites) or pooled to a composite sample (Zierenberg site).

Water flux via stemflow for Zierenberg and Solling has been estimated to be 15% of the total throughfall flux. At the Solling site, a value of 15% is close to the mean value of stemflow flux estimated by Benecke (1984) for the period 1969–1975. For Göttinger Wald, stemflow fluxes were obtained from regression functions between stemflow volume and throughfall (Gerke 1987).

Analytical methods are described by Fassbender and Ahrens (1977) and König and Fortmann (1996a–d) for the Solling and Göttinger Wald sites. In short, pH was measured potentiometrically. Sodium (Na^+), potassium (K^+), magnesium (Mg^{2+}), calcium (Ca^{2+}), and manganese (Mn^{2+}) were determined by AAS until 1989, and since 1990 by ICP-AES. NH_4^+, NO_3^- and chloride (Cl^-) were analysed colorimetrically with a continuous flow system. Sulphate (SO_4^{2-}) was measured by precipitation with Ba and by potentiometric titration of excess Ba with EDTA until 1982, from 1983 to 1992 by the methyl-thymol-blue method, and since 1993 by ICP-AES. Organic N (N_{org}) is calculated as the difference between total N (N_{tot}) (measured after digestion) and the sum of NH_4^+ and NO_3^-.

Because no independent estimates of interception deposition are available for the study sites over the whole observation period, we have used the canopy budget model developed by Ulrich (1994) for the calculation of total deposition. Annual fluxes of ions have been used for the calculation procedure. The canopy budget model of Ulrich (1994) estimates the interception deposition for element species A (ID_A), which are not adsorbed or leached from the canopy, from the difference of stand precipitation (SP_A) and open field precipitation (wet deposition OF_A). Interception deposition ID_A is the sum of particulate ($ID_{part,\,A}$) and gaseous deposition ($ID_{gas,\,A}$). Total deposition (TD_A) is the sum of wet deposition and interception deposition.

$$TD_A = OF_A + ID_A, \tag{15.1}$$

$$ID_A = ID_{part,A} + ID_{gas,A}, \tag{15.2}$$

$$ID_A = SP_A - OF_A. \tag{15.3}$$

$$A = \text{Na, Cl, SO}_4.$$

Particulate interception deposition ($ID_{part,\,A}$) is estimated from the ratio between interception deposition and open field precipitation of Na assuming that Na^+ is only deposited by wet deposition and particulate interception. It is assumed that the particle size distribution of all deposited substances is similar, resulting in a similar deposition velocity. Another assumption is that particulate interception is caused to a large degree by fog droplets. In contrast to earlier formulations of the model by Ulrich (1983) and Bredemeier (1988), Ulrich (1994) extended the model for NH_4^+ and NO_3^- implying the assumption of similar deposition velocities holds for particulates containing NH_4^+ and NO_3^- particles. The assumption of similar deposition velocities is highly questionable especially for N-compounds (Spranger 1992):

$$f_{Na} = \frac{ID_{Na}}{OF_{Na}}, \tag{15.4}$$

$$ID_{part,A} = f_{Na}\, OF_A. \tag{15.5}$$

$$A = \text{H, K, Mg, Ca, Mn, Al, Fe, NH}_4\text{, Cl, SO}_4\text{, NO}_3.$$

Gaseous deposition ($ID_{gas,\,A}$) of metal cations is assumed to be negligible.

$$ID_{gas,A} = 0. \tag{15.6}$$

$$A = \text{Na, K, Mg, Ca, Mn, Al, Fe}.$$

Gaseous deposition ($ID_{gas,\,A}$) of SO_2, HCl, HNO_3 and NH_3 is estimated from the difference of interception deposition and particulate deposition:

$$ID_{gas,A} = ID_A - ID_{part,A} = SP_A - OF_A - ID_{part,A}. \tag{15.7}$$

$$A = \text{NH}_4\text{, Cl, SO}_4\text{, NO}_3.$$

Gaseous deposition of SO_2, HCl and HNO_3 causes an equivalent input of protons, gaseous deposition of NH_4^+ a consumption of protons:

$$ID_{gas,H} = ID_{gas,SO_2} + ID_{gas,Cl} + ID_{gas,NO_3} - ID_{gas,NH_4}. \tag{15.8}$$

If particulate deposition of NH_4^+ and NO_3^- is higher than the difference between SP_A and OF_A, no gaseous deposition can be calculated.

The difference between total deposition and stand precipitation is interpreted as canopy budget (CB_A). Positive values are interpreted as leaching, negative values as uptake by the canopy:

$$CB_A = SP_A - TD_A = SP_A - OF_A - ID_A. \tag{15.9}$$

From the calculation scheme, it arises that either gaseous deposition (if $ID_{part} < SP_B - OF_B$) or uptake by the canopy (if $ID_{part} > SP_B - OF_B$) is calculated for NH_4^+ and NO_3^-. As both processes can occur concurrently (Veithen 1996; Garten et al. 1998), total deposition of NH_4^+ and NO_3^- is underestimated by the model. Moreover, the processes of canopy uptake and leaching are highly seasonal and the use of annual budgets disregards the seasonal nature of these processes. The assumptions involved in the model were not tested on these three sites and will need due consideration during the interpretations of model results.

Despite the limitations mentioned above, the canopy model of Ulrich (1994) has been applied to NH_4^+ and NO_3^-. Total deposition of nitrogen (N_{tot}) has been calculated as the sum of total deposition of NH_4^+, total deposition of NO_3^- and open field deposition of N_{org} (Ulrich 1994).

15.2 Precipitation Chemistry

For characterisation of the chemical composition of atmospheric deposition, data for the period 1993–1998 were selected as all three study sites had information for this period. As precipitation chemistry has changed significantly during the last few decades, only the pattern for the selected period is described.

At the Solling site, open field precipitation chemistry is dominated by NH_4^+ (44% of cations on equivalent basis) and Na^+ (23%), and H^+, Ca^{2+}, Mg^{2+} and K^+ are only of minor importance in that order. Anions of significance are SO_4^{2-} (43%), NO_3^- (35%) and Cl^- (22%). Concentrations of NH_4^+ are higher than concentrations of NO_3^- (Table 15.1). The contribution of N_{org} to N_{tot} is about 7%.

At the Göttinger Wald site, 44% of the sum of cations in open field precipitation is NH_4^+. Na^+ (19%), Ca^{2+} (15%), and H^+ (10%) also have some quantitative importance. The contribution to the sum of anions is 16% for Cl^-, 39% for NO_3^- and 44% for SO_4^{2-}.

At the Zierenberg site, NH_4^+ (30%) is the most abundant cation in open field precipitation, but the relative contributions of Ca^{2+} (27%) and Mg^{2+} (14%) are

Table 15.1 Mean concentrations and standard deviation (in parentheses) of total N (N_{tot}) and relative contribution of NH_4^+, NO_3^-, and N_{org} to N_{tot} in open field precipitation, throughfall and stemflow from 1993 to 1998 at the study sites

Site		N_{tot} (mmol l^{-1})	NH_4^+ %	NO_3^- %	N_{org} %
Solling	Open field deposition	125 (1)	52	41	7
	Throughfall	260 (18)	51	37	12
	Stemflow	216 (28)	41	40	19
Göttinger Wald	Open field deposition	170 (1)	50	43	7
	Throughfall	293 (5)	49	40	11
	Stemflow	303 (43)	44	42	15
Zierenberg	Open field deposition	174	50	41	9
	Throughfall	454	39	51	10
	Stemflow	310	45	35	20

much higher than at the Solling and Göttinger Wald sites. This may be related to higher input of dust particles and to other factors relating to methodology of collection and analysis. The contribution of Na^+ to the sum of cations is 16%. The most abundant anion is SO_4^{2-} (45%), which is followed by NO_3^- (34%) and Cl^- (21%). Various anions have similar fractions in the open field precipitation on the three sites with values following the order: SO_4^{2-} (43–45%) > NO_3^- (34–39%) > Cl^- (16–22%).

N_{tot} concentrations are generally highest in throughfall and lowest in open field precipitation (Table 15.1). After passing through the canopy, concentrations of N_{tot} in the precipitation increased by 108% at Solling, 72% at Göttinger Wald and 161% at Zierenberg compared to open field precipitation. This indicates a much higher interception of dry deposition at Zierenberg. The increase of concentrations of N_{tot} in stemflow was very similar at the three study sites (73–78%).

The relative contribution of the N species to N_{tot} is similar at the three study sites (Table 15.1). NH_4^+ concentrations in open field deposition are higher than NO_3^- at all sites indicating that N-inputs are influenced to a large degree by animal husbandry. In throughfall and stemflow, the relative contribution of N_{org} to N_{tot} is generally higher than in open field precipitation indicating leaching of N_{org} from leaves and bark of the trees or from other sources of N_{org} in the canopy. At the Zierenberg site, the enrichment of the N-compounds in throughfall as compared to open field precipitation is much higher than at the Solling and Göttinger Wald sites. The enrichment of NH_4^+ in throughfall is higher than that of NO_3^- at the Solling and Göttinger Wald sites and lower at the Zierenberg site. The strong enrichment of NO_3^- relative to NH_4^+ in throughfall at Zierenberg may be explained by nitrification of NH_4^+ in the canopy (Papen et al. 2002). (Table 15.1). However, Eichhorn (1995) attributed the enrichment of NO_3^- in throughfall to its leaching from the canopy. At the Solling and Göttinger Wald sites, the relative contribution of NH_4^+ in stemflow is lower than in throughfall. This pattern may be partly explained by nitrification of NH_4^+ at branches and stems.

Mean pH in throughfall is higher than in open field deposition at the Göttinger Wald and Zierenberg sites reflecting the buffering of acids in the canopy, but lower at Solling. pH is generally lower in stemflow than in open field deposition.

Sea spray is a major source for Cl^-, Na^+, and Mg^{2+} in open field precipitation. An influence of road salt can be excluded since the sampling sites are far away from roads. Higher concentrations of Cl^- during the winter season are related to more frequent storm events. At the Solling site, the contribution of sea spray is 100% for Cl^-, 77% for Na^+ and 44% for Mg^{2+} (calculated with Cl^- as index element). At the Göttinger Wald site, the influence of sea spray is somewhat lower than at the Solling site with values of 100% for Cl^-, 72% for Na^+ and 29% for Mg^{2+}. Sea spray is an important source at the Zierenberg site for Cl^- (100%) and Na^+ (80%) whereas a low value for Mg^{2+} (10%) points to sources other than sea spray being more important, e.g. soil dust due to agricultural activities or Mg containing particles from different industrial processes such as coal burning or handling of bulk cargo.

The covariance analysis of the concentrations of solutes in open field precipitation, throughfall and stemflow at the three sites was undertaken by employing principal component analysis (SPSS version 6.1.2). Principal component analysis was used to find the least linear combinations of the parameters which were required to explain as much of the total variance of the data as possible. The major ions Na^+, K^+, Mg^{2+}, Ca^{2+}, H^+, $SO_4{}^{2-}$, Cl^-, NH_4^+ and NO_3^- and N_{org} were included for the analyses. Principal component analysis has been frequently used for assigning different sources to various solutes in the precipitation (Gorham et al. 1984; Feger 1986) and to describe the predominant processes occurring in ecosystems (Christophersen and Hooper 1992). Varimax rotation has been performed to find out the contribution of different processes.

As an example, results for open field precipitation at Solling are given in Table 15.2 for principal components with eigenvalues >1.0. Three components could be differentiated. Component 1, which explains 46% of the variance, has high loadings of NH_4^+, NO_3^-, $SO_4{}^{2-}$ and Ca^{2+}. These ions are negatively correlated to the amount of precipitation. NH_4^+, NO_3^- and SO_4^{2-} can be ascribed to the emission of

Table 15.2 Matrix of factor loadings (varimax rotation) and communalities of principal component analysis for concentrations of ions in open field deposition at Solling. Factor loadings above |0.6| are given in bold

	Component 1	Component 2	Component 3	Communality
Na^+	0.13	**0.97**	0.02	0.96
K^+	0.48	−0.15	**0.64**	0.67
Mg^{2+}	0.38	**0.84**	0.20	0.89
Ca^{2+}	**0.75**	0.27	0.27	0.71
H^+	0.21	−0.25	**−0.71**	0.62
$SO_4{}^{2-}$	**0.94**	0.21	0.08	0.93
Cl^-	0.09	**0.96**	−0.10	0.93
$NH_4{}^+$	**0.90**	0.14	0.10	0.84
$NO_3{}^-$	**0.93**	0.09	0.02	0.93
N_{org}	0.29	−0.08	**0.79**	0.72
Explained variance (%)	46.0	22.4	13.7	82.0

NH_3, NO_x and SO_2. NH_3 emissions can be attributed mainly to intensive farming practices such as animal husbandry and field application of faecal materials. NO_x is emitted to a high degree by vehicles, whereas SO_2 emissions can be ascribed mainly to large power plants. Ca^{2+} and K^+ can be attributed partly to the emission of dust. Thus, component 1 describes components of air pollution in open field precipitation.

Component 2 explains about one-quarter of the variance and has high loadings of Cl^-, Na^+, and Mg^{2+}. This component can be ascribed to the influence of sea spray. Cl^-, Na^+ and Mg^{2+} are independent of the amount of precipitation. The third component explains 14% of the variance and has high loadings of N_{org} and K^+ and a high negative loading of H^+, which can be ascribed to plant-based organic substances.

The first principal component of stemflow at Solling explains over 60% of the variance, and has high loadings of Mg^{2+}, Ca^{2+}, H^+ and $SO_4{}^{2-}$ and considerable loadings of K^+ and NO_3^- (Table 15.3). Mg^{2+}, Ca^{2+} and K^+ ions are leached from the vegetation, when buffering of H^+ occurs whereas $SO_4{}^{2-}$ and NO_3^- are involved in the charge balance. Thus, the first component can be interpreted as describing the input of acid depositions and their subsequent buffering by ion exchange. The second component in Table 15.3 explains 18% of the variance and has high loadings of NH_4^+, NO_3^- and N_{org} which may account for the transformation of N-species and their interaction with the different tree compartments and canopy epiphytes. The third component explains 10% of the variance and has high loadings of Na^+ and Cl^-, which can be interpreted as sea spray.

The principal component analysis for open field precipitation at the Göttinger Wald and Zierenberg sites gives similar results as for the Solling site (Table 15.4). Also for throughfall data at Solling, principal components provided similar interpretation of results. For throughfall data at the Göttinger Wald and Zierenberg sites, canopy interactions are a major source affecting the variance of the data (Table 15.4).

For stemflow data at the Göttinger Wald and Zierenberg sites similar results as for the Solling site have been obtained (Table 15.4). In contrast to open field deposition, where the components can be interpreted as different sources of the

Table 15.3 Matrix of factor loadings (varimax rotation) and communalities of principal component analysis for concentrations of ions in stemflow at Solling. Factor loadings above |0.6| are given in bold

	Component 1	Component 2	Component 3	Communality
Na^+	0.34	0.10	**0.93**	0.99
K^+	0.49	0.53	0.10	0.52
Mg^{2+}	**0.83**	0.19	0.49	0.96
Ca^{2+}	**0.92**	0.20	0.26	0.96
H^+	**0.92**	0.20	0.26	0.93
$SO_4{}^{2-}$	**0.81**	0.50	0.26	0.97
Cl^-	0.30	0.17	**0.92**	0.97
$NH_4{}^+$	0.07	**0.93**	0.19	0.90
$NO_3{}^-$	0.48	**0.76**	0.29	0.91
N_{org}	0.13	**0.93**	−0.03	0.72
Explained variance (%)	62.4	18.1	9.6	90.1

Table 15.4 Explained variance of principal components and sum of explained variance of principal component analysis (varimax rotation) for concentrations of ions in open field deposition (OF), throughfall (TF) and stemflow (SF) at Solling, Göttinger Wald and Zierenberg (interpretation of principal components is given by letters; the same interpretation means that the same elements have high loadings in certain components)

Site	Flux	Component 1 (%)	Component 2 (%)	Component 3 (%)	Σexplained variance (%)
Solling	OF	48.9[a]	24.2[b]	12.2[c]	85.3
	TF	50.4[a]	17.7[b]	12.3[c]	80.4
	SF	62.4[d]	18.1[e]	9.6[b]	90.1
Göttinger Wald	OF	53.3[a]	16.5[b]	13.2[c]	83.0
	TF	44.7[d]	21.1[a]	16.5[b]	82.4
	SF	52.8[d]	19.5[e]	16.1[b]	88.4
Zierenberg	OF	56.2[a]	13.1[b]	9.8[c]	79.2
	TF	59.6[ba]	16.4[d]	8.3[f]	84.3
	SF	57.7[d]	17.7[e]	10.0[b]	85.5

[a]Air pollution
[b]Sea spray
[c]Organic deposition
[d]Canopy leaching
[e]N mineralisation processes
[f]H^+ buffering

solutes, transformation and interaction processes of the solutes with the vegetation seems to play an important role for the variance of the stemflow data.

15.3 Element Fluxes

We present the element fluxes with open field deposition, throughfall, stemflow, stand precipitation and total deposition separately for three periods of roughly a decade each (Tables 15.5–15.7). The period from 1969 to 1980 is only available for the Solling site. In 1981, the Göttinger Wald site was established and represents the beginning of the second period from 1981 to 1989. The Zierenberg site was established in 1990, which is the beginning of the third period from 1990 to 2002 and represents the period after emission control in Germany. Comparing the period 1990–2002 between the sites indicates higher fluxes of SO_4^{2-}, Cl^- and Na^+ at the Solling site than at the Göttinger Wald and Zierenberg sites due to higher precipitation rates at the Solling site. N-fluxes in open field deposition were lowest at the Zierenberg site and highest at the Solling site, whereas they were quite similar in stand precipitation and total deposition at the three study sites. However, fluxes of K^+, Mg^{2+} and $Ca^{2+}$are highest at the Zierenberg site.

Fluxes of most elements have decreased during the last 22 years at the Solling and Göttinger Wald sites (Tables 15.5 and 15.6). Decreasing trends have been detected for open field deposition, throughfall, stemflow and stand precipitation as

Table 15.5 Mean annual rates of open field deposition (*OF*), throughfall (*TF*), stemflow (*SF*), stand precipitation (*SP*), and total deposition (*TD*) at the Solling site during the periods 1969–1980 (1971–1980 for NH_4^+, NO_3^-, N_{tot}, N_{org}), 1981–1989, and 1990–2002 (Data source: Nordwestdeutsche Forstliche Versuchsanstalt)

Flux	Period	H_2O (mm)	Na^+	K^+	Mg^{2+}	Ca^{2+}	H^+	Mn^{2+}	SO_4^{2-}	Cl^-	NH_4^+	NO_3^-	N_{tot}	N_{org}
			($mmol_c$ m^{-2} per year)										(mmol m^{-2} per year)	
OF	69–80	985	33	9	15	51	80	1.3	148	48	84	58	175	34
	81–89	1,145	36	9	11	36	78	1.4	123	47	85	69	181	28
	90–02	1,184	40	5	7	16	17	0.5	59	33	63	51	126	12
TF	69–80	729	45	48	26	96	88	10.7	207	67	75	63	201	61
	81–89	768	46	50	23	77	57	10.0	183	66	99	80	202	27
	90–02	812	46	49	14	30	13	4.0	82	48	74	59	157	24
SF	69–80	121	16	23	8	28	51	3.4	115	25	13	10	52	25
	81–89	122	16	20	7	24	44	2.8	82	22	15	15	42	16
	90–02	119	10	10	2	5	8	0.5	20	11	8	8	20	5
SP	69–80	850	61	72	33	124	138	14.1	322	92	87	73	253	86
	81–89	890	61	70	30	101	101	12.8	264	88	114	95	244	42
	90–02	931	55	59	16	35	21	4.5	102	59	82	67	178	29
TD	69–80	850	61	17	29	95	199	2.4	322	92	157	109	300	64
	81–89	890	61	15	19	62	204	2.5	264	88	143	117	289	47
	90–02	931	55	8	10	22	55	0.8	102	59	94	73	179	21

Table 15.6 Mean annual rates of open field deposition (OF), throughfall (TF), stemflow (SF), stand precipitation (SP) and total deposition (TD) at the Göttinger Wald site during the periods 1981–1989 and 1990–2002 (Data source: Nordwestdeutsche Forstliche Versuchsanstalt)

Flux	Period	H_2O (mm)	Na^+	K^+	Mg^{2+}	Ca^{2+}	H^+	Mn^{2+}	SO_4^{2-}	Cl^-	NH_4^+	NO_3^-	N_{tot}	N_{org}
			($mmol_c$ m^{-2} per year)										(mmol m^{-2} per year)	
OF	81–89	750	28	9	12	56	40	0.5	108	32	61	61	141	22
	90–02	684	24	3	7	19	11	0.2	48	20	47	45	100	8
TF	81–89	516	25	38	21	88	16	1.2	128	39	63	57	137	21
	90–02	519	28	42	13	37	5	0.5	59	32	59	50	126	17
SF	81–89	119	11	24	7	32	28	0.4	82	19	21	22	50	8
	90–02	97	7	14	2	7	3	0.1	20	8	10	11	25	4
SP	81–89	636	36	62	28	121	44	1.6	210	58	84	79	187	28
	90–02	615	35	56	15	44	9	0.6	79	41	70	61	151	22
TD	81–89	636	36	10	16	75	132	0.7	210	58	89	84	187	31
	90–02	615	35	5	9	27	37	0.4	79	41	75	68	151	13

Table 15.7 Mean annual rates of open field deposition (OF), throughfall (TF), stemflow (SF), stand precipitation (SP) and total deposition (TD) at the Zierenberg site during the period 1990–2002 (Data source: Nordwestdeutsche Forstliche Versuchsanstalt)

Flux	Period	H_2O (mm)	Na^+	K^+	Mg^{2+}	Ca^{2+}	H^+	Mn^{2+}	SO_4^{2-}	Cl^-	NH_4^+	NO_3^-	N_{tot}	N_{org}
			(mmol$_c$ m^{-2}per year)										(mmol m^{-2}per year)	
OF	90–02	752	26	7	19	38	18	0.2	45	22	39	36	85	9
TF	90–02	527	46	67	44	72	8	0.5	75	54	54	69	143	21
SF	90–02	77	7	13	5	9	3	0.1	16	9	8	7	20	5
SP	90–02	604	53	80	49	82	11	0.5	91	63	62	76	164	25
TD	90–02	604	53	14	39	77	61	0.4	91	63	81	76	171	19

well as for total deposition. Significant trends can be observed for SO_4^{2-}, H^+, Cl^-, Mg^{2+}, Ca^{2+}, Mn^{2+} and N-species (Table 15.8). In open field deposition, stemflow and total deposition, decreasing trends are also visible for K^+. Fluxes of SO_4^{2-} decreased by 81–94% during the last 22 years and reflect the general trend of S emissions in Germany (Fig. 15.1) (Umweltbundesamt 2000; Gauger et al. 2002). Decreasing deposition rates of Mg^{2+} and Ca^{2+} can be attributed to reductions of dust emission in Central Europe. At the Zierenberg site, similar trends were observed although due to a shorter observation period the relative reduction of deposition was less than for the Solling and Göttinger Wald sites (Table 15.7). Significant negative trends can be observed for SO_4^{2-}, H^+, Ca^{2+} and Mn^{2+} at Solling even though water fluxes increased during the period 1990–2002. For Cl^-, increasing fluxes have been measured under the canopy at the Zierenberg site. Decreasing deposition rates for many elements have been observed also for other forest ecosystems in Lower Saxony (Meesenburg et al. 1995) and Hesse (Balazs 1998) as well as for most parts of Germany (Gauger et al. 2002; Matzner et al. 2004).

N deposition at the Solling site showed different trends during the periods 1971–1985 and 1985–2002. During the first period, fluxes of NH_4^+, NO_3^- and N_{tot} showed no change or slightly increasing trend (Fig. 15.2). However, since 1985, a slightly decreasing trend was evident. N_{tot} fluxes have reduced by 30–55% during the period of 1981–2002 from about 36 to 22 kg ha^{-1} per year. Fluxes of NH_4^+ and NO_3^- with stemflow decreased even more. For N_{org}, high flux rates occurred in the 1970s, whereas low rates were measured afterwards. Temporal trends of N_{org} fluxes should be treated cautiously, since N_{org} is calculated as the difference between N_{tot} and the inorganic N-components and any alteration in the analytical methods of any one of the three components may have substantially affected the N_{org} estimates.

A decrease in N-fluxes was observed at Göttinger Wald during the whole observation period (Table 15.8, Fig. 15.3). However, for NH_4^+, N_{org}, and N_{tot} in throughfall, these trends were not significant. NH_4^+ fluxes at the different pathways decreased by 30–50% within the last two decades. Reduction of NO_3^- was about 40% for stand precipitation. N_{tot} fluxes have reduced by about 35% (from about 29–17 kg ha^{-1} per year). For stemflow, relative reductions have been generally higher.

N fluxes during the period 1990–2002 were generally very similar at the study sites and the interannual variations showed the same pattern (Fig. 15.3, Tables 15.5–15.7). Open field N-deposition was highest at Solling and lowest at Zierenberg, whereas N-values in stand precipitation and total deposition were somewhat lower at Göttinger Wald than at Solling and Zierenberg. For a spruce stand at Solling, Ibrom et al. (1995) calculated by use of micro-meteorological methods a total deposition of N of 460 mmol m^2 per year, whereas total deposition at the same stand calculated with the model of Ulrich (1994) was only 285 mmol m^{-2} per year. Marques et al. (2001) showed that dry deposition (particulate and gaseous) contributed 75% to total deposition at the Solling spruce stand. These results and similar results from other locations (Harrison et al. 2000; Zimmerling et al. 2000; Meesenburg et al. 2005) suggest that the canopy model may be underestimating total deposition of N (see Horn et al. 1989; Harrison et al. 2000).

Table 15.8 Trends of solute fluxes with open field deposition (*OF*), throughfall (*TF*), stemflow (*SF*), stand precipitation (*SP*), and total deposition (*TD*, according to Ulrich 1994) at the Solling and Göttinger Wald sites from 1981 to 2002 and at the Zierenberg site from 1990 to 2002 (test of trend with correlation coefficient after Pearson; o no trend, + significant increasing trend $p \leq 0.05$, ++ highly significant increasing trend $p \leq 0.01$, – significant decreasing trend $p \leq 0.05$, – – highly significant decreasing trend $p \leq 0.01$)

Site	Flux	H_2O	Na^+	K^+	Mg^{2+}	Ca^{2+}	H^+	Mn^{2+}	NH_4^+	NO_3^-	N_{tot}	N_{org}	SO_4^{2-}	Cl^-
Solling	OF	o	o	– –	– –	–	– –	–	–	–	–	–	–	–
	TF	o	o	o	– –	–	–	–	o	–	–	o	–	–
	SF	o	– –	– –	– –	–	–	–	–	–	–	–	–	–
	SP	o	–	o	– –	–	–	–	–	–	–	o	–	–
	TD	o	–	– –	– –	–	–	–	–	–	–	o	–	–
Göttinger Wald	OF	o	–	– –	– –	–	–	–	–	–	–	–	–	–
	TF	o	o	o	– –	–	–	–	o	–	o	o	–	–
	SF	–	–	– –	– –	–	–	–	–	–	–	–	–	–
	SP	o	o	o	– –	–	–	–	–	–	–	–	–	–
	TD	o	o	– –	– –	–	–	–	–	–	–	–	–	–
Zierenberg	OF	++	o	–	o	o	–	–	o	o	o	o	–	o
	TF	++	o	o	o	–	–	–	o	o	o	o	–	o
	SF	++	o	++	o	–	–	–	o	o	o	++	o	+
	SP	++	o	o	o	–	–	–	o	o	o	+	–	+
	TD	++	o	o	o	o	–	–	o	o	o	o	–	+

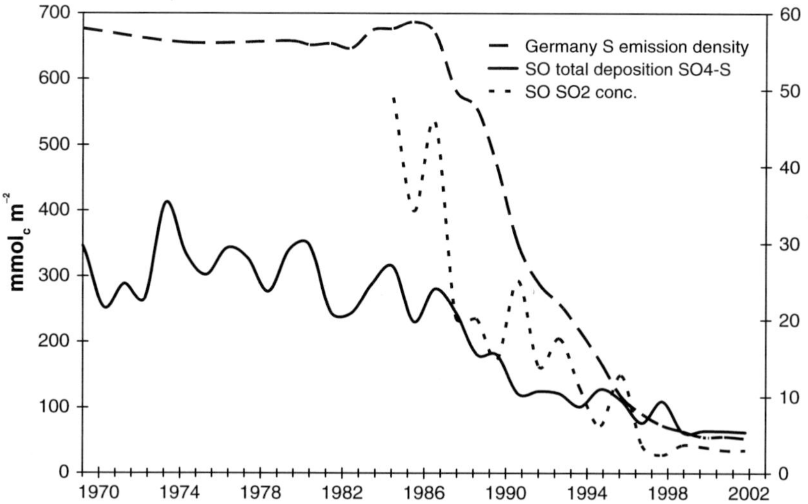

Fig. 15.1 Time series of emission density of sulphur (mmol$_c$ S m^{-2}) in Germany, total deposition (mmol$_c$ S m^{-2}) and SO_2 concentration (μg SO_2 m^{-3}) in ambient air at Solling site (*SO*)

Mean Na^+ fluxes in stand precipitation have been higher than in open field precipitation during the period 1990–2002 by a factor (ID/OF) of 0.49 at Göttinger Wald, 0.41 at Solling and 1.02 at Zierenberg sites. This factor (f_{Na}, see data collection and evaluation procedures) is used to calculate the particulate interception which is higher at the Zierenberg site than at the Solling and Göttinger Wald sites and may be caused by the high agricultural activity surrounding the Zierenberg site, and by higher aerodynamic roughness of the forest stand, which is situated at the slope of a relatively isolated mountain.

A high fraction of total acid inputs was contributed by NH_4^+, where the acid inputs are given by the sum of H^+, Mn^{2+} and $NH_4{}^+$ depositions. Despite decreasing NH_4^+ fluxes at the Solling and Göttinger Wald sites during the last two decades, the relative contribution of NH_4^+ to the fluxes of acids has increased significantly because of a considerable decrease in free acidity fluxes. During 1990–2002, the contribution of NH_4^+ to total acid deposition was 50–85% at Solling, 75–95% at Göttinger Wald and 40–65% at Zierenberg, whereas at the beginning of the 1970s, NH_4^+ contributed from 10 to 40% at the Solling site.

Annual element fluxes for open field deposition, throughfall and stemflow at Solling, Göttinger Wald, and Zierenberg are documented in Annex Tables 15.11–15.19.

15.4 Canopy Rain Interactions

Various tree compartments (leaves, twigs, branches and bark) act as sources or sinks for solutes in precipitation when they pass through the canopy. The canopy budget is commonly estimated by subtracting stand precipitation from total deposition

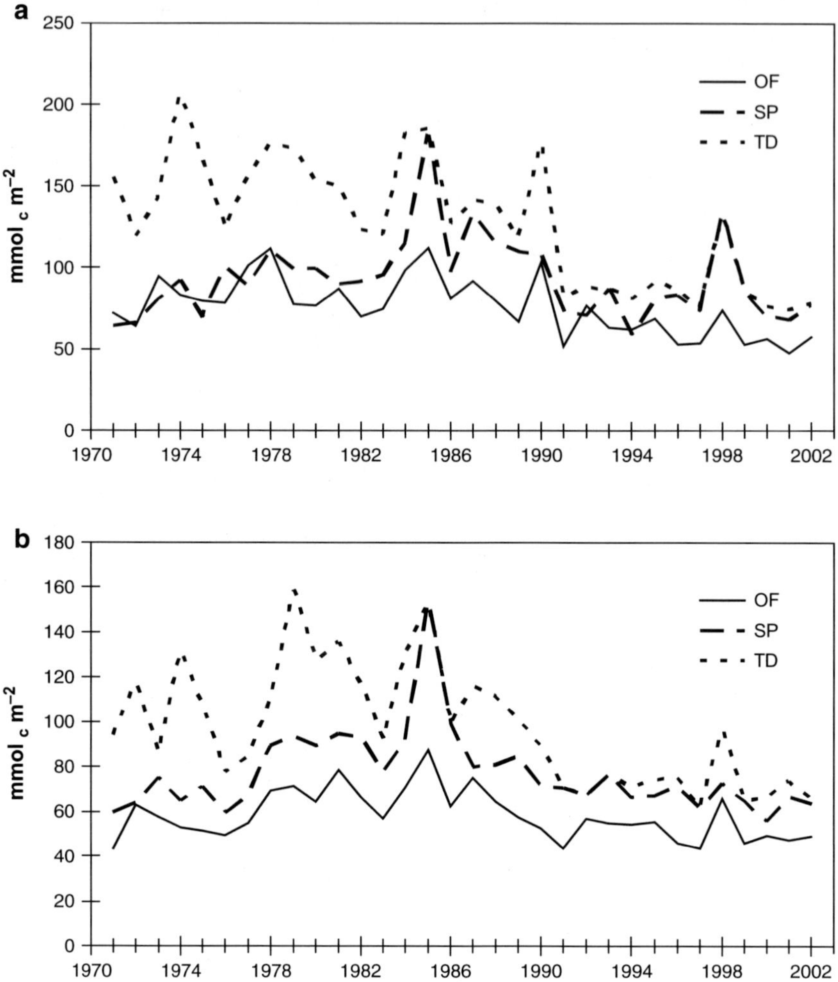

Fig. 15.2 Time series of annual fluxes of (**a**) NH_4^+ and (**b**) NO_3^- with open field deposition (*OF*), stand precipitation (*SP*) and total deposition (*TD*, according to Ulrich 1994) at Solling site

(canopy model by Ulrich 1994). Negative values can be interpreted as indication of leaching from the leaves (Langusch et al. 2003) and of dissolution of dry deposited particles, whereas positive values indicate a sink function, e.g. uptake by the canopy. Plant leaching may occur as a diffusion of organically complexed cations or as an exchange process, where nutrient cations such as K^+, Mg^{2+}, Ca^{2+} and Mn^{2+} are exchanged against H^+ or NH_4^+ (Klemm et al. 1989; Draaijers and Erisman 2005).

Mean values for canopy budgets have been negative for K^+, Mg^{2+}, Ca^{2+} and Mn^{2+}, whereas they have been positive for H^+, NH_4^+ and NO_3^- (Table 15.9). At the Zierenberg site, no retention of NO_3^- has been observed during the observation period. Retention of nitrogen by the canopy can be caused by (1) uptake by

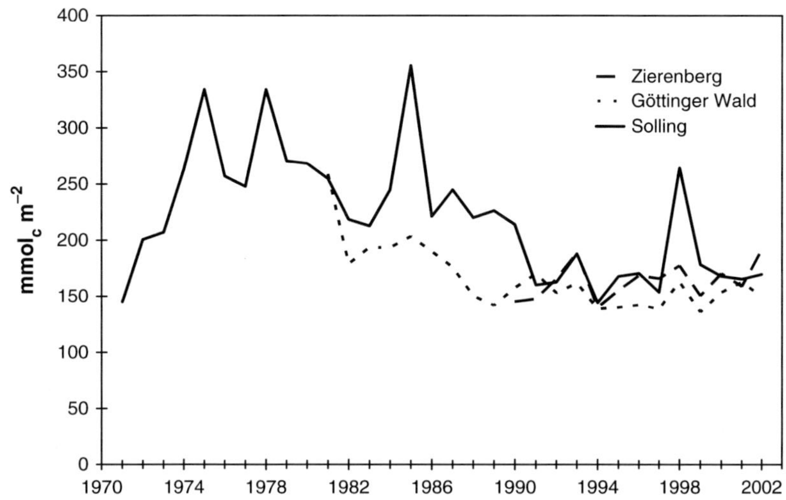

Fig. 15.3 Time series of annual N_{tot} fluxes with stand precipitation at the Solling, Göttinger Wald and Zierenberg sites

Table 15.9 Average canopy budgets (total deposition−stand precipitation) for a number of elements during the periods 1969–1980 (Solling only, 1971–1980 for NH_4^+, NO_3^-), 1981–1989 (Solling and Göttinger Wald), and 1990–2002 at the three sites

Site	Period	K^+	Mg^{2+}	Ca^{2+}	H^+	Mn^{2+}	NH_4^+	NO_3^-
				(mmol$_c$ m^{-2} per year)				
Solling	1969–1980	−55	−5	−29	65	−11.8	70	36
	1981–1989	−56	−12	−39	103	−10.3	29	22
	1990–2002	−51	−6	−14	34	−3.7	12	6
Göttinger Wald	1981–1989	−50	−12	−46	88	−0.9	5	5
	1990–2002	−52	−6	−17	29	−0.2	6	7
Zierenberg	1990–2002	−65	−10	−5	50	−0.1	18	0

epiphytic algae and lichens, (2) immobilisation by micro-organisms (Stadler and Michalzik 2000), and (3) assimilation into the leaves (Garten et al. 1998). The last process is likely to be the most relevant one for the retention of N by the canopy (Lovett and Lindberg 1993).

Mean NH_4^+ retention rates by the canopy decreased during the study period at the Solling site. For the measurement period of 1990–2002, NH_4^+ retention by the canopy was estimated as: 12 mmol$_c$ for Solling, 6 mmol$_c$ for Göttinger Wald, and 18 mmol$_c$ m^{-2} per year for Zierenberg. The corresponding figures for NO_3^- were: 6 mmol$_c$ (Solling), 7 mmol$_c$ (Göttinger Wald) and 0 mmol$_c$ m^{-2} per year (Zierenberg), and for N_{tot}: 18 mmol$_c$ (Solling), 13 mmol$_c$ (Göttinger Wald) and 18 mmol$_c$ m^{-2} per year (Zierenberg). As mentioned above, these estimates are uncertain and are most probably an underestimation. For the sites of the Integrated Forest Study (IFS), mean inorganic N-retention rates of 10–160 mmol$_c$ m^{-2} per year were found (Lovett and Lindberg 1993), which brackets the values found for the sites of

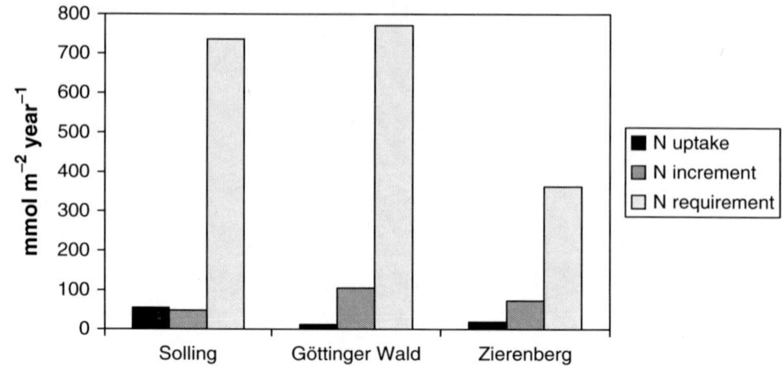

Fig. 15.4 Canopy retention (N-uptake) of total N compared to requirement for growth increment and forest requirement (growth and turnover) at Solling, Göttinger Wald and Zierenberg (data for N increment and N requirement from Rademacher et al. (Chap. 8)

this study. Horn et al. (1989) calculated N-retention rates by the canopies of a healthy and a declining spruce stand at Fichtelgebirge, Bavaria, of 89 and 185 $mmol_c$ m^{-2} per year, respectively.

If all the assumptions for calculating the canopy retention of N were valid, the canopy absorption would cover 7% of N-demand for growth and turnover at the Solling stand which has been calculated to be 740 mmol m^{-2} (Chap. 8). Similar values of N-demand calculated for the Göttinger Wald (770 mmol m^{-2} per year) and Zierenberg (360 mmol m^{-2} per year) stands (Chap. 8), of which about 1.5% and 5% will be taken up by the canopy from atmospheric deposition (Fig. 15.4). Compared to the N required for the growth increment, the Solling stand can cover its N-demand completely by uptake within the canopy, whereas the relative contribution of crown uptake is low for the Göttinger Wald (11%) and Zierenberg stands (25%). Annual N-retention for forest growth has been estimated as 43 mmol m^{-2} for the stands at Solling, 38–63 mmol m^{-2} for Göttinger Wald and 126 mmol m^{-2} for Zierenberg (Chap. 8).

The assimilation of NH_4^+ and NO_3^- from precipitation has been experimentally confirmed by Brumme et al. (1992); Veithen (1996); Garten et al. (1998); and Harrison et al. (2000); Brumme et al. (1992) found that 6–12% of ^{15}N applied to the aboveground parts of 3 to 9-years-old beech plants from the Solling site over 4 months was allocated to the roots. By the use of ^{15}N labelled NH_4^+ and NO_3^-, they found higher uptake rates for NH_4^+ as compared to NO_3 indicating preferential uptake of NH_4^+. In a similar study, Garten et al. (1998) found a retention of ^{15}N labelled wet deposition of 12–26% for deciduous trees at Walker Branch Watershed, Tennessee, USA. Veithen (1996) used washing procedures at leaves from the Solling and Göttinger Wald stands for the study of canopy interactions and also found higher uptake rates for NH_4^+ as compared to NO_3^-. The preferential uptake of NH_4^+ is confirmed for the Solling and Zierenberg sites from the data of this study.

At the Solling site, uptake of NH_4^+ and NO_3 by the canopy decreased during the last 18 years (Table 15.10). At the Göttinger Wald and Zierenberg sites, no such

Table 15.10 Trends of canopy budgets at Solling and Göttinger Wald sites from 1981 to 2002 and at Zierenberg site from 1990 to 2002 (test of trend with correlation coefficient after Pearson; o no trend, + significant increasing trend $p \leq 0.05$, ++ highly significant increasing trend $p \leq 0.01$, – significant decreasing trend $p \leq 0.05$, – – highly significant decreasing trend $p \leq 0.01$)

Site	K^+	Mg^{2+}	Ca^{2+}	H^+	Mn^{2+}	NH_4^+	NO_3^-
Solling	o	+	++	– –	++	–	–
Göttinger Wald	o	++	++	– –	++	o	o
Zierenberg	o	o	o	– –	o	o	o

trend was observed. Brumme et al. (1992) and Veithen (1996) showed that NH_4^+ uptake is dependent on the NH_4^+ concentration in precipitation. Thus, decreasing uptake rates of NH_4^+ may be attributed to decreasing NH_4^+ concentrations in deposition. Veithen (1996) estimated the compensation point for NH_4^+, where NH_4^+ uptake turns to NH_4^+ leaching, to be between 28 and 46 $\mu mol_c\ l^{-1}$. As mean NH_4^+ concentrations in throughfall and stemflow at the Solling and Göttinger Wald sites are currently well above the compensation point found by Veithen (1996), NH_4^+ uptake is likely to take place at these stands.

Positive canopy budgets of H^+ indicated proton buffering at all sites. The buffering can occur as an exchange process where H^+ is exchanged against cations such as K^+, Mg^{2+} or Ca^{2+} (Lindberg et al. 1986; Bredemeier 1988; Klemm et al. 1989, Matzner and Meiwes 1994; Draaijers and Erisman 1995). Mean proton buffering rates for 1990–2002 were calculated at 34 $mmol_c$ (Solling), 29 $mmol_c$ (Göttinger Wald) and 50 $mmol_c\ m^{-2}$ per year (Zierenberg). Between 1981 and 2002, proton buffering rates decreased significantly at the Solling and Göttinger Wald sites accompanied by a simultaneous decrease in the leaching rates of Mg^{2+} and Ca^{2+} (Table 15.10). Proton buffering rates decreased significantly at the Zierenberg site between 1990 and 2002.

Buffering of total acidity in the canopy has been calculated from the sink functions of H^+ and NH_4^+. At the study sites, 13–59% of the load of total acidity is buffered in the canopy on an annual basis (Fig. 15.5). In addition to decreasing loads of total acidity, the degree of buffering of acidity in the canopy has also decreased during the last 22 years. For the Göttinger Wald site, this might be explained by the more or less constant NH_4^+ uptake, which causes a production of H^+ in the canopy (Ulrich 1994). At the Solling site, the decrease of H^+ buffering has been twice as high as the decrease of NH_4^+ uptake. Hence, the H^+ production by NH_4^+ uptake has become more important. At the Zierenberg site, the degree of acid buffering in the canopy was generally higher compared to the other sites that also decreased during the study period (Fig. 15.5).

Leaching rates of K^+ from the canopy have been similar at the three study sites, whereas leaching of Mn^{2+} has occurred only at the Solling site probably due to the higher Mn^{2+} availability in this acid soil (Table 15.9). The higher amount of leaching of Mn^{2+} from the canopy at the Solling site was confirmed by Veithen (1996). Leaching of Mg^{2+} was highest at the Zierenberg site because of high Mg foliar content and also high Mg^{2+} concentration in soil solution resulting from the high Mg^{2+} content of the magmatic bedrock. Leaching of Ca^{2+} from the canopy was

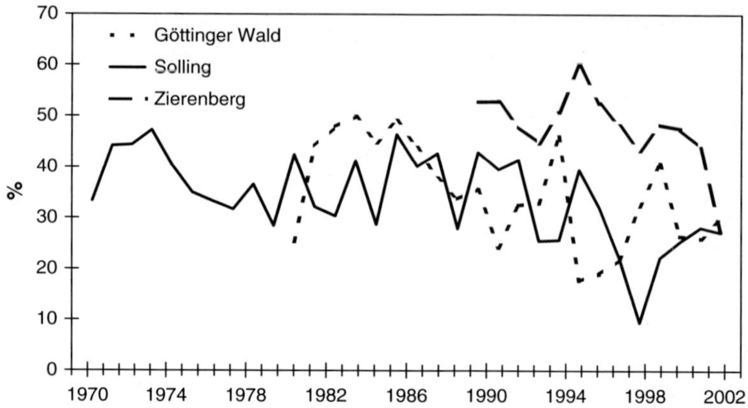

Fig. 15.5 Time series of buffering of acids in the canopy relative to total acidity load at the Zierenberg, Solling and Göttinger Wald sites

similar at the Solling and Göttinger Wald sites, but lower at the Zierenberg sites. Contrary to these results, Veithen (1996) found higher leaching rates for Ca^{2+} from the canopy at the Göttinger Wald site than at the Solling site, which were attributed to a better Ca^{2+} nutrition of the stand growing on calcareous substrate. Surprisingly, the Ca^{2+} availability of the soils and also foliar contents of the three sites were not reflected by the Ca^{2+} leaching rates from the canopy. Mohr et al. (2005) found a positive relation between leaching rates of Ca^{2+} and Mg^{2+} and foliar nutrient concentrations. Time series of canopy budgets indicate that leaching rates of Ca^{2+}, Mg^{2+} and Mn^{2+} decreased significantly during 1981 and 2002 at the Solling and Göttinger Wald sites (Table 15.10). A decrease in the leaching rates of nutrient cations from the canopy may be related to decreasing deposition rates of H^+. No significant trends of canopy budgets have been observed at the Zierenberg site for these ions.

15.5 Discussion

Atmospheric deposition of many major elements can be attributed almost quantitatively to human activities. The only important natural source of salts is of marine origin. Main sources of anthropogenic air pollutants are: combustion processes, industrial processes and agriculture. Combustion and industrial processes are mainly responsible for the emission of SO_2 and NO_x, whereas dust particles are released mainly through industrial processes. NH_3 is mainly emitted through agricultural activities related to animal farming (Umweltbundesamt 2000). Deposition rates at the study sites are moderate when compared to other study sites in Germany (Gauger et al. 2002), but are relatively high compared to other regions in Europe (Hauhs et al. 1989; de Vries et al. 2001). Compared to spruce stands, deposition at beech stands is generally lower (Meesenburg et al. 1995; Balazs 1998; Rothe et al. 2002, Matzner et al. 2004); Eichhorn et al. (2001) evaluated data from 49 beech plots across Europe with average total N-deposition of 136 $mmol_c$ m^{-2} per year (min–max: 59–210 $mmol_c$ m^{-2} per year). The study sites are with 151–179 $mmol_c$ m^{-2} per year among the European beech sites with above-average N-deposition. Compared to 144 intensive monitoring plots in Europe, Na^+ scaling factors (f_{Na}) at the study sites are above average (medium value 0.34, de Vries et al. 2001), suggesting that the contribution of dry deposition to total deposition is more important than at most other forests in Europe.

Deposition rates of the most major elements decreased during the last two decades which is evident at the Solling and Göttinger Wald sites with long observation periods. Atmospheric deposition is primarily characterised by NH_4^+, NO_3^- and SO_4^{2-}. During the 1970s and 1980s, H^+also played a major role in precipitation chemistry. During the last two decades deposition of free acidity has decreased but that of potential acidity has remained high due to high NH_4^+ deposition. The reductions of SO_4^{2-} deposition are related to reductions of SO_2 emission in western Germany (Umweltbundesamt 2000). Because H^+ is mainly generated through the oxidisation of SO_2 to SO_4^{2-}, the H^+ deposition decreased simultaneously. The reduction of acid depositions to central German forest ecosystems is attributed to the reduction of emissions, which became effective through legislation and the closing down of industrial units in eastern Germany after the reunion. There has been a reduction in dust emissions, decreasing the depositions of Ca^{2+} and Mg^{2+}. This decease in cations may have consequences for the buffering of acids in precipitation and for the nutrition of the forest stands. N-deposition decreased only slightly, which is also in agreement with constantly high N-emissions in western Germany (Umweltbundesamt 2000). N-deposition at the three sites is currently well above the amount retained for the forest increment.

Estimates of total deposition of different N-species are very uncertain because of the involvement of several different deposition and transformation processes in the canopy (Marques et al. 2001). N-deposition into forests occurs as rainwater, as fog or in gaseous form and with several different N-species. The deposited N-compounds interact with the canopy including their assimilation into leaves and transformations of the N-species. Garten et al. (1998) argued that the assimilation of gaseous N-compounds was the most effective uptake process, whereas Harrison et al. (2000) estimated the uptake from wet deposition to be more important. N-leaching from the canopy is possible at certain growth phases. Due to the complex biochemical processes, estimates of total N-deposition with a simple approach such as the model of Ulrich (1994) should be viewed with due care. Because N-uptake and gaseous deposition cannot be independently calculated, an underestimation of total deposition is likely by this model. Alternative canopy budget models are more specific with respect to nitrogen, but require independent input parameters, which are sometimes difficult to estimate (Horn et al. 1989; Draaijers and Erisman 1995). Independent measurements with micro-meteorological methods at a spruce stand adjacent to the Solling beech stand suggested an undestimation of N-deposition of almost 50% by the Ulrich model (Ibrom et al. 1995; Marques et al. 2001). Gauger et al. (2002) compared deposition estimates from inferential modelling with estimates from throughfall measurements and canopy budget modelling and found an underestimation of about 50% by canopy budget modelling for Level II monitoring plots in Germany stocked with spruce. For Swiss long-term forest monitoring sites, Schmitt et al. (2005) found on an average 17 $mmol_c m^{-2}$ per year higher N-deposition rates estimated with an inferential method than by throughfall measurements. Nevertheless, estimates of total deposition for the study sites are plausible but are regarded as lower limits of true values.

Acid deposition on the canopy will induce interaction between rainwater and foliage. Leaching of nutrients from the canopy may induce nutrient deficiencies depending on the capacity of the trees to replenish the nutrient pools (Lindberg et al. 1986). When large quantities of N are assimilated in the canopy, the N-uptake by roots may be reduced. This may increase nutrient imbalances due to the spatial decoupling of N and base cation uptake (Harrison et al. 2000). Finally, the major changes in the deposition have also affected canopy-rain interactions. A decrease in the atmospheric emissions has reduced interaction between precipitation and canopy. As precipitation acidity is a major driving force for exchange of nutrient cations at foliage surfaces and their subsequent leaching, a reduced acidity input results in decreased exchange rates (Klemm 1989).

15.6 Conclusions

- N concentrations in open field precipitation, throughfall and stemflow at three beech stands in the northwest German low mountain ranges differ substantially between sites and pathways. However, the contributions of the different N-species NH_4^+, NO_3^-, and N_{org} to the deposition fluxes are relatively similar between sites.
- The sources of solutes in the deposition pathways can be related to anthrogenic air pollution, sea spray, deposition of organic compounds and canopy processes.
- Atmospheric deposition of most major components has declined over the last two decades. Reductions of deposition of acids and S have been more pronounced than reduction of N-input.
- Estimation of total N-deposition with the canopy budget model of Ulrich (1994) most probably underestimates real input rates. However, total deposition estimates can be regarded as lower limits of true values.
- According to the canopy budget model, N is taken up by the canopy of the beech stands. The uptake of NH_4^+ is higher than the uptake of NO_3^-.
- The N-requirement of beech stands for growth increment may be fulfilled to a substantial part by canopy uptake.
- Due to the decline of atmospheric deposition, the interactions between rainwater and the canopy have also reduced.

Annex Tables

See Tables 15.11to15.19.

Table 15.11 Rates of open field deposition at Solling during the period 1969–2002 (Data source: Nordwestdeutsche Forstliche Versuchsanstalt)

Year	H_2O (mm)	Na^+	K^+	Mg^{2+}	Ca^{2+}	H^+	Mn^{2+}	SO_4^{2-}	Cl^-	NH_4^+	NO_3^-	N_{tot}	N_{org}
		($mmol_c\ m^{-2}$)										($mmol\ m^{-2}$)	
1969	1,064	28	7.2	15.6	41	68.5	0.7	150	44				
1970	1,479	37	7.4	10.7	68	109.1	1.1	170	58				
1971	810	26	11.8	32.1	71	70.4	0.7	138	36	72	44	135	19
1972	910	22	9.2	14.8	109	91.3	0.4	157	30	64	63	183	54
1973	1,037	53	10.0	16.5	38	99.2	2.6	132	72	95	57	165	13
1974	1,235	32	6.1	11.5	32	65.5	1.8	145	49	83	53	164	29
1975	884	28	8.2	10.7	37	82.4	1.1	158	40	80	51	166	35
1976	688	43	9.0	14.8	40	60.5	1.1	143	55	79	49	163	35
1977	897	42	14.3	18.1	58	84.3	3.3	158	56	101	55	198	42
1978	950	39	11.5	14.0	47	75.4	1.1	150	61	111	69	223	43
1979	845	21	6.7	10.7	34	85.3	1.1	122	30	78	71	188	39
1980	1,018	30	8.4	13.2	35	68.5	0.7	152	39	77	64	169	28
1981	1,544	43	13.3	18.1	46	89.3	2.9	140	50	87	79	208	42
1982	867	29	7.5	12.0	52	64.5	2.5	146	36	70	66	154	17
1983	1,029	44	10.6	12.2	37	61.5	1.3	131	54	75	57	156	24
1984	1,217	35	10.3	10.6	38	126.0	1.0	135	76	99	71	202	33
1985	1,071	30	10.0	9.5	47	85.3	1.6	145	38	112	87	209	17
1986	1,209	36	7.8	8.5	35	114.1	0.6	99	48	81	62	195	52
1987	1,204	32	6.8	7.0	25	58.5	0.6	122	39	92	75	207	40
1988	1,214	46	8.0	11.4	24	57.6	1.8	101	46	80	64	157	15
1989	949	32	5.4	9.0	23	41.7	0.8	85	40	67	57	138	15
1990	1,040	51	11.5	13.0	22	14.9	0.8	78	42	103	52	181	26
1991	837	40	5.0	9.7	22	7.9	1.0	56	30	52	43	103	9
1992	1,219	62	8.3	12.2	29	5.0	0.7	83	41	77	57	154	20
1993	1,198	56	5.9	11.2	24	6.3	0.4	72	33	63	55	136	15

(*continued*)

Table 15.11 (continued)

Year	H_2O (mm)	Na^+	K^+	Mg^{2+}	Ca^{2+}	H^+	Mn^{2+}	SO_4^{2-}	Cl^-	NH_4^+	NO_3^-	N_{tot} (mmol m^{-2})	N_{org}
		(mmol$_c$ m^{-2})											
1994	1,386	41	4.9	8.4	16	20.5	0.5	69	38	62	54	125	8
1995	1,321	46	4.6	7.4	12	25.8	0.5	69	46	69	55	131	8
1996	928	26	2.8	4.2	10	17.9	0.3	49	23	53	46	103	2
1997	1,008	29	3.1	5.2	13	17.7	0.4	48	28	54	43	105	7
1998	1,575	39	5.9	6.3	15	27.8	0.6	69	39	74	66	147	7
1999	1,109	37	4.6	5.3	11	18.0	0.4	44	33	53	46	106	7
2000	1,131	33	4.6	4.4	10	16.6	0.4	46	28	57	49	119	13
2001	1,187	29	3.3	4.5	10	19.5	0.9	41	27	48	48	112	17
2002	1,451	27	4.5	4.3	10	25.3	0.2	46	28	58	49	122	15

Table 15.12 Throughfall fluxes at Solling during the period 1969–2002 (Data source: Nordwestdeutsche Forstliche Versuchsanstalt)

Year	H_2O (mm)	Na^+	K^+	Mg^{2+}	Ca^{2+}	H^+ (mmol$_c$ m^{-2})	Mn^{2+}	SO_4^{2-}	Cl^-	NH_4^+	NO_3^-	N_{tot} (mmol m^{-2})	N_{org}
1969	793	40	46.8	24.7	104	77.4	10.9	195	61				
1970	1,035	51	46.8	28.0	129	113.1	8.4	258	86				
1971	541	42	32.7	30.5	106	90.3	8.4	179	61	58	52	129	19
1972	623	31	47.1	22.2	126	76.4	6.6	202	51	52	54	159	54
1973	759	67	46.8	28.8	93	83.3	10.9	188	95	77	69	185	39
1974	832	48	66.5	26.3	85	60.5	16.4	144	74	76	57	191	58
1975	632	44	47.8	23.0	87	89.3	10.9	204	63	48	60	203	81
1976	550	50	76.0	29.6	96	79.4	10.6	247	78	91	51	229	86
1977	745	47	47.1	25.5	93	116.1	15.3	231	68	76	56	204	72
1978	770	44	49.4	26.3	93	100.2	12.7	238	67	100	79	271	92
1979	710	35	28.1	18.9	65	84.3	7.3	180	35	85	80	221	56
1980	762	43	45.5	23.9	74	85.3	10.6	220	65	84	76	216	56
1981	1,050	53	46.6	26.3	82	83.3	13.1	202	64	79	81	201	41
1982	562	37	39.0	22.8	97	51.4	8.5	187	59	81	84	185	20
1983	652	54	52.1	22.6	70	46.4	9.6	151	80	78	64	164	21
1984	832	50	57.7	25.3	81	93.7	11.8	222	69	102	78	210	30
1985	692	39	56.6	27.2	100	83.3	13.3	238	70	156	133	303	24
1986	838	42	39.0	18.6	68	56.9	6.5	150	60	79	73	170	19
1987	795	38	72.7	24.2	76	35.4	11.3	208	58	120	67	212	26
1988	844	57	42.7	23.7	71	37.3	8.9	165	78	104	69	186	26
1989	647	41	45.9	19.2	52	24.6	7.0	122	55	95	73	191	33
1990	737	62	74.7	23.9	49	17.8	6.2	119	67	95	60	180	25
1991	603	54	46.2	18.8	50	10.3	7.4	97	56	64	61	140	16
1992	834	66	77.7	21.3	47	8.8	4.5	114	58	68	64	154	22
1993	842	57	31.5	17.5	42	10.8	3.7	100	44	82	69	171	21

(continued)

Table 15.12 (continued)

Year	H_2O	Na^+	K^+	Mg^{2+}	Ca^{2+}	H^+	Mn^{2+}	SO_4^{2-}	Cl^-	NH_4^+	NO_3^-	N_{tot}	N_{org}
	(mm)	(mmol$_c$ m^{-2})										(mmol m^{-2})	
1994	889	44	28.8	14.1	29	26.0	2.9	83	46	53	58	127	16
1995	863	51	85.8	16.1	25	17.1	4.2	106	59	73	58	148	16
1996	652	34	36.2	10.8	24	14.5	3.8	84	38	73	61	146	12
1997	690	35	30.7	10.3	24	12.3	3.3	63	40	66	54	136	15
1998	1,068	47	48.7	13.4	26	12.6	3.6	94	53	126	64	246	56
1999	736	41	39.4	10.1	20	10.3	3.7	50	44	75	54	153	25
2000	747	35	57.7	10.1	17	10.7	2.3	54	39	63	48	149	37
2001	912	37	30.8	9.6	22	11.5	3.6	54	42	62	58	146	26
2002	982	30	44.2	7.8	18	8.9	2.6	53	36	69	56	150	25

Table 15.13 Stemflow fluxes at Solling during the period 1969–2002 (Data source: Nordwestdeutsche Forstliche Versuchsanstalt)

Year	H_2O (mm)	Na^+	K^+	Mg^{2+}	Ca^{2+}	H^+	Mn^{2+}	SO_4^{2-}	Cl^-	NH_4^+	NO_3^-	N_{tot} (mmol m^{-2})	N_{org}
		(mmol$_c$ m^{-2})											
1969	138	16	27.4	7.4	30	77.4	3.6	180	18				
1970	92	12	22.3	7.4	25	52.6	2.5	88	21				
1971	105	13	13.6	4.1	23	49.6	1.8	74	15	6	7	16	3
1972	118	10	17.4	4.9	34	32.7	2.2	87	29	14	10	41	17
1973	134	12	25.8	8.2	12	37.7	2.9	79	20	4	6	22	11
1974	160	31	35.8	14.0	43	66.5	7.3	268	38	17	8	73	48
1975	115	15	28.9	8.2	35	47.6	5.8	130	23	21	11	131	41
1976	89	17	18.7	8.2	24	30.8	2.2	56	37	10	8	29	21
1977	116	18	24.3	7.4	27	54.6	3.3	112	24	13	11	44	20
1978	112	17	18.2	5.8	22	51.6	2.5	90	22	11	10	63	41
1979	112	13	21.2	7.4	26	49.6	2.9	97	33	14	14	50	22
1980	162	16	27.1	8.2	32	55.6	3.6	120	23	15	14	53	24
1981	199	21	39.1	9.9	35	64.5	5.1	145	27	11	13	54	29
1982	66	14	13.4	5.2	18	27.0	2.0	58	20	10	9	34	14
1983	136	18	23.7	8.3	34	45.6	3.2	93	25	17	14	49	19
1984	127	15	16.8	6.7	21	57.5	2.3	68	21	14	14	35	7
1985	101	10	12.1	5.2	19	42.7	2.4	77	16	26	20	52	6
1986	125	14	16.2	6.1	21	53.6	2.0	81	22	18	26	51	7
1987	119	12	19.7	6.4	21	34.7	2.4	74	17	13	12	33	17
1988	126	23	23.3	8.1	24	38.7	3.2	81	29	12	12	34	22
1989	97	15	18.0	6.8	20	31.7	2.5	60	22	14	12	35	20
1990	110	25	19.5	7.4	20	30.8	2.8	61	31	13	11	34	20
1991	91	9	8.1	3.2	8	6.0	0.6	25	10	9	9	20	3
1992	72	5	4.4	1.4	3	4.0	0.2	12	6	3	4	9	3
1993	140	9	7.6	2.3	6	8.4	0.5	22	7	6	8	17	4

(*continued*)

Table 15.13 (continued)

Year	H_2O	Na^+	K^+	Mg^{2+}	Ca^{2+}	H^+	Mn^{2+}	SO_4^{2-}	Cl^-	NH_4^+	NO_3^-	N_{tot}	N_{org}
	(mm)						($mmol_c\ m^{-2}$)					($mmol\ m^{-2}$)	
1994	133	10	6.0	2.2	4	10.0	0.3	19	10	6	8	17	4
1995	129	11	14.7	2.5	4	7.1	0.4	23	13	8	8	20	4
1996	98	8	14.0	2.2	5	9.9	0.5	28	10	10	10	24	5
1997	104	6	7.0	1.4	4	5.0	0.3	14	6	7	8	18	3
1998	160	10	10.7	1.8	4	5.0	0.3	16	11	7	8	19	4
1999	110	9	8.6	1.7	3	4.6	0.3	14	9	10	11	25	4
2000	112	9	8.8	1.6	3	3.6	0.2	11	11	6	8	19	5
2001	137	8	11.1	1.4	3	2.1	0.3	11	9	6	8	20	5
2002	147	7	10.6	0.9	2	2.0	0.2	10	9	8	7	20	5

Table 15.14 Rates of open field deposition at Göttinger Wald during the period 1981–2002 (Data source: Nordwestdeutsche Forstliche Versuchsanstalt)

Year	H_2O (mm)	Na^+	K^+	Mg^{2+}	Ca^{2+}	H^+	Mn^{2+}	SO_4^{2-}	Cl^-	NH_4^+	NO_3^-	N_{tot}	N_{org}
		($mmol_c\ m^{-2}$)										($mmol\ m^{-2}$)	
1981	1,252	35	20.2	17.6	97	52.5	0.7	160	50	73	62	169	34
1982	550	18	8.6	9.6	46	20.4	0.7	82	23	43	41	112	28
1983	635	32	7.1	10.4	43	31.9	0.6	96	32	59	51	129	23
1984	754	21	6.2	11.3	55	49.4	0.8	113	25	59	59	148	30
1985	621	22	8.9	9.4	49	26.2	0.5	92	27	68	51	142	23
1986	784	25	6.9	9.9	45	43.7	0.4	100	32	58	66	151	28
1987	900	26	9.3	13.0	76	54.8	0.3	129	33	75	73	162	15
1988	763	45	7.4	15.9	58	45.3	0.4	118	45	66	84	155	15
1989	495	24	3.4	9.5	36	38.2	0.2	85	27	51	58	99	5
1990	546	39	3.0	11.3	32	19.6	0.3	70	30	53	54	115	9
1991	496	37	3.2	11.2	35	12.4	0.2	67	29	51	54	113	8
1992	686	38	3.6	12.5	37	17.0	0.3	84	31	67	66	142	9
1993	715	30	3.6	8.2	24	3.4	0.3	53	15	46	43	102	12
1994	758	18	2.8	4.4	15	17.5	0.3	52	15	49	44	96	13
1995	598	20	4.1	4.8	11	0.8	0.2	42	21	42	35	82	5
1996	589	14	2.7	3.2	10	1.2	0.2	36	12	42	37	85	5
1997	664	21	2.8	8.2	25	9.9	0.4	44	18	44	40	89	5
1998	893	21	2.7	4.7	14	21.6	0.3	47	19	52	50	109	7
1999	635	18	2.6	3.7	11	11.7	0.2	31	14	42	38	87	6
2000	719	18	3.3	4.1	14	8.1	0.3	32	13	39	37	85	9
2001	718	18	2.7	4.0	11	11.7	0.1	29	17	35	36	83	12
2002	868	18	3.2	4.4	15	14.7	0.1	39	21	51	45	107	9

Table 15.15 Throughfall fluxes at Göttinger Wald during the period 1981–2002 (Data source: Nordwestdeutsche Forstliche Versuchsanstalt)

Year	H_2O (mm)	Na	K	Mg	Ca	H	Mn	SO_4^{2-}	Cl	NH_4	NO_3	N_{tot} (mmol m^{-2})	N_{org}
		(mmol$_c$ m^{-2})											
1981	843	34	40.0	26.5	124	27.7	1.7	185	52	92	69	188	32
1982	366	19	31.4	20.3	89	8.1	1.0	112	33	54	53	129	22
1983	428	30	44.1	22.7	78	14.0	1.5	119	48	61	53	132	18
1984	543	23	39.3	22.0	100	16.7	1.4	147	36	62	59	149	28
1985	417	22	36.2	21.1	89	14.6	1.5	127	39	63	60	144	21
1986	556	23	34.2	18.8	87	19.2	1.1	122	38	49	65	144	29
1987	621	22	53.0	20.3	89	20.1	1.0	146	39	66	53	127	13
1988	522	28	25.9	15.2	66	15.0	0.8	99	36	52	51	115	15
1989	351	23	34.7	18.6	71	10.4	0.9	95	33	64	47	109	9
1990	415	36	50.6	18.7	51	3.6	0.7	76	41	60	42	113	12
1991	329	32	31.0	17.2	57	2.6	0.8	73	34	65	50	129	15
1992	501	33	50.3	19.9	53	4.6	0.7	79	37	63	47	132	22
1993	590	34	26.2	15.3	47	5.2	0.5	77	28	59	55	134	20
1994	562	26	32.7	10.9	33	11.5	0.5	60	30	47	51	112	14
1995	502	29	65.7	18.4	35	6.9	0.6	65	35	58	45	117	14
1996	450	19	31.4	9.3	29	5.9	0.3	58	23	60	51	123	11
1997	463	26	32.8	10.6	33	4.4	0.3	48	31	54	50	116	12
1998	647	29	46.4	11.5	32	5.6	0.4	57	34	68	55	142	20
1999	497	28	36.3	9.5	27	6.0	0.3	42	33	51	52	116	14
2000	547	25	51.8	9.9	26	3.9	0.3	41	29	58	48	134	28
2001	612	27	47.1	10.9	31	4.1	0.3	45	35	68	52	145	25
2002	627	23	43.2	8.6	27	3.6	0.3	45	31	60	51	131	20

Table 15.16 Stemflow fluxes at Göttinger Wald during the period 1981–2002 (Data source: Nordwestdeutsche Forstliche Versuchsanstalt)

Year	H_2O (mm)	Na^+	K^+	Mg^{2+}	Ca^{2+}	H^+	Mn^{2+}	SO_4^{2-}	Cl^-	NH_4^+	NO_3^-	N_{tot} (mmol m^{-2})	N_{org}
		(mmol$_c$ m^{-2})											
1981	183	14	24.1	8.1	41	53.6	0.5	120	25	33	27	70	10
1982	90	8	18.8	5.8	28	16.9	0.7	65	16	20	23	50	7
1983	132	22	37.6	12.3	49	43.7	0.9	124	39	22	30	61	10
1984	137	11	30.9	8.7	40	33.7	0.4	102	20	16	19	45	10
1985	91	9	24.3	6.7	28	25.8	0.4	78	16	27	24	59	9
1986	102	9	21.7	7.7	34	24.8	0.4	75	17	16	23	46	7
1987	159	10	31.6	7.2	35	32.7	0.3	85	17	20	23	49	6
1988	95	8	16.0	4.4	18	7.9	0.2	48	12	19	13	36	4
1989	86	9	13.9	4.2	18	12.9	0.2	43	13	15	16	33	5
1990	104	15	22.5	6.0	19	9.9	0.3	43	19	18	21	45	6
1991	71	8	15.4	3.4	12	5.0	0.1	31	11	20	15	41	6
1992	95	9	15.3	3.3	11	6.0	0.1	27	11	7	9	21	5
1993	103	6	13.2	2.1	8	5.4	0.1	26	6	12	12	28	5
1994	156	9	17.8	2.5	10	8.0	0.1	30	11	11	14	27	5
1995	86	8	15.8	2.0	6	3.2	0.1	21	10	10	9	23	3
1996	78	4	12.3	1.3	5	2.3	0.1	17	5	9	8	20	3
1997	90	5	10.9	1.1	4	2.0	0.0	14	6	10	10	23	3
1998	108	5	11.5	0.9	4	1.1	0.1	11	5	9	9	21	3
1999	80	5	8.7	1.0	4	1.3	0.1	9	6	8	9	20	3
2000	99	7	12.2	1.4	5	1.0	0.0	10	8	6	9	20	4
2001	91	4	14.3	1.5	4	0.2	0.0	11	5	7	6	18	5
2002	95	5	15.2	1.3	4	0.2	0.0	12	6	7	8	19	4

Table 15.17 Rates of open field deposition at Zierenberg during the period 1990–2002 (Data source: Nordwestdeutsche Forstliche Versuchsanstalt)

Year	H_2O (mm)	Na^+	K^+	Mg^{2+}	Ca^{2+}	H^+	Mn^{2+}	SO_4^{2-}	Cl^-	NH_4^+	NO_3^-	N_{tot}	N_{org}
		($mmol_c$ m^{-2})										($mmol$ m^{-2})	
1990	630	29	10.3	18.7	39	19.9	0.2	51	17	32	34	78	11
1991	529	21	6.9	17.4	39	29.2	0.2	49	17	28	35	72	9
1992	742	34	9.1	17.6	37	34.2	0.3	63	25	44	37	91	10
1993	742	28	7.7	30.4	61	18.3	0.3	60	21	52	42	103	9
1994	780	35	14.5	34.1	67	16.7	0.3	57	29	61	39	111	10
1995	721	29	6.9	19.1	33	17.9	0.3	46	22	42	30	82	10
1996	688	17	5.7	15.4	28	14.4	0.2	43	17	45	37	91	9
1997	707	18	4.3	16.8	32	18.8	0.1	42	22	31	32	69	7
1998	890	28	6.3	18.8	35	18.3	0.2	42	27	35	36	80	9
1999	802	27	5.1	16.1	30	13.2	0.1	32	23	30	31	73	11
2000	800	25	7.5	18.4	35	10.5	0.1	39	23	45	42	98	11
2001	758	24	3.6	13.8	27	8.7	0.1	26	22	24	33	65	7
2002	981	23	5.0	14.8	27	8.9	0.2	36	21	41	40	89	9

Table 15.18 Throughfall fluxes at Zierenberg during the period 1990–2002 (Data source: Nordwestdeutsche Forstliche Versuchsanstalt)

Year	H_2O (mm)	Na^+	K^+	Mg^{2+}	Ca^{2+}	H^+ (mmol$_c$ m^{-2})	Mn^{2+}	SO_4^{2-}	Cl^-	NH_4^+	NO_3^-	N_{tot} (mmol m^{-2})	N_{org}
1990	424	42	61.6	39.9	73	9.9	0.6	77	35	40	70	123	13
1991	355	35	55.5	44.6	79	6.9	0.7	83	36	42	64	122	16
1992	508	54	114.4	55.7	81	18.8	0.7	94	61	50	67	144	26
1993	497	46	51.3	49.7	89	5.5	0.4	96	51	68	80	168	19
1994	537	50	49.2	53.6	97	10.2	0.3	76	54	49	62	125	14
1995	483	57	70.0	58.2	79	6.5	0.7	88	64	50	61	141	30
1996	491	33	61.0	38.8	62	5.8	0.3	90	44	60	75	152	18
1997	495	42	48.2	42.1	68	11.5	0.5	76	62	55	80	152	18
1998	646	54	75.2	48.8	78	4.8	0.5	70	68	65	73	164	26
1999	555	48	46.0	33.8	57	7.024	0.3	49	57	48	71	135	16
2000	555	49	80.6	42.3	66	6.65	0.2	57	59	61	62	147	24
2001	549	46	66.7	34.4	54	2.383	0.4	51	59	43	65	128	20
2002	755	43	86.4	35.9	56	3.617	0.4	62	52	70	67	165	28

Table 15.19 Stemflow fluxes at Zierenberg during the period 1990–2002 (Data source: Nordwestdeutsche Forstliche Versuchsanstalt)

Year	H_2O (mm)	Na^+	K^+	Mg^{2+}	Ca^{2+}	H^+	Mn^{2+}	SO_4^{2-}	Cl^-	NH_4^+	NO_3^-	N_{tot} (mmol m^{-2})	N_{org}
		(mmol$_c$ m^{-2})											
1990	58	7	10	5	10	3	0.1	17	5	9	11	22	3
1991	50	7	11	5	12	5	0.1	22	7	10	12	26	4
1992	68	8	12	5	10	14	0.1	18	8	8	10	22	4
1993	74	6	10	5	10	4	0.1	16	6	11	7	21	3
1994	79	6	8	5	10	3	0.0	11	7	8	4	15	3
1995	68	8	9	7	11	3	0.1	23	10	6	5	15	4
1996	70	5	10	4	8	2	0.1	19	6	7	6	17	4
1997	72	5	10	4	8	3	0.0	16	7	5	4	14	4
1998	96	8	14	5	9	1	0.1	15	10	5	3	14	6
1999	85	9	13	5	8	1	0.0	12	11	5	5	16	6
2000	83	11	19	5	10	1	0.0	14	14	10	7	23	7
2001	82	9	23	5	9	0	0.1	18	13	13	9	31	9
2002	113	7	19	4	6	0	0.0	12	9	14	6	27	8

References

Aber JD (2002) Nitrogen saturation in temperate forest ecosystems: current theory, remaining questions and recent advances. Plant Soil 247:179–188

Aber J, McDowell W, Nadelhoffer K, Magill A, Berntson G, Kamakea M, McNulty S, Currie W, Rustad L, Fernandez I (1998) Nitrogen saturation in temperate forest ecosystems: hypotheses revisited. BioScience 48:921–934

Balazs A (1998) 14 Jahre Niederschlagsdeposition in Hessischen Waldgebieten: Ergebnisse von den Meßstationen der Waldökosystemstudie Hessen. Hessische Landesanstalt für Forsteinrichtung, Waldforschung und Waldökologie Forschungsbericht 25

Benecke P (1984) Der Wasserumsatz eines Buchen- und eines Fichtenwaldökosystems im Hochsolling. Schr. Forstl. Fak. Univ. Göttingen Nieders. Forstl. Versuchsanstalt 77, JD Sauerländer, Frankfurt, Germany

Binkley D, Högberg P (1997) Does atmospheric deposition of nitrogen threaten Swedish forests? Forest Ecol Manage 92:119–152

Brechtel HM, Hammes W (1984) Aufstellung und Betreuung des Niederschlagssammlers "Münden". Meßanleitung 3, Hess Forstl Versuchsanstalt, 2. Auflage

Bredemeier M (1988) Forest canopy transformation of atmospheric deposition. Water Air Soil Pollut 40:121–138

Brumme R, Leimcke U, Matzner E (1992) Interception and uptake of NH_4 and NO_3 from wet deposition by above-ground parts of young beech (*Fagus silvatica* L.) trees. Plant Soil 142:273–279

Christophersen N, Hooper RP (1992) Multivariate analysis of stream water chemical data: the use of principal components analysis for the end-member mixing problem. Water Resour Res 28:99–107

De Vries W, Reinds GJ, van der Salm C, Draaijers GPJ, Bleeker A, Erisman JW, Auee J, Gundersen P, van Dobben H, de Zwart D, Derome J, Voogd JCH, Vel EM (2001) Intensive monitoring of forest ecosystems in Europe: technical report 2001. EC-UN/ECE 2001, Brussels, pp 166

Draaijers GPJ (1999) Canopy budget models applicable for use within the intensive monitoring program. In: United Nations Economic Commission for Europe, Convention on Long–Range Transboundary Air Pollution, International Co-operative Programme on Assessment and Monitoring of Air Pollution Effects on Forests (eds) Manual on methods and criteria for harmonized sampling, assessment, monitoring and analysis of the effects of air pollution on forests Part VI. Measurement of deposition and air pollution. UNECE, CLRTAP, ICPAMA-PEF, Geneva, Switzerland, pp 40–52

Draaijers GPJ, Erisman JW (1995) A canopy budget model to assess atmospheric deposition from throughfall measurements. Water Air Soil Pollut 85:2253–2258

Draaijers GPJ, Erisman JW, Spranger T, Wyers GP (1996) The application of throughfall measurements for atmospheric deposition monitoring. Atmos Environ 30

Draaijers GPJ, Bleeker A, van der Veen D, Erisman JW, Möls H, Fonteijn P, Geusenbroek M (2001) Field intercomparison of throughfall, stemflow and precipitation measurements performed within the framework of the Pan-European Intensive Monitoring Program of EU/ICP forests. TNO rep. R 2001/140

Eichhorn J (1995) Stickstoffsättigung und ihre Auswirkungen auf das Buchenwaldökosystem der Fallstudie Zierenberg. Ber Forschungszentrum Waldökosysteme A 124

Eichhorn J, Haußmann T, Paar U, Reinds GJ, de Vries W (2001) Assessments of impacts of nitrogen deposition on beech forests. Scientific World Journal 1(Suppl 3) 423–432

Fassbender HW, Ahrens E (1977) Laborvorschriften und Praktikumsanleitung: Zur chemischen Untersuchung von Vegetations-, Boden- und Wasserproben im Institut für Bodenkunde und Waldernährung der Universität Göttingen. Göttinger Bodenkd Ber 47

Feger K-H (1986) Biogeochemische Untersuchungen an Gewässern im Schwarzwald unter besonderer Berücksichtigung atmogener Stoffeinträge. Freiburger Bodenkundl Abh 17

Galloway NJ (1995) Acid deposition: perspectives in time and space. Water Air Soil Pollut 85:15–24

Garten CT, Schwab AB, Shirshac TL (1998) Foliar retention of 15N tracers: implications for net canopy exchange in low- and high-elevation forest ecosystems. Forest Ecol Manage 103:211–216

Gauger T, Anshelm F, Schuster H, Erisman JW, Vermeulen AT, Draaijers GPJ, Bleeker A, Nagel H-D (2002) Mapping of ecosystem specific long-term trends in deposition loads and concentration of air pollutants in Germany and their comparison with critical loads and critical levels, Part 1: deposition loads 1990–1999. Inst. f. Navigation, Stuttgart, pp 1–207

Gerke H (1987) Untersuchungen zum Wasserhaushalt eines Kalkbuchenwald-Ökosystems und zur Wasserbewegung in flachgründigen Böden und im durchwurzelten Kalkgestein als Grundlage zur Modellentwicklung. Ber Forschungszentrum Waldökosysteme/Waldsterben A27

Godbold DL, Fritz H-W, Jentschke G, Meesenburg H, Rademacher P (2003) Root turnover and root necromass accumulation of Norway spruce (*Picea abies*) are affected by soil acidity. Tree Physiol 23:915–921

Gorham E, Martin FB, Litzau JT (1984) Acid rain: ionic correlations in the eastern United States, 1980–1981. Science 225:407–409

Harrison AF, Schulze E-D, Gebauer G, Bruckner G (2000) Canopy uptake and utilization of atmospheric pollutant nitrogen. In: Schulze ED (ed) Carbon and nitrogen cycling in European forest ecosystems. Ecol Stud 142:171–188

Hauhs M, Rost-Siebert K, Raben G, Paces T, Vigerust B (1989) Summary of European data. In: Malanchuk JL, Nilsson J (eds) The role of nitrogen in the acidification of soils and surface waters. NORD, København, Denmark, 5–1–5–37

Horn R, Schulze E-D, Hantschel R (1989) Nutrient balance and element cycling in healthy and declining Norway spruce stands. Biogeochemistry of forested catchments in a changing environment: a German case study. Ecol Stud 77:444–455

Ibrom A (1993) Die Deposition und die Pflanzenauswaschung (leaching) von Pflanzennährstoffen in einem Fichtenbestand im Solling. Ber Forschungszentrum Waldökosysteme A 105

Ibrom A, Oltchev A, Constantin J, Marques M, Gravenhorst G (1995) Die Stickstoffimmission und – deposition in Wäldern. IMA-Querschnittseminar Wirkungskomplex Stickstoff und Wald, Umweltbundesamt Texte 28/95:20–29

Johnson DW, Cresser MS, Nilsson SI, Turner J, Ulrich B, Binkley D, Cole DW (1991) Soil changes in forest ecosystems: evidence for and probable causes. Proc Royal Soc Edinburgh 97B:81–116

Klemm O (1989) Leaching and uptake of ions through above-ground Norway spruce tree parts. In: Schulze E-D, Lange O, Oren R (eds) Forest decline and air pollution. A study or spruce on acid soils. Ecol Stud 77:210–237

König N, Fortmann H (1996a) Probenvorbereitungs-, Untersuchungs- und Elementbestimmungs-Methoden des Umweltanalytik-Labors der Niedersächsischen Forstlichen Versuchsanstalt und des Zentrallabor II des Forschungszentrums Waldökosysteme. Teil 1: Elementbestimmungsmethoden A-M. Ber Forschungszentrum Waldökosysteme B 46

König N, Fortmann H (1996b) Probenvorbereitungs-, Untersuchungs- und Elementbestimmungs-Methoden des Umweltanalytik-Labors der Niedersächsischen Forstlichen Versuchsanstalt und des Zentrallabor II des Forschungszentrums Waldökosysteme. Teil 2: Elementbestimmungsmethoden N-Z und Sammelanhänge. Ber Forschungszentrum Waldökosysteme B 47

König N, Fortmann H (1996c) Probenvorbereitungs-, Untersuchungs- und Elementbestimmungs-Methoden des Umweltanalytik-Labors der Niedersächsischen Forstlichen Versuchsanstalt und des Zentrallabor II des Forschungszentrums Waldökosysteme. Teil 3: Gerätekurzanleitungen und Gerätekurzanleitungen Datenverarbeitung. Ber Forschungszentrum Waldökosysteme B 48

König N, Fortmann H (1996d) Probenvorbereitungs-, Untersuchungs- und Elementbestimmungs-Methoden des Umweltanalytik-Labors der Niedersächsischen Forstlichen Versuchsanstalt und des Zentrallabor II des Forschungszentrums Waldökosysteme. Teil 4: Probenvorbereitungs- und Untersuchungsmethoden, Qualitätskontrolle und Datenverarbeitung. Ber Forschungszentrum Waldökosysteme B 49

Langusch J-J, Borken W, Armbruster M, Dise NB, Matzner E (2003) Canopy leaching of cations in Central European forest ecosystems: a regional assessment. J Plant Nutr Soil Sci 166: 168–174

Lindberg SE, Lovett GM, Richter DD, Johnson DW (1986) Atmospheric deposition and canopy interactions of major ions in a forest. Science 231:141–145

Lovett GM, Lindberg SE (1993) Atmospheric deposition and canopy interactions of nitrogen in forests. Can J Forest Res 23:1603–1616

Marques MC, Gravenhorst G, Ibrom A (2001) Input of atmospheric particles into forest stands by dry deposition. Water Air Soil Pollut 130:571–576

Matzner E, Meiwes KJ (1994) Long-term development of element fluxes with bulk precipitation and throughfall in two German forests. J Environ Qual 23:162–166

Matzner E, Murach D (1995) Soil changes induced by air pollutant deposition and their implication for forests in central Europe. Water Air Soil Pollut 85:63–76

Matzner E, Zuber T, Alewell C, Lischeid G, Moritz K (2004) Trends in deposition and canopy leaching of mineral elements as indicated by bulk depostion and throughfall measurements. In: Matzner E (ed) Biogeochemistry of forested catchments in a changing environment: a German case study. Ecol Stud 172:233–250

Meesenburg H, Meiwes KJ, Rademacher P (1995) Long-term trends in atmospheric deposition and seepage output in northwest German forest ecosystems. Water Air Soil Pollut 85:611–616

Meesenburg H, Meiwes KJ, Schulze A, Rademacher P (1997) Bodendauerbeobachtungsflächen auf forstlich genutzten Böden (BDF-F). Arb-H Boden 2/1997:77–94

Meesenburg H, Mohr K, Dämmgen U, Schaaf S, Meiwes KJ, Horvath B (2005) Stickstoff-Einträge und – Bilanzen in den Wäldern des ANSWER-Projektes – eine Synthese. Landbauforschung Völkenrode, Sonderh 279:95–108

Meiwes KJ, Hauhs M, Gerke H, Asche N, Matzner E, Lamersdorf N (1984) Die Erfassung des Stoffkreislaufes in Waldökosystemen: Konzept und Methodik. Ber Forschungszentrum Waldökosysteme/Waldsterben 7:68–142

Meiwes KJ, Meesenburg H, Rademacher P (1999) Wirkung von Stickstoff auf Waldökosysteme. Mitt NNA 1/99:118–121

Mohr K, Schaaf S, Horvath B, Meesenburg H, Dämmgen U (2005) Stoff- und Energieflüsse der im ANSWER-Projekt untersuchten Waldbestände. In: Bestimmung von Ammoniak-Einträgen aus der Luft und deren Wirkung auf Waldökosysteme (ANSWER-Projekt), Landbauforschung Völkenrode, Sonderheft 279:69–94

Papen H, Gessler A, Zumbusch E, Rennenberg H (2002) Chemolithoautotrophic nitrifiers in the phyllosphere of a spruce ecosystem receiving high atmospheric nitrogen input. Curr Microbiol 44:56–60

Rothe A, Huber C, Kreutzer K, Weis W (2002) Deposition and soil leaching in stands of Norway spruce and European beech. Results from the Höglwald research in comparison with other European case studies. Plant Soil 240:1–14

Schmitt M, Thöni L, Waldner P, Thimonier A (2005) Total deposition of nitrogen on Swiss long-term forest ecosystem research (LWF) plots: comparison of the throughfall and the inferential method. Atmos Environ 39:1079–1091

Spranger T (1992) Erfassung und ökosystemare Bewertung der atmosphärischen Deposition und weiterer oberirdischer Stoffflüsse im Bereich der Bornhöveder Seenkette. EcoSys Beitr Ökosystemforschung (Suppl 4)

Stadler B, Michalzik B (2000) Effects of phytophagous insects on micro-organisms and throughfall chemistry in forest ecosystems: herbivores as switches for the nutrient dynamics in the canopy. Basic Appl Ecol 1:109–116

Sutton MA, Fowler D, Burkhardt JK, Milford C (1995) Vegetation atmosphere exchange of ammonia: canopy cycling and the impacts of elevated nitrogen inputs. Water Air Soil Pollut 85:2057–2063

Tyler G (1983) Does acidification increase metal availability and thereby inhibit decomposition and mineralisation processes in forest soils? In: National Swedish Environment Protection Board (ed) Ecological effects of acid deposition. Report and Backgroundpapers 1982 Stockholm conference on the acidification of the environment, Expert meeting I – Report PM 1636:245–256

Ulrich B (1983) Interaction of forest canopies with atmospheric constituents: SO_2, alkali and alkali earth cations and chloride. In: Ulrich B, Pankrath J (eds) Effects of accumulation of air pollutants in forest ecosystems. Dordrecht, Reidel publ Co, pp 33–45

Ulrich B (1994) Nutrient and acid/base budget of central European forest ecosystems. In: Hüttermann A, Godbold DL (Hrsg) Effects of acid rain on forest processes. Wiley, New York, pp S1–S43

Umweltbundesamt (2000) Daten zur Umwelt, 7th edn. Erich Schmidt, Berlin, p 378

Van Leeuwen EP, Draaijers GPJ, de Jong PGH, Erisman JW (1996) Mapping dry deposition of acidifying components and base cations on a small scale in Germany. National Institute of Public Health and the Environment (RIVM). Report no. 722108012

Veithen C (1996) Ionentransporte durch Cuticlen von Buchenblättern und Fichtennadeln. Ber Forschungszentrum Waldökosysteme A 136

Wright RF, Roelofs JGM, Bredemeier M, Blanck K, Boxman AW, Emmett BA, Gundersen P, Hultberg H, Kjønaas OJ, Moldan F, Tietema A, van Breemen N, van Dijk HFG (1995) NITREX: responses of coniferous forest ecosystems to experimentally changed deposition of nitrogen. Forest Ecol Manage 71:163–169

Zimmerling R, Dämmgen U, Haenel H-D (2000) Flüsse versauernd und eutrophierend wirkender Spezies zwischen Atmosphäre und Wald- und Forstökosystemen. In: Dämmgen U (ed) Versauernde und eutrophierende Luftverschmutzung in Nordost-Brandenburg, Landbauforschung Völkenrode Sonderheft 213:95–128

Chapter 16
Changes in Soil Solution Chemistry, Seepage Losses, and Input–Output Budgets at Three Beech Forests in Response to Atmospheric Depositions

R. Brumme, H. Meesenburg, M. Bredemeier, C. Jacobsen,
E. Schönfelder, K.J. Meiwes, and J. Eichhorn

16.1 Introduction

Atmospheric depositions have increased substantially since industrialisation and have affected many forest properties. Forest soils of low buffering to the acid load as those of sandy or silty texture significantly declined in soil pH and some essential nutrients during the last decades (see Chap. 21) causing a controversial discussion about the stability of such forest ecosystems (Ulrich 1981, 1987, 1992, 1994b). Of 1,700 soils studied in the forest soil survey programme of Germany, 60% of all soils have pH(H_2O) <4.2 (Wolff and Riek 1997a, b). N deficiency, which has been a common feature of forest stands in the temperate region (Tamm 1991), does not occur any more due to high N deposition rates in Europe (except of north and east Europe). On the contrary, the N supply has increased as indicated by high N contents of the tree foliage and surface organic layer samples (Tietema and Beier 1995; Alewell et al. 2000b; McNulty et al. 1991; Wolff and Riek 1997a). Nitrate losses with seepage water have increased at some locations (Dise and Wright 1995; Bredemeier et al. 1998; Matzner et al. 2004; Borken and Matzner 2004), which led to a discussion about the saturation of forests with nitrogen (e.g. Ågren and Bosatta 1988; Tamm 1991; Gundersen et al. 1998; Aber et al. 1998). However, political actions by European countries have led to a noticeable decline in atmospheric depositions of sulphur and nitrogen in the 1980s and 1990s (Ferrier et al. 2001) raising questions about the recovery of forest ecosystems from soil acidification. A general recovery of alkalinity of lakes and streams was observed in all regions of Europe in the 1990s (Stoddard et al. 1999) except at many sites in central Europe where a significant delay in aquatic recovery from acidification (Alewell et al. 2000a) was observed. Recovery of water acidification can take decades, because of the release of previously stored sulphate from soils with a high storage capacity for sulphate continues leading to acidification of aquatic systems (Matzner 2004). Moreover, many of the European forest stands are experiencing changes due to the so-called "climate change" phenomenon. For example, significant changes in the climate at the Solling site with

R. Brumme and P.K. Khanna (eds.), *Functioning and Management of European Beech Ecosystems*, Ecological Studies 208,
DOI: 10.1007/978-3-642-00340-0_16, © Springer-Verlag Berlin Heidelberg 2009

increased air temperature and precipitation have been recorded since the Solling project was established in 1966 (see Chap. 2). All these impacts on forest ecosystems constitute the basis for extensive research for which the most promising tool is the long-term monitoring of element budgets of representative ecosystems. This allows one to follow forest development under changing environmental conditions.

Three beech forests covering a range of soil properties were considered in this study to evaluate the effects of acid and nitrogen depositions, especially on carbon and nutrient turnover and pools. Soil solution chemistry was studied at different depths to observe the reactions of deposited elements with soil matrix. For quantifying the element fluxes with seepage, water output was estimated by modelling the water balance on each site. The response of soil solution to decreasing atmospheric deposition was studied by trend analysis. Input–output (I/O) budgets were used to assess the overall increase or loss of elements for the stands. The annual I/O budgets were established for three periods of roughly a decade each: (1) from 1969 to 1980, when the atmospheric deposition of SO_x and NO_x increased to peak emissions in Europe, (2) from 1981 to 1989, when deposition started to decline, and (3) from 1990 to 2002 when deposition further declined (Ferrier et al. 2001) as a result of an international agreement in Europe under the Convention on Long Range Transboundary Air Pollution (CLRTAP) of the United Nations Economic Council for Europe (UN–ECE). The selected beech forests were intensively studied during the observational periods and the results have been previously presented in many publications. An update of the results will be presented here. For the most extensive publications on input–output analysis of preceding periods refer to Matzner (1989), Ulrich (1994a), Meesenburg et al. (1995) and Eichhorn (1994, 1999).

Our intention was to address the following questions in this chapter. (1) How do these ecosystems with different soil properties respond to acid depositions? (2) Do these ecosystems accumulate or lose N, and are they linked to the soil chemical status? (3) How do these ecosystems respond to decreasing atmospheric depositions of sulphuric acid and base cations? (4) Are there any indications of recovery from high depositions of acid and N?

16.2 Soil Solution Collection and Statistical Analysis

Monitoring of soil solution (measurement of ion concentrations) started at the Solling site in 1969 at the interface of the organic layer and mineral soil (depth 0 cm) and in the subsoil (−90 cm depth). From 1981 onwards, the monitoring was extended with soil solution sampling at several other soil depths, e.g. −20 cm depth. At the Göttinger Wald site, soil solution monitoring started in 1981, and at the Zierenberg site, in 1989 at several soil depths. The monitoring is still being continued.

Soil solution is collected by suction plates and suction cups (tension lysimeters) which sample soil solution at the same place producing a time series with short sampling intervals during several years. A detailed description of the technique applied is given by Meiwes et al. (1984), and Meesenburg et al. (1997) for the Solling and Göttinger Wald sites and by Jochheim (1992) and Eichhorn (1995) for Zierenberg site. The lysimeters were installed with three to ten replicates at fixed soil depths of 0, 20 and 90 cm at Solling, 5, 20 and 100 cm at Göttinger Wald, and 20, 60 and 100 cm at Zierenberg. Soil solution is collected at weekly or biweekly intervals. Combined samples of two to three lysimeters of each depth are sampled weekly at Solling and Göttinger Wald, which are pooled to monthly samples for chemical analysis. Biweekly samples from Zierenberg are analysed for each lysimeter separately. Analytical methods are described by Fassbender and Ahrens (1977) and König and Fortmann (1996a–d).

Linear regression analyses were used to describe the temporal variability of solute concentrations and to detect trends. Analyses have been performed for annual mean concentrations or solute fluxes for the period 1981–2002 for Solling and Göttinger Wald and for 1990–2002 for Zierenberg. Statistical analyses have been performed with the software package SPSS 11.5 (SPSS).

16.3 Modelling of Water Fluxes

We used the hydrology model SIMPEL for the estimation of evapotranspiration and water fluxes beneath the rooting zone (Hörmann 1997). Although other hydrology models are available (Hörmann and Schmidt 1995; Tiktak and van Grinsven 1995), we chose SIMPEL because it has been successfully calibrated to various forest ecosystems including beech stands (Hörmann and Meesenburg 2000), and because of its limited requirement for input data.

SIMPEL is a very simple hydrologic model. A detailed description of SIMPEL is given by Hörmann (1997, 1998). Thus, only a short overview of the basic principles is given here. SIMPEL uses a capacity approach for the simulation of water fluxes. Three consecutive storages (ecosystem layers) are defined: the canopy of the stand, the litter layer and the soil layer. Potential evapotranspiration within SIMPEL can be calculated with different methods ranging from simple (e.g. Thornthwaite 1948) to more physically based equations (e.g. Penman 1948/Monteith 1965). Interception is simulated as a simple bucket model with canopy water storage capacity as the only parameter. Canopy water storage capacity can be estimated by using leaf area index (LAI) which is considered to change in time according to phonological phases. The soil water storage is defined by field capacity, permanent wilting point and maximum rooting depth. Outflow from the soil water storage above field capacity is the difference between infiltration and transpiration. Below field capacity, seepage is simulated according to Glugla (1969). The reduction from potential to actual evapotranspiration is calculated using a linear reduction function where

Table 16.1 Input parameters and parameter values for the soil module of SIMPEL

SIMPEL input parameter	Unit	Göttinger Wald	Zierenberg	Solling
Field capacity	Vol.%	25	25	25
Permanent wilting point	Vol.%	12	12	12
Reduction limit	Vol.%	20	19	24
Maximum rooting depth	Cm	180	180	180

transpiration starts to be restrained at the reduction limit and ceases at the permanent wilting point.

In the present case, the canopy layer and the litter layer have been omitted from the model. Measured values of stand precipitation were used as infiltration data into the soil storage. For the estimation of potential evapotranspiration, a derivation of the Penman/Monteith formula with constant values for albedo, aerodynamic and canopy resistance has been used. Global radiation, air temperature and relative humidity are required as site-specific input parameters for the calculation of potential evapotranspiration. No site-specific climatic data are available for the whole simulation period of the study sites. Thus, data from nearby climate stations of the German Weather Service (DWD) have been used for the simulations. Site-specific transfer functions for the extrapolation of the DWD data were derived from short-term measurements at the study sites. Global radiation has been calculated from sunshine duration using the Angström formula (DVWK 1996). A common set of parameter values for the three study sites was used for the soil module (Table 16.1). Only the reduction limit served as a calibration parameter. The model has been run on a daily time step.

16.4 Calculation of Input–Output Balances of Various Elements and Acid Production

Element fluxes at 100 cm soil depth (soil output) were obtained by multiplying the element concentrations in soil solution with the respective water flux for that period. For periods with missing values of soil solution concentrations (no soil solution sample), a linear interpolation between the preceding and the following periods was applied.

Element budgets of the ecosystems and the production of acidity were calculated using the approach suggested by Ulrich (1994a). The budget of any solute is the difference between total deposition and seepage output. Additional input of base cations by mineral weathering was estimated by using the PROFILE model (Sverdrup and Warfvinge 1992) (Table 16.2). The mineral composition of the soils used for weathering estimates is described in Chap. 3. The weathering rates for Ca and Mg are similar to the rates obtained by the so-called "historical" method which was applied for the Solling site (Wesselink et al. 1994). ZrO was used as an

Table 16.2 Net element retention for annual increment by trees (from Chap. 8) and element weathering (mmol$_c$ m^{-2} per year) on the three beech sites. H is the negative equivalent of the sum of base cations (Na, K, Mg, Ca)

	Flux	Na	K	Mg	Ca	H	N_{tot}
Solling	Uptake	0	8.2	6.8	21	−36	43
	Weathering	26	52	27	8	−113	0
Göttinger Wald	Uptake	0	22	18	151	−190	126
	Weathering	7	20	27	1,179	−1,233	0
Zierenberg	Uptake[a]	0	7.5–12	4.6–13	33–55	−45 to −79	38–63
	Weathering	5	56	202	67	−331	0

[a]Uptake is given as *x* –SD and *x* +SD (*x* mean, SD standard deviation). The mean was calculated based on the measured breast diameter and height at Zierenberg site (Chap. 8), regression analysis for the different biomass compartments and the basal areas from the Göttinger Wald and Solling sites (Peter Rademacher, personal communication), and element concentrations (mean ± SD) of a literature review by Jacobsen et al. (2003) of which the concentrations for noncalcareous soils were taken

internal standard to calculate historical weathering rates by comparing the mineral pools at 60 cm depth with the composition in the reference C horizon at 180 cm depth. Unfortunately, the Solling soil does not contain loess material in the C horizon, so a loess profile southeast of Solling was used which has a mineralogical composition similar to that of the Solling site in the topsoil. While the historical weathering rates for Ca (8 mmol$_c$ m^{-2} per year) and Mg (30 mmol$_c$ m^{-2} per year) were identical, the rates for Na (5 mmol$_c$ m^{-2} per year) and K (6 mmol$_c$ m^{-2} per year) were much lower than the estimates obtained with the PROFILE model indicating methodological problems in estimating realistic weathering rates. N release by weathering was assumed to be negligible. Export of cations by biomass was estimated as net accumulation in plant biomass by Rademacher et al. and included an aboveground increment (>7 cm diameter) and coarse roots (>2 mm) (Chap. 8).

The acid load of the ecosystems by deposition and system-internal H production was calculated (see Sect. 16.10) according to Ulrich (1994a) and is the sum of following values:

(1) Acid input: positive budget values (input > output) of Ma cations (H, Mn, Al, Fe) (input =total deposition + the negative equivalent of the sum of cations taken up for plant increment; output =leaching losses).
(2) N transformation: NH_4 input $-NH_4$ output $-NO_3$ input $+NO_3$ output.
(3) S release: negative budget values of SO_4. Dissolution of aluminum sulphate represents the mobilisation of previously stored H_2SO_4 from atmospheric deposition in the solid soil phase (Prenzel 1983).
(4) Weak acidity (carbonic and organic acids) is calculated as the charge balance difference. Cl is only taken into account in the calculation if a negative budget cannot be attributed to the loss of neutral salts, i.e. if the budget is not balanced by Mb cations.
(5) Mb retention: positive budgets of Mb cations (Na, K, Mg, Ca) corrected for neutral salt accumulation. Mb cations are bound to acidic groups releasing Ma cations (e.g. protons).

16.5 Solution Composition From Soils of Different Buffer Systems

Sulphate is the dominant anion in the soil solution at the Solling site (Table 16.3) which is related to its high atmospheric depositions. For this highly acid soil Al and H occur in all soil layers. A strong relation between Al and sulphate and nitrate in soil solution was found to be typical for soils with base saturation values of <25% and pH values of <4.5 for 192 sites spread over 12 European countries (de Vries et al. 2003). In soils with base saturation values of >25% and pH values of >4.5, there was a strong relationship between Ca and strong acid anions, indicating that the acidity in these soils is mainly neutralised by the release of base cations. Soils of Göttinger Wald and Zierenberg fall in this group with mean base saturation values of >93% (Chap. 3). High H concentration in soil solution below the surface organic layer at Solling (Table 16.3) has atmospheric deposition and nitrification processes as primary sources (Chaps. 13 and 15). Since buffer substances are absent in the surface organic layer most of H is leached and buffered in the mineral soil as indicated by low pH values of soil solution (pH 4.1–4.5). Owing to the low content of silicates such as feldspars and illites in the Solling soil and the advanced level of acidification in this soil, silicate weathering does not buffer all the protons completely. Therefore, Al occurs in considerable concentrations making up more than 50% of the sum of cations in solution of the mineral soil layers. However, at >80 cm soil depth proton buffering capacity increases due to a higher content of illites causing slightly lower concentrations of H and Al in the soil solution. Al occurs predominantly in monomeric form in the soil solution collected at >20 cm depth which is supposed to be toxic for tree roots whereas less toxic soluble organic Al complexes occur in solutions from below the surface organic layer (Dietze and Ulrich 1992).

At the less acid Göttinger Wald and Zierenberg sites, acidity generated by the system internal processes or added through atmospheric depositions is buffered by easily weatherable silicates or carbonates, as indicated by solution pH values of 6.1–8.0 and high concentrations of Ca and Mg in soil solution. Buffering of protons by Al silicates occurs only in the acidified surface mineral soil where carbonate is absent as indicated by Al concentration in soil solution. Increased concentrations of Ca and Mg indicate buffer reactions with carbonates at >5 cm soil depth (Göttinger Wald) and at >60 cm soil depth (Zierenberg). HCO_3 is the accompanying anion of Ca and Mg. HCO_3 was not measured but could be calculated as the excess of cations over anions in soil solutions where organic anions are negligible ($\Sigma+ - \Sigma-$, Table 16.3). High Mg concentration in the soil solution from Zierenberg originates from dissolved Mg-rich minerals like olivine, pyroxene or hornblende of the basaltic debris (Chap. 3).

Concentrations of SO_4 and Cl increased with soil depth due to their discrimination by plant roots during water uptake. However, increase in the concentration of SO_4 with depth was smaller than that of Cl during the period of 1969–1980 at Solling (Table 16.3) which was explained by the SO_4 retention in this highly acidic

Table 16.3 Mean concentrations (unweighted) standard deviation of spatial replicates of major ions ($\mu mol_c\ L^{-1}$) in soil solution at different sampling depths and for different sampling periods at the study sites ($\Sigma+ - \Sigma-$ = sum of cations −sum of anions)

Depth	Period	Na	K	Mg	Ca	Al	Mn	H	NH_4	NO_3	SO_4	Cl	$\Sigma+ - \Sigma-$
Mean, Solling													
0 cm	69–80	81	95	68	222	176	50	178	84	221	411	116	218
	81–89	80	86	58	187	130	42	211	51	280	329	106	130
	90–02	90	47	34	72	93	15	146	18	95	141	90	189
20 cm	81–89	95	12	37	60	452	29	60	7	125	410	134	82
	90–02	102	4	21	25	228	14	40	8	29	213	113	87
90 cm	69–80	104	17	47	88	314	36	78	9	18	426	163	87
	81–89	75	14	35	39	361	21	49	7	18	437	110	38
	90–02	85	7	24	21	191	10	35	6	12	252	91	24
Mean, Göttinger Wald													
5 cm	81–89	69	52	74	1131	20	1.2	0.53	7	313	593	124	324
	90–02	79	20	41	560	41	0.3	0.49	7	51	244	101	353
20 cm	81–89	80	9	86	2396	6	0.5	0.05	6	298	634	152	1,501
	90–02	91	5	69	2132	3	0.4	0.04	6	190	276	104	1,736
100 cm	81–89	222	20	126	5004	7	0.4	0.02	13	286	771	218	4,117
	90–02	187	14	108	5095	3	0.3	0.01	17	99	431	148	4,746
Mean, Zierenberg													
20 cm	90–02	166	29	745	604	40	0.3	1.02	6	683	408	185	316
60 cm	90–02	202	13	844	729	7	0.3	0.21	3	739	486	219	355
100 cm	90–02	250	14	711	1378	4	0.3	0.07	2	536	511	217	1,096
Standard deviation, Solling													
0 cm	69–80	3	17	5	40	27	10	14	36	89	19	7	
	81–89	5	14	4	26	80	9	57	33	57	23	3	
	90–02	9	20	4	21	42	5	62	5	64	9	8	

(*continued*)

Table 16.3 (continued)

Depth	Period	Na	K	Mg	Ca	Al	Mn	H	NH_4	NO_3	SO_4	Cl	Σ+ − Σ−
20 cm	81–89	6	1	3	7	40	3	2	1	38	19	2	
	90–02	3	0	3	2	5	2	3	0	2	16	1	
90 cm	69–80	4	5	2	2	5	1	2	1	1	14	17	
	81–89	7	5	3	3	25	0	2	0	3	19	12	
	90–02	6	1	2	1	22	0	2	0	3	23	8	
Standard deviation, Göttinger Wald													
5 cm	81–89	7	51	15	280	5	0	2	2	121	135	23	
	90–02	7	10	4	74	12	0	1	2	14	24	11	
20 cm	81–89	6	4	13	240	1	0	0	7	88	78	15	
	90–02	4	2	12	520	1	0	0	1	78	46	8	
100 cm	81–89	27	4	24	377	1	0	0	6	58	26	19	
	90–02	22	2	25	578	1	0	0	4	52	84	16	
Standard deviation, Zierenberg													
20 cm	90–02	38	20	306	194	18	0	1.1	8	1,798	235	44	
60 cm	90–02	60	10	390	268	6	0	0.1	3	2,204	246	58	
100 cm	90–02	33	11	212	861	3	0	0.1	1	785	178	49	

soil (Ulrich et al. 1980; Meiwes et al. 1980; Khanna et al. 1987; Alewell et al. 2000b). The retained SO_4 was released into the soil solution after 1985 as indicated by the higher increase in concentration of SO_4 than that of Cl during 1990–2002. This also resulted in a negative sulphur budget (Sect. 16.9). Retention or loss of SO_4 in acid soils is explained by the formation and dissolution of aluminum hydroxyl sulphate ($AlOHSO_4$, e.g. Jurbanite) (Meiwes et al. 1980; Prenzel 1983), by adsorption processes (Alewell et al. 2000b) and by retention in biomass. $AlOHSO_4$ may exist as an intermediate product on the surface of other minerals in soils, and is dissolved when pH values increase to produce Al and SO_4 in soil solution.

The nitrate concentration in soil solution of the three sites followed the order: Solling < Göttinger Wald < Zierenberg (Table 16.3). High concentrations of nitrate in the solution collected below the litter layer indicated high mineralisation rates in the surface organic layer at the Solling site. But relatively high concentrations of NH_4 suggested that N was only partly nitrified in the acid surface organic layer. Controlled incubation studies showed that nitrification rate was low in the F horizon of the surface organic layer which was explained by the absence of autotrophic nitrifiers and low activity of heterotrophic nitrifiers (Chap. 13). A decrease in the concentrations of ammonium and nitrate with soil depth at the Solling site suggested their transformation or uptake at deeper depths. Ammonium concentrations are generally low due to high nitrification rates in the less acid soils of Zierenberg and Göttinger Wald (Chap. 13). In these soils, mineralisation processes occurred in the mineral soil and were maintained by bioturbation of organic matter by earthworms that resulted in high nitrate concentrations in the Ah horizon. Very high nitrate concentrations below 100 cm depth at Zierenberg indicated high N losses with seepage water.

Concentrations of dissolved organic nitrogen (DON) were high, its values lying between those of ammonium and nitrate concentrations (Tables 16.3 and 16.4). DON and dissolved organic carbon (DOC) values decreased with increasing soil depth as has been observed in several studies (Kaiser and Zech 2000; Michalzik et al. 2001). The ratios between DOC and DON slightly decreased with depth at the Göttinger Wald and Zierenberg sites and increased at the Solling site. However, the values of these ratios (13–56), when calculated from long-term mean concentrations in Table 16.4 were in the range as reported by Michalzik et al. (2001) who compiled such data for temperate forests.

16.6 Temporal Trends in Soil Solution Chemistry

Stricter emission control measures in Europe and structural changes in the industry, i.e. reduction in the size of heavy industry and changes in the energy consumption and fuel mixtures used for energy production, significantly reduced atmospheric deposition since mid 1980s in Europe (Ferrier et al. 2001). Atmospheric depositions peaked around 1985 across Europe. The long-term observations at the Solling and Göttinger Wald sites showed a decrease for sulphur (−60% at Solling and Göttinger

Table 16.4 Mean pH, mean concentrations and standard deviation of spatial replicates (μmol L^{-1}) of total nitrogen (N_{tot}), dissolved organic N (DON), and dissolved organic C (DOC) in soil solutions collected from different sampling depths and for different sampling periods at the three study sites (n.d., not determined)

Depth	Period	pH	N_{tot}	DON	DOC
Mean, Solling					
0 cm	69–80	3.8	406	98	n.d.
	81–89	3.7	433	103	n.d.
	90–02	3.8	188	78	3,490
20 cm	81–89	4.2	168	34	n.d.
	90–02	4.4	49	13	729
90 cm	69–80	4.1	44	19	n.d.
	81–89	4.3	24	0	n.d.
	90–02	4.5	17	0	182
Mean, Göttinger Wald					
5 cm	81–89	6.4	403	84	1,427
	90–02	6.3	120	62	1,886
20 cm	81–89	7.3	375	69	983
	90–02	7.5	243	52	1,053
100 cm	81–89	7.8	369	68	863
	90–02	8.0	152	36	827
Mean, Zierenberg					
20 cm	90–02	6.1	813	83	1,879
60 cm	90–02	6.7	821	66	954
100 cm	90–02	7.2	590	37	747
Standard deviation, Solling					
0 cm	69–80	n.d.	122	23	n.d.
	81–89	n.d.	75	18	n.d.
	90–02	n.d.	72	23	1,012
20 cm	81–89	n.d.	34	0	n.d.
	90–02	n.d.	3	3	121
90 cm	69–80	n.d.	25	26	n.d.
	81–89	n.d.	2	3	n.d.
	90–02	n.d.	4	2	34
Standard deviation, Göttinger Wald					
5 cm	81–89	n.d.	120	11	262
	90–02	n.d.	11	2	84
20 cm	81–89	n.d.	100	12	217
	90–02	n.d.	93	25	294
100 cm	81–89	n.d.	62	8	142
	90–02	n.d.	65	10	164
Standard deviation, Zierenberg					
20 cm	90–02	n.d.	432	35	639
60 cm	90–02	n.d.	519	32	408
100 cm	90–02	n.d.	195	23	196

Wald), H (−72% at Solling and Göttinger Wald), and nitrogen depositions (−38% for Solling and −20% for Göttinger Wald) for the periods 1981–1989 and 1990–2002 (Chap. 15). Emission control measures such as air filtration on large combustion plants have led to the reduction in depositions of other essential nutrients such as base cations which have decreased by about 60% at the Solling and Göttinger Wald sites (Chap. 15). These reductions in the atmospheric deposition have changed the soil solution chemistry significantly as indicated by long-term monitoring data of the Solling and Göttinger Wald sites. However, these trends are less distinct at the Zierenberg site due to the short observational period. Trend analysis of soil solution concentrations performed by linear regression analysis showed declining trends for several elements (Table 16.5) which is most consistent for sulphate at all the three beech sites (Fig. 16.1). The effect of reduced atmospheric depositions on soil solution chemistry was most pronounced at Solling where a significant reduction in the concentrations of most elements was observed. Mb cations (Na +K +Mg +Ca) also showed a dramatic decrease in soil solutions collected at 90 cm depth (Fig. 16.2). The amount of exchangeable base cations have decreased from 35 to <20 $kmol_c\ ha^{-1}$ during 1969 to 2001 (Chap. 3).

At the Göttinger Wald and Zierenberg sites, this reduction in atmospheric depositions had only a small effect on soil solution chemistry. There was a decreasing trend in base cations and an increasing Al content indicating still ongoing acidification of the surface soil at the Göttinger Wald site (Table 16.5) despite the decrease in acid depositions. There was a higher reduction of nitrate concentration in soil solution than in total atmospheric depositions pointing to a reduction in mineralisation, which was probably induced by the increased acidification in the topsoil at the Göttinger Wald site. The Zierenberg site did not show any effect in the surface 20 cm soil except for K and Mn. However, an increase in Al and H at a depth of below 20 cm pointed to acidification in relation to high net nitrification. Fluxes of Al, H and nitrate increased with time at 100 cm soil depth (Table 16.7).

Table 16.5 Temporal trends of concentrations in soil solution at different depths (cm) of the Solling (*SO*) and Göttinger Wald (*GW*) sites from 1981 to 2002 and of the Zierenberg (*ZB*) site from 1990 to 2002

Site	Depth	Na	K	Mg	Ca	H	Mn	Al	TIC	NO_3	N_{tot}	DON	SO_4	Cl
SO	0	○	– –	– –	– –	○	– –	– –		– –	– –	– –	– –	–
	20	○	– –	– –	– –	– –	– –	– –		– –	– –	– –	– –	–
	90	○	– –	– –	– –	– –	– –	– –		○	○	○	– –	– –
GW	5	○	– –	– –	– –	○	○	++	○	– –	– –	○	– –	–
	20	○	–	– –	○	○	○	–	++	○	–	– –	– –	– –
	100	○	○	–	○	– –	– –	– –	++	–	–	○	– –	– –
ZB	20	○	– –	○	○	○	– –	○		○	○	–	– –	○
	60	○	– –	○	○	○	– –	++		○	○	○	– –	○
	100	○	○	○	○	++	– –	++		○	○	○	– –	○

Test of trend with correlation coefficient after Pearson; ○ no trend, + significant increasing trend $p \leq 0.05$, ++ highly significant increasing trend $p \leq 0.01$, – significant decreasing trend $p \leq 0.05$, – – highly significant decreasing trend $p \leq 0.01$

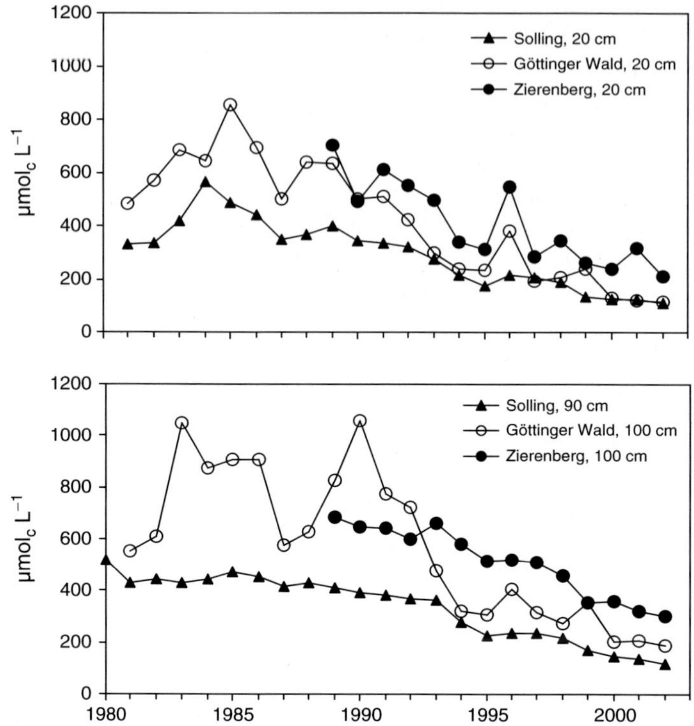

Fig. 16.1 Time series of sulphate concentrations (μmol$_c$ L^{-1}) in soil solutions from 20 cm and 90 or 100 cm depths at three beech sites

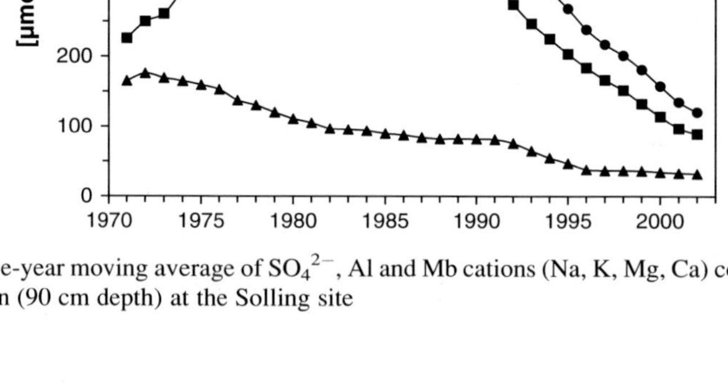

Fig. 16.2 Five-year moving average of SO_4^{2-}, Al and Mb cations (Na, K, Mg, Ca) concentrations in soil solution (90 cm depth) at the Solling site

There was a decrease in nitrate concentration in soil solution which could be observed clearly below the surface organic layer and this decrease followed the declining trend in N depositions (Chap. 15). Nitrate concentrations increased during 1970s, stayed constant during 1980s, and then declined to values observed at the beginning of the 1970s (Fig. 16.3). The magnitude of changes in nitrate concentration decreased with increasing soil depth (Fig. 16.4; Table 16.5) at the Solling and Göttinger Wald sites but no temporal trend was observed at the Zierenberg site probably because of the short observation period. Since deposited ammonium is either immobilised in the surface organic layer as was indicated by ^{15}N studies (Chap. 14) or is nitrified, ammonium concentration in the solutions did not show any temporal pattern.

Toxic effects of monomeric Al on roots and soil fauna are commonly expressed through molar ratios of base cations (BC) to Al (Sverdrup and Warfvinge 1992; Cronan and Grigal 1995; de Vries et al. 2003). (Mg+Ca)/Al ratio of less than 1.0–3.0 in soils with low BC concentrations as proposed by Sverdrup and Warfvinge (1992) indicated a high degree of stress on plant roots which was the case for the Solling site (Fig. 16.5). In such situations, the rhizosphere activity may mitigate the adverse effects of Al (Högberg and Jensén 1994).

The alkalinity of the solution is used as a measure of the recovery of lakes, streams or soils from acidification (Reuss and Johnson 1986; Alewell 2001) and was calculated for the Solling site as (Na +K +Ca +Mg +NH_4) –(Cl +NO_3 +SO_4). The alkalinity was negative when soil solution sampling started at Solling in 1969 (Fig. 16.5). The temporal pattern of alkalinity inversely followed the concentrations of sulphate, for example from 1975 to 1989 at the Solling site concentration of sulphate was high but alkalinity was very low. Since the beginning of the 1990s, the alkalinity has increased to values which are slightly higher than those at the beginning of the 1970s though still negative. A similar increase in alkalinity of soil solution was reported by Matzner et al. (2004) who interpreted it as a slight but significant recovery of soil solution from acidification.

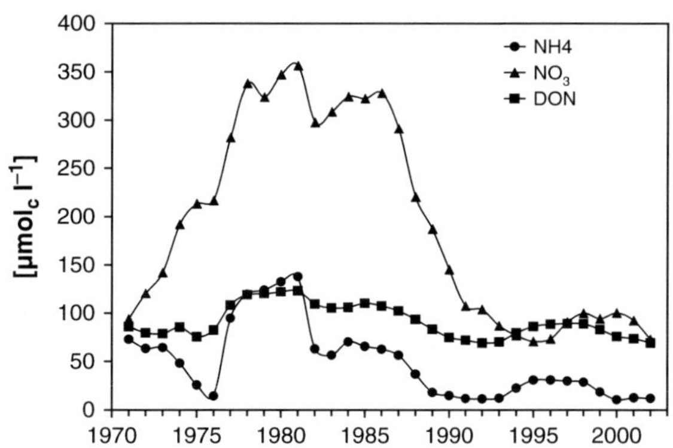

Fig. 16.3 Five-year mooving average of NH_4, NO_3 and DON concentrations in soil solution beneath the surface organic layer at the Solling site

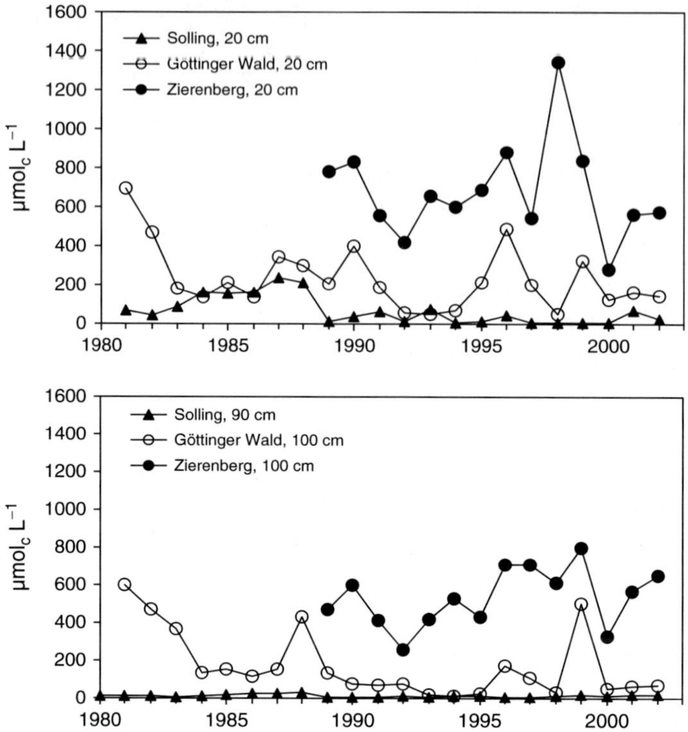

Fig. 16.4 Time series of nitrate concentrations ($\mu mol_c\ L^{-1}$) at 20 cm and 90 or 100 cm depths of the three beech sites

16.7 Hydrologic Regime

For the determination of solute outputs, estimates of soil water fluxes have to be coupled with nutrient concentrations obtained by the sampling of soil solution. For none of the study sites, direct measures of all water fluxes, especially the drainage (e.g. by the use of calibrated watersheds) are available. Thus, a number of water fluxes were estimated by using simulation model (see Sect. 16.3).

During 1990 to 2002, simulated mean potential evapotranspiration (interception not included) was highest for the Göttinger Wald site and lowest for the Solling site (Table 16.6). Differences in actual evapotranspiration among the three sites followed the same order but were smaller in magnitude. The highest evapotranspiration at the Göttinger Wald site is related to higher air temperatures at this site (Chap. 2). Due to very different precipitation and interception values on these sites, the outflow flux from soils is also very different with Solling having the highest value. Mean output water flux of 638 mm for the Solling site was similar to the value of 588 mm given by Benecke (1984) for 1969 to 1975 and of 572 mm by Ellenberg et al. (1986) for 1976–1981. The present simulation with the SIMPEL model yielded an average

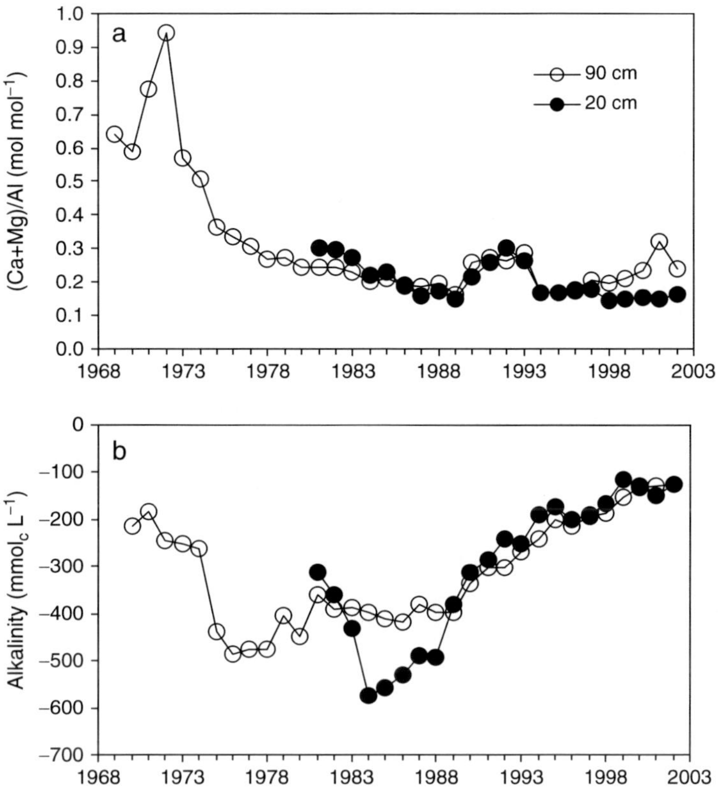

Fig. 16.5 Time series of (**a**) the ratio between concentrations of base cations (Ca +Mg) and Al in soil solution (mol$_c$ mol$_c$ L^{-1}) and (**b**) the soil solution alkalinity ((Na +K +Ca +Mg +NH_4) − (Cl +NO_3 +SO_4)) at 20 and 90 cm depths at the Solling site

Table 16.6 Mean annual values of measured (precipitation, interception) and simulated water balance components (mm) for the study sites for the period 1990–2002 (evapotranspiration does not include crown interception)

	Göttinger Wald	Zierenberg	Solling
Precipitation	684	752	1,184
Interception	68[a]	148	253
Potential evapotranspiration	616	553	457
Actual evapotranspiration	387	368	293
Output water flux	229	236	638

[a]Low interception probably relate to an underestimation of precipitation

output flux of 568 mm for 1976–1981. Mean annual soil water fluxes for the Zierenberg and Göttinger Wald sites were calculated to be lower than those observed for the Solling site with large year to year variation in the values

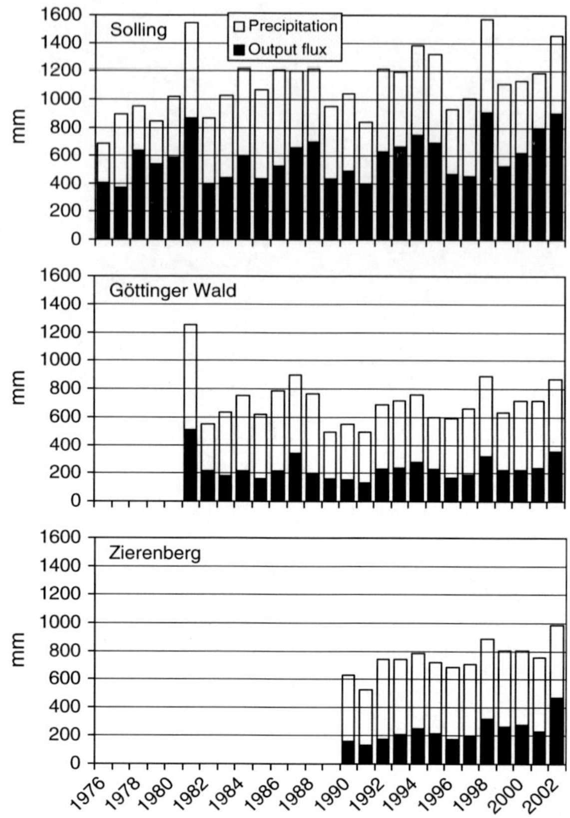

Fig. 16.6 Annual precipitation rates and output water fluxes (mm) at the three beech sites

(Fig. 16.6). At Solling, the lowest and highest values of annual output fluxes occurred in 1977 with 370 and in 1998 with 908 mm. At Göttinger Wald, the highest annual values of output flux (505 mm) occurred in 1981, and the lowest value of 134 was calculated for 1991. The driest year at Zierenberg was also 1991 with 134 mm of output fluxes and the highest flux during the 1990–2002 observation period for Zierenberg occurred in 2002 with 471 mm.

Seasonal pattern of monthly output fluxes for water during the 1990–2002 (Fig. 16.7) indicated that the highest values occurred in January and the lowest values in August. Simulation with the SIMPEL model indicates that the outflow from these soils occurred during the whole year, though at low rates during the summer months.

The long-term observation at the Solling site revealed an increase in precipitation and temperature (Chap. 2) and raised the question about future changes in the hydrologic regime. However, energy input from global radiation remained more or less constant and evapotranspiration did not change, so any surplus water from increased precipitation will increase output water fluxes.

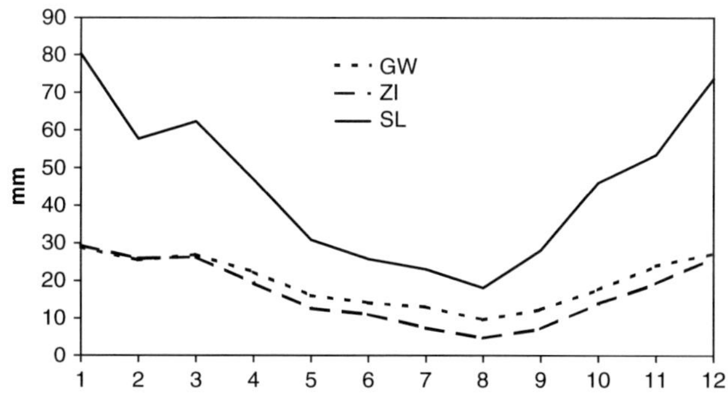

Fig. 16.7 Simulated values of mean monthly output water fluxes (1990–2002) at Göttinger Wald (*GW*), Zierenberg (*ZI*), and Solling (*SL*) sites

16.8 Element Seepage Losses

Among the cations, the amount of Al loss was the highest at the Solling site (Table 16.7). Losses of sulphate exceeded those of Al and caused losses of base cations such as Mg and Ca. At the Göttinger Wald and Zierenberg sites, Ca is the main cation in the soil solution (also Mg at the Zierenberg site) which is lost in large amounts. Sulphate and HCO_3 are the main anions accompanying Ca and Mg with seepage water. Nitrogen losses are low or moderate at Solling (12 mmol m^{-2} per year) and Göttinger Wald (32 mmol m^{-2} per year) in contrast to very high losses of 148 mmol m^{-2} per year at Zierenberg.

Linear trend analysis of the mean solute fluxes indicated that flux trends were somewhat different from those obtained for concentrations at the three beech sites (Table 16.8). The negative trends indicated a decrease in solute fluxes and thus in groundwater contamination and followed the pattern of emission reduction in the 1990s at the Solling and Göttinger Wald sites (the Zierenberg site was an exception). Decreasing trends in the fluxes of sulphate (highly prominent) and of Al, Ca, Mg, Mn and K (significant) were observed at the Solling site, whereas the negative trend of nitrogen at the Göttinger Wald site was not very clear because of very high water fluxes at the beginning of the project in 1981 (see Fig. 16.5). Losses of nitrate and DOC at the Zierenberg site increased in association with high output water fluxes and this also led to higher losses of Mg, Na, K, Cl and DON.

16.9 Element Budgets

Annual mean values of total deposition, drainage losses and plant uptake for increment were used to develop input–output budgets for a number of elements for the three sites, which are presented as mean values of different periods in Table 16.9.

Table 16.7 Mean seepage losses at 90 cm (Solling) and 100 cm depth (Göttinger Wald and Zierenberg). Values are given in mmol$_c$ m^{-2} per year except mmol m^{-2} per year for *DON*, *N$_{tot}$*, and *DOC*

Period	mm	NO_3	DON	N_{tot}	DOC	Cl	SO_4	Ca	Mg	K	Na	Al	Mn	H
Solling														
76–80	508	10	7	19	n.d	66	276	32	24	9	48	204	17	50
81–89	560	12	3	14	n.d.	60	248	22	20	7	42	205	12	29
90–02	638	8	1	12	101	54	154	12	14	4	51	117	6	22
Göttinger Wald														
81–89	244	80	13	97	164	50	181	1251	32	5	50	1	0	0
90–02	229	23	6	32	157	31	89	1130	23	3	39	1	0	0
Zierenberg														
90–02	236	140	8	148	181	52	112	314	167	4	60	1	0	0

Table 16.8 Mean solute fluxes (90 cm depth (Solling) or 100 cm depth (Göttinger Wald and Zierenberg)) and linear trends (slope of the regression functions) for the study sites for the periods 1981–2002 (Solling and Göttinger Wald) and 1990–2002 (Zierenberg)

	Unit	Solling		Göttinger Wald		Zierenberg	
		Mean	Trend	Mean	Trend	Mean	Trend
H_2O	mm	588		235		236	16.74
H	$mmol_c\ m^{-2}$ per year	29		0.003	−0.0001	0.03	0.005
Ca	$mmol_c\ m^{-2}$ per year	19	−1.0	1,179		314	
Mg	$mmol_c\ m^{-2}$ per year	18	−0.6	27		167	10.9
Na	$mmol_c\ m^{-2}$ per year	47		43		60	4.73
K	$mmol_c\ m^{-2}$ per year	6	−0.2	4		4	0.36
Al	$mmol_c\ m^{-2}$ per year	162	−7.6	1	−0.07	1.2	0.24
Mn	$mmol_c\ m^{-2}$ per year	10	−0.5	0.1	−0.004	0.06	
Cl	$mmol_c\ m^{-2}$ per year	58		39	−1.8	52	2.65
SO_4	$mmol_c\ m^{-2}$ per year	208	−8.6	126	−8.4	112	
NO_3	$mmol_c\ m^{-2}$ per year	10		46	−4.6	140	14.31
DON	$mmol\ m^{-2}$ per year	1		9	−0.5	8	0.71
N_{tot}	$mmol\ m^{-2}$ per year	14		59	−5.4	148	15.03
DOC	$mmol\ m^{-2}$ per year	93		160		181	19.16

Only values for significant trends $p \leq 0.05$ are given

16.9.1 Solling Site

The input–output budgets of this ecosystem (TD-SL) are positive for most of the elements except aluminum and manganese in the first two decades, indicating their retention within the ecosystem (Table 16.9). Losses of Al and Mn are due to high acidity levels in this soil. I/O budgets for S and Mg indicate their losses during the recent periods.

The long-term sulphate budget is driven by strongly decreasing input rates of atmospheric deposition from 264 to 102 $mmol_c\ m^{-2}$ per year (Table 16.9), accompanied by seepage rates which have decreased from 248 to 154 $mmol_c\ m^{-2}$ per year during the last two decades (Table 16.7). Thus, a delay in response of seepage outputs to strongly decreased deposition results in negative values of balance in the budget for sulphur. The temporal variation is high (Fig. 16.8) due to large fluctuations of seepage water fluxes. Source of sulphate is the amount of S which has accumulated in the soil during high atmospheric input conditions (Meiwes et al. 1980; Khanna et al. 1987; Alewell et al. 2000b). A net release of sulphate from acid forest soils after emission reduction has recently been reported from other sites in Germany (Matzner et al. 2004). Studies at the Fichtelgebirge mountains and the Steigerwald hillside in South Germany indicated a net release of 4 kg in the period of 1995–2001 and 19 kg S ha^{-1} per year for 1993–2001 similar to the Solling site with losses of 8 kg S ha^{-1} per year in the period 1990–2002. Catchments studies conducted in Europe showed a similar reduction in sulphate output. However, they differed in their response to reduced sulphur depositions. Whereas catchments in

Table 16.9 Total deposition (*TD*) and element budgets (*TD-SL* and *TD-SL-PI*) in $mmol_c\ m^{-2}$ per year ($mmol\ m^{-2}$ per year for *DON*, N_{tot}, *DOC*) where *SL* is seepage losses and *PI* plant increment (from Table 16.2) for consecutive periods for the Solling, Göttinger Wald, and Zierenberg sites

	Na	K	Mg	Ca	N_{tot}	DON	DOC	SO_4	Cl	H	Mn	AL	Σ+ - Σ−
Solling, 1976–1980													
TD	60	17	25	75	311	66	n.d.	318	91	205	2	n.d.	46
TD–SL	12	8	1	43	292	60	n.d.	42	25	155	−15	−204	−17
TD–SL–PI	12	0	−6	22	249	60	n.d.	42	25	191	−15	−204	−17
Solling, 1981–1989													
TD	61	15	19	62	289	47	n.d.	264	88	204	2	n.d.	37
TD–SL	20	8	−2	40	275	49	n.d.	16	28	175	−10	−205	16
TD–SL–PI	20	−1	−8	19	232	49	n.d.	16	28	211	−10	−205	16
Solling, 1990–2002													
TD	55	8	10	22	186	17	n.d.	102	59	55	1	n.d.	9
TD–SL	4	3	−4	10	174	17	−101	−52	5	32	−5	−117	−6
TD–SL–PI	4	−5	−11	−11	131	17	−101	−52	5	68	−5	−117	−6
Göttinger Wald, 1981–1989													
TD	36	12	16	75	196	31	36	210	58	132	1	n.d.	8
TD–SL	−14	6	−16	−1,176	99	18	−133	29	9	132	1	−1	−1,027
TD–SL–PI	−14	−15	−34	−1,326	−28	18	−133	29	9	322	1	−1	−1,027
Göttinger Wald, 1990–2002													
TD	35	5	9	27	158	13	173	79	41	37	0	n.d.	2
TD–SL	−4	2	−14	−1,102	126	7	16	−10	10	37	0	−1	−1,054
TD–SL–PI	−4	−20	−32	−1,253	0	7	16	−10	10	227	0	−1	−1,054
Zierenberg 1990–2002													
TD	53	14	39	77	171	19	n.d.	91	63	61	0	n.d.	96
TD–SL	−7	11	−128	−237	23	18	−181	−21	11	61	0	−1	−147
TD–SL–PI	−7	−1 to −3	−132 to −140	−271 to −292	−15 to −40	18	−181	−21	11	106 – 140	0	−1	−147

Positive numbers of TD-SL and TD-SL-PI indicate net retention of a component, negative numbers net release. Σ+ - Σ− is the difference between the sum of cations and anions (n.d., not determined)

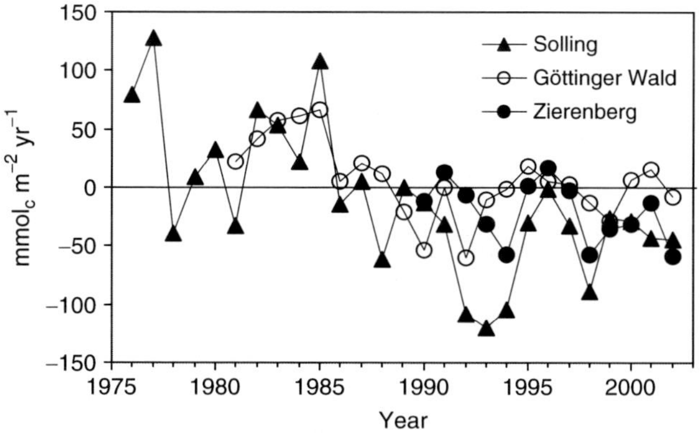

Fig. 16.8 Long-term annual budgets of sulphate (total deposition – leaching losses) at the three beech sites at Solling, Göttinger Wald and Zierenberg

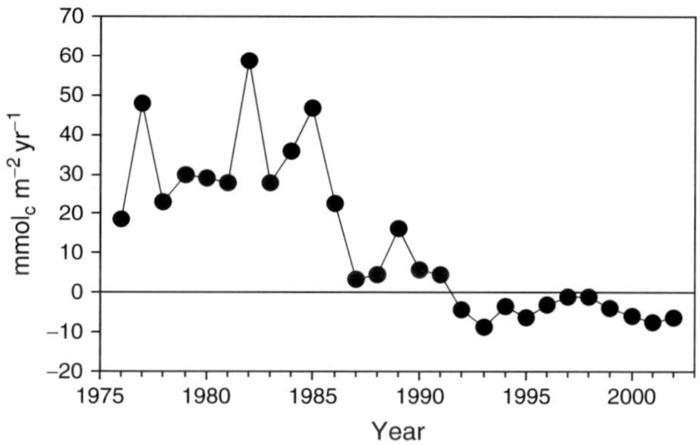

Fig. 16.9 Long-term values of annual ecosystem budget of Ca (total deposition + weathering – seepage output – plant increment) at the Solling beech site

Scandinavia and the Czech Republic/Slovakia with shallow soils and relatively small sulphate storage capacity responded quickly to decreased S input, the deeply weathered soils in Germany with high sulphate storage capacity responded more slowly to decreased deposition and this response was considered to be the results of the release of stored sulphate (Alewell 2001; Prechtel et al. 2001).

When the amount of elements retained by plant increments (PI) is included in the output value (TD-SO-PI) then the obtained budgets indicated that the Solling soil which was a sink for Ca became a net source during the three decades of observation period (Table 16.9). The net source amounted to 11 mmol$_c$ m^{-2} per year and is higher than that calculated for mineral weathering (8 mmol$_c$ m^{-2} per year; Sect. 16.4). When the amount of Ca released by mineral weathering is included in the input, most of the Ca values remained under the zero line since 1992 (Fig. 16.9) indicating a net loss of Ca. Both atmospheric depositions (TD) and seepage outputs of Ca have decreased during this period. However, decrease in TD was higher than the decrease in seepage losses. This pattern is hence quite similar to the one observed for sulphate. Under these conditions, sustainable forest utilisation seems to be no longer possible with respect to forest nutrition.

The long-term observation at Solling site indicated that exchangeable Ca decreased from about 14 $kmol_c\ ha^{-1}$ in 1966 to 7 $kmol_c\ ha^{-1}$ in 1983 (0–50 cm, Chap. 3), and since then had stayed at this low level indicating a significant decline of plant available Ca. This may partially be due to the retention of Ca in the surface organic layer for which the amount increased from 4.5 to 8.8 $kmol_c\ ha^{-1}$ from 1966 to 2001 (Chap. 4). As the mass of surface organic layer at the Solling site increased, it retained C, N, Ca and other elements. Foliage Ca levels showed Ca deficiency (Chap. 5) without any prospect of recovery since input by weathering and TD did not cover the plant demand for growth increment. The recent decrease in TD and exchangeable Ca and its retention in the surface organic layer will have some long-term consequences on the nutrition and vitality of trees in future.

Mg showed similar decreasing trends in total deposition and leaching losses as Ca. The decrease in TD of Mg (from 19 to 10 $mmol_c\ m^{-2}$ per year, Table 16.9) exceeded the decline in leaching losses during the last observation decades (from 20 to 14 $mmol_c\ m^{-2}$ per year, Table 16.7), thus TD did not cover the amount of leaching losses. Considering the additional amount of Mg retained in plant increment (7 $mmol_c\ m^{-2}$ per year) the total amount required annually in addition to that received through TD at the Solling site, was about 11 $mmol_c$ Mg m^{-2} per year (Table 16.9). Through mineral weathering about 27 $mmol_c$ Mg m^{-2} per year would be released (Table 16.2) which may cover the requirements at the Solling site of 11 $mmol_c$ Mg m^{-2} per year as well as the additional amount required for humus accumulation in surface organic layer (6 $mmol_c\ m^{-2}$ per year). Even though the overall ecosystem budget is positive the content of Mg in leaves pointed to its deficiency (Chap. 11). Weathering rates might be realistic because two independent methods estimated similar values for Mg (and Ca, see Sect. 16.4). Thus, a reduction in Mg availability through interactions with Al in the apoplast of roots (Godbold and Jentschke 1998) might be a plausible explanation for Mg deficiency at this acid soil.

A negative K budget (TD-SL-PI, −5 $mmol_c\ m^{-2}$ per year, Table 16.9) and high weathering rates of 52 $mmol_c\ m^{-2}$ per year as calculated by PROFILE model of Sverdrup and Warfvinge (1992) (Table 16.2) suggested that the overall budget was balanced. The main assumption is that the weathering rate for K is realistic. This may not be the case because the amount of exchangeable K did not show any increase during the observational period of more than three decades at the Solling site (Chap. 3). A much lower weathering rate of 6 $mmol_c$ K m^{-2} per year was calculated by the ZrO method (see Sect. 16.4) which would just cover the negative budget. However, K content in beech leaves revealed an increasing trend of K deficiency during the last decade (Chap. 11). Thus, there was an overestimation of K weathering by the PROFIL model. The same result is revealed for Na. The ecosystem budget of +4 $mmol_c\ m^{-2}$ per year is just balanced by the weathering rate calculated by the ZrO method (5 $mmol_c\ m^{-2}$ per year) whereas a much higher value was estimated by the PROFILE model (26 $mmol_c\ m^{-2}$ per year). Losses of Al and

Mn decreased in time (Table 16.9) which indicated that, due to low inputs of H through atmospheric depositions, the amount of protons buffered through the release of Al was also reduced.

High amount of nitrogen inputs from atmospheric depositions was retained at the Solling site, which was more than 281 $mmol_c\ m^{-2}$ per year (about 39 kg N ha^{-1} per year) during the first two observation periods (Table 16.9). During the last decade, N retention decreased markedly to 174 $mmol_c\ m^{-2}$ per year (24 kg N ha^{-1} per year). As the amount of N lost by leaching is low at this site and the annual amount retained by plant increment is constant and low (43 $mmol_c m^{-2}$ per year or 6 kg N ha^{-1} per year), the main reduction in N retention occurred in the soil which has decreased from about 33 to 18 kg N ha^{-1} per year (Fig. 16.10). Most important for N retention at the Solling site is the surface organic layer where on average 21 kg N ha^{-1} per year were accumulated during the 36-year period (Chap. 3). This increase was assigned to the increase in mass of surface organic layer, though the C/N ratio did not change. According to the extensive studies by Berg and McClaugherty (2003), high N content in freshly fallen litter would decrease its decomposition in the long-term and would thus increase the amount of organic matter remaining in the surface organic layer. Berg (2003) calculated from the data on humus accumulation at Solling site and decomposition studies of litter from another 26 European beech forests that incomplete litter decomposition may sequester about 27 kg N ha^{-1} per year at the Solling site which is in the range of the value obtained for other sites.

16.9.2 Göttinger Wald Site

Temporal changes in the input–output (I/O) budgets at Göttinger Wald site do not indicate any clear trends for most of the elements (Table 16.9). However, for some individual years, the N budget values was negative (Fig. 16.10). At this site, N is not

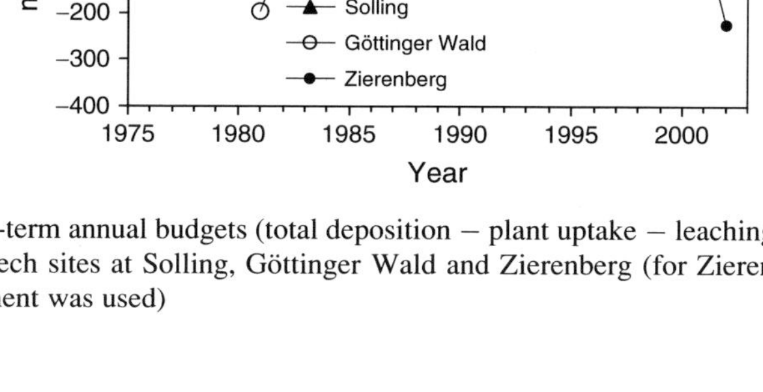

Fig. 16.10 Long-term annual budgets (total deposition − plant uptake − leaching losses) of total N at the three beech sites at Solling, Göttinger Wald and Zierenberg (for Zierenberg, the mean uptake for increment was used)

retained by the soil because of the low amount of organic matter in the surface organic layer and high biological activity. The amount of N retained by plant increment is about three times higher than at the Solling site (Table 16.2). I/O budgets of Ca, Mg, K and Na are negative indicating high amount of outputs by seepage water and plant increment. Weathering rates of Mg and K would just meet the demand for considering the relationship TD-SL-PI, by assuming weathering rates were correctly estimated by the PROFIL model. Foliage nutrient levels indicated sufficient levels for Mg, an optimal level of nutrition for K and a luxuriant nutrient level for Ca. Budgets for aluminum and sulphur generally exhibited small values when compared to the Solling site. However, a shift from sulphur retention (period 1981 to 1989) to a release of stored sulphur from the soil (1990–2002, Table 16.9) was observed which is expected to cause an increased acid load for the soil.

16.9.3 Zierenberg Site

At the Zierenberg site, where the observation period was the shortest of the three, a general trend towards more negative values of I/O budgets for nutrients was observed, despite a strong seasonal character of the values. The unique feature of the Zierenberg site is the continuing high leaching loss of N. The amount of N in seepage output and sequestration by growth increment was 15–40 mmol$_c$ m^{-2} per year more than total N deposition at this site indicating an annual loss of 2–6 kg N ha^{-1} per year from the soil (Table 16.9). Figure 16.10 indicates an increasing trend of nitrate fluxes which was positively correlated with precipitation rates ($r^2 = 0.73$) and water output rates ($r^2 = 0.82$), though the amount of N deposition did not change much. The negative I/O budgets for N suggested a decrease of organic N pools in the soil (Eichhorn 1995; Eichhorn and Hüttermann 1994, 1999; Fried 2001). This loss of N through humus degradation would assume high level of acidity or the presence of aluminum in soil solution. OH-groups of phenolic substances could be masked by Al-cation forming stable Al-complexes which is assumed to prevent further polymerisation processes to larger organic molecules as proposed by Ulrich (1981). In general, organo-metallic complexes are regarded to be stabilising soil humus (Blaser and Klemedson 1987); however, dissolved organic carbon as a precursor of humus showed a similar reaction with Al. Using lignin as a model, the enzymatically catalysed polymerisation of phenols was clearly inhibited in the presence of Al-ions (Jentschke 1984; Haars et al. 1989; Hüttermann et al. 1989). Aluminum and nitrates were released into seepage water through acidification for which increasing positive trends (Table 16.8) were observed. In addition to Al, other elements as K, Ca and Mg were released by silicate weathering and carbonate dissolution, of which Ca and Mg occurred in appreciable amounts in the output fluxes. Increased precipitation rates and temperatures during the nonvegetation period of recent decades were observed as a general trend (Chap. 2). This would increase the leaching losses of nitrates at high nitrifying sites similar to Zierenberg during the nonvegetation period, especially those with additional litter input from N-rich herbaceous vegetation like *Urtica dioica*.

16.10 Acid Loading: Atmospheric Depositions and System-Internal H Production

Production and consumption of H ions in ecosystems are determined by the input and output of other ions as the electro-neutrality has to be maintained at any macroscopic scale in soil solutions. During their uptake and losses electro-neutrality is preserved through transfer of protons among different components of a system (Ulrich 1981, 1994a). Thus, the total ecosystem proton production rates can be inferred from the input-output budgets of major ions (Bredemeier et al. 1990; Ulrich 1994a).

N transformation processes may play an important role for consumption and production of protons. For each mole of NH_4 input, one mole of H is released if taken up or 2 moles of H will be produced if completely nitrified (according to the overall stoichiometric equation $NH_4^+ + 2O_2 \rightarrow NO_3^- + 2H^+ + H_2O$). Thus, depending upon the sources and fate of N, the total proton production can be calculated. For the methodology to calculate acid budgets refer to Sect. 16.4.

The overall total acid production rates were highest at Göttinger Wald (>1 mol m^{-2} per year in both decades, where a significant part of the total protons was weak acidity from carbonic acid dissociation (73% and 81% of the total protons produced in the two consecutive periods) (Table 16.10). This acidity is primary buffered through dissolution of limestone (carbonate buffer range) and the release of Ca into the solution phase. This process is therefore related to the amount of precipitation as indicated by positive correlation between water output and production of weak acidity (r^2 = 0.94) and losses of Ca (r^2 = 0.94) and Mg (r^2 = 0.77) with seepage water. The contribution of atmospheric deposition and uptake of base cations for plant increment to the total production of acidity has decreased from 23 to 18% (of which plant increment contributed to about 14%) reflecting the decline in atmospheric depositions.

Table 16.10 Mean annual acid production rates in mmol m^{-2} per year calculated from input/output balances (considering total deposition, plant increment, and leaching losses), given separately for the consecutive decades, and relative contributions of four different processes to the total at the Solling, Göttinger Wald and Zierenberg sites, calculated according to Ulrich 1994a

Site	Period	Total H^+ production (mmol m^{-2} per year)	S release (% of total)	N transformation (% of total)	Acid input (% of total)	Weak acidity release (% of total)
Solling	76–80	284	3	18	67	8
	81–89	280	4	12	75	4
	90–02	150	34	17	46	8
Göttinger Wald	81–89	1,400	0	6	23	73
	90–02	1,300	1	2	18	81
Zierenberg	90–02	407–416	6	35–36	23–26	35–36

H production by Mb retention is not shown. Acid input refers to the sum of positive budgets of H, Mn, Al and Fe. Weak acidity refers to carbonic and organic acids

At Zierenberg, the weak acids contributed 35–36% to the total proton load which occurred in deeper soil layers close to the boundary of tertiary basaltic debris overlaying Triassic limestone. The proportion of protons produced from N transformation processes was 35–36% of the total proton load at the Zierenberg site, where nitrification was highest among the three sites. The input of acid contributed to 23–26% of the total load of which plant increment added 19%. Proton buffering occurred through consumption of protons for weathering of primary silicates or carbonates causing the release of Mb cations of which Ca and Mg dominated the soil solution.

Due to the acid conditions at the Solling site the production of weak acidity was low and consisted almost exclusively of organic acid deprotonation (4–8% of the total load). The primary sources of acidity in this ecosystem were the inputs by H deposition and the negative equivalent of base cations taken up for plant increment. The total amount of proton load decreased from 280 to 150 mmol m^{-2} per year during the last two decades. The relative proportion of acid input to the total load decreased from 75 to 46% (Table 16.10), of which 14–24% were contributed by cation retention by plants. Contribution of N transformations to acid load at the Solling site was primarily related to the retention of deposited ammonium whose relative proportion to total acid load increased from 12 to 17% during the last two decades, despite a decrease in acidity produced by N transformation from 35 to 25 mmol m^{-2} per year. The most important component of the current acid load at Solling is the release of stored sulphate increasing from 12 to 52 mmol m^{-2} per year and contributing about 4% and 34% to the total acid load during the last two decades. This resulted in delaying the recovery from soil acidification. A delay in recovery from soil acidification due to release of sulphate present in the deep weathered acid soils has been shown to be widespread in central Europe (Alewell et al. 2000a). Total acid load at Solling would have been higher than 150 mmol m^{-2} per year if the acid production for the uptake and accumulation of base cations in the surface organic layer were included in the calculation. Assuming that Na, K, and Mg were retained in the same proportion as Ca, an additional potential of H load of 30 mmol m^{-2} per year has to be considered. Buffering of acidity at Solling occurred predominantly by the leaching of Al which contributed about 54% of the sum of cations in the output flux.

Declining N depositions suggested a decreasing trend in acidity production. H generation by transformation of N at Solling decreased from 51, 35, to 25 mmol m^{-2} per year during the three decades (Fig. 16.11). The Göttinger Wald site showed a small decreasing trend in protons, if the initial value was considered as an outlier in the data. The H production by N transformations at Zierenberg, however, showed an increasing trend which is driven by increased precipitation and water output during the observation period. A higher loss of N with increased precipitation might occur at sites where an excess of nitrates exist in the soil which is leached at higher rates of water flow before N is retained by plants or microorganisms. The highest value of acid production was observed in 2002 and corresponded to an acidification rate of 360 mmol $Hm^{-2.}$ per year by N transformation alone, and this in turn makes up 47% of total proton production in that year.

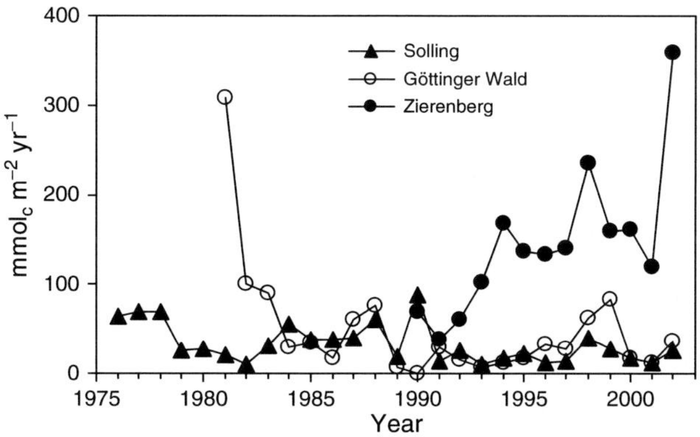

Fig. 16.11 Comparison among the three sites of time series of acidity production by transformation of deposited N compounds

16.11 Final Discussion

The long-term observation at the Solling and Göttinger Wald sites indicated that total depositions had decreased for sulphur (−60% at Solling and Göttinger Wald), H (−72% at Solling and Göttinger Wald), and nitrogen (−38% at Solling, −20% at Göttinger Wald) and base cations (−60% at Solling and Göttinger Wald) between two periods of 1981–1989 and 1990–2002. Emissions of SO_2 have significantly decreased since the mid-1980s in Europe (Ferrier et al. 2001). Deposition measurements from other studies in Europe confirmed a decline by 38–82% for sulphate and approximately 20% for inorganic nitrogen during the 1990s (Prechtel et al. 2001; Wright et al. 2001). This poses the question as to what were the effects of these changes on soils of different soil chemical states? The soil chemical properties determine biochemical and biological properties and thus contribute to the stability of an ecosystem (Ulrich 1994b). The three beech forests, Solling, Göttinger Wald and Zierenberg, represent the range of soil properties in Europe, with calcareous, intermediate and very acid soils, and thus could be regarded as typical representatives of three contrasting forest ecosystem types with respect to element budgets. Owing to differences in soil properties the response of these ecosystems to decreasing atmospheric depositions differed widely.

16.11.1 Response of Acid–Base Soil Properties to Decreasing Atmospheric Deposition

The decreasing sulphate depositions caused a decrease in sulphate export from all three beech sites. Owing to the small contribution of deposited acidity to the total H production at the carbonate-rich Göttinger Wald site, the response of soil solution chemistry to decreasing atmospheric deposition was small. Solute concentration of sulphate, N and base cations in the surface 5 cm soil decreased, but that of Al increased suggesting that soil acidification may have continued in this layer despite the reduction in acid load. Similarly, it was not possible to assess if this decrease in acid deposition has changed soil solution chemistry in the surface soil at the Zierenberg site due to lack of solution sampling within 0–20 cm layer. At lower depths at the Zierenberg site there was an increase in the concentration of Al and H suggesting that soil processes significantly modified the concentration of solute in atmospheric depositions. In addition, high precipitation and soil water drainage during 1990–2002 increased nitrate losses at Zierenberg and thus contributed further to increased soil acidification.

In the highly acid Solling soil, a decrease in leaching losses of almost all ions in soil solution was observed indicating the importance of decreasing atmospheric acid load for soil solution chemistry on similar sites. Though the total production of acidity decreased during the last 10 years at the Solling site, changes in soil solution composition were still of a minor nature. For example, the base-cations/Al ratio remained low and the alkalinity of soil solution increased only slightly indicating that the recovery of soil solution acidity was a very slow process in such soils. One of the main reasons is desorption and the release of previously stored sulphate from the solid phases (Stoddard et al. 1999; Alewell et al. 2000a). The net release of sulphate from the soil profile amounted to about 8 kg ha^{-1} per year and contributed to 34% of total acidity production at the Solling site during the last decade. The soil capacity to adsorb sulphate is large, the modelled adsorbed pool in 2003 down to 100 cm was 1,030 kg S ha^{-1} for a "clean-rain" treatment in the Solling roof project adjacent to the Solling beech forest (Martinson et al. 2005). Model prediction shows no recovery, based on the criteria of Bc/Al ratio above 1 in the rooting zone, before the year 2050, independent of future deposition cuts. Another reason for the delayed recovery is related to low levels of base cations in the atmospheric depositions. The decrease in calcium fluxes in total deposition exceeded the decrease in leaching losses. The Ca budget clearly indicated that the ecosystem at the Solling site has changed from being a sink to a net source. Foliage chemistry also showed deficiency of Ca and Mg in the leaves (see Chap. 11). Thus, a further significant recovery in acidity of soil solution is expected in future depending on the amount of sulphur stored in the soil. Similar studies by Matzner et al. (2004) reported an annual net release of S from the soil profile of 4 and 19 kg ha^{-1} as an average between 1993–1995 and 2001 at two sites. Long-term trends in sulphate concentrations and fluxes in throughfall and streamwater of 20 European catchments revealed a similar trend (Prechtel et al. 2001). However, it is highly unlikely that the Ca supply of the trees will improve without liming.

16.11.2 Response of N and C Budgets of Soils to Decreasing Atmospheric Depositions

N depositions have also decreased at all sites but the decrease was only detectable at sites where long time series (Solling and Göttinger Wald) were available. The 13-year observation period at Zierenberg was too short to detect such changes. A significant decreasing trend in nitrate concentrations was detectable only in the surface soil layers of the Solling and Göttinger Wald sites. Most of the 21 kg ha^{-1} per year N atmospheric inputs at Göttinger Wald were retained for annual increment by plants (18 ha^{-1} per year, considering above ground biomass and coarse roots) and the remaining amount leached with seepage water (4.5 kg N ha^{-1}) while the soil N pool was constant during 1990–2002. Thus, N deposition at this site meets the demand for the annual growth and any excess amount is leached with seepage water indicating that this forest is near (quasi-) steady state. Such systems are defined by organic matter production in the soil that equals the respiration, a closed nutrient cycle, and amount of organic matter stored in the soil being at a maximum level under given environmental conditions (Odum 1969, 1983; Ulrich 1992).

The Zierenberg site showed two interesting features. Firstly, nitrate concentrations showed no clear trends in time and the annual seepage fluxes were significantly increasing during the 13 years of observation period mainly due to increasing water fluxes. This pattern was not observed on the other two sites. Secondly, an interesting feature of N dynamics at the Zierenberg site was a negative N budget. Output by seepage water (21 kg N ha^{-1}) and retention by plant increment (5–9 kg N ha^{-1}) exceeded the input by total deposition (24 kg N ha^{-1}) by about 2–6 kg N ha^{-1} annually. This may be attributed to a process described as humus degradation (Ulrich 1981; Eichhorn 1994, 1995). The theory of humus degradation assumes the presence of aluminum in the soil solution causing inhibition of the repolymerisation of organic components to form humus. High nitrification leads to the production of acidity and free Al in the soil triggering an amplifying effect where organic N in the soil is further nitrified and lost. In this soil, organic matter production and respiration processes are not balanced. According to the ecosystem theory by Ulrich (1992), such a forest is in a transient state and will continue losing carbon, nitrogen and other nutrients from the mineral soil until a new (quasi-) steady state with lower C and N contents in the mineral soil under high soil acidity is achieved.

At the Solling site, most of the N received from atmospheric depositions was retained in the soil. However, this retention decreased to about 18 kg N ha^{-1} per year during the last 10 years following a similar decrease in N inputs. Leaching losses of N were low and only a small amount was retained for growth increment, which did not change with the decrease in N depositions. Most of the N was retained in the surface organic layer in association with an accumulation of organic matter. C/N ratio of the soil did not change. C/N ratios of >25 in the surface organic layer have been found as critical values for the ability of forest ecosystems to retain N (Matzner and Grosholz 1997; McDonald et al. 2002). But in the surface organic layer at the Solling site this value was much lower than that critical value. An annual increase in mass of surface organic layer of 350 kg C and 21 kg N ha^{-1} occurred during the last 36-year period (see Chap. 4 for more details). Whether organic matter accumulation is controlled by high N depositions as suggested by Berg and McClaugherty (2003) and/or by acid depositions as suggested by Persson and Wirén (1993) is not clear from the data. Matzner et al. (2004) reported a relatively low N retention in a spruce forest at the Fichtelgebirge (Coulissenhieb site) with an acid podzolic soil and mor humus type litter layer. At this site, an average 80% of the mineral N in throughfall was leached as nitrate with seepage water.

16.12 Conclusions

Political action by European countries has led to a noticeable decline in atmospheric depositions of sulphuric acid and to a lesser extent of nitrogen since the end of the 1980s. This decline has raised the question: how do forest ecosystems with different soil properties respond and recover? Data presented from the three sites provided the following conclusions:

- In the highly acid Solling soil, a decrease in leaching losses of almost all ions in soil solution was observed, although base cations/Al ratio and alkalinity still remained low or increased only slightly. This was related to desorption and release of previously stored sulphate from the solid phases which contributed to one-third of total acidity produced or added to the system. A decline in atmospheric deposition of base cations changed this ecosystem from a sink to a net source of Ca. To improve Ca supply of the trees, liming will be needed.
- Solute concentrations in the carbonate-rich Göttinger Wald soil decreased in the carbonate-free surface soil layer whereas solute concentrations in deeper soil layers which were rich in carbonates and silicates were not affected. Highly significant increases of Al in solution samples from 5 cm depth suggested that soil acidification may have continued despite the reduction in acid load.
- In the base-rich and carbonate-free Zierenberg soil, high values of Al and H concentrations were observed in solutions collected from 20 cm soil depths. This was not related to atmospheric depositions but to degradation of soil organic matter, probably initiated by formerly high levels of acid load.
- A significant decrease in nitrate concentrations was observed in soil solutions of the surface mineral soils of the Göttinger Wald and Solling sites. At the Göttinger Wald site, high annual increment meets N deposition and excess amounts are leached indicating that this forest is near (quasi-) steady state with respect to nutrients (a closed nutrient cycle) and constant C and N pools in the soil.
- At the Zierenberg site, a net loss of N from soil organic matter was observed, where acidification of the mineral soil promoted humus degradation suggesting a transient ecosystem state. The acid Solling soil retained most of the N received from atmospheric deposition by increasing the surface organic layer. This moder humus soils has probably a high potential for N retention in future.

References

Aber J, McDowell W, Nadelhoffer K, Magill A, Berntson G, Kamakea M, McNulty S, Currie W, Rustad L, Fernandez I (1998) Nitrogen saturation in temperate forest ecosystems: hypotheses revisited. BioScience 48:921–934

Ågren GI, Bosatta E (1988) Nitrogen saturation of terrestrial ecosystems. Environ Pollut 54:185–197

Alewell C, Manderscheid B, Meesenburg H, Bittersohl J (2000a) Is acidification still an ecological threat? Nature 407:856–857

Alewell C, Manderscheid B, Gerstberger P, Matzner E (2000b) Effects of reduced atmospheric deposition on soil solution chemistry and elemental contents of spruce needles in NE-Bavaria, Germany. J Plant Nutr Soil Sci 163:509–516

Alewell C (2001) Predicting reversibility of acidification: the European sulfur story. Water Air Soil Pollut 130:1271–1276

Benecke P (1984) Der Wasserumsatz eines Buchen- und eines Fichtenwaldökosystems im Hochsolling. Schr. Forstl. Fak. Univ. Göttingen u. Nieders. Forstl. Versuchsanst 77

Berg B (2003) Sequestration rates for C and N in soil organic matter at four N-polluted temperate forest stands. In: Matzner E (ed) Biogeochemistry of forested Catchments in a Changing Environment, Ecological Studies 172. Springer Berlin, pp 361–376

Berg B, McClaugherty C (2003) Plant litter. Springer, Berlin, Germany

Blaser P, Klemedson JO (1987) Die Bedeutung von hohen Aluminiumgehalten für die Humusanreicherung in sauren Waldböden. Z Pflanzenern Bodenk 150:334–341

Borken W, Matzner E (2004) Nitrate leaching in forest soils: an analysis of long-term monitoring sites in Germany. J Plant Nutr Soil Sci 167:277–283

Bredemeier M, Blanck K, Xu YJ, Tietema A, Boxman AW, Emmett B, Moldan F, Gundersen P, Schleppi P, Wright RF (1998) Input-output budgets at the NITRES sites. For Ecol Manage 101:57–64

Bredemeier M, Matzner E, Ulrich B (1990) Internal and external proton load to forest soils in northern Germany. J Environ Qual 19:469–477

Cronan CS, Grigal DF (1995) Use of calcium/aluminum ratios as indicators of stress in forest ecosystems. J Environ Qual 24:209–226

de Vries W, Reinds GJ, Vel E (2003) Intensive monitoring of forest ecosystems in Europe 2: atmospheric deposition and its impacts on soil solution chemistry. For Ecol Manage 174:97–115

Dietze G, Ulrich B (1992) Aluminium speciation in acid soil water and ground water. In: Mathess G, Frimmel F, Hirsch P, Schulz HD, Usdowski HE (eds) Progress in hydrogeochemistry. Springer, Berlin, pp 269–281

Dise NB, Wright RF (1995) Nitrogen leaching from European forests in relation to nitrogen deposition. For Ecol Manage 71:153–161

DVWK Deutscher Verband für Wasserwirtschaft und Kulturbau e.V. (1996) Ermittlung der Verdunstung von Land- und Wasserflächen. DVWK-Merkblätter zur Wasserwirtschaft 238

Eichhorn J, Hüttermann A (1994) Humusdisintegration and N-mineralisation. In: Godbold DL, Hüttermann A (eds) Effects of acid rain on forest processes. Wiley, New York, USA, pp 129–162

Eichhorn J (1995) Stickstoffsättigung und ihre Auswirkungen auf das Buchenwaldökosystem der Fallstudie Zierenberg. Ber Forschungszentrum Waldökosysteme A 124

Eichhorn J, Hüttermann A (1999) Mechanisms of humus dynamics and nitrogen mobilisation. In: Rastin N, Bauhus J (eds) Going underground, ecological studies in forest soils. Research Signpost, Trivandrum, India, pp 239–277

Fassbender HW, Ahrens E (1977) Laborvorschriften und Praktikumsanleitung: Zur chemischen Untersuchung von Vegetations-, Boden- und Wasserproben im Institut für Bodenkunde und Waldernährung der Universität Göttingen. Göttinger Bodenkundliche Berichte 47

Ferrier RC, Jenkins A, Wright RF, Schöpp W, Barth H (2001) Assessment of recovery of European surface waters from acidification 1970–2000: an introduction to the Special Issue. Hydrol Earth Syst Sci 5:274–282

Fried R, Eichhorn J, Paar U (2001) Exploratory analysis and a stochastic model for humusdisintegration. Environ Monit Assess 68:273–295

Glugla G (1969) Berechnungsverfahren zur Ermittlung des aktuellen Wassergehalts und Gravitationswasserabflusses im Boden. Albrecht-Thaer-Arch 13:371–376

Godbold DL, Jentschke G (1998) Aluminum accumulation in root cell walls coincides with inhibition of root growth but not with inhibition of magnesium uptake in Norway spruce. Physiol Plant 102:553–560

Gundersen P, Emmett BA, Kjǿnaas OJ, Koopmans CJ, Tietema A (1998) Impact of nitrogen deposition on nitrogen cycling in forests: synthesis of NITREX data. For Ecol Manage 101:37–55

Haars A, Karazipour H, Zanker B, Hüttermann A (1989) Room-temperature curing adhesives based on lignin and phenoloxidases. In: Hemingway RW, Conner AH, Branhams SJ (eds) Adhesives from renewable resources. ACS-Symposium Series No 385. American Chemical Society, Washington DC, pp 126–134

Högberg P, Jensén P (1994) Aluminium and uptake of base cations by tree roots: a critique of the model proposed by Sverdrup et al. Water Air Soil Pollut 75:121–125

Hörmann G (1997) SIMPEL – Ein einfaches, benutzerfreundliches Bodenwassermodell zum Einsatz in der Ausbildung. Dtsch Gewässerkundl Mitt 41:67–72

Hörmann G (1998) SIMPEL – Speichermodelle zum Bodenwasserhaushalt. User reference, unpubl

Hörmann G, Meesenburg H (2000) Die Erfassung und Modellierung des Wasserhaushaltes im Rahmen des Level II-Programmes in Deutschland. Forstarchiv 71:70–75

Hörmann G, Schmidt J (1995) Dokumentation von Wasserhaushaltsmodellen. Ber Forschungsz Waldökosysteme B 42

Hüttermann A, Haars A, Trojanowski J, Milstein O, Kharazipour A (1989) Enzymatic modification of lignin for its technical use – strategies and results. In: Glasser WG, Sarkanen S (eds) New polymeric materials from lignin. ACS-Symposium Series No 397. American Chemical Society, Washington DC, pp 361–371

Jentschke G (1984) Untersuchungen zum biochemischen Mechanismus der Humusdisintegration – Versuch einer Falsifizierung. Ber d Forschz Waldökosysteme Göttingen 5:175–256

Jacobsen C, Rademacher P, Meesenburg H, Meiwes KJ (2003) Gehalte und chemische Elemente in Baumkompartimenten – Literaturstudie und Datensammlung. Studie im Auftrage des BMELF, FZW Reihe B, Bd 69

Jochheim H (1992) Raum/zeitliche Variabilität des chemischen Bodenzustandes in der Phase des Stickstoff-Vorratsabbaus. Abschlussbericht des Teilprojektes P 6.3.2.2 im BMFT-Projekt Stabilitätsbedingungen von Waldökosystemen. GH Kassel

Kaiser K, Zech W (2000) Sorption of dissolved organic nitrogen by acid subsoil horizons and individual mineral phases. Eur J Soil Sci 51:403–411

Khanna PK, Prenzel J, Meiwes KJ, Ulrich B, Matzner E (1987) Dynamics of sulfate retention by acid forest soils in an acidic deposition environment. Soil Sci Soc Am J 51:446–452

König N, Fortmann H (1996a) Probenvorbereitungs-, Untersuchungs- und Elementbestimmungs-Methoden des Umweltanalytik-Labors der Niedersächsischen Forstlichen Versuchsanstalt und des Zentrallabors II des Forschungszentrums Waldökosysteme. Teil 1: Elementbestimmungsmethoden A-M. Ber Forschungszentrum Waldökosysteme B 46

König N, Fortmann H (1996b) Probenvorbereitungs-, Untersuchungs- und Elementbestimmungs-Methoden des Umweltanalytik-Labors der Niedersächsischen Forstlichen Versuchsanstalt und des Zentrallabors II des Forschungszentrums Waldökosysteme. Teil 2: Elementbestimmungsmethoden N-Z und Sammelanhänge. Ber Forschungszentrum Waldökosysteme B 47

König N, Fortmann H (1996c) Probenvorbereitungs-, Untersuchungs- und Elementbestimmungs-Methoden des Umweltanalytik-Labors der Niedersächsischen Forstlichen Versuchsanstalt und des Zentrallabors II des Forschungszentrums Waldökosysteme. Teil 3: Gerätekurzanleitungen und Gerätekurzanleitungen Datenverarbeitung. Ber Forschungszentrum Waldökosysteme B 48

König N, Fortmann H (1996d) Probenvorbereitungs-, Untersuchungs- und Elementbestimmungs-Methoden des Umweltanalytik-Labors der Niedersächsischen Forstlichen Versuchsanstalt und des Zentrallabors II des Forschungszentrums Waldökosysteme. Teil 4: Probenvorbereitungs- und Untersuchungsmethoden, Qualitätskontrolle und Datenverarbeitung. Ber Forschungszentrum Waldökosysteme B 49

Martinson L, Lamersdorf N, Warfvinge P (2005) The Solling roof revisited – slow recovery from acidification observed and modeled despite a decade of 'clean-rain' treatment. Environ Pollut 135:293–302

Matzner E (1989) Acidic precipitation: case study Solling. In: Adriano DC, Havas M (eds) Advances in environmental science, acidic precipitations, Volume I Case studies. Springer, Berlin, pp 39–84

Matzner E (2004) Biogeochemistry of forested catchments in a changing environment. Ecological Studies 172. Springer, Berlin

Matzner E, Grosholz C (1997) Beziehung zwischen NO_3-Austrägen, C/N Verhältnissen der Auflage und N-Einträgen in Fichtenwald (*Picea abies* Karst.)-Ökosystemen Mitteleuropas. Forstwissenschaftliches Centralblatt 116:39–44

Matzner E, Zuber T, Lischeid G (2004) Response of soil solution chemistry and solute fluxes to changing deposition rates. In: Matzner (ed) Biogeochemistry of forested catchments in a changing environment, Ecological Studies 172. pp 339–360

MacDonald JA, Dise NB, Matzner E, Armbruster M, Gundersen P, Forsius M (2002) Nitrogen input together with ecosystem nitrogen enrichment predict nitrate leaching from European forests. Global Change Biol 8:1028–1033

McNulty SG, Aber JD, Boone RD (1991) Spatial changes in forest floor and foliar chemistry of spruce-fir forests across New England. Biogeochemistry 14:13–29

Meesenburg H, Meiwes KJ, Rademacher P (1995) Long term trends in atmospheric deposition and seepage output in northwest German forest ecosystems. Water Air Soil Pollut 85:611–616

Meesenburg H, Meiwes KJ, Schulze A, Rademacher P (1997) Bodendauerbeobachtungsflächen auf forstlich genutzten Böden (BDF-F). Arb-H Boden 2/1997:77–94

Meiwes KJ, Khanna PK, Ulrich B (1980) Retention of sulphate by an acid brown earth and its relationship with the atmospheric input of sulphur to forest vegetation. Z Pflanzenernaehr Bodenkd 143:402–411

Meiwes KJ, Hauhs M, Gerke H, Asche N, Matzner E, Lamersdorf N (1984) Die Erfassung des Stoffkreislaufes in Waldökosystemen: Konzept und Methodik. Ber Forschungszentrum Waldökosysteme/Waldsterben 7:68–142

Michalzik B, Kalbitz K, Park JH, Solinger S, Matzner E (2001) Fluxes and concentrations of dissolved organic carbon and nitrogen: a synthesis for temperate forests. Biogeochemistry 52:173–205

Montheith JL (1965) Evaporation and Environment. In: Fogg GE (ed) The state and movement of water in living organisms. Sypm Soc Experimental Biol 19. The University Press, Cambridge, pp 205–234

Odum EP (1969) The strategy of ecosystem development. Science 164:262–370

Odum EP (1983) *Grundlagen der Ökologie*. Georg Thieme, Stuttgart, Germany

Penman HL (1948) Natural evaporation from open water, bare soil and grass. Proc Roy Soc Lond A 193:120–145

Persson T, Wirén A (1993) Effects of experimental acidification on C and N mineralization in forest soils. Agric Ecosyst Environ 47:159–174

Prechtel A, Alewell C, Armbruster M, Bittersohl J, Cullen J, Evans C, Helliwell R, Kopacek J, Marchetto A, Matzner E, Meesenburg H, Moldan F, Moritz K, Vesely J, Wright RF (2001) Response of sulfur dynamics in European catchments to decreasing sulfate deposition. Hydrol Earth System Sci 5:311–325

Prenzel J (1983) A mechanism for storage and retrieval of acid in acid soils. In: Ulrich B, Pankrath J (eds) Effects of Accumulation of air pollutants in forest ecosystems. D. Reidel, Dordrecht, The Netherlands, pp 157–170

Reuss JO, Johnson DW (1986) Acidic deposition and the acidification of soils and waters. Springer, New York

Stoddard JL, Jeffries DS, Lükewille A, Clair TA, Dillon PJ, Driscoll CT, Forsius M, Johannessen M, Kahl JS, Kellogg JH, Kemp A, Mannio J, Monteith DT, Murdoch PS, Patrick S, Rebsdorf A, Skjelkvåle BL, Stainton MP, Traaen T, van Dam H, Webster KE, Wieting J, Wilander A (1999) Regional trends in aquatic recovery from acidification in North America and Europe. Nature 401:575–578

Sverdrup H, Warfvinge P (1992) A model for the impact of soil solution Ca:Al ratio, soil moisture and temperature on tree base cation uptake. Water Air Soil Pollut 61:365–383

Tamm CO (1991) Nitrogen in terrestrial ecosystems. Ecological Studies 81. Springer, Berlin

Thornthwaite CW (1948) An approach toward a rational classification of climate. Geogr Rev 38:55–94
Tietema A, Beier C (1995) A correlative evaluation of nitrogen cycling in the forest ecosystems of the EC projects NITREX and EXMAN. For Ecol Manage 71:143–151
Tiktak A, van Grinsven HJM (1995) Review of sixteen forest-soil-atmosphere models. Ecol Model 83:35–53
Ulrich B, Mayer R, Khanna PK (1980) Chemical changes due to acid precipitation in a loess-derived soil in Central Europe. Soil Sci 130:193–199
Ulrich B (1981) Theoretische Betrachtung des Ionenkreislaufs in Waldökosystemen. Z Pflanzenernährung Bodenkunde 144:647–659
Ulrich B (1987) Stability, elasticity and resilience of terrestrial ecosystems with respect to matter balance. In: Schulze ED, Zwölfer H (eds) Ecological Studies 61. Springer, Berlin, pp 11–49
Ulrich B (1992) Forest ecosystem theory based on material balance. Ecol Model 63:163–183
Ulrich B (1994a) Nutrient and acid/base budget of central European forest ecosystems. In: Godbold DL, Hüttermann A (eds) Effects of acid rain on forest processes. Wiley, New York, pp 1–43
Ulrich B (1994b) Process hierarchy in forest ecosystems: an integrative Ecosystem theory. In: Godbold DL, Hüttermann A (eds) Effects of acid rain on forest processes. Wiley, New York, pp 343–398
Wesselink LG, van Grinsven JJM, Grosskurth G (1994) Measuring and modeling mineral weathering in an acid forest soil, Solling, Germany. In: Bryant RB, Arnold RW (eds) Quantitative modeling of soil forming processes. Soil Science Society of America Special Publication Number 39, pp 91–110
Wolff B, Riek W (1997a) Deutscher Waldbodenbericht 1996, Band 1. Bundesministerium für Ernährung, Landwirtschaft und Verbraucherschutz, Bonn, Germany
Wolff B, Riek W (1997b) Deutscher Waldbodenbericht 1996, Band 2. Bundesministerium für Ernährung, Landwirtschaft und Verbraucherschutz, Bonn, Germany
Wright RF, Alewell C, Cullen JM, Evans CD, Marchetto A, Moldan F, Prechtel A, Rogora M (2001) Trends in nitrogen deposition and leaching in acid-sensitive streams in Europe. Hydrol Earth System Sci 5:299–310

Chapter 17
Soil Respiration

R. Brumme, W. Borken, and J. Prenzel

17.1 Introduction

The recent discussion about global climate change has intensified the role of soils which may act as a sink or a source of carbon under changing environmental conditions (Schulze 2000; Valentini 2003). A decrease of organic carbon by 10% in the whole world soils would be equivalent to all the anthropogenic CO_2 emitted over 30 years (Kirschbaum 2000). It is generally argued that with an increase in global warming both net primary production (NPP) and decomposition of soil organic carbon (SOC) will increase (Kirschbaum 1995, 2000). The issue is which of these processes are stimulated by increasing temperature. Soil respiration includes heterotrophic respiration and autotrophic respiration, and their temperature sensitivities will largely determine the effect of a warmer world on net carbon flux between soils and atmosphere (Boone et al. 1998). However, there is increasing evidence that soil respiration is determined more by carbohydrate supply than by temperature (Högberg et al. 2001).

A reliable technique of measuring respiration that provides temporal and spatial variations in the soil and autotrophic respiration is required for assessing ecosystem levels effects. Another issue is whether forest soils, which have been affected by acid and nitrogen deposition or liming, will react differently to changing climate. We will therefore consider: (1) methodological aspects of assessing soil respiration by using chambers, (2) ways to obtain reliable flux estimates with consideration of temporal and spatial variations, (3) evaluation of annual CO_2 fluxes from three beech forest soils, (4) the contribution of autotrophic and heterotrophic respiration to soil respiration, (5) the temperature sensitivity of autotrophic respiration and heterotrophic respiration, and (6) the effect of liming and N-fertilisation and stand harvesting by the formation of small gaps on soil respiration.

Method. For studying the *temporal and spatial variation of CO_2 emissions* we used 25 static, double-wall, closed chambers (12 cm inside diameter surrounded by a buffer zone to prevent wind effects as reported by Matthias et al. 1980) on 30 occasions from May 1991 to September 1991 at the B2 plot at Solling (about 500 m

R. Brumme and P.K. Khanna (eds.), *Functioning and Management of European Beech Ecosystems*, Ecological Studies 208,
DOI: 10.1007/978-3-642-00340-0_17,

from the long-term monitoring plot B1). Five subplots (1 × 1 m) with five closed chambers on each subplot were established. The subplots were located in the middle and at the corners of a 60 × 60 m plot. This experimental design allowed us to study the small scale (1 × 1 m) and the large scale variation (60 × 60 m) of soil respiration under field conditions. After the experiment was over, the surface organic layer and 0–5 cm of the mineral soil beneath each chamber were collected and analysed for weights of the surface organic layer and the fine root contents, the C- and N-content, and the C/N ratios in the surface organic layer and the mineral soil. For studying the *CO_2 emission from beech forests*, closed chambers were also used at the long-term trace gas plot B1 at the Solling site (*n* =3; Brumme and Beese 1992), the control plot for a litter exchange experiment (*n* =3; see Brumme et al. 1999), and a control plot used by Borken et al. (2002) (*n* =5) from 1995 to 1999 which was used to calculate a mean annual CO_2 flux for the beech site at Solling. *Effects of forest management on CO_2 emission* were studied with three chambers at the limed (BK plot) and, the fertilised plot (BN plot) from 1993 to 1995 at the Solling site. A forest gap at the Solling site was used to distinguish between autotrophic and heterotrophic respiration (B2 plot, see Bartsch et al. 2002). Three automatic chambers were placed in the middle of the gap and the forest stand from May to October 1991 (Brumme 1995), and three manual closed chambers were placed at the same locations from August 1993 to March 1995. At the Zierenberg site, three chambers were placed between *Urtica dioica* from 1991 to 1992 (*n* = 3). Gas samples were taken with evacuated glass flasks (100 mL) and automatically analysed in the laboratory (Loftfield et al. 1997; Brumme and Beese 1995).

17.2 Soil Chamber Design Affects Soil Respiration Measurements

Chamber systems are simple to use but it has been reported that they often underestimate or overestimate trace gas fluxes depending on the conditions of measurements. Two chamber types are commonly used for studies in trace gas emissions (Mosier 1989). (1) Dynamic, closed chambers where a constant air flow through the chamber is established and gas flux can be calculated from concentration difference, flow rate, volume and area covered by the chamber. (2) Static, closed chambers where the air is not circulated and gas flux can be calculated by periodically collecting gas samples from the chamber and measuring the change in gas concentration with time during the period of linear concentration change. Both types of chambers have disadvantages which should be considered while using a specific chamber technique. Static, closed chambers may underestimate fluxes by lowering the concentration gradient between soil and atmosphere which may lead to an increase in gas concentrations in the upper part of the soil profile (Conen and Smith 2000). The resultant error which depends on several factors was estimated to range from 10 to 15% (Rayment 2000; Norman et al. 1997) and might be negligible when chambers are kept closed for a short period. Dynamic, closed chambers tend to overestimate gas fluxes. Air movement inside the chambers is thought to disrupt the laminar boundary layer over the soil and increase fluxes by 30–50% (Le Dantec et al. 1999). The dynamic, closed chamber system led to 10 and 36% higher fluxes than those obtained by soda lime or eddy correlation techniques, respectively (Janssens et al. 2000). Another problem with dynamic, closed chambers is the pressure anomalies induced by air circulation and during chamber closure. For greater than ambient chamber pressure, soil CO_2 flux decreases (Lund et al. 1999). Independent of the systems used, CO_2 fluxes obtained by direct insertion of a chamber into the soil were significantly higher compared to those where a collar was installed before measurement (Ngao et al. 2006). An increase in vertical diffusivity coefficient measured directly after insertion may be related to the cutting of fine roots during collar installation (Wang et al. 2005; Ngao et al. 2006). However, there are large discrepancies among methods indicating the need for a careful comparison with other techniques (Lankreijer et al. 2003) or to employ means for obtaining an absolute calibration of the method (Widén and Lindroth 2003).

We used three different types of static, closed chambers at Solling, which allowed us to compare the effects of chamber design on soil respiration. We started in 1987 with an automatic chamber system and placed three automatic, double-walled chambers (inner frame: 50 × 50 cm; outer frame: 70 × 70 cm) side by side (Loftfield et al. 1992), and measured CO_2 and N_2O emissions five times a day until 1988 (Brumme and Beese 1992). Double-walled chambers were used to reduce wind effects which would be expected if the soil–chamber seal between chamber and soil surface was not tight as described by Matthias et al. (1980). We continued measurements with manual, quadratic-shaped chambers (50 × 50 cm) (U-shaped water-traps with a 2-cm collar at the bottom) from 1991 to 1995. They were then replaced by round chambers (70 cm diameter) from 1996 to 1998 to facilitate the top-up procedure of the water-traps. All chambers had a height of 15–20 cm. Large chambers were used to overcome the spatial variations. However, our study on spatial variation of soil respiration (see below) showed that this approach could account for small-scale spatial variation only, but not for the large-scale spatial variation.

There were only small differences in soil respiration between the chamber types when soil temperatures of >10°C were recorded (Fig. 17.1). The quadratic chamber type underestimates the soil respiration measured at <5°C as indicated in Fig. 17.1. The difference might be due to a poor seal by the 2-cm collar at the bottom of the U-shaped water-trap which was insufficient for a horizontal installation on an uneven ground surface. The two other chambers types showed similar rates indicating that the seal of the round chamber was similar as the double-walled chamber. The double-walled chamber is assumed to have the best seal in forests because any leakage of the seal between the chamber frame and surface organic layer would primary influence the outer but not the inner compartment of the chamber where the gas fluxes were measured (Loftfield et al. 1992). Perfect sealing of chambers in forests remains a problem because a deep insertion would cut tree roots in contrast to agricultural systems, where the whole plant-root system can be included in the chamber. This is primarily the case in acid forest soils with moder or mor humus where the amount of fine roots in the F and H horizons could be high. A

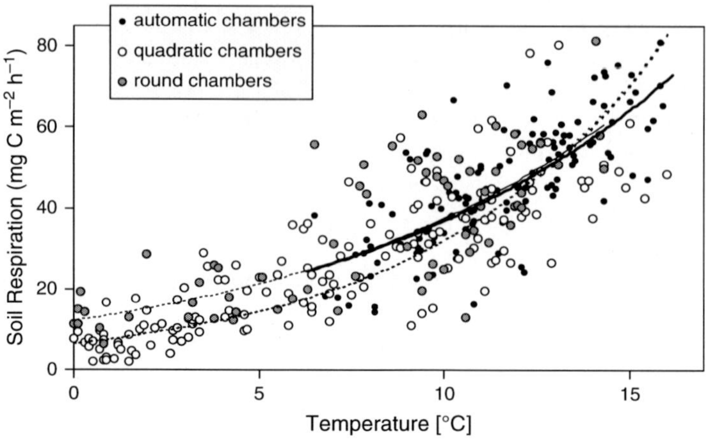

Fig. 17.1 Soil respiration measured with an automatic double-walled chamber (1987–1988), a quadratic type water-sealed manual chamber (1991–1995) and a round chamber (1996–1998). All chamber types were installed at the same plots in succession at the beech stand B1 at Solling. Exponential regressions are given for automatic (*solid line*, $r^2 = 0.53$), round (*upper dotted line*, $r^2 = 0.67$) and quadratic chambers (*lower dotted line*, $r^2 = 0.71$)

deeper insertion of chamber frames could also increase the moisture content and decomposition of cut roots. Moreover, higher water content may increase soil respiration as indicated in Fig. 17.4. However, autotrophic respiration is reduced. Thus, a compromise between chamber sealing and root disturbance has to be made during chamber installation in acid forests soils.

17.3 Temporal Variation of Soil Respiration

Soil respiration showed a distinct diurnal response in values similar to those found for soil temperature (Fig. 17.2) with maximum values between 1200 and 1600 hours, when measured with double-walled automatic chambers (Loftfield et al 1992). Diurnal variation in soil temperature between 0.2 and 2°C resulted in an increase of soil respiration between 10 and 220% as measured in August 1988 at the Solling B1 plot (Brumme and Beese 1992). This diurnal variation is much higher than that reported by Matteucci et al. (2000) for a European transect. Tang et al. (2005) demonstrated that, beside soil temperature and moisture, photosynthesis drives soil respiration on a diurnal time scale. In our long-term study, the average flux of daily cycles occurred between 0630 and 1130 hours. This has been shown by comparing 700 measurements (five measurements per day with one automatic chamber, nine chambers in total over 140 days) and 30 selected measurements (one measurement per week between 0630 and 1130 hours) (Brumme and Beese 1992). The "one measurement per week" estimate for CO_2 was similar to the "five

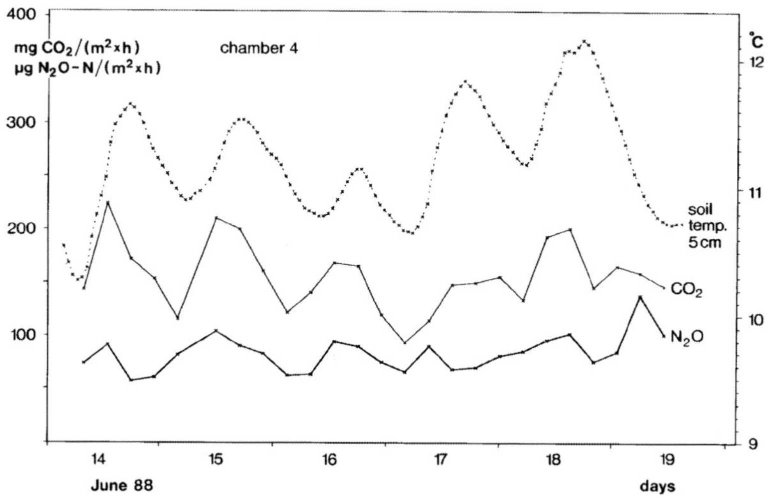

Fig. 17.2 Diurnal and day-to-day variation of N_2O and CO_2 emission rates (five flux calculations per day) and soil temperature (5 cm depth) for June 1988 measured by an automatic chamber at the Solling B1 plot (Loftfield et al. 1992)

measurements per day" estimate. The average was less than +1% indicating that one weekly CO_2 measurement between 0630 and 1130 hours would overcome diurnal, day-to-day and seasonal variations.

The trend in seasonal variation in soil respiration followed the changes in soil temperature which explained 83% of the variation in soil respiration, using a linear regression analysis, but excluding measurements when soil water suction values were higher than 200 hPa. Values of soil water suctions >200 hPa reduced the soil respiration at the Solling site (Figs. 17.3, 17.4, and 17.6). High and low soil moisture may reduce soil respiration by limiting oxygen availability or causing water stress on micro-organisms (Linn and Doran 1984). Heterotrophic respiration is highly sensitive to soil water tension as was observed in a laboratory study with sieved agricultural soil (Orchard and Cook 1983). Heterotrophic respiration decreased linearly with an increase in water tension from 100 to 15,000 hPa. The use of field measurements for calculating temperature and water effects on soil respiration should be considered with care as the measured CO_2 efflux may be derived from different soil layers with different temperature and water regimes. Mean Q_{10} values, the relative increase in decomposition rate per 10°C difference, increased asymptotically with soil depth from 1.87 at 0 cm to 3.46 at 10 cm depth in European forest ecosystems, indicating a large uncertainty in the prediction of the temperature dependency of soil respiration (Borken et al. 2002). Moreover, plant physiological processes can significantly contribute to the seasonality of soil respiration by belowground C-allocation and, hence, can influence the temperature sensitivity of soil respiration (Högberg et al. 2001; Yuste et al. 2004) similar to the model used for the calculation of Q_{10} (Davidson and Janssens 2006).

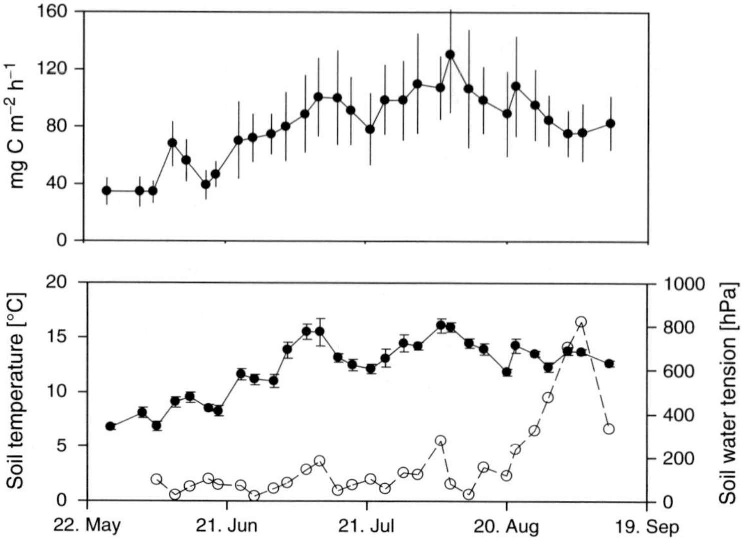

Fig. 17.3 Seasonal variation of CO_2 emission rates ($n = 25$), soil temperature (*closed circle*) (5 cm depth) and soil water tension (*open circle*) (hPa) (5 cm below surface organic layer) from May to September 1991 at Solling on the B2 plot

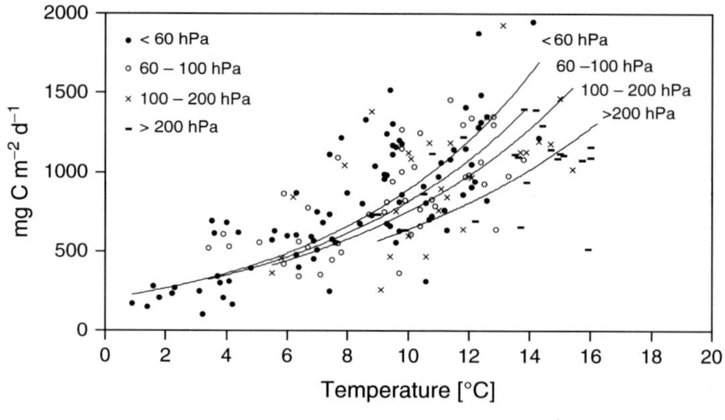

Fig. 17.4 Soil respiration at the Solling B1 plot ($n = 3$) in relation to soil temperature (5 cm depth) and soil water tension (hPa) (5 cm below surface organic layer) from 1991 to 2000 (exponential regressions for the three highest moisture classes were forced through the same point on the *y*-axis as obtained for <60 hPa)

17.4 Spatial Variation of Soil Respiration

A spatial variation in soil respiration, expressed as the coefficient of variation (cv) of 25 chambers, ranged from 18 to 39% (average 27%) during the 107-day period in the 60 × 60 m plot. The cv showed no clear temporal pattern. However, in the summer period, cv of >30% were observed when soil dryness occurred. A lower cv value for soil respiration was found for the cumulative rates. Within the five subplots ($n = 5$) and between the 25 chambers, average cv values of 17% and 19% were calculated. Therefore, the small scale spatial variation (1 × 1 m) was similar (cv =17%, $n = 5$) as the large spatial scale (60 × 60 m) variation (cv = 19%, $n = 25$). Other studies indicated a higher spatial variation in soil respiration. For example, Fang et al. (1998) measured high spatial variation in soil respiration with 12 dynamic chambers within a 25 × 25 m plot (cv of 55%). The high cv value for soil respiration was related to fine root biomass and the amount of litter and humus in the surface organic layer. The variation was smaller after plots with and without understory were grouped (21% in each group). Nakayama (1990) reported that soil CO_2 efflux rates varied by 100% between locations 1 m apart. Rayment and Jarvis (2000) found that soil respiration at locations separated by 1 m showed as much intra-site variability as at locations separated by up to 6 m.

To determine the factors affecting small and large scale spatial variation in soil respiration, linear regressions were performed between cumulative soil respiration and surface organic layer mass, fine root biomass (dead and live roots) in the surface organic layer and 0–5 cm mineral soil, the C- and N-content of the surface organic layer, and the C/N ratios in the surface organic layer and the mineral soil. The small scale variation of soil respiration within the 1 × 1 m subplots showed $r^2 > 0.5$ only for %C in the surface organic layer in two subplots. The best correlation between soil respiration on the larger spatial scale (25 chambers in a 60 × 60 m plot) was found for the C/N ratio in the surface organic layer ($p = 0.01$) and the fine root biomass (<2 mm) ($p = 0.04$) (Fig. 17.5a). The negative correlation between soil respiration and the C/N ratio of surface organic layer suggested that the quality of the substrate affects heterotrophic respiration (Fig. 17.5b). Multiple regression of soil respiration with fine root biomass and C/N explained 32% of the variation. Soil respiration increased from 1.9 to 2.4 Mg C ha^{-1} per 107 days with increasing root biomass from 0.2 to 3.7 g per chamber indicating the role of root biomass in controlling soil respiration at the larger spatial scale. Fine root biomass was the most important variable in a mull type humus forest in central Germany, explaining 30–40% of spatial variation in soil respiration rates (Søe and Buchmann 2005).

17.5 Soil Respiration in Temperate Beech Forests

Soil respiration was higher at Zierenberg (4.5 Mg C ha^{-1}, SD 0.31, for 23 sampling dates) than at the Solling site (3.7 Mg C ha^{-1}, SD 1.0, for 218 sampling dates). The Zierenberg site was mostly covered by herbs (stinging nettle, *Urtica dioica*) whereas the Solling site had sparse understorey vegetation (Chap. 5 and 10). The understorey at the Zierenberg site may have increased both autotrophic and heterotrophic respirations. The Solling site accumulated about 350 kg C ha^{-1} per year in the surface organic layer (see Chap. 4) and had reduced the emission of CO_2 from the soil. The large differences in respiration rates summarised by Raich and Schlesinger

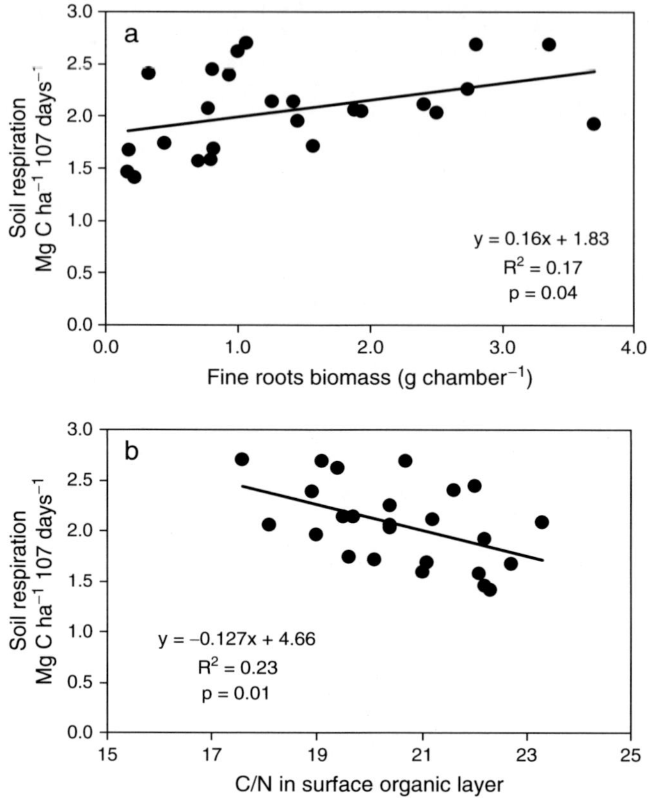

Fig. 17.5 Relationship between cumulative soil respiration (Mg C ha^{-1} per 107 days) and fine root biomass (g) (**a**) and C/N ratio in the organic layer (**b**) of 25 soil chambers within an 60 × 60 m area at the B2 plot at Solling

(1992) for 29 temperate deciduous forests, which ranged from 3 to 14 Mg C ha^{-1} per year (mean 6.47 ± 0.51 Mg C ha^{-1} per year), may result from a number of factors, e.g. site-to-site variation, spatial and temporal variation and chamber design (Lankreijer et al. 2003). Beech sites with calcareous parent material annually produced 8.8 Mg C ha^{-1} (Collelongo, Italy; Matteucci et al. 2000) and 7.5 Mg C ha^{-1} (south Germany; Dannenmann et al. 2007), those on acid soils fluxes 5.8 Mg C ha^{-1} (north Germany; Borken et al. 2002) and 8.2 Mg C ha^{-1} (Ardennes, Belgium; Jonard et al. 2007) were reported.

17.6 Autotrophic Respiration

Estimates of autotrophic respiration are commonly made using a number of methods: (1) comparison between soil respiration with and without roots, (2) direct respiration measurement of living roots, (3) application of stable or radioactive isotopes, (4) the use of tree-girdling (Hanson et al. 2000; Högberg et al. 2001). We used a forest gap and forest stand at Solling to calculate autotrophic respiration by subtracting CO_2 emission of the gap from that of the forest stand.

The forest gap (30 m diameter) was originally established in October 1989 at Solling (Bartsch et al. 2002). The stand was free of ground vegetation (percentage cover of less than 5%) and significant establishment did not occur until the fourth vegetation period after the establishment of the gap. In the fourth and fifth year after formation of the gap, when soil respiration was measured, percentage cover of ground vegetation was 7 and 20%, respectively (Bartsch et al. 2002). Soil temperatures did not differ between the plots under the gap and the neighbouring forest stand due to lack of direct solar radiation on the soil surface (Brumme 1995). Ingrowth-core experiments showed that growth of beech fine roots at ingrowth–core experiments gap centre was negligible during the second and third year after gap creation, indicating that although fine roots from stumps stayed alive, they did not grow (Bauhus and Bartsch 1996). Roots were not active in water uptake in the centre of the gap indicated by soil water potential measurements (Bauhus and Bartsch 1995). Eight years after harvesting there was no change in C- and N-content in the entire soil profile and no effect on mass loss of F material in litter bags (Bauhus et al. 2004) was noted.

Gas measurements, conducted with soil chambers free of ground vegetation in the fourth and fifth years after gap creation, showed that autotrophic respiration and soil respiration followed a seasonal trend (Fig. 17.6). The increase in soil temperature at the beginning of April increased decomposition of soil organic matter whereas autotrophic respiration increased later with the bud burst in May. Autotrophic respiration was insignificant between December and April and approached zero at 0°C in contrast to heterotrophic respiration (Fig. 17.7). Autotrophic respiration started to increase at the end of April, reached a value of 10 mg C m^{-2} h^{-1} in May and a maximum of 20 mg C m^{-2} h^{-1} in July and September. Autotrophic respiration decreased during leaf litter fall in October to 13 mg C m^{-2} h^{-1} and to 8 mg C m^{-2} h^{-1} in November. In July–August, the soil moisture was low (high soil

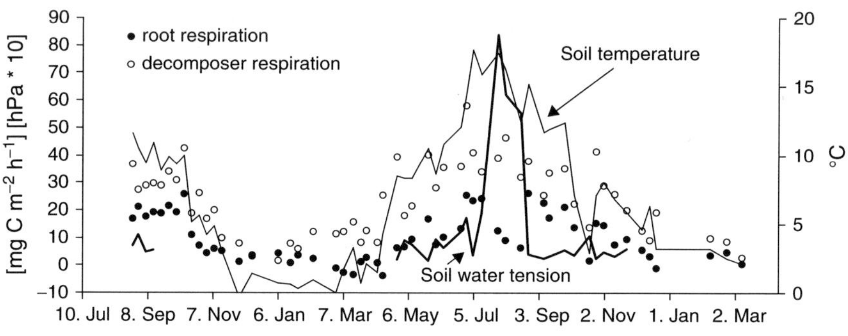

Fig. 17.6 Soil respiration (mg C m^{-2} h^{-1}) of a forest gap without roots (heterotrophic respiration), difference between an adjacent forest and the gap (autotrophic respiration), soil temperature (5 cm depth from the top soil), and soil water tension (5 cm depth in the mineral soil) from August 1993 to March 1995 at Solling

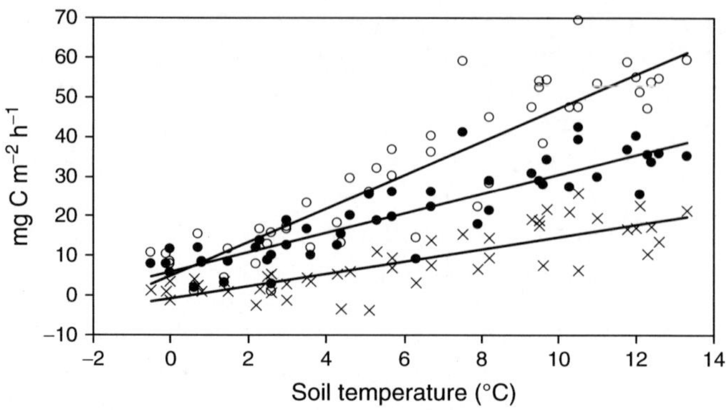

Fig. 17.7 Linear regression between soil temperature (5 cm depth from the top soil) and soil respiration (*open circle*, $r^2 = 0.83$), autotrophic ($\times$, $r^2 = 0.67$) and heterotrophic respiration (*closed circle*, $r^2 = 0.79$) (mg C m^{-2} h^{-1}) from August 1993 to March 1995 at Solling

water tension) which reduced CO_2 emission from the rooted plot but not from the non-rooted plot resulting in apparent low autotrophic respiration (Fig. 17.6). However, the effect of soil dryness on autotrophic respiration might be of less importance. Partitioning of soil respiration using radiocarbon signatures of CO_2 revealed a less strong decline in autotrophic respiration with increasing drought stress in a mixed hardwood forest (Borken et al. 2006).

Soil temperature explained more of the variation in autotrophic respiration ($r^2 = 0.67$; Fig. 17.7) than the air temperature ($r^2 = 0.47$) if respiration values during soil dryness were excluded. In a trenching experiment in a beech forest in France, 86% of variance could be explained by soil temperature (Epron et al. 2001). The respiration of excised roots has been found to increase exponentially with temperature in ten forested sites across North America (Burton et al. 2002) and indicates that root activity may at least partly be controlled by soil temperature.

The contribution of autotrophic respiration to soil respiration was calculated for the second (Brumme 1995) and fifth year after formation of the gap, showing similar values for the two years (Fig. 17.8). The monthly fraction of autotrophic respiration to total soil respiration ranged from 24 to 50% with the highest value in August. Throughout the growing season autotrophic respiration amounted to 35 and 30% in the second and fifth year after formation of the gap, respectively, and the value for the entire year 1994 was 27%. The reduction in soil respiration after gap creation was accompanied by a decrease in microbial biomass and might have resulted from a reduction of ecto-mycorrhizal hyphae (Bauhus and Bartsch 1995). Harvesting would reduce litterfall affecting the litter layer but root biomass would stay alive long after trees were cut (Bauhus and Bartsch 1996).

A similar autotrophic respiration was observed after girdling of a boreal Scots pine forest. Girdling reduced soil respiration within 1–2 months by about 54% relative to the respiration on ungirdled control plots (Högberg et al. 2001). During the second year after girdling, autotrophic respiration increased up to 65% and averaged about 50% from June to October (Singh et al. 2003). Andersen et al. (2005)

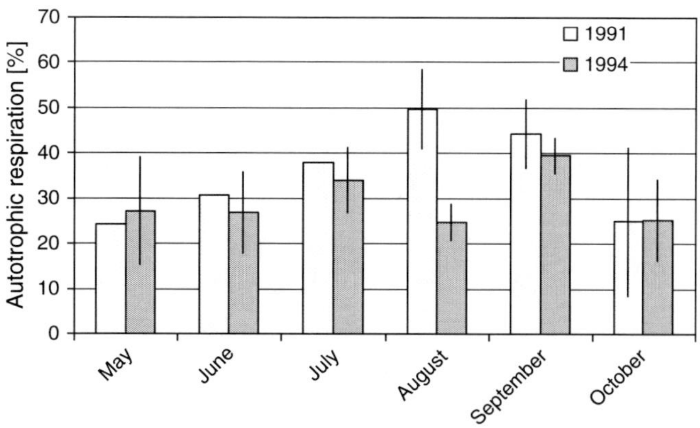

Fig. 17.8 Contribution of autotrophic respiration to soil respiration in the second and fifth year after formation of a gap at the Solling site

observed a decrease in soil respiration of 50% close to the bole of beech trees, during the 3 months following girdling. Autotrophic respiration decreased with increasing distance from the base of the trees (1 m and 1.5–2 m). Several factors might influence autotrophic respiration like tree species, climate, soil properties or spatial variation of root biomass which have not been studied so far. Moreover, a reliable estimate of autotrophic respiration is difficult, because it is an indirect measure of respiration values in a tree-girdled area or in gap studies where several counteracting factors are operative. Starch reserves were calculated to be of less significance for the soil CO_2 efflux 2 months after tree girdling (Högberg et al. 2001). Microbial biomass was reduced by 41 and 23% 1 and 3 months after girdling, respectively, (Högberg and Högberg 2002) and proteomic diversity changed drastically (Schulze et al. 2005). This indicated the importance of carbohydrate allocation to roots as a substrate for heterotrophic respiration. Other processes like decomposition of fine roots and a decrease in litter-fall and their decomposition might be important in the longer term (Bauhus and Bartsch 1995; Bauhus and Bartsch 1996) in both girdling and gap studies. A similar autotrophic respiration in the second and fifth year after formation of the forest gap at Solling (Fig. 17.8) indicated that many counteracting effects are less important during this period after clearing.

17.7 Effects of Additional N-inputs and Liming on Soil Respiration

Soil respiration in the fifth and sixth years on ammonium fertilisation (BN) plot was about 16% higher than on the control plot B1 at Solling (Brumme and Beese 1992). At the end of 11 years of continuous fertilisation, the difference between the two plots for soil respiration was reduced to 4%, with rates of 2.1 Mg C ha^{-1} per year at the BN plot (SD 0.16 Mg C ha^{-1} per year) and 2.01 Mg C ha^{-1} per year (SD 1.04 Mg C ha^{-1} per year) at the control plot B1 between 1993 and 1995. This may indicate that N-fertilisation had no long-term effect on soil respiration at Solling. Effects of fertilisation on soil respiration are difficult to interpret because the effects on heterotrophic respiration may differ from those on the autotrophic respiration. Root studies indicate a similar fine root biomass and turnover compared to the control plot in the third to fourth years after fertilisation (Chap. 9) which may indicate that fertilisation did not change heterotrophic respiration in the long-term. However, other studies have been shown to decrease heterotrophic respiration after N-fertilisation (Berg and Matzner 1997).

Soil respiration in the fifth and sixth years after liming at the BK plot was higher by about 31% than on the control plot B1 at Solling (Brumme and Beese 1992). A lower fine root biomass in the fourth and fifth years after liming (Chap. 9) indicated that liming may have stimulated microbial activity. In the 11th and 12th years, this effect of liming reduced to 7% with rates of 2.1 Mg C ha^{-1} per year (SD 0.56 Mg C ha^{-1} year) at BK plot and 2.01 Mg C ha^{-1} year (SD 1.04 Mg C ha^{-1} year) at the control plot B1 between 1993 and 1995. Liming did not show any clear long-term effect on soil respiration in four forests receiving 6–43 Mg ha^{-1} (Borken and Brumme 1997) of lime. Liming and fertilisation did not change the temperature sensitivity of soil respiration as indicated by linear regression analysis. However, a constant soil respiration does not indicate that heterotrophic respiration was not effected by liming or fertilisation since half of the soil respiration is sustained by current photosynthate (Högberg and Read 2006) which makes it difficult to separate the effects on heterotrophic and autotrophic respiration.

17.8 Summary and Implications for Management Options

- Soil chamber design, chamber installation procedure and the spatial variation in soil respiration are important factors influencing soil respiration on a site, and may cause a high intra- and inter-site variation in soil respiration values. Interpretations of soil respiration data therefore warrant special care.
- A reliable estimate of soil respiration has to consider the spatial variation of root biomass and the C/N ratio in the surface organic layer, which explained 32% of the spatial variation of soil respiration at the Solling site. An adequate number of replicates of chambers to cover spatial variation are more important than temporal resolution of more than one measurement per week.
- Soil temperature explained 83% of the temporal variation in soil respiration. Soil water suction reduced soil respiration at values higher than 200 hPa in the upper mineral soil.
- The fraction of autotrophic respiration to the total soil respiration changed seasonally and contributed up to 50% in August and averaged to 30–35% over the growing season.

- Forest harvesting by creating gaps in the stands did not increase decomposition processes since soil temperature did not change by direct solar radiation in small gaps. Liming increased and fertilisation reduced soil respiration for the short-term while such an effect was not observed in the long-term.

References

Andersen CP, Nikolov I, Nikolova P, Matyssek R, Haberle KH (2005) Estimating "autotrophic" belowground respiration in spruce and beech forests: decreases following girdling. Eur J For Res 124:155–163

Bartsch N, Bauhus J, Vor T (2002) Effects of management practice on ecosystem processes in European Beech forests. In: Dohrenbuch A, Bartsch N (eds) Forest development, succession, environmental stress and forest management. Springer, Berlin, pp 109–137

Bauhus J, Bartsch N (1995) Mechanisms for carbon and nutrient release and retention in beech forest gaps 1. Microclimate, water-balance and seepage water chemistry. Plant Soil 168:579–584

Bauhus J, Bartsch N (1996) Fine-root growth in beech (*Fagus sylvatica*) forest gaps. Can J For Res 26:2153–2159

Bauhus J, Vor T, Bartsch N, Cowling A (2004) The effects of gaps and liming on forest floor decomposition and soil C and N dynamics in a *Fagus sylvatica* forest. Can J For Res 34: 509–518

Berg B, Matzner E (1997) Effect of N deposition on decomposition of plant litter and soil organic matter in forest systems. Environ Rev 5:1–25

Boone RD, Nadelhoffer KJ, Canary JD, Kaye JP (1998) Roots exert a strong influence on the temperature sensitivity of soil respiration. Nature 396:570–572

Borken W, Brumme R (1997) Liming practise in temperate forest ecosystems and the effects on CO_2, N_2O and CH_4 fluxes. Soil Use Manage 13:251–257

Borken W, Xu YJ, Davidson EA, Beese F (2002) Site and temporal variation of soil respiration in European beech, Norway spruce, and Scots pine forests. Glob Change Biol 8:1205–1216

Borken W, Savage K, Davidson EA, Trumbore SE (2006) Effects of experimental drought on soil respiration and radiocarbon efflux from a temperate forest soil. Glob Change Biol 12:177–193

Brumme R (1995) Mechanisms for carbon and nutrient release and retention in beech forest gaps. 3. Environmental regulation of soil respiration and nitrous oxide emissions along a microclimatic gradient. Plant Soil 168–169:593–600

Brumme R, Beese F (1992) Effects of Liming and Nitrogen Fertilization on Emissions of CO_2 and N_2O from a Temporate Forest. J Geophys Res 97:12851–12858

Brumme R, Beese F (1995) Automated monitoring of biological trace gas production and consumption. In: Alef K, Nannipiere P (eds) Methods in applied soil microbiology and biochemistry. Academic, London, pp 468–472

Brumme R, Borken W, Finke S (1999) Hierarchical control on nitrous oxide emission in forest ecosystems. Glob Biochem Cycles 13:1137–1148

Burton AJ, Pregitzer KS, Ruess RW, Hendrik RL, Allen MF (2002) Root respiration in North American forests: effects of nitrogen concentration and temperature across biomes. Oecologia 131:559–568

Conen F, Smith KA (2000) An explanation of linear increases in gas concentration under closed chambers used to measure gas exchange between soil and the atmosphere. Eur J Soil Sci 51:111–117

Dannenmann M, Gasche R, Ledebuhr A, Holst T, Mayer H, Papen H (2007) The effect of forest management on trace gas exchange at the pedosphere-atmosphere interface in beech (*Fagus sylvatica* L.) forests stocking on calcareous soils. Eur J For Res 126:331–346

Davidson EA, Janssens IA (2006) Temperature sensitivity of soil carbon decomposition and feedbacks to climate change. Nature 440:165–173

Davidson EA, Belk E, Boone RD (1998) Soil water content and temperature as independent or confounded factors controlling soil respiration in a temperate mixed hardwood forest. Glob Change Biol 4:217–227

Epron D, Le Dantec V, Dufrene E, Granier A (2001) Seasonal dynamics of soil carbon dioxide efflux and simulated rhizosphere respiration in a beech forest. Tree Physiol 21:145–152

Fang C, Moncrieff JB, Gholz HL, Clark KL (1998) Soil CO_2 efflux and its spatial variation in a Florida slash pine plantation. Plant Soil 205:135–146

Hanson PJ, Edwards NT, Garten CT, Andrews JA (2000) Separating root and soil microbial contributions to soil respiration: a review of methods and observations. Biogeochemistry 48:115–146

Högberg MN, Högberg P (2002) Extramatrical ectomycorrhizal mycelium contributes one-third of microbial biomass and produces, together with associated roots, half the dissolved organic carbon in a forest soil. New Phytol 154:791–795

Högberg P, Read DJ (2006) Towards a more plant physiological perspective on soil ecology. Trends Ecol 21:548–554

Högberg P, Nordgren A, Buchmann N, Taylor A, Ekblad A, Högberg MN, Nyberg G, Ottosson-Löfvenius M, Read DJ (2001) Large-scale forest girdling shows that current photosynthesis drives soil respiration. Nature 411:789–792

Janssens IA, Kowalski AS, Longdoz B, Ceulemans R (2000) Assessing forest soil CO_2 efflux: an in situ comparison of four techniques. Tree Physiol 20:23–32

Jonard M, Andre F, Jonard F, Mouton N, Proces P, Ponette Q (2007) Soil carbon dioxide efflux in pure and mixed stands of oak and beech. Ann For Sci 64:141–150

Kirschbaum MUF (1995) The temperature dependence of soil organic matter decomposition, and the effect of global warming on soil organic C storage. Soil Biol Biochem 27(6):753–760

Kirschbaum MUF (2000) Will changes in soil organic carbon act as a positive or negative feedback on global warming? Biogeochemistry 48:21–51

Lankreijer H, Janssens IA, Buchmann N, Longdoz B, Epron D, Dore S (2003) Measurement of soil respiration. In: Valentini (ed) Fluxes of Carbon, Water and Energy of European Forests. Ecological Studies 163. Springer, Berlin, pp 37–54

Le Dantec V, Epron D, Dufrêne E (1999) Soil CO_2 efflux in a beech forest: comparison of two closed dynamic systems. Plant Soil 214:125–132

Linn DM, Doran JW (1984) Effect of water-filled pore space on carbon dioxide and nitrous oxide production in tilled and nontilled soils. Soil Sci Soc Am J 48:1267–1272

Loftfield N, Brumme R, Beese F (1992) An automated high resolution chamber method for monitoring N_2O and CO_2 flux from forest soil. Soil Sci Soc Am J 56:1147–1150

Loftfield N, Flessa H, Augustin J, Beese F (1997) Automated gas chromatographic system for rapid analysis of the atmospheric trace gases methane, carbon dioxide, and nitrous oxide. J Environ Qual 26:560–564

Lund CP, Riley WJ, Pierce LL, Field CB (1999) The effects of chamber pressurization on soil-surface CO_2 flux and the implications for NEE measurements under elevated CO_2. Glob Change Biol 5:269–281

Matteucci G, Dore S, Stivanello S, Rebmann C, Buchmann N (2000) Soil respiration in beech and spruce forests in Europe: trends, controlling factors, annual budgets and implications for the ecosystem carbon balance. In: Schulz ED (ed) Carbon and nitrogen cycling in European forest ecosystems. Ecological studies 142. Springer, Berlin, pp 217–236

Matthias AD, Blackmer AM, Bremner JM (1980) A simple chamber technique for field measurement of emission of nitrous oxide from soil. J Environ Qual 9:251–256

Mosier AR (1989) Chamber and Isotope Techniques. In: Andreae MO, Schimel DS (eds) Exchange of Trace Gases between Terrestrial Ecosystems and the Atmosphere. Wiley, New York, pp 175–187

Nakayama FS (1990) Soil respiration. Remote Sensing Rev 5:311–321

Ngao J, Longdoz B, Perrin D, Vincent G, Epron D, Le Dantec V, Soudani K, Aubinet M, Willm F, Granier A (2006) Cross-calibration functions of soil CO_2 efflux measurement systems. Ann For Sci 63:477–484

Norman JM, Kucharik CJ, Gower ST, Baldocchi DD, Crill PM, Rayment M, Savage K, Striegl RG (1997) A comparison of six methods for measuring soil-surface carbon dioxide fluxes. J Geophys Res 102(D24):28771–28777

Orchard VA, Cook FJ (1983) Relationship between soil respiration and soil moisture. Soil Biol Biochem 15:447–453

Raich JW, Schlesinger WH (1992) The global carbon dioxide flux in soil respiration and its relationship to vegetation and climate. Tellus 44B:81–99

Rayment MB (2000) Closed chamber systems underestimate soil CO_2 efflux. Eur J Soil Sci 51:107–110

Rayment MB, Jarvis PG (2000) Temporal and spatial variation of soil CO_2 efflux in a Canadian boreal forest. Soil Biol Biochem 32:35–45

Schulze ED (2000) Carbon and Nitrogen Cycle in European Forest Ecosystems. Ecological Studies 142. Springer, Berlin

Schulze WX, Gleixner G, Kaiser K, Guggenberger G, Mann M, Schulze ED (2005) A proteomic fingerprint of dissolved organic carbon and of soil particles. Oecologia 142:335–343

Singh B, Nordgren A, Lofvenius MO, Hogberg MN, Mellander PE, Hogberg P (2003) Tree root and soil heterotrophic respiration as revealed by girdling of boreal Scots pine forest: extending observations beyond the first year. Plant Environ 26:1287–1296

Søe ARB, Buchmann N (2005) Spatial and temporal variations in soil respiration in relation to stand structure and soil parameters in an unmanaged beech forest. Tree Physiol 25:1427–1436

Tang J, Baldocchi DD, Xu L (2005) Tree photosynthesis modulates soil respiration on a diurnal time scale. Glob Change Biol 11:1298–1304

Valentini R (2003) Fluxes of Carbon, Water and Energy of European Forests. Ecological Studies 163, Springer, Berlin

Wang WJ, Zu YG, Wang HM, Hirano T, Takagi K, Sasa K, Koike T (2005) Effect of collar insertion on soil respiration in a larch forest measured with a LI-6400 soil CO_2 flux system. J For Res 10:57–60

Widén B, Lindroth A (2003) A calibration system for soil carbon dioxide-efflux measurement chambers: Description and application. Soil Sci Soc Am J 67:327–334

Yuste JC, Janssens IA, Carrara A, Ceulemans R (2004) Annual Q(10) of soil respiration reflects plant phenological patterns as well as temperature sensitivity. Glob Change Biol 10:161–169

Chapter 18
N_2O Emission from Temperate Beech Forest Soils

R. Brumme and W. Borken

18.1 Introduction

The interest on N_2O emission has increased since the late 1980s after realizing that N_2O is an important greenhouse gas (Lashof and Ahuja 1990; Bouwman 1990a) which destroys ozone in the stratosphere by catalytic reactions (Crutzen 1970). The high global warming potential (GWP) of N_2O has increased the scientific research effort on assessing N_2O fluxes from soils of terrestrial ecosystems (Andreae and Schimel 1989; Bouwman 1990a; Granli and Bøckman 1994) because soils are the largest natural source of N_2O (IPCC 2001). Studies in the 1980s suggested that tropical forests are larger sources for N_2O than temperate and boreal forests, whereas recent studies have indicated that beech (*Fagus Sylvatica* L.) forests can have N_2O fluxes similar to those observed in tropical forests (Brumme and Beese 1992; Papen and Butterbach-Bahl 1999; Zechmeister-Boltenstern et al. 2002). Beech forests with high annual N_2O fluxes have a seasonal emission pattern with high N_2O fluxes in summer and low N_2O fluxes in winter. However, most temperate forests (beech, spruce, oak) have low background N_2O emissions during the year which lack any seasonal trend (Brumme et al. 1999). There are some questions which need to be answered to understand the importance of forests with a seasonal emission pattern for the global balance of N_2O (Brumme et al. 2005). In this chapter, temporal and spatial variations of N_2O emissions from a beech forest ecosystem with a seasonal emission pattern will be provided, leading to a discussion on the mechanisms and processes responsible for seasonal and background patterns of N_2O emissions. Attempts will be made to assess the effect of temperature change, forest management practices (harvesting, liming, soil compaction), and nitrogen inputs on N_2O emissions.

R. Brumme and P.K. Khanna (eds.), *Functioning and Management of European Beech Ecosystems*, Ecological Studies 208,
DOI: 10.1007/978-3-642-00340-0_18,

18.2 Method

To study the *temporal and spatial variation* of N_2O emissions we used 25 static, double-walled, closed chambers which had 12 cm inside diameter and were surrounded by a buffer zone to prevent wind effects as reported by Matthias et al. (1980). The outside diameter of the chamber was 25 cm. Thirty measurements from May 1991 to September 1991 were taken at the B2 plot at the Solling site (about 500 m from the long-term monitoring plot B1). Five subplots (1 × 1 m), each with five closed chambers, were established. The subplots were located in the middle and at the corners of a 60 × 60 m plot. This experimental design allowed us to study the small- and large-scale spatial variations of N_2O emissions. After gas measurements, the surface organic layer and 0–5 cm of the mineral soil under each chamber were collected and analyzed for the weight of different layers, fine root mass, the C and N contents, and the C/N ratios. To study the *N_2O emissions from beech forests* large (0.25 m^2) closed chambers ($n = 3$) were installed for gas measurements at the Göttinger Wald and Zierenberg sites, and on the following different plots at the Solling site: the control plot (long-term trace gas plot B1), the limed plot (BK, 30 tons of dolomitic limestone applied in 1982; see site description in Chap. 3), and the fertilized plot (BN, 140 kg N ha^{-1} applied annually from 1983 to 1993 as $(NH_4)_2SO_4$; see site descriptions in Chap. 3).

Effects of forest management on N_2O emissions were studied in a limed (3 tons ha^{-1} dolomite applied in autumn 1989) and an unlimed forest gap (30 m in diameter, established in autumn 1989) in the B2 plot at the Solling site. Another gap was established at the Zierenberg site in autumn 1990. Gas samples were taken between January and December 1994 at the Solling site and from June 1991 to June 1992 at the Zierenberg site with three closed chambers at each site using evacuated glass bottles (100 mL) and an automatic gas chromatograph (Loftfield et al. 1997).

18.3 Temporal Variation of N_2O Emission

Diurnal changes in soil temperature affected N_2O emissions much more strongly than soil respiration (Chap. 17). A daily increase in soil temperature between 0.2 and 2°C increased N_2O emissions by 10–470% as measured with automatic chambers in August 1988 at the B1 plot (Solling) (Brumme and Beese 1992). In order to find the most suitable time to take measurements during the day that would avoid diurnal variations, 700 measurements (five measurements per day) were taken during 140 days in each of nine chambers, and compared with the measurements taken during 0630 and 1130 hours once a week (30 measurements at each chamber taken during 140 days). One time gas samples taken between 0630 and 1130 hours showed an overestimation of N_2O emissions by values of +3% and +49% in single chambers with a mean increase of +21%. Similar results were obtained by Smith and Dobbie (2001) who used an automated system that provided flux data at 8-h intervals when compared with manual sampling conducted at intervals of 3–7 days. Integrated flux values based on the more intensive measurements were on average no more than

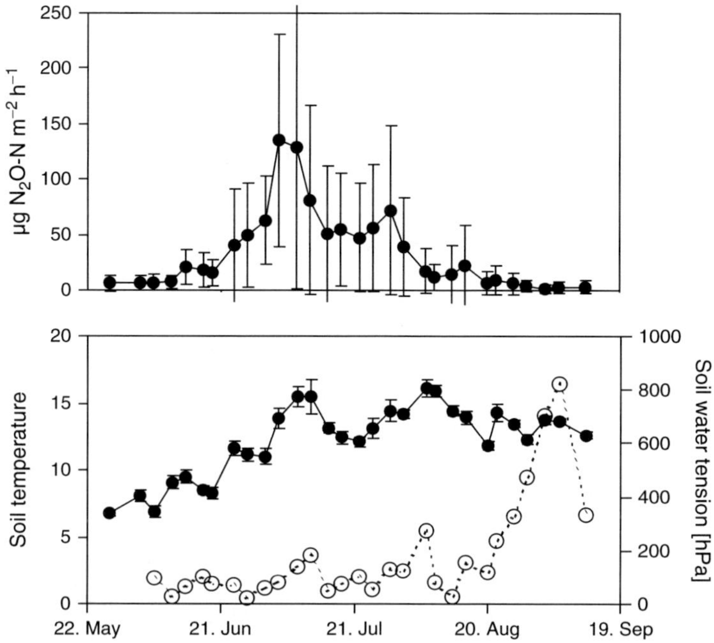

Fig. 18.1 Seasonal variation of N_2O emission rates ($n = 25$), soil temperature (*filled circle*) (5 cm depth) and soil water tension (*open circle*) (hPa) (5 cm below surface organic layer) and standard deviation from May to September 1991 at Solling on the B2 plot

14% greater than those based on manual sampling. However, Flessa et al. (2002) showed that fluxes could be underestimated by less intensive measurements during short periods of extremely high N_2O emissions as would occur during frost–thaw cycles.

The seasonal variations of N_2O emissions were pronounced at the Solling beech site (Fig. 18.1) and were observed since the measurements started in 1987 (Loftfield et al. 1992; Brumme and Beese 1992; Brumme 1995). Temperatures of >10°C increased the N_2O emissions which resulted in high summer fluxes until the soil dried with water tension values exceeding 200 hPa (Figs. 18.1, 18.7). Low precipitation in the summer months therefore resulted in high water tensions and low annual fluxes of N_2O emissions, in contrast to those years when summer precipitation was high, as shown for the years 1994 and 1993 in Fig. 18.8.

At the Göttinger Wald and Zierenberg sites, a low background emission pattern of N_2O emissions was measured during the year and the seasonal emission pattern was lacking. The emission values did not exceed 15 μg N m^{-2} h^{-1} at the Göttinger Wald site and 30 μg N m^{-2} h^{-1} at the Zierenberg site. Seasonal emission patterns were often observed in beech forests with moder humus, whereas the background emission patterns occurred in those with mull humus (Brumme et al. 1999). High rates of N_2O emissions were observed during short periods of freezing–thawing cycles in all forests whether they showed seasonal emission pattern or background emission pattern.

18.4 Spatial Variation of N_2O Emissions

The spatial variation, expressed as the coefficient of variation (cv) of mean N_2O emissions of 25 chambers in a 60 × 60 m plot, ranged from 64 to 228% with a mean value of 119% for 30 sampling dates at the Solling site. Values of coefficient of variation of <125% occurred in spring and summer months and those of >125% in autumn months indicating primarily the irregular spatial changes in soil moisture content in autumn. A lower cv value for N_2O emission was found within the five 1 × 1 m subplots (35–101%, mean 59%) and between the five 1 × 1 m subplots (49%).

Mean N_2O emissions of the 25 chambers within the 60 × 60 m plot correlated significantly with the fine root mass (Fig. 18.2). With increasing fine root biomass from 0.2 to 3.7 g per chamber N_2O emission decreased from 2.5 to 0.1 kg N ha^{-1} per 107 days. A higher root density reduced the water and nitrate concentration and increased the oxygen consumption. There is, however, an interaction between a lowering of water content and an increase in the oxygen diffusivity. In this case, the reduction of N_2O emission indicated that the higher oxygen demand of roots was overcompensated by higher oxygen diffusivity. Nitrate concentration was probably not the limiting factor in this high N soil. Within the five subplots where small scale variation was measured, the mean N_2O emissions correlated significantly with the moisture content at three subplots (R^2 = 0.47, 0.54, 0.97) and with the C content (R^2 = 0.67) at one subplot.

18.5 Landscape Control on N_2O Emissions

Annual value of N_2O emission at the beech forest at the Solling B1 plot was 1.92 kg N ha^{-1} per year for a 10-year measuring period (1995–2000) (Table 18.1). This N_2O emission rate was three and eight times higher than at the Zierenberg and Göttinger Wald sites where amounts of 0.41 and 0.16 kg N ha^{-1} per year were recorded, respectively.

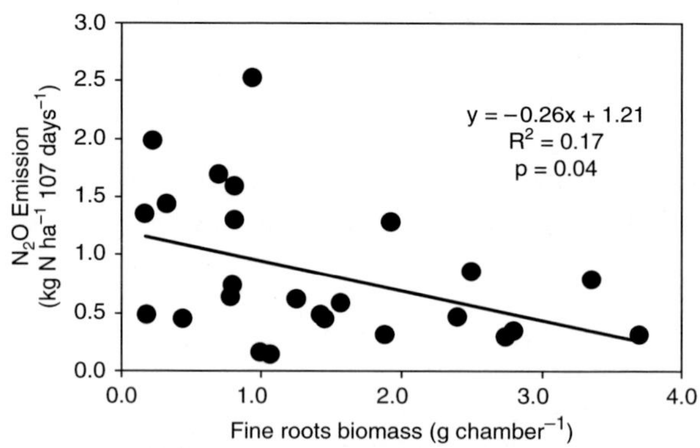

Fig. 18.2 Relationship between cumulative values of N_2O emission for 107 days and fine root biomass (O-horizon and 0–5 cm depth per chamber) within an 60 × 60 m of the B2 plot at Solling

Table 18.1 N_2O emission (kg N ha^{-1} per year) at Göttinger Wald, at the main research area at Zierenberg (ZB1) and at a control and harvested site at Zierenberg (ZB2 control and ZB2 gap), at Solling (B1, B2 sites) and for a fertilized (BN) and limed plots (BK) adjacent to B1 and a gap and limed gap adjacent to B2 at Solling site (standard deviation values in parenthesis)

Site	Plot	Sampling dates	Period	N_2O emissions
Solling	B1, control	319	1990–2000	1.92 (0.63)
	BN, fertilized	72	1993–1995	2.91 (0.90)
	BK, limed	71	1993–1995	0.41 (0.16)
	B2, control	55	1994	1.34 (0.30)
	B2, gap	55	1994	4.19 (1.28)
	B2, limed gap	55	1994	3.29 (1.98)
Göttinger Wald		63	1993–1995	0.16 (0.002)
Zierenberg	ZB1	23	1991–1992	0.41 (0.12)
	ZB2, control	23	1991–1992	0.39 (0.08)
	ZB2, gap	23	1991–1992	0.27 (0.13)

A literature review by Brumme et al. (1999) indicated that only 5 of 29 forests in the temperate biome where whole-year N_2O emissions measurements were carried out, showed high seasonal emission pattern with annual emission rates of >1 kg N ha^{-1} per year. One key factor responsible for high seasonal emission pattern values was the low air diffusivity causing reducing conditions in soils. At the beech forest at the Solling site, low diffusivity values were observed in the surface organic layer and the mineral soil as compared to the Göttinger Wald site (Fig. 18.3). The broad leaves of beech litter are often tightly packed on the surface of a moder type humus causing long diffusion pathways for oxygen (Fig. 18.4). When air diffusivity of the surface organic layer at the beech site was compared with that on the spruce site (both sites have moder form of humus, but the Solling spruce site had low background emissions), the values were about 70% lower at the beech site (Ball et al. 1997). A litter fall exchange experiment at the Solling site, where litter falls at the beech and spruce stands were collected above the gas chambers and replaced across the sites, revealed the importance of different structure of tree litter for N_2O emissions (Brumme et al. 1999). The N_2O emission decreased at the beech stand and increased at the spruce site during the following 3–4 years of the experiment.

Low air diffusivity in the mineral soil layers caused by temporary water-logged soil is another important factor at the ecosystem level. Permanently water-logged soils (Histosol) are low sources of N_2O while a drained site showed very high losses (Brumme et al. 1999). A study on well-drained *Cambisols* and water-logged *Gleysols* and *Histosols* showed the following ranges in N_2O emission: from 0.3–0.8 kg N ha^{-1} per year (*Cambisols*), 1.4–2.7 kg N ha^{-1} per year (*Gleysols*), and 3.7–4 kg N ha^{-1} per year (*Histosols*) of a Norway spruce stand in Germany (Jungkunst et al. 2004; Lamers et al. 2007).

In a majority (21 out of 29) of forests where whole-year measurements of N_2O emissions were made, low background emission patterns of < 1 kg N ha^{-1} per year with a mean of 0.39 ± 0.27 kg N ha^{-1} per year were reported (Brumme et al. 1999).

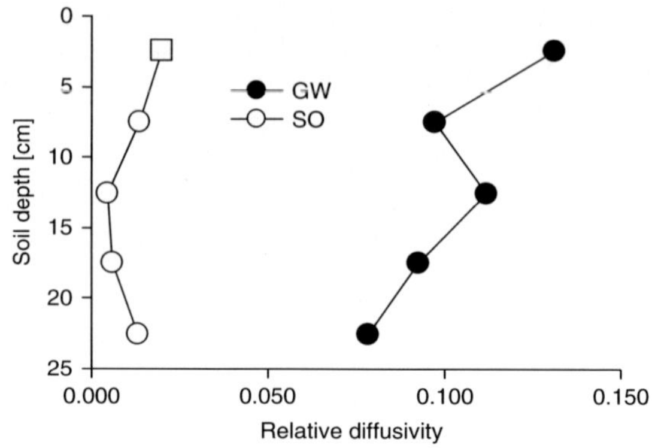

Fig. 18.3 Relative diffusivity (diffusivity in the soil related to diffusivity in air) in different soil depths at the beech sites Göttinger Wald (*GW*) and Solling (*SO*) (Ball et al. 1997) (the *square* for the *SO* site indicates the relative diffusivity in the moder humus)

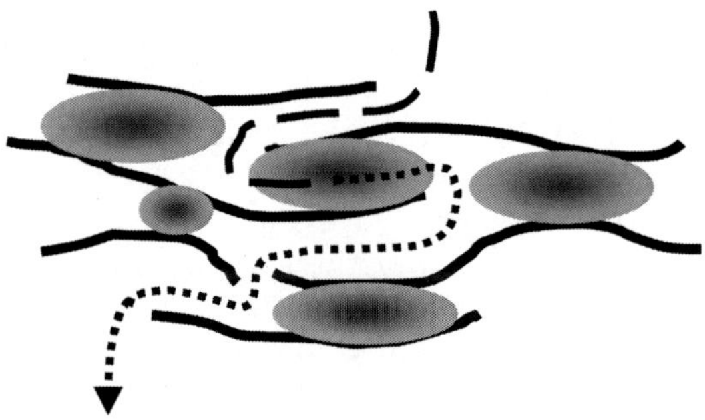

Fig. 18.4 Diffusion pathway of oxygen around beech leaves in the air and water-filled (*ellipsis*) pore spaces of a moder type humus

Similar values (0.34 ± 0.21 kg N ha^{-1} per year, $n = 10$) were calculated for beech forests with background emission pattern of N_2O with values ranging from 0 to 0.9 kg N ha^{-1} per year. Coniferous forests with moder type humus and deciduous forests with mull type humus in most cases show this type of background emission pattern. Mull type humus soils are common for base rich soils like Göttinger Wald or Zierenberg. These soils showed a high earthworm activity (Chap. 7) which created better air diffusivity conditions by preventing litter accumulation on the surface of the mineral soils, and by creating a high proportion of macro pores in deeper soil layers (Fig. 18.3). Most of these sites with background emission pattern of N_2O showed that emissions increased with increasing C content of the surface organic layer (Fig. 18.5). Hence, the thickness of the surface organic layer could serve as a simple indicator for the magnitude of N_2O emissions in deciduous forest ecosystems, and was used to produce a regional-based estimate of N_2O emissions from German forests (Schulte-Bisping et al. 2003). Mean annual emission of 0.32 kg N ha^{-1} per year from German forest soils was calculated after stratifying forests in to seasonal emission type (2.0 kg N ha^{-1} per year) and background emission types (deciduous: 0.37 kg N ha^{-1} per year; conifers: 0.17 kg N ha^{-1} per year). With

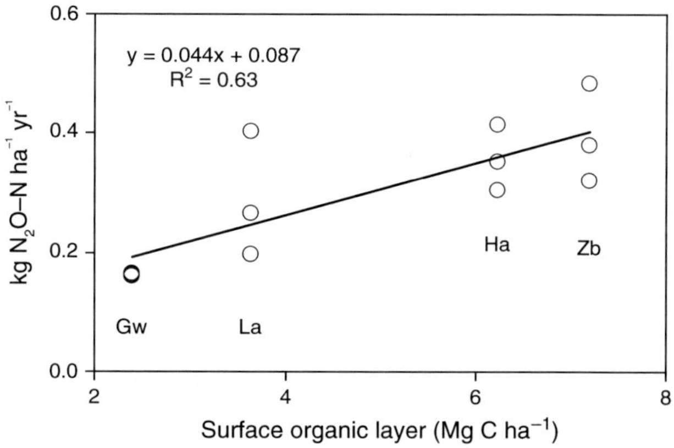

Fig. 18.5 Relationship between background emission pattern of N_2O ($n = 3$) and C-amount in the surface organic layer of four beech forests (*Gw* Göttinger Wald, *Zb* Zierenberg; for *Ha* Harste and *La* Lappwald, from Brumme et al. 1999)

this approach, the emissions from forests in the temperate biome were estimated to be 0.43 kg N ha^{-1} per year (Brumme et al. 2005). However, the results from a process-oriented model (PnET-N-DNDC) by Butterbach-Bahl et al. (2001, 2004), which was validated for a forest site with seasonal emission pattern, has resulted in much higher emission rates for southern Germany and Saxony with values of approximately 2 kg N ha^{-1} per year. This estimate was questioned by Kesik et al. (2005) using a new version of PnET-N-DNDC for calculating the European inventory of N_2O and NO for forest soils. The source strength for German forest soils was considerably lower and amounted to 0.7 kg N ha^{-1} per year.

Field studies with ^{15}N labelled nitrate and ammonium indicated that denitrification was the main source of high seasonal N_2O emissions at Solling (Wolf and Brumme 2002). However, in spring or autumn, heterotrophic nitrification may be involved. In laboratory experiments where high soil moisture conditions (100% WHC) and high nitrate concentrations were maintained, high N_2O emissions were observed for the Göttinger Wald soil. This soil usually had low emissions under field conditions, indicating that the laboratory conditions increased the N_2O emissions up to the level of seasonal emission pattern (Wolf and Brumme 2003). Similar experiments which were carried out with ^{15}N enrichment in nitrate and N_2O indicated that denitrification was the primary process responsible for the formation of N_2O at the Göttinger Wald site. This leads us to hypothesize that high seasonal emissions of N_2O are derived from denitrification, and low background emissions may probably be associated with heterotrophic nitrification. The dominance of denitrification for N_2O formation in forest soils was also suggested by a laboratory study on undisturbed soil cores from 11 forests in Europe (Ambus et al. 2006). In this study, nitrate was the dominant substrate for N_2O with an average contribution of 62% to N_2O formation.

18.6 Relationship Between Temperature and N_2O Emission

Forests with seasonal emission pattern of N_2O showed a high temperature sensitivity for N_2O (Q_{10} value of 9.0; Fig. 18.6, Table 18.2). Low values of soil moisture contents (soil water tensions of >200 hPa) reduced the temperature sensitivity of N_2O emissions (Fig. 18.6) and were excluded from the calculation. High Q_{10} values of N_2O emission of up to 14 have been observed at a forest gap at the Solling site where emissions were not restricted by low soil moisture content (Brumme 1995). A high temperature response (Q_{10} = 6.5) was observed at the beech forest at Höglwald, Germany, a site with seasonal emission pattern of N_2O (Papen and Butterbach-Bahl 1999). At forests with background emission pattern of N_2O, for example at the Göttinger Wald site, N_2O emissions were not affected by soil temperature. However, at the Zierenberg site, a small temperature response of N_2O emissions were observed. Fertilization decreased the temperature sensitivity (Q_{10} of 6.8) as compared to the control plot, and liming drastically reduced the N_2O emissions lowering the temperature response to a value of zero. The literature

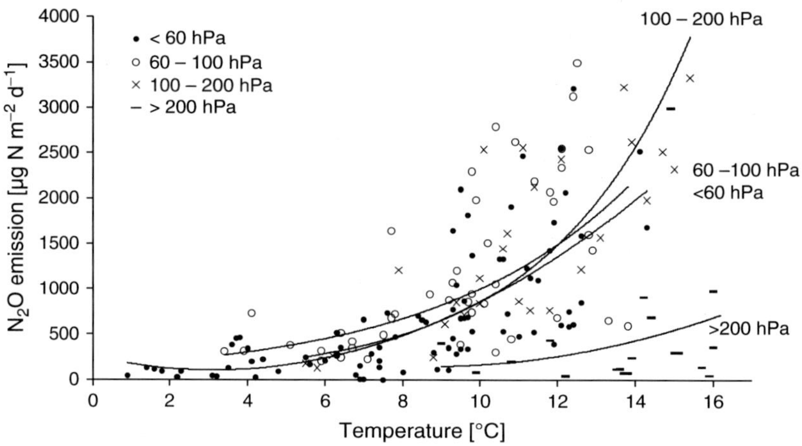

Fig. 18.6 Nitrous oxide emissions with changes in soil temperature (5 cm depth) and soil water tension (hPa) (5 cm below surface organic layer) at the Solling B1 plot (n = 3) during 1991–2000

Table 18.2 Temperature sensitivity expressed as apparent Q_{10} values of N_2O emissions at Göttinger Wald, Zierenberg, and Solling (control B1, limed BK, and fertilized BN) sites

Site	Plot	Treatment	Q_{10}
Solling	B1	Control	9.0
	BK	Limed	0
	BN	Fertilized	6.8
Göttinger Wald			0
Zierenberg			1.5

reviews by Smith (1997) and Granli and Bøckman (1994) showed that high Q_{10} values of N_2O emission have often been reported. The rule of vant't Hoff means that temperature normally doubles biological reaction by a temperature increase of 10°C ($Q_{10} = 2$). Q_{10} values greater than 2–3 indicate that additional to temperature positive feedback mechanisms are involved (Brumme 1995). High temperatures increase the oxygen consumption by micro-organisms, roots, and mycorrhizae and reduce the oxygen concentration in soils under conditions of low oxygen supply. Root detritus and exudates may increase with temperature exerting a positive feedback effect on N_2O emissions. Partially anoxic conditions in soil aggregates usually increase exponentially because of a strong nonlinear increase of anaerobic soil volume in soil aggregates with temperature, and would result in high Q_{10} (Smith 1997) values even under laboratory conditions without the positive feedback mechanisms by roots. We conclude that reliable assessment of the temperature sensitivity of the emission of N_2O is hardly possible since the oxygen availability, the most important regulator for the production of N_2O, changed with temperature. Thus, the values presented for the temperature sensitivity of N_2O emission are apparent Q_{10} values and reflect the temperature response of the whole ecosystem.

18.7 Implications of Forest Management for N_2O Emissions

Thinning and harvesting practices in forests usually reduce root content and may increase soil temperature, soil moisture, nitrate concentrations, and radiation of the surface organic layer (Likens and Bormann 1995; Bauhus and Bartsch 1995), causing an increase in N_2O emissions. Measurements with a full automatic chamber system on a transect from the center of a forest gap (30 m diameter; see Bartsch et al. 2002) to the stand at the Solling site (B2 plot) showed that seasonal N_2O emissions were higher in the gap and existed for a prolonged period in the summer and autumn months of 1991 (1.5 years after clearing; Brumme 1995). The effect of harvesting continued at least up to 1994 (4.5 years after clearing) as indicated in Fig. 18.7. On an annual basis, N_2O emissions increased by 460% in 1991 and by 220% in 1994 in the center of the gap compared to the stand. A temperature effect was not observed because of shading by the surrounding trees. The surface organic layer was not covered by ground vegetation before harvesting and the removal of trees did not effect the growth in 1991 (Bartsch et al. 2002). In 1994, only 20% of the land surface was covered by ground vegetation. Consequently, soil water tension did not exceed 200 hPa and nitrate concentration was several times higher than on the control plot and may have caused the high N_2O emissions in the gap.

In contrast to the Solling site, harvesting did not affect N_2O emission at the Zierenberg site (Table 18.1). The ground vegetation grew back quickly after harvesting (Godt 2002) which probably reduced soil moisture levels and nitrate concentration. Similar results were shown by Dannenmann et al. (2007) in a thinning study in three beech forests on calcareous soils covered with mull type humus. One of three forests in that trial showed increased N_2O emission in the first

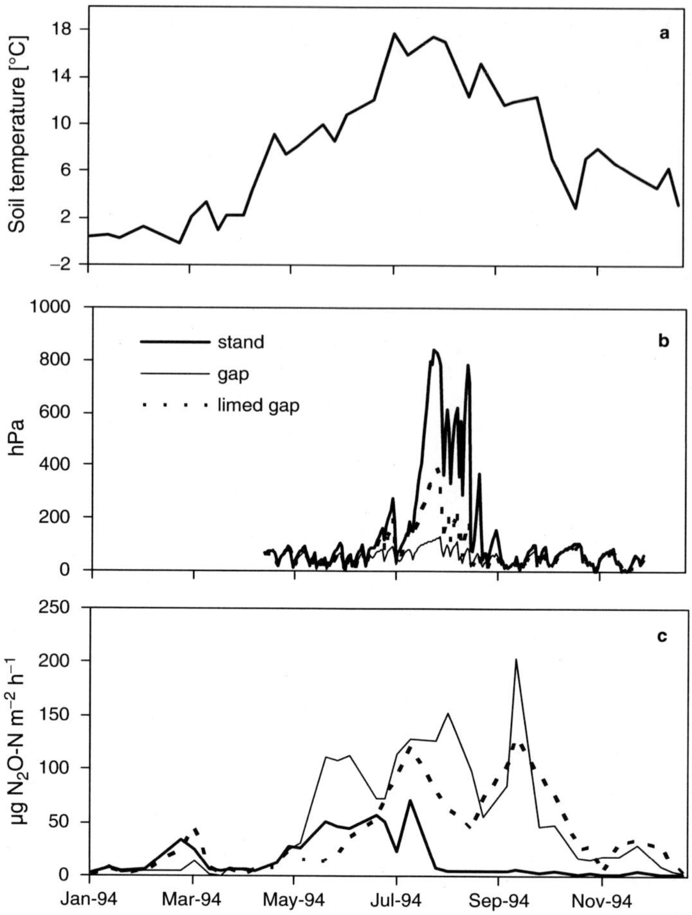

Fig. 18.7 Soil temperature in the OH layer (**a**), soil water tension (hPa) at 10 cm soil depth (**b**), and N_2O emission (μg N m^{-2} h^{-1}) at the Solling control plot B2 and a adjacent limed gap and a control gap (**c**) for January to December 1994

two summers after thinning and declined in the third year, which was accompanied by the development of understorey vegetation.

Nitrate is one of the most important controlling factors affecting N_2O emissions from agricultural soils (Bouwman 1990b) whereas it is of less importance on forest land (Schmidt et al. 1988; Bowden et al. 1991; Mogge et al. 1998; Brumme et al. 1999; Papen and Butterbach-Bahl 1999; Ambus and Robertson 2006). A fertilizer experiment at the Solling site (BN plot) confirmed this result. After fertilization, the mean nitrate concentration was 20 mg N L^{-1} in contrast to 0.7 mg N L^{-1} at the control B1 plot while the emission of N_2O amounted to 2.9 kg N ha^{-1} per year (Fig. 18.8, Table 18.1) and was not significantly higher than observed for the same period at the control plot (2.3 kg N ha^{-1} per year, SD 0.57). The differences between forest and agricultural land are probably related to high nitrification rates

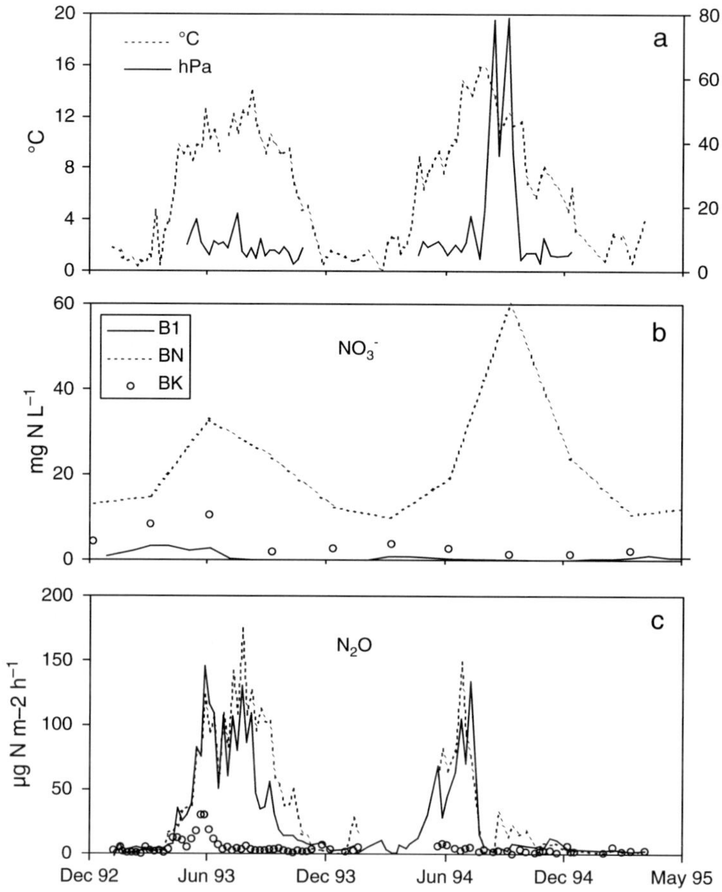

Fig. 18.8 Soil temperature in the H layer and soil water tension at 10 cm soil depth at the Solling control plot B1 (**a**), soil nitrate concentrations at 10 cm depth (**b**) and N_2O emissions (**c**) at the beech site at Solling (*B1*), a fertilized plot (*BN*), and a limed plot (*BK*) at the Solling site for January 1993 to March 1995

after fertilization when compared to soils under forests, and also to more microsites with low oxygen diffusivity (compacted soil) under agricultural conditions.

Liming of acid forest soils is very common in Germany (Meiwes 1994; AFZ 1995) and has been found to reduce the N_2O emissions by 74% after 5 years of applying 30 tons of dolomitic limestone at the BK plot when compared to the control B1 plot at the Solling site (Brumme and Beese 1992). In the 11th and 12th years after liming, the N_2O emissions were still low (83% or 0.41 kg N ha^{-1} per year) (Table 18.1, Fig. 18.8). A reduction of N_2O emissions by liming was also observed in a forest gap at the Solling site where 3 tons of dolomite ha^{-1} was worked into the upper soil after harvesting (Fig. 18.7). In contrast to the unlimed gap where a ground vegetation cover of <5% occurred, the limed gap showed about 90% of the area covered with ground vegetation after 2 years of clear-cut (Bartsch et al. 2002). Dense ground vegetation would lead to a decrease in soil moisture, nitrate, and N_2O emissions during the growing season. Moreover, a high population of earthworms in the mineral soil (Theenhaus and Schaefer 1995) might have contributed to the reduction in N_2O emissions by increasing macropores. Borken and Brumme (1997) reported that 9–82% reduction of N_2O emissions in separate field experiments occurred where lime was applied 5–19 years ago to beech and spruce forests. At the Höglwald site, the effect of a higher diffusivity (Schack-Kirchner and Hildebrand 1998) after the establishment of earthworms in the surface organic layer after liming (Ammer and Makeschin 1994) increased the N_2O emission (Papen and Butterbach-Bahl 1999). In this study, the effect of a higher diffusivity may have been overcompensated by an increase in net nitrification rates.

Soil compaction during harvesting caused a considerable increase in N_2O emissions, with values elevated by up to 40 times of those on the uncompacted soil (Teepe et al. 2004). These changes were caused by a reduction in macropore volume and an increase of the water-filled pore space. N_2O emissions were altered in the trafficked soil and in the adjacent areas. Despite the significant changes in N_2O emission, the cumulative effect on the atmosphere was small with respect to the soil trafficking. The fractional area on the skid trails to total area as calculated by Teepe et al. (2004) was 13%, considering a distance of 20 m between skid trails which is commonly used in Germany. The cumulative values of increase due to compaction were approximately 0.1 kg N_2O–N ha^{-1} per year for the sandy loam and silty clay loam sites and 0.3 kg N_2O–N ha^{-1} per year for the silt site.

18.8 Conclusions

- The N_2O emissions from the acid soil at Solling (B1 plot) was up to 15 times higher than from the less acid soils at Zierenberg (0.41 kg N ha^{-1} per year) and Göttinger Wald (0.16 kg N ha^{-1} per year). The high values of N_2O emissions at the Solling site were mainly related to a strong seasonal pattern of N_2O emissions at that site. These high N_2O emission values at the Solling site were attributed to reduced oxygen diffusion in the surface organic layer (densely packed leaves acted as a diffusion barrier) and in the mineral soil (high bulk density).
- Soil temperature was the main factor determining N_2O emissions at the Solling site except for the periods when soils were dry (soil water tensions >200 hPa) in summer and autumn months. High temperatures caused a seasonal pattern of N_2O emissions, as compared to the commonly observed background patterns.
- Forests with seasonal N_2O emissions (as at the Solling site) responded much more strongly to temperature increase (apparent Q_{10} = 9) than forests with background N_2O emissions such as the Göttinger Wald or Zierenberg sites. The strong response to temperature increase was explained by higher oxygen consumption and increased anaerobic zones in the soil. Forests with a seasonal

emission pattern may therefore increase the global N_2O budget if the temperature increases.

- High spatial variations of N_2O emissions were partly related to the heterogeneous distribution of fine root biomass on the Solling site. High numbers of replicates are required to assess mean N_2O emissions from a forest site.
- Forest clearing may affect N_2O production in forests by restricting oxygen diffusion as it was observed in the moder-like humus soil at the Solling site. High soil moisture and nitrate content in a forest gap increased the period of high N_2O emissions in summer and autumn months and led to 3–5 times higher annual fluxes. Forest harvesting did not affect N_2O emissions at the Zierenberg site. This may be related to the well aerated soil and the low background N_2O emissions at this site, and to the fast development of a dense ground vegetation, reducing nitrate and water content in the soil.
- Liming reduced the N_2O emissions by improving oxygen diffusion in soils through an increase in the faunal activity that improved bioturbation in soils (removing the diffusion barrier in the surface organic layer). Liming increased ground vegetation cover and further reduced N_2O emissions by lowering moisture content in the soil. The recent liming to forests may have long-term effects on the N_2O emissions as is indicated by low values in up to 19-year-old liming plots.
- In order to reduce the N_2O emissions from beech forests, it is important to reduce the seasonal N_2O emissions through adequate soil aeration which may be achieved in many situations by liming and by keeping soil compaction low during forest operations.

References

AFZ (1995) 10 Jahre Waldkalkung. Allgemeine Forst Zeitschrift 17:928–949

Ambus P, Robertson GP (2006) The effect of increased N deposition on nitrous oxide, methane and carbon dioxide fluxes from unmanaged forest and grassland communities in Michigan. Biogeochemistry 79:315–337

Ambus P, Zechmeister-Boltenstern S, Butterbach-Bahl K (2006) Sources of nitrous oxide emitted from European forest soils. Biogeosciences 3:135–145

Ammer S, Makeschin F (1994) Auswirkungen experimenteller saurer Beregnung und Kalkung auf die Regenwurmfauna (*Lumbircidae, Oligochaeta*) und die Humusform in einem Fichtenaltbestand (Höglwaldexperiment). Forstw Cbl 113:70–85

Andreae MO, Schimel DS (1989) Exchange of trace gases between terrestrial ecosystems and the atmosphere. Wiley, New York

Ball BC, Smith KA, Klemedtsson L, Brumme R, Sitaula BK, Hansen S, Prieme A, MacDonald J (1997) The influence of soil gas transport properties on methane oxidation in a selection of northern European soils. J Geophys Res 102:23309–23317

Bartsch N, Bauhus J, Vor T (2002) Effects of group selection and liming on nutrient cyling in an European beech forest on acidic soil. In: Dohrenbusch A, Bartsch N (eds) Forest development. Springer, Berlin, pp 109–142

Bauhus J, Bartsch N (1995) Mechanisms for carbon and nutrient release and retention in beech forest gaps. I. Microclimate, water balance and seepage water chemistry. Plant Soil 168–160:579–584

Borken W, Brumme R (1997) Liming practise in temperate forest ecosystems and the effects on CO_2, N_2O and CH_4 fluxes. Soil Use Manage 13:251–257

Bouwman AF (1990a) Soils and the Greenhouse Effect. Wiley, New York

Bouwman AF (1990b) Exchange of greenhouse gases between terrestrial ecosystems and the atmosphere. In: Bouwman AF (ed) Soils and the greenhouse effect. Wiley, New York, pp 61–127

Bowden RD, Melillo JM, Steudler PA, Aber JD (1991) Effects of nitrogen addition on annual nitrous oxide fluxes from temperate forest soils in the northeastern United States. J Geophys Res 96:9321–9328

Brumme R (1995) Mechanisms for carbon and nutrient release and retention in beech forest gaps. III. Environmental regulation of soil respiration and nitrous oxide emissions along a microclimatic gradient. Plant Soil 168–169:593–600

Brumme R, Beese F (1992) Effects of liming and nitrogen fertilization on emissions of CO_2 and N_2O from a temporate forest. J Geophys Res 97:12 851–12 858

Brumme R, Borken W, Finke S (1999) Hierarchical control on nitrous oxide emission in forest ecosystems. Glob Biochem Cycles 13:1137–1148

Brumme R, Verchot LV, Martikainen PJ, Potter CS (2005) Contribution of trace gases nitrous oxide (N_2O) and methane (CH_4) to the atmospheric warming balance of forest biomes. In: Griffiths H, Jarvis PG (eds) The carbon balance of forest biomes. Taylor and Francis Group, Oxon, New York, pp 293–318

Butterbach-Bahl K, Stange F, Papen H, Li CS (2001) Regional inventory of nitric oxide and nitrous oxide emissions for forest soils of southeast Germany using the biogeochemical model PnET-N-DNDC. J Geophys Res Atmos 106:34155–34166

Butterbach-Bahl K, Kesik M, Miehle P, Papen H, Li C (2004) Quantifying the regional source strength of N-trace gases across agricultural and forest ecosystems with process based models. Plant Soil 260:311–329

Crutzen PJ (1970) The influence of nitrogen oxide on the atmospheric ozone content. Q J R Meteorol Soc 96:320–325

Dannenmann M, Gasche R, Ledebuhr A, Holst T, Mayer H, Papen H (2007) The effect of forest management on trace gas exchange at the pedosphere-atmosphere interface in beech (*Fagus sylvatica* L.) forests stocking on calcareous soils. Eur J For Res 126:331–346

Flessa H, Ruser R, Schilling R, Loftfield N, Munch JC, Kaiser EA, Beese F (2002) N_2O and CH_4 fluxes in potato fields: automated measurement, management effects and temporal variation. Geoderma 105:307–325

Godt J (2002) Canopy disintegration and effects on element budgets in a nitrogen-saturated beech stand. In: Dohrenbusch A, Bartsch N (eds) Forest development. Springer, Berlin, pp 143–165

Granli T, Bøckman OC (1994) Nitrous oxide from agriculture. Norw J Agric Sci 12

IPCC (2001) Intergovernmental panel on climate change. Climate change 2001: mitigation. Contribution of working group III to the third assessment report. Cambridge University Press, Cambridge

Jungkunst HF, Fiedler S, Stahr K (2004) N_2O emissions of a mature Norway spruce (*Picea abies*) stand in the Black Forest (southwest Germany) as differentiated by the soil pattern. J Geophys Res Atmos 109:D07302

Kesik M, Ambus P, Baritz R, Bruggemann NB, Butterbach-Bahl K, Damm M, Duyzer J, Horvath L, Kiese R, Kitzler B, Leip A, Li C, Pihlatie M, Pilegaard K, Seufert G, Simpson D, Skiba U, Smiatek G, Vesala T, Zechmeister-Boltenstern S (2005) Inventories of N_2O and NO emissions from European forest soils. Biogeosciences 2:353–375

Lamers M, Ingwersen J, Streck T (2007) Nitrous oxide emissions from mineral and organic soils of a Norway spruce stand in South-West Germany. Atmos Environ 41:1681–1688

Lashof DA, Ahuja DR (1990) Relative contribution of greenhouse gas emissions to global warming. Nature 344:529–531

Likens GE, Bormann H (1995) Biogeochemistry of a forested ecosystem. Springer, New York

Loftfield N, Brumme R, Beese F (1992) An automated high resolution chamber method for monitoring N_2O and CO_2 flux from forest soil. Soil Sci Soc Am J 56:1147–1150

Loftfield N, Flessa H, Augustin J, Beese F (1997) Automated gas chromatographic system for rapid analysis of the atmospheric trace gases methane, carbon dioxide, and nitrous oxide. J Environ Qual 26:560–564
Matthias AD, Blackmer AM, Bremner JM (1980) A simple chamber technique for field measurement of emission of nitrous oxide from soil. J Environ Qual 9:251–256
Meiwes KJ (1994) Kalkungen. In: Matschullat J, Heinrichs H, Schneider J, Ulrich B (eds) Gefahr für Ökosysteme und Wasserqualität. Springer, Berlin, pp 415–431
Mogge B, Kaiser EA, Munch JC (1998) Nitrous oxide emissions and denitrification N-losses from forest soils in the Bornhöved lake region (Northern Germany). Soil Biol Biochem 30:703–710
Papen H, Butterbach-Bahl K (1999) A 3-year continuous record of nitrogen trace gas fluxes from untreated and limed soil of a N-saturated spruce and beech forest ecosystem in Germany. J Geophys Res 104:18487–18503
Schack-Kirchner H, Hildebrand EE (1998) Changes in soil structure and aeration due to liming and acid irrigation. Plant Soil 199:167–176
Schmidt J, Seiler W, Conrad R (1988) Emission of nitrous oxide from temperate forest soils into the atmosphere. J Atmos Chem 1:95–115
Schulte-Bisping H, Brumme R, Priesack E (2003) Nitrous oxide emission inventory of German forest soils. J Geophys Res 108(D4):4132
Smith KA (1997) The potential for feedback effects induced by global warming on emissions of nitrous oxide by soils. Glob Chang Biol 3:327–338
Smith K, Dobbie K (2001) The impact of sampling frequency and sampling times on chamber-based measurements of N_2O emissions from fertilized soils. Glob Chang Biol 7:933–945
Teepe R, Brumme R, Beese F, Ludwig B (2004) Nitrous oxide emission and methane consumption following compaction of forest soils. Soil Sci Soc Am J 68:605–611
Theenhaus A, Schaefer M (1995) The effects of clear-cutting and liming on the soil macrofauna of a beech forest. For Ecol Manage 77:35–51
Wolf I, Brumme R (2002) Contribution of nitrification and denitrification sources for seasonal N_2O emissions in an acid German forest soil. Soil Biol Biochem 34:741–744
Wolf I, Brumme R (2003) Dinitrogen and nitrous oxide formation in beech forest floor and mineral soils. Soil Sci Soc Am J 67:1862–1868
Zechmeister-Boltenstern S, Hahn M, Meger S, Jandl R (2002) Nitrous oxide emissions and nitrate leaching in relation to microbial biomass dynamics in a beech forest soil. Soil Biol Biochem 34:823–832

Chapter 19
Methane Uptake by Temperate Forest Soils

W. Borken and R. Brumme

19.1 Introduction

Net uptake of atmospheric methane by methanotrophic bacteria in forest soils is generally less than 20 kg CH_4 ha^{-1} per year and therefore has almost no impact on the carbon budget of forest ecosystems. However, terrestrial soils are the most important biological sink for atmospheric methane, consuming between 20 and 45 Tg CH_4 per year (Smith et al. 2000; Dutaur and Verchot 2007). The methane concentration in the atmosphere increased from about 700 to 1,774 ppb (by volume) during the last 150 years, though the concentration has possibly stabilised during the past decade (IPCC 2007). Methane currently contributes approximately 18% to the anthropogenic greenhouse effect and has a global warming potential 25 times higher than that of CO_2 based on a time horizon of 100 years (IPCC 2007).

Human activities have reduced the soil sink for atmospheric methane by converting natural forests and grasslands to coniferous plantations, agricultural and urban land (Ojima et al. 1993; Dobbie et al. 1996; Smith et al. 2000; Borken et al. 2003; Borken and Beese 2006) and by causing acid deposition in temperate forests (Brumme and Borken 1999). Undisturbed forest soils generally consume much higher rates of atmospheric methane than agricultural and urban soils, although the methane uptake rates may vary considerably within and among forests over all climatic regions. Comparisons of adjacent forest and agricultural soils showed that converting forest land to agricultural use reduced methane uptake rates by two-thirds in both temperate and tropical regions (Smith et al. 2000). The mechanisms for long-term reduction in methane uptake are not completely understood. Disturbances of the soil structure as well as changes in vegetation, micro-climate, and nutrient status may have diminished the population of methane-utilising bacteria in several ways after land conversion to agricultural use. It is believed that these still unidentified bacteria have an extremely slow growth rate due to the small atmospheric methane concentration. Methanotrophs have a high affinity for methane as indicated by low K_m values and low threshold concentrations (Bender and Conrad 1993). Smith et al. (2000) pointed out that it probably takes more than 200 years for

R. Brumme and P.K. Khanna (eds.), *Functioning and Management of European Beech Ecosystems*, Ecological Studies 208,
DOI: 10.1007/978-3-642-00340-0_19,

methane utilising bacteria to recover after reversion of agricultural land to forest or grassland.

Since the industrial revolution in the nineteenth century, there have been anthropogenic changes in temperate forests of Central Europe and North America, mainly in association with high atmospheric depositions and intensive forest management practices (Puhe and Ulrich 2001). Soil acidification, high nitrogen loads, uniform plantations of foreign tree species and liming may all have contributed to an altered methane sink in forest soils. To counteract soil acidification, forest soils have been limed and SO_2 emissions have been greatly decreased since the 1990s. However, the nitrogen input by atmospheric deposition from fuel combustion and large-scale animal husbandry is still very high and may acidify forest soils. Various studies have demonstrated a negative effect of N addition on CH_4 uptake in forest soils (e.g. Steudler et al. 1989; Sitaula et al. 1995; Gulledge et al. 2004) and some authors reported a negative effect of soil acidification on CH_4 uptake (Borken and Brumme 1997; Brumme and Borken 1999).

In this study, we focus on the effects of soil pH, nitrogen addition, liming, compaction and forest harvesting on methane uptake rates at Solling and Göttinger Wald, two forest soils of different acidity but similar atmospheric N-inputs (for detailed description of the sites see Chap. 1). We measured methane uptake rates using static chambers (Chap. 17) in control plots over a 2-year period from March 1993 to April 1995. At the Solling beech forest methane uptake was measured in limed and N-fertilised plots (for detailed description of soils, see Chap. 3). The effect of nitrogen addition on methane uptake was investigated at Göttinger Wald from April 1994 to April 1995. Nitrogen compounds, i.e. 30 kg N ha^{-1} of KNO_3, $NaNO_3$, NH_4Cl, $(NH_4)_2SO_4$, NH_4NO_3, NH_4-acetate or urea, were added four times a year. Another plot received 120 kg N ha^{-1} as $(NH_4)_2SO_4$ once per year. An automated chamber system (Brumme and Beese 1995) was used to investigate the short-term effect of nitrogen addition in an adjacent spruce forest at Solling from August 1993 to December 1994. The spruce soil was treated four times with 30 kg N ha^{-1}, added as $NaNO_3$, NH_4Cl, $(NH_4)_2SO_4$, NH_4NO_3 or urea. Additionally, results from other untreated beech and spruce sites were included in the analysis of site factors governing CH_4 uptake (see Brumme and Borken 1999 for detailed site description). The effects of forest harvesting were studied in gaps at the Solling beech site, which were established in autumn 1989 (Bauhus and Bartsch 1995; Bartsch et al. 2002). Two gaps were limed with 3 Mg ha^{-1} of fine dolomitic limestone. Methane uptake was measured in the unlimed gap, the limed gap and the control plot with three closed chambers (0.25 m^2) from January to December 1994.

19.2 Moisture and Temperature Effects

Figures 19.1 and 19.2 show the temporal dynamics of CH_4 uptake rates for control plots of the beech forests at Solling and Göttinger Wald from April 1993 to April 1995. CH_4 uptake rates were extremely low at Solling compared to Göttinger

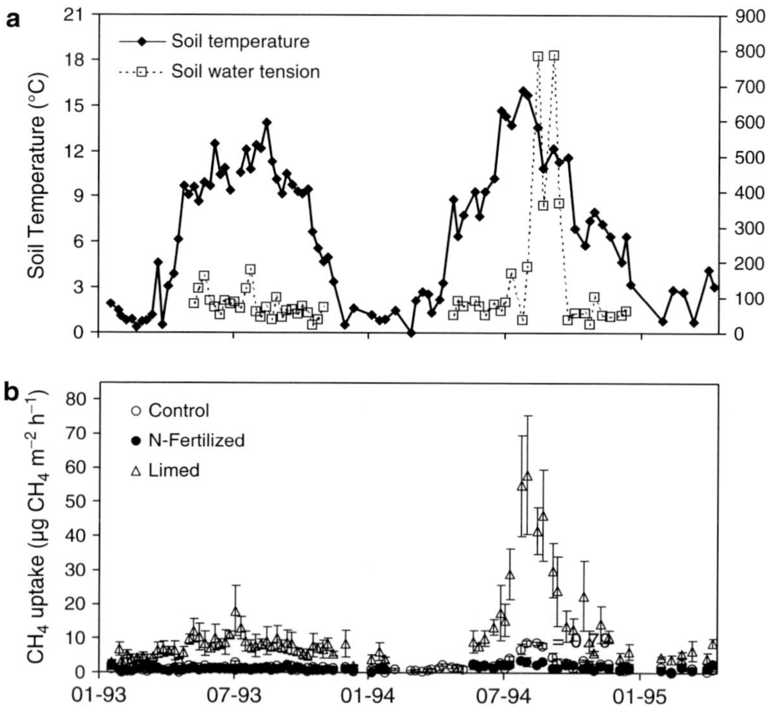

Fig. 19.1 (**a**) Soil temperature, soil water tension in the control plot and (**b**) CH_4 uptake ($\pm$ standard error of the mean) in the control, N-fertilised and limed plots of the beech forest at the Solling from January 1993 to April 1995

Wald and varied from 0 to 8.8 μg CH_4 m^{-2} h^{-1} (Solling) and from 9.1 to 72 μg CH_4 m^{-2} h^{-1} (Göttinger Wald) over the entire period. The highest rates were measured in the summer of 1994 when the top mineral soil was relatively dry and soil water tension increased to >800 hPa. In contrast, 1993 was a rather wet year and soil water tension did not exceed 200 hPa during the growing season. This clearly demonstrates the importance of soil moisture on temporal variation in CH_4 uptake. Soil moisture controls the mass flow of air and diffusion of atmospheric CH_4 into the soil by changing the air-filled pore space of soils (Ball et al. 1997). In poorly drained forest soils, anaerobic production may exceed the consumption of CH_4 and turn the soil into a net CH_4 source under wet conditions (Fiedler et al. 2005).

Diffusion of CH_4 in water is 10^4 times slower than in air (Marrero and Mason 1972). Consequently, soil water may affect the temporal dynamics as well as spatial variation in CH_4 uptake at the stand and regional scales (Castro et al. 1995; Potter et al. 1996; Priemé and Christensen 1997). Future changes in rainfall amount and distribution could alter the soil sink of temperate forest for atmospheric CH_4. The simulation of extended summer droughts by exclusion of throughfall increased CH_4 uptake rates by two to three times in a spruce forest at Solling (Borken et al. 2000b).

Based on the Arrhenius function, apparent Q_{10} values of 2.8 and 2.1 have been calculated for CH_4 uptake in the beech forests at Solling and Göttinger Wald.

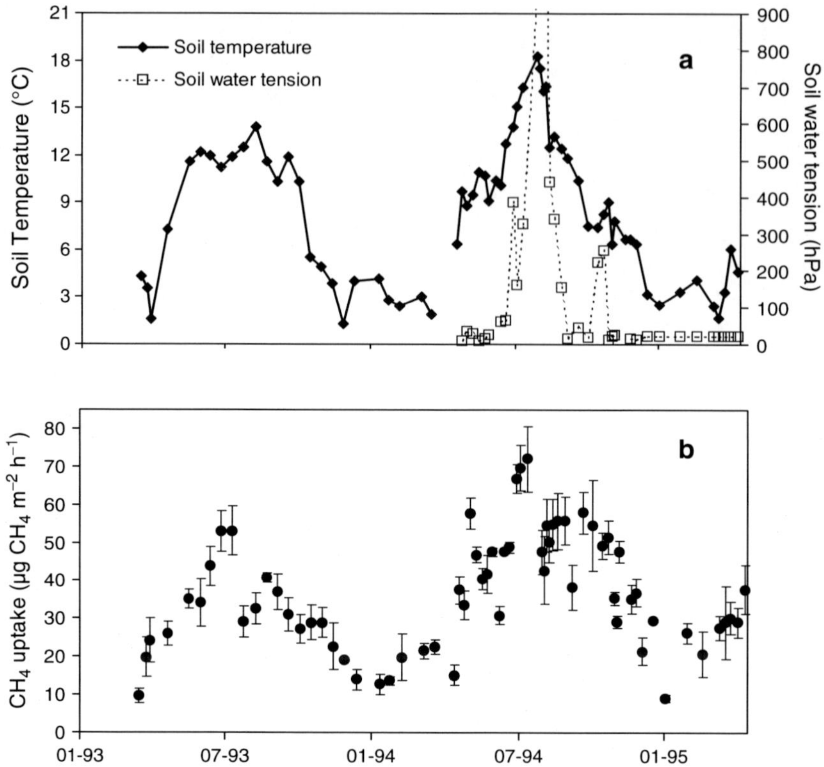

Fig. 19.2 (**a**) Soil temperature, soil water tension and (**b**) CH_4 uptake (± standard error of the mean) in the beech forest at the Göttinger Wald from April 1993 to April 1995

Contrary to our results, some studies have shown that soil temperature has little effect on CH_4 uptake, as indicated by small apparent Q_{10} values in the range of 1.1–1.5 (Born et al. 1990; King and Adamsen 1992; Priemé and Christensen 1997; Borken et al. 2000b). Observed differences in Q_{10} values within a year have been explained by growth of methane oxidising bacteria (Crill 1991). Crill found Q_{10} values of 2.0 for the period from April to July and of 1.2 for August to December and suggested that the growth of methane oxidisers were responsible for the higher temperature sensitivity from April to July. Another reason for the differences in Q_{10} values is that soil moisture and soil temperature may be confounding factors during the growing season, when evaporation and transpiration increase with increasing temperature (Borken et al. 2000b). Thus, the relatively high Q_{10} at our sites could be related to changes in soil moisture during the growing season.

Some authors have pointed out that high Q_{10} values from field and laboratory studies would overestimate the intrinsic temperature dependency of CH_4 uptake. One reason is that the atmospheric CH_4 concentration of 1.7 µl l^{-1} is too low for saturation of the key enzyme methane monooxygenase as indicated by a low half-saturation constant K_m (Bender and Conrad 1993). Thus, the enzymatic activity is mainly controlled by CH_4 concentration and to a lesser extent by temperature. A typical parabolic temperature function was found when pure cultures of *M. rubra*, a CH_4 oxidising species, were incubated at a CH_4 concentration of 10,000 μl l^{-1} while only a weak temperature dependency was observed at a CH_4 concentration of 100 μl l^{-1} (King and Adamsen 1992). Similarly, Whalen et al. (1990) reported a parabolic temperature response for a landfill cover soil at a CH_4 concentration of 9,500 μl l^{-1} whereby the optimum temperature for CH_4 uptake was 31°C and the average Q_{10} was 1.9 for a temperature range between 5 and 26°C.

King and Adamsen (1992) argue that the diffusion of atmospheric CH_4 into the mineral soil is also a temperature-dependent process whereby the temperature dependencies of CH_4 diffusion coefficients in soil air and soil water are different. They calculated increases of 21% for diffusion of CH_4 in soil air and of 150% in soil water, when temperature increased from 0 to 30°C. Consequently, the apparent temperature dependency of CH_4 uptake should be higher in wet soils than in dry soils.

This discussion suggests that the impact of future global warming on the soil CH_4 sink is still uncertain. The direct effect of a temperature increase on CH_4 uptake is probably negligible, whereas the indirect effects such as interaction between temperature and soil moisture may be more important and should be considered for predicting the soil CH_4 sink.

19.3 Site Factors Determine Methane Uptake

Atmospheric CH_4 uptake occurs in all terrestrial soils with aerobic conditions although site-to-site variation can be very high (Smith et al. 2000). Physical processes involved in the regulation of CH_4 uptake such as gas diffusivity in soils are mostly understood, but little is known about the effects of soil chemical parameters on biological and physical soil properties. Our study on CH_4 uptake at three spruce and four beech forests indicates that soil acidity is an important factor for site variation in CH_4 uptake from 0.1 to 2.5 kg CH_4 ha^{-1} per year (Fig. 19.3). Several authors reported high methane uptake rates in soils with low pH (Crill 1991; Borken et al. 2006), indicating that other factors may obscure the intrinsic effect of soil pH on methanotrophs. Saari et al. (2004) found pH optima for CH_4 oxidation between 4 and 7.5 for different forest soils and some optima were above the natural pH. Negative effects of soil acidification on soil physical parameters and bioturbation have the potential to reduce CH_4 uptake in forest soils.

Of the seven forests studied in northern Germany, those at Solling and Göttinger Wald have the most different soils regarding their ability to oxidise CH_4 (Fig. 19.4). Göttinger Wald had the highest CH_4 uptake rate of 2.5 kg CH_4 ha^{-1} per year and Solling had the lowest rate of 0.1 kg CH_4 ha^{-1} per year (Table 19.1). An important parameter that affects soil porosity is the soil pH. The soil at Göttinger Wald has a higher soil pH than at Solling, a 2.4-times higher air-filled pore space at field capacity and six times higher relative gas diffusivity in the top 10 cm of the mineral

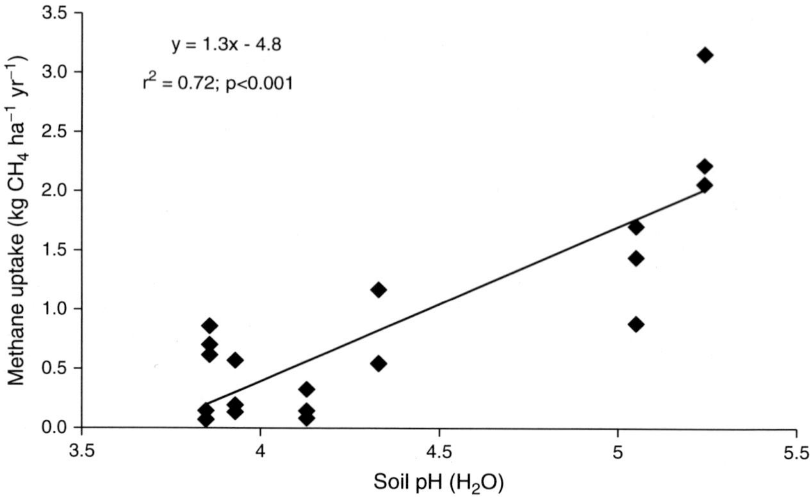

Fig. 19.3 Relationship between annual CH_4 uptake rates and soil pH at 0–10 cm depth from different beech and spruce forests. For site description, see Brumme and Borken (1999)

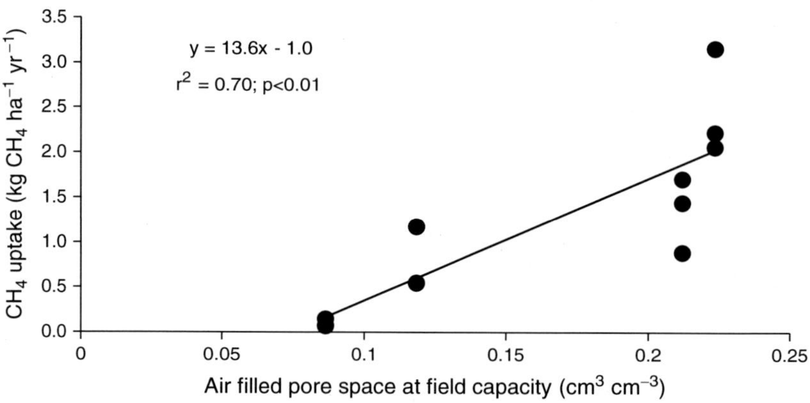

Fig. 19.4 Correlation between annual CH_4 uptake rates and air filled pore space at field capacity of the top 10 cm mineral soil from four different beech forests in Germany. For site description, see Brumme and Borken (1999)

soil (Ball et al. 1997). The strong linear correlation between annual CH_4 uptake rates and air-filled pore space at field capacity ($r^2 = 0.72$) at four different beech forests illustrates the impact of soil porosity on CH_4 uptake (Fig. 19.4). Air-filled soil pores at field capacity have a diameter of >50 µm and are often created by soil-burrowing earthworms (Edwards and Bohlen 1996) or may result from decomposed roots. However, most of the indigenous earthworms are not adapted to acid soils. The relatively low earthworm population at Solling with 19 individuals m^{-2} compared to 205 individuals m^{-2} at Göttinger Wald (Schaefer and Schauermann

Table 19.1 Mean annual CH_4 uptake rates (± standard deviation) and some soil properties at 0–10 cm depth of different beech stands in northern Germany

Study site	Treatment	Soil pH	Organic C-concentration ($g\ g^{-1}$)	CH_4 uptake (kg CH_4 ha^{-1} per year)
Solling	Control	3.9	0.04	$0.17 \pm 0.06^{a,b}$
	Limed	5.3	0.05	0.87 ± 0.39^{c}
	$(NH_4)_2SO_4$			$0.11 \pm 0.06^{a,b}$
	Control gap			0.92 ± 0.10^{c}
	Gap			0.14 ± 0.03^{a}
	Limed gap			$0.71 \pm 0.51^{b,c}$
Göttinger Wald	Control	5.2	0.05	2.48 ± 0.59^{d}
Lappwald	Control	5.1	0.05	1.34 ± 0.42^{c}
Harste	Control	4.3	0.02	0.75 ± 0.36^{c}

At Solling, the control, limed and N-fertilised plots were measured over 2 years from 1993 to 1995, the 'gap' plots from January to December 1994 and all other plots from April 1993 to April 1994. Differences in CH_4 uptake between the sites and the treatments were tested for same time periods using one-way ANOVA. Different letters indicate significant differences at $\alpha = 0.05$

1990) are related to the low soil pH at Solling. The soil at Göttinger Wald is inhabited by endogeic and anecic earthworm species. Both ecological groups loosen the mineral soil by burrowing activity, but most important for gas diffusivity seems to be *L. terrestris*, an anecic earthworm species. It forms large vertical borrows which are open to the atmosphere and allow rapid exchange of gases between the atmosphere and the soil. In contrast, only epigeic earthworm species were found at Solling (Schaefer and Schauermann 1990). These species live in the O horizon and have little or no effect on the porosity of the mineral soil. High and low densities of endogeic and anecic earthworms were also found respectively at the beech stands Lappwald with high (5.1) and Harste with low (4.3) soil pH (Fig. 19.3 and Table 19.1). Therefore, our results suggest that a high activity of endogeic and anecic earthworms increase soil CH_4 uptake by creation of macro-pores.

We speculate that the invasion of endogeic and anecic earthworm species could increase CH_4 uptake of forest soils in the long run. However, the inoculation with *L. terrestris* in the subsoil of B1 resulted in a reduction in net CH_4 uptake (Borken et al. 2000a); the CH_4 uptake of the subsoil was extremely low in the experiment and the density of *L. terrestris* (113 individuals m^{-2}) was rather high. CH_4 was obviously produced in anaerobic micro-sites of the gut and/or the fresh faeces of *L. terrestris*. Despite the observed CH_4 production, endogeic and anecic earthworms could increase CH_4 uptake of acid forest soils in the long-term by loosening the mineral soil, provided that the soil pH has increased to the adaptation range of the earthworms.

Because CH_4 uptake is positively correlated with soil pH (Fig. 19.3), high acid depositions could have reduced the CH_4 uptake by many forests of Central and North Europe by triggering changes in the biological and physical status of the soils during recent decades. Phylogenetic studies suggest an impact of soil pH on the community structure of methanotrophs in forest soils (Knief et al. 2003; Kolb et al. 2005). Whether these different methanotrophic communities have different affinities for CH_4 oxidation is still an object of actual research.

In German forests, soils with pH 3–4 have decreased by 0.4–0.6 pH units, and those with pH 5.5–6 by 2 pH units within the last 20–30 years (Veerhoff et al. 1996). A reduction in soil pH from 5.5 to 4.5 would decrease the uptake rate by 1.3 kg CH_4 ha^{-1} per year (Brumme and Borken 1999). Our results are based on soils with high clay and silt contents but we did not include any sandy soils, which may not have changed their physical properties by soil acidification.

Soil acidity may have also affected CH_4 uptake through accumulation of litter in the O horizon following the reduction or disappearance of anecic earthworms. Anecic earthworms prevent the accumulation of organic matter in the O horizon by incorporating litter into the mineral soil. O horizons barely consume atmospheric CH_4 although CH_4 supply is not restricted by diffusion compared to the mineral soil (Borken et al. 2006; Saari et al. 1998). We found a strong negative exponential relationship (r^2 = 0.92) between annual CH_4 uptake rates and thickness of O horizons of four beech forest soils (Fig. 19.5). The increase in CH_4 uptake rates by 25–171% after removal of the O horizon from nine different forest soils indicate that the O horizon may act as a diffusion barrier for atmospheric CH_4 at high water contents (Borken and Brumme 1997). Moreover, the laminar structure of beech litter influences the resistance to CH_4 transport. Repeated addition of beech litter from the beech plot at Solling to the O horizon of an adjacent spruce stand significantly reduced the annual CH_4 uptake rate (Brumme and Borken 1999). An opposite effect was found when beech litter was periodically replaced by spruce litter in the beech stand. CH_4 uptake was about twice as high as in the beech control plot during the third year of treatment. This experiment demonstrated that the

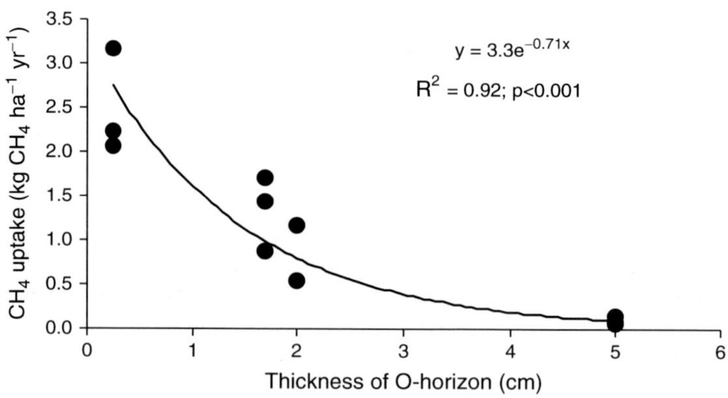

Fig. 19.5 Relationship between annual CH_4 uptake rates and thickness of O horizons from four different beech forests in Germany. The CH_4 uptake on all sites were measured weekly to biweekly from April 1993 to April 1994. For site description, see Brumme and Borken (1999)

morphological structure of litter affects the transport of atmospheric CH_4 into the mineral soil.

The direct effects of acid rain on CH_4 uptake observed in laboratory studies are contradictory. Increasing volumes and concentrations of added acid to soils from a mixed coniferous-hardwood forest decreased CH_4 uptake, whereas nitric acid was more inhibitory than sulphuric acid at equivalent soil pH (Benstead and King 2001). The optimum CH_4 uptake was between pH 4.8 and 6.0, and strong reductions were found when soil pH decreased to <pH 4.4 or increased to >pH 6.0. Benstead and King (2001) treated the soil within 3 days with 1.28 mmol H^+ cm^{-2}, which is 320-times higher than the annual H^+-input by precipitation in the study area. This strong acid application might have killed methanotrophic bacteria and therefore did not simulate field conditions. In contrast, CH_4 uptake of a spruce forest soil increased when acidic water with pH = 3 was applied, whereas water with pH 4.0 and 5.5 had no effect on CH_4 uptake (Sitaula et al. 1995). In another study, high inputs of sulphuric acid stimulated CH_4 uptake in a temperate deciduous woodland, while those of nitric acid had no effect (Bradford et al. 2001). These authors concluded that H_2SO_4 could increase the activity or population of methanotrophs. A 10-year reduction in sulphuric and nitric acid in rainwater to pre-industrial concentrations had no effect on CH_4 uptake of a spruce forest soil at Solling (Borken et al. 2002).

19.4 Effects of Liming and Forest Harvesting

In order to decrease soil acidity, surface applications of 3–4 tons ha^{-1} of lime have been employed in many forests of Germany since the mid–1980s. These small amounts of lime increase the pH in the O horizons while the pH of mineral soils mostly remains unaffected. A return of strongly acidic forest soils to the pre-industrial level requires lime applications of approximately 30 tons ha^{-1} (Beese and Meiwes 1995). Little is known about the effects of lime applications on CH_4 uptake in forest soils, but liming could influence CH_4 uptake by changes in physical, chemical and biological soil properties.

We studied the long-term effect of liming on CH_4 uptake in four different forest stands, which previously received 6–43 tons lime ha^{-1} (Borken and Brumme 1997). In three stands, mean CH_4 uptake rates increased by 26–560% compared with the control plots; the exception was the spruce stand at Solling treated with 6 tons lime ha^{-1}, which showed a decrease in CH_4 uptake of 24%. CH_4 uptake in the limed beech stand at Solling significantly increased from 0.11 to 0.87 kg CH_4 ha^{-1} per year (Fig. 19.1 and Table 19.1). We attributed the increase in the three stands to changes in the morphology of the O horizon by increased density of epigeic earthworms. The earthworms have probably diminished the diffusion resistance for atmospheric CH_4 by loosening the O horizons. Nevertheless, the correlation between mineral soil pH and annual CH_4 uptake rates including untreated and limed soils was weak ($r^2 = 0.40$).

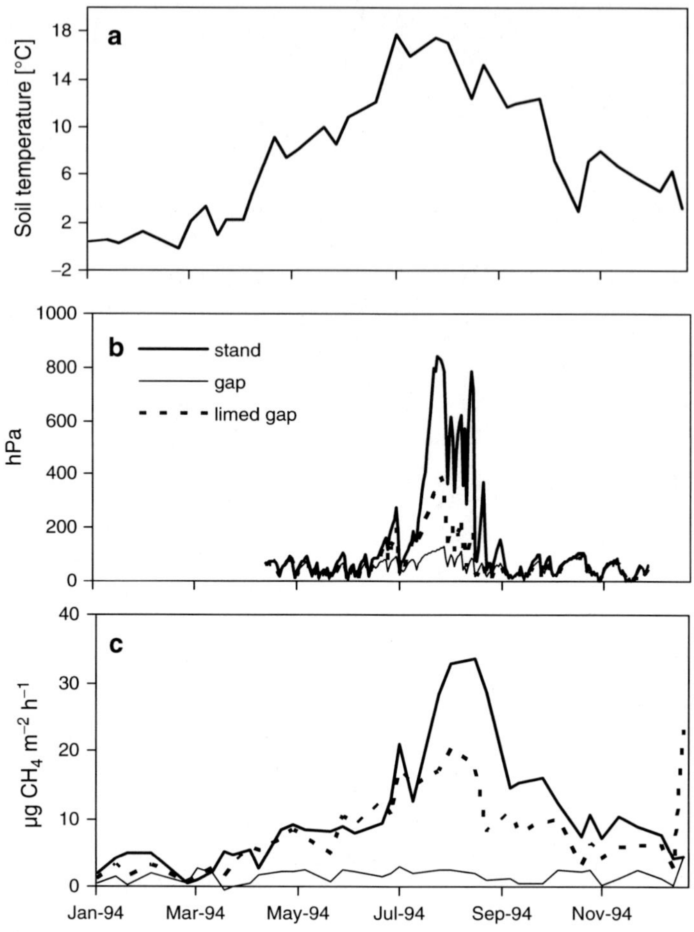

Fig. 19.6 (**a**) Soil temperature in the Oa horizon, (**b**) soil water tension (hPa) at 10 cm soil depth, and (**c**) CH_4 uptake in a control, limed gap and control gap plot of a beech stand at Solling from January to December 1994

Contradicting results were also found in the literature. An insignificant reduction in CH_4 uptake was observed in a hardwood forest 2 years after liming using 10 tons $CaCO_3$ ha^{-1} (Yavitt et al. 1993). A small reduction in CH_4 uptake was observed in a limed spruce stand (4 tons dolomite ha^{-1}), which was attributed to elevated N-mineralisation in the O horizon (Butterbach-Bahl and Papen 2002). In accordance to our results, Kasimir-Klemedtsson and Klemedtsson (1997) found an increase in CH_4 uptake of 14% by liming of a Norway spruce stand. Although the direct and indirect effects of acid rain on CH_4 uptake are not clear, our results indicate that large lime additions keep the potential of acid forest soils to consume atmospheric CH_4 constant.

Forest harvesting may change CH_4 uptake by increased soil temperature and moisture or by increased nitrification rates as a result of a failing root uptake. Temperature has not been found to be an important factor for the uptake of CH_4 in soils (Sect. 19.2) and did not increase in the centre of a gap (30 m diameter, established in autumn 1989 at the Solling site (see Chap. 22). Methane uptake in the gap, measured with closed chambers ($n = 3$) from January to December 1994, never exceeded 5 μg CH_4 m^{-2} h^{-1} in contrast to the stand where rates up to 34 μg CH_4 m^{-2} h^{-1} were measured in August at maximum soil water tension (Fig. 19.6). Annual uptake rates significantly declined from 0.92 kg CH_4 ha^{-1} per year in the stand to 0.14 kg CH_4 ha^{-1} per year in the centre of the gap (Table 19.1). Nitrate losses increased after harvesting if compared to the stand (Bartsch et al. 2002) and might be involved in the reduction of CH_4 oxidation in the soil (see Sect. 19.5). However, probably more important is the higher moisture content in the soil (Fig. 19.6b), which controls the diffusion of atmospheric CH_4 into the soil. Application of lime to improve forest regeneration within another gap had a positive effect on CH_4 uptake (Fig. 19.6 and Table 19.1). The increase in CH_4 uptake in the limed gap is likely a result of reduced soil moisture following water consumption by the dense ground vegetation. In agreement, Dannenmann et al. (2007) reported reduced CH_4 uptake by increasing soil moisture in a beech forest after thinning.

Soil compaction during soil harvesting caused a reduction in macro-pore volume and an increase of the water-filled pore space and reduced the CH_4 consumption at skid trails established by applying two passes with a forwarder in three beech stands (Teepe et al. 2004). The extent of reduction in the cumulative value of CH_4 oxidation during the growing season at different sites was −92% (sandy loam) and −77% (silt). A silty clay loam became a source of CH_4 when water-filled pore space increased to approximately 100%.

19.5 Effects of Nitrogen Input

High atmospheric inputs of inorganic nitrogen are considered to substantially reduce the soil CH_4 sink in different terrestrial ecosystems. A decrease in CH_4 uptake due to ammonium or nitrate fertilisation was first reported for a mixed hardwood and a pine forest in Massachusetts, USA (Steudler et al. 1989). Nitrogen fertilisers may reduce soil CH_4 oxidation due to competitive inhibition of methane monooxygenase by NH_3 or NH_4^+ (Dunfield and Knowles 1995). The similar molecular size of CH_4 and NH_4^+ permit both compounds to compete for enzyme binding, whereby the affinity of methane monooxygenase for CH_4 is higher than for NH_4^+. Recent studies, however, challenge this mechanism (Whalen 2000; Saari et al. 2004; Tate et al. 2007) since most findings were obtained from fertilisation experiments.

Observed reductions in CH_4 uptake by nitrogen fertilisation do not necessarily reflect a direct inhibitory effect of NH_4^+ on methanotrophs. Indirect effects of N-treatments such as high osmotic pressure or toxicity of specific ions could have killed methanotrophs. From laboratory studies it is known that non-N-salts can reduce CH_4 uptake rates of soils. Whalen (2000) found decreasing first-order rate k constants for atmospheric CH_4 uptake with increasing amounts of NH_4Cl, $NaNO_3$ and NaCl (0–20 μmol N g^{-1} soil), indicating that the reduction in k resulted from salt concentration rather than from inhibition of methane monooxygenase by NH_4^+. Moreover, similar decreases in k with NH_4Cl and NaCl and less decrease in k with $NaNO_3$ indicate a strong inhibitory effect of Cl^-. From Whalen's (2000) study, it may be deduced that reductions in CH_4 uptake by high amendments of N-compounds result from non-specific salt effects.

The long-term effect of repeated annual $(NH_4)_2SO_4$ addition (140 kg N ha^{-1} per year) on CH_4 uptake was investigated in the N-fertilised beech plot at Solling from January 1993 to April 1995 (Fig. 19.1). CH_4 uptake rates were 0.08 kg CH_4 ha^{-1} per year in 1993 and 0.14 kg CH_4 ha^{-1} per year in 1994, which is slightly below the rates of the control plot (Table 19.1). The difference in the reduction between the years resulted mainly from the growing season in 1994 when CH_4 uptake rates were generally higher due to lower soil moisture. Mean NH_4^+ concentration in the soil solution at the 10 cm depth was 7 mg N l^{-1} higher than in the control plot from 1983 to 1994 (Meiwes et al. 1998). Despite the long-term treatment of 140 kg NH_4–N ha^{-1} per year over 13 years, CH_4 uptake of this soil was not totally inhibited.

The effect of repeated $(NH_4)_2SO_4$, NH_4-acetate, urea, NH_4Cl, $NaNO_3$, NH_4NO_3 and KNO_3 treatments on CH_4 uptake was studied in the beech forest at Göttinger Wald from April 1994 to April 1995. Insignificant changes in the range of −11 to 10% reveal that four N-treatments each of 30 kg N ha^{-1} had little or no effect on CH_4 uptake within a year (Fig. 19.7). The only exception is the 21% reduction in the plot which received 120 kg N ha^{-1} as $(NH_4)_2SO_4$ in a single application.

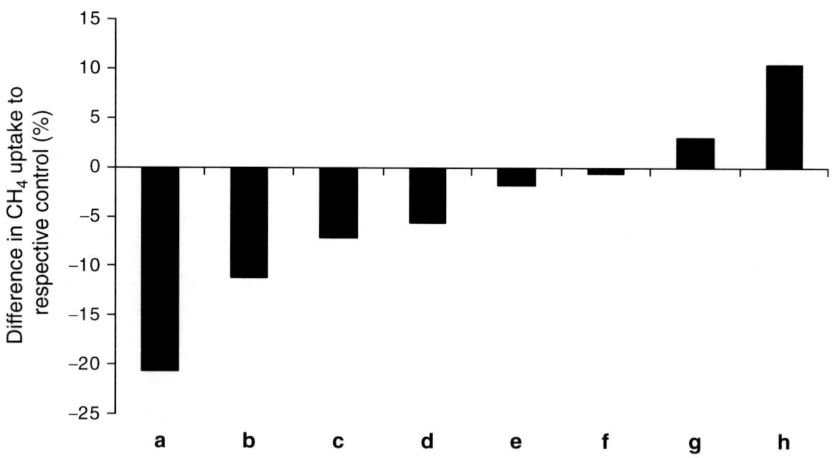

Fig. 19.7 Proportional deviation of N-treated plots to the control plot in CH_4 uptake at Göttinger Wald measured from April 1994 to April 1995 (n = 41) (**a** $(NH_4)_2SO_4$, **b** $(NH_4)_2SO_4$, **c** NH_4-acetate, **d** urea, **e** NH_4Cl, **f** $NaNO_3$, **g** NH_4NO_3, **h** KNO_3). Plot **a** received 120 kg N ha^{-1} as $(NH_4)_2SO_4$ on 1 June 1994. All other plots received 30 kg N ha^{-1} on 1 June 1994, on 4 August 1994, on 28 November 1994 and on 27 February 1995

The effects of repeated $(NH_4)_2SO_4$, NH_4NO_3, NH_4Cl, urea and $NaNO_3$ treatments on CH_4 uptake were measured using automated chambers in a spruce stand adjacent to the beech stand at Solling from August 1993 to December 1994. CH_4 uptake rates at the control plot were 1.3 kg CH_4 ha^{-1} in 1993 and 1.6 kg CH_4 ha^{-1} in 1994 (Borken et al. 2000b). All N-fertilisers reduced CH_4 uptake at the spruce stand by 20–44% during the period from November 1993 to December 1994 (Fig. 19.8). The greatest reduction in CH_4 uptake resulted from the $NaNO_3$ treatment (−44%), followed by urea (−34%), NH_4Cl (−25%), NH_4NO_3 (−23%), and $(NH_4)_2SO_4$ (−20%). Concentrations of ammonia, nitrate and dissolved organic nitrogen were periodically measured in soil solution at 10 cm depth and indicated a higher ammonium and nitrate concentration at each treatment compared to the control plot (Borken 1996). Mean NH_4^+ concentrations increased after fertilisation from 1.5 mg N l^{-1} in the control plot to 5.1 (urea), 5.5 (NH_4NO_3), 6.0 ($(NH_4)_2SO_4$) and 14.8 mg N l^{-1} (NH_4Cl), while the NH_4^+ concentration of 1.1 mg N l^{-1} in the $NaNO_3$ plot was lower than the control. NO_3^- concentration increased only in the $NaNO_3$ plot, from 5.9 mg N l^{-1} (control) to 26.1 mg N l^{-1}.

The N-treatments in the spruce stand at Solling had a stronger effect on CH_4 uptake than those in the beech forest at Göttinger Wald, although the forms and amounts of added N-compounds were not different. The insignificant effect of NH_4 treatments at Göttinger Wald could result from high nitrification or immobilisation rates that would protect methanotrophs from exposure to NH_4^+. It is likely that the reported high nitrification rate at Göttinger Wald (Chap. 13) led to rapid and complete removal of applied NH_4^+. By contrast, nitrification was incomplete in the control plot of the beech stand at Solling (Mochoge and Beese 1983). The rather high NH_4^+ concentrations in the soil solution from the NH_4^+ treated spruce plot at Solling (Borken 1996) suggest that low nitrification rates may have contributed to the decrease in CH_4 uptake (Fig. 19.8). Gulledge et al. (1997) found no inhibitory

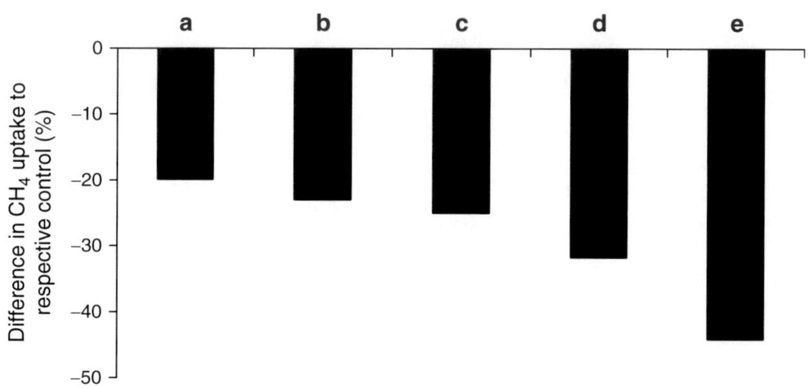

Fig. 19.8 Proportional deviations of N-treated plots to the control plot in CH_4 uptake in the spruce forest at Solling using automated chambers from August 1993 to December 1994 (**a** $(NH_4)_2SO_4$, **b** NH_4NO_3, **c** NH_4Cl, **d** urea, **e** $NaNO_3$). All fertilised plots were treated four times with 30 kg N ha^{-1} on 4 November 1993, on 22 May 1994, on 12 July 1994 and on 14 September 1994

effect after $(NH_4)_2SO_4$ fertilisation of a spruce soil but a delayed inhibition for a birch soil in upland taiga in Alaska. Low nitrification rates and high NH_4^+ concentrations were found in both soils. They concluded that different responses of the fertilised soils resulted from physiologically distinct CH_4 oxidiser communities in the two study sites.

In accordance with our results, Castro et al. (1995) observed that CH_4 uptake rates decreased by 25–62% in annually repeated NH_4NO_3 fertilised plots (150 kg N ha^{-1} per year) of a red pine plantation, while CH_4 uptake of an adjacent mixed hardwood stand was reduced by 15–36% following the same treatment over 6 years (Castro et al. 1995). However, the annual repeated application of 50 kg N ha^{-1} per year as NH_4NO_3 reduced CH_4 uptake by only 17–27% in both stands, with no temporal trend during 5 years of treatment.

Because of indirect salt effects, it is not evident whether heavy N additions simulate the long-term effect of high N-deposition on CH_4 uptake. Only a few field studies were performed to assess the in situ long-term effect of moderate nitrogen inputs on the soil CH_4 sink of forest soils. Kasimir-Klemedtsson and Klemedtsson (1997) reported a strong decrease in CH_4 uptake from 1.6 to 0.6 kg CH_4 ha^{-1} per year in a well-drained coniferous forest soil in Sweden, which received additionally 35 kg NH_4NO_3–N ha^{-1} per year, applied weekly throughout the year. However, the poorly drained site treated with the same amount of NH_4NO_3 showed higher CH_4 uptake rates than the respective control plot. The total N-input of the fertilised plots was 47 kg N ha^{-1} per year including the ambient atmospheric deposition of 12 kg N ha^{-1} per year (Klemedtsson et al. 1997). In contrast, no effect on CH_4 uptake was observed from elevated inputs of 11 kg HNO_3–N ha^{-1} per year and 18 kg $(NH_4)_2SO_4$–N ha^{-1} per year added to throughfall in a mixed deciduous woodland (Bradford et al. 2001). A 10-year reduction in nitrogen and proton input by throughfall water to pre-industrial concentrations did not increase soil CH_4 uptake in a spruce stand at Solling (Borken et al. 2002).

From these studies it may be concluded that elevated atmospheric input of nitrogen had no effect on CH_4 uptake during recent decades. Alternatively, the soil CH_4 sink was strongly reduced by nitrogen deposition during past decades and the recovery of methanotrophic bacteria needs some decades or centuries.

19.6 Conclusions and Indications of Human Impacts

- CH_4 uptake is controlled by multiply factors, which require special attention when any prediction is made. Temperate forest soils act generally as a sink for atmospheric CH_4, but both CH_4 oxidation and production coexist in soils and may determine the net CH_4 uptake rate.
- Soil moisture mainly controls the dynamics of CH_4 uptake and also has an effect on the spatial variability of CH_4 uptake. Extended summer droughts may increase annual CH_4 uptake rates in temperate forests, as indicated by large

inter-annual variation in CH_4 uptake at our study sites. In contrast, temperature has little effect on the oxidation of CH_4 per se, but it may affect soil moisture and thus the CH_4 uptake rate.

- Soil acidity has diminished the soil CH_4 sink in this study by decreasing bioturbation and accumulation of organic matter in the O-horizons. Because of the low ability to oxidise CH_4, particularly thick O horizons may act as diffusion barriers for atmospheric CH_4 at high moisture contents.
- Forest harvesting decreased CH_4 oxidation under conditions where root uptake is reduced and high soil moisture content is established during growing season at a gap study at Solling. Soil compaction during harvesting with harvesters and forwarder reduced CH_4 oxidation primary by decreased macro-pore volume. Liming maintains or increases the potential of acid forest soils to act as a sink for atmospheric CH_4.
- Inputs of nitrogen by atmospheric deposition have probably not directly affected the soil CH_4 sink of forests in Central Europe. The observed reductions in CH_4 uptake by N-fertilisation may result from non-specific salt effects or from toxic effects of accompanying ions on methanotrophic bacteria.

References

Ball BC, Smith KA, Klemedtsson L, Brumme R, Sitaula BK, Hansen S, Priemé A, MacDonald J, Horgan GW (1997) The influence of soil gas transport properties on methane oxidation in a selection of northern European soils. J Geophys Res 102:23309–23317

Bartsch N, Bauhus J, Vor T (2002) Effects of group selection and liming on nutrient cyling in an European beech forest on acidic soil. In: Dohrenbusch A, Bartsch N (eds) Forest development. Springer, Berlin, pp 109–142

Bauhus J, Bartsch N (1995) Mechanisms for carbon and nutrient release and retention in beech forest gaps. I. Microclimate, water balance and seepage water chemistry. Plant Soil 168–169:579–584

Beese F, Meiwes KJ (1995) 10 Jahre Waldkalkung: Stand und Perspektiven. AFZ 17:946–949

Bender M, Conrad R (1993) Kinetics of methane oxidation in oxic soils. Chemosphere 26: 687–696

Benstead J, King GM (2001) The effect of soil acidification on atmospheric methane uptake by a Maine forest soil. FEMS Microb Ecol 34:207–212

Borken W (1996) Methan-Aufnahme und Kohlendioxid-Freisetzung von Waldböden. Berichte des Forschungszentrums Waldökosysteme Univ Göttingen A137, pp 1–182

Borken W, Brumme R (1997) Liming practice in temperate forest ecosystems and the effects on CO_2, N_2O and CH_4 fluxes. Soil Use Manage 13:251–257

Borken W, Gründel S, Beese F (2000a) Potential contribution of *Lumbricus terrestris L* to carbon dioxide, methane and nitrous oxide fluxes from a forest soil. Biol Fertil Soils 32:142–148

Borken W, Xu YJ, Brumme R (2000b) Effects of prolonged soil drought on CH_4 oxidation in a temperate spruce forest. J Geophys Res 105:7079–7088

Borken W, Beese F, Brumme R, Lamersdorf N (2002) Long-term reduction in nitrogen and proton inputs did not affect atmospheric CH_4 uptake and N_2O emission from a German spruce forest soil. Soil Biol Biochem 34:1815–1819

Borken W, Xu YJ, Beese F (2003) Conversion of hardwood forests to spruce and pine plantations strongly reduced soil methane sink in Germany. Glob Chang Biol 9:956–966

Borken W, Beese F (2006) Methane and nitrous oxide fluxes of soils in pure and mixed stands of European beech and Norway spruce. European Journal of Soil Science 57: 617–625.

Borken W, Davidson EA, Savage K, Sundquist ET, Steudler P (2006) Effect of summer throughfall exclusion, summer drought, and winter snow cover on methane fluxes in a temperate forest soil. Soil Biol Biochem 38:1388–1395

Born M, Dörr H, Levin I (1990) Methane consumption in aerated soils of the temperate zone. Tellus 42:2–8

Bradford MA, Wookey PA, Ineson P, Lappin-Scott HM (2001) Controlling factors and effects of chronic nitrogen and sulphur deposition on methane oxidation in a temperate forest soil. Soil Biol Biochem 33:93–102

Brumme R, Beese F (1995) Automated monitoring of biological trace gas production and consumption. In: Alef K, Nannipieri P (eds) Methods in applied soil microbiology and biochemistry. Academic, San Diego, CA, pp 468–472

Brumme R, Borken W (1999) Site variation in methane oxidation as affected by atmospheric deposition and type of temperate forest ecosystem. Glob Biogeochem Cycles 13:493–501

Butterbach-Bahl K, Papen H (2002) Four years continuous record of CH_4-exchange between the atmosphere and untreated and limed soil of a N-saturated spruce and beech forest ecosystem in Germany. Plant Soil 240:77–90

Castro MS, Steudler PA, Melillo JM, Aber JD, Bowden RD (1995) Factors controlling atmospheric methane consumption by temperate forest soils. Glob Biogeochem Cycles 9:1–10

Crill PM (1991) Seasonal patterns of methane uptake and carbon dioxide release by a temperate woodland soil. Glob Biogeochem Cycles 5:319–334

Dannenmann M, Gasche R, Ledebuhr A, Holst T, Mayer H, Papen H (2007) The effect of forest management on trace gas exchange at the pedosphere-atmosphere interface in beech (*Fagus sylvatica L.*) forests stocking on calcareous soils. Eur J For Res 126:331–346

Dobbie KE, Smith KA, Priemé A, Christensen S, Degorska A, Orlanski P (1996) Effect of land use on the rate of methane uptake by surface soils in northern Europe. Atmos Environ 30: 1005–1011

Dunfield P, Knowles R (1995) Kinetics of inhibition of methane oxidation by nitrate, nitrite, and ammonium in a humisol. Appl Environ Microbiol 61:3129–3135

Dutaur L, Verchot LV (2007) A global inventory of the soil CH_4 sink. Glob Biogeochem Cycles 21:GB4013

Edwards CA, Bohlen PJ (1996) Biology and ecology of earthworms. Chapman and Hall, NY

Fiedler S, Höll BS, Jungkunst HF (2005) Methane budget of a Black Forest spruce ecosystem considering soil pattern. Biogeochemistry 76:1–20

Gulledge J, Doyle AP, Schimel JP (1997) Different NH_4^+-inhibition patterns of soil CH_4 consumption – a result of distinct CH_4-oxidizer populations across sites. Soil Biol Biochem 29:13–21

Gulledge J, Hrywna Y, Cavanaugh C, Seudler PA (2004) Effects of long-term nitrogen fertilization on the uptake kinetics of atmospheric methane in temperate forest soils. Soil Biol Biochem 49:389–400

IPCC (2007) Climate change 2007: the physical basis. In: Solomon S, Qin D, Manning M, Chen Z, Marquis M, Averyt KB, Tignor M, Miller HL (eds) Contribution of Working Group I to the fourth assessment report of the intergovernmental panel on climate change. Cambridge University Press, Cambridge, p 996

Kasimir-Klemedtsson A, Klemedtsson L (1997) Methane uptake in Swedish forest soil in relation to liming and extra N-deposition. Biol Fertil Soils 25:296–301

King GM, Adamsen APS (1992) Effects of temperature on methane consumption in a forest soil and in pure cultures of the methanotroph *Methylomonas rubra*. Appl Environ Microbiol 58:2758–2763

Klemedtsson L, Kasimir-Klemedtsson A, Moldan F, Weslien P (1997) Nitrous oxide emission from Swedish forest soils in relation to liming and simulated increased N-deposition. Biol Fertil Soils 25:290–295

Knief C, Lipski A, Dunfield PF (2003) Diversity and activity of methanotrophic bacteria in different upland soils. Appl Environ Microbiol 69:6703–6714

Kolb S, Knief C, Dunfield PF, Conrad R (2005) Abundance and activity of uncultured methanotrophic bacteria involved in the consumption of atmospheric methane in two forest soils. Environ Microbiol 7:1150–1161

Marrero TR, Mason EA (1972) Gaseous diffusion coefficients. J Chem Ref Data 1:3–118

Matzner E (2004) Biogeochemistry of forested catchments in a changing environment. Ecol Stud 172:1–498

Meiwes KJ, Merino A, Beese FO (1998) Chemical composition of throughfall, soil water, leaves and leaf litter in a beech forest receiving long term application of ammonium sulphate. Plant Soil 201:217–230

Mochoge B, Beese F (1983) The behavior of nitrogen fertilizers in neutral and acid loess soils. II. Distribution and balances of 15N tagged nitrogen. Z Pflanzern Bodenk 146:504–515

Ojima DS, Valentine DW, Mosier AR, Parton WJ, Schimel DS (1993) Effect of land use change on methane oxidation in temperate forest and grassland soils. Chemosphere 26:675–685

Potter CS, Davidson EA, Verchot LV (1996) Estimation of global biogeochemical controls and seasonality in soil methane consumption. Chemosphere 32:2219–2246

Priemé A, Christensen S (1997) Seasonal and spatial variation of methane oxidation in a Danish spruce forest. Soil Biol Biochem 29:1165–1172

Puhe J, Ulrich B (2001) Global climate change and human impacts on forest ecosystems. Ecol Stud 143:1–592

Saari A, Heiskanen J, Martikainen PJ (1998) Effect of the organic horizon on methane oxidation and uptake in soil of a boreal Scots pine forest. FEMS Microbiol Ecol 26:245–255

Saari A, Rinnan R, Martikainen PJ (2004) Methane oxidation in boreal forest soils: kinetics and sensitivity to pH and ammonium. Soil Biol Biochem 36:1037–1046

Schaefer M, Schauermann J (1990) The soil fauna of beech forests: comparison between a mull and a moder soil. Pedobiologia 34:299–314

Sitaula BK, Bakken LR, Abrahamsen G (1995) CH_4 uptake by temperate forest soil: effect of N input and soil acidification. Soil Biol Biochem 27:871–880

Smith KA, Dobbie KE, Ball BC, Bakken LR, Sitaula BK, Hansen S, Brumme R, Borken W, Christensen S, Priemé A, Fowler D, MacDonald JA, Skiba U, Klemedtsson L, Kasimir-Klemedtsson A, Degorska A, Orlanski P (2000) Oxidation of atmospheric methane in Northern European soils, comparison with other ecosystems, and uncertainties in the global terrestrial sink. Glob Chang Biol 6:791–803

Steudler PA, Bowden RD, Mellilo JM, Aber J (1989) Influence of nitrogen fertilization on methane uptake in temperate forest soils. Nature 341:314–315

Tate KR, Ross DJ, Saggar S, Hedley CB, Dando J, Singh BK, Lambie SM (2007) Methane uptake in soils from *Pinus radiata* plantations, a reverting shrubland and adjacent pasture: effects of land use change, and soil texture, water and mineral nitrogen. Soil Biol Biochem 39:1437–1449

Teepe R, Brumme R, Beese F, Ludwig B (2004) Nitrous oxide emission and methane consumption following compaction of forest soils. Soil Sci Soc Am J 68:605–611

Veerhoff M, Roscher S, Brümmer GW (1996) Ausmaß und ökologische Gefahren der Versauerung von Böden unter Wald. Berichte des Umweltbundesamtes 1, Erich Schmidt Verlag, Berlin, S1–S364

Whalen SC (2000) Influence of N and non–N salts on atmospheric methane oxidation by upland boreal forest and tundra soils. Biol Fertil Soils 31:279–287

Whalen SC, Reeburgh WS, Sandbeck KA (1990) Rapid methane oxidation in a landfill cover soil. Appl Environ Microbiol 56:3405–3411

Yavitt JB, Simmons JA, Fahey TJ (1993) Methane fluxes in a northern hardwood forest ecosystem in relation to acid precipitation. Chemosphere 26:721–730

Part C
Forest Management and Regional Scale Issues Concerning C and N

Chapter 20
Microbial Biomass in Broad-Leaved Forest Soils

T.-H. Anderson

20.1 Introduction

Long-term analysis of ecosystem processes involves, to a great extent, inventories of element and energy transfers between biotic and abiotic compartments. In terrestrial ecosystems, the heterotrophic activity of soil micro-organisms plays a central role in this element and energy transfer as it involves the release of organic nutrients and mineral elements during microbial mineralisation activities of organic residues from the primary producers. Chronic detrimental impacts would directly affect this vital microbial compartment with subsequent changes in the element and energy budget of a system. For that reason, it is an obligatory necessity to develop a sound knowledge about the size of this microbial pool and the microbial biomass of total fungal and bacterial cells, together with an understanding of its controlling mechanisms. It is most likely that adverse changes in an ecosystem will more easily and at an early stage be detectable at the microbial community level.

The objectives of this study were (1) to survey a large number of different forest sites in Lower Saxony for microbial biomass content, microbial growth or activity indices, in order to (2) provide the underlying principles of microbial development in forest soils. All the methods used are well described in the pertaining literature.

20.2 Regional Distribution of Microbial Carbon of Broad-Leaved Forest Soils in Lower Saxony

More than 20 years ago, three methods were developed for microbial biomass estimations: the fumigation incubation method (Jenkinson and Powlson 1976), the substrate-induced respiration method (Anderson and Domsch 1978) and the fumigation-extraction method (Vance et al. 1987). A detailed description of the methods

R. Brumme and P.K. Khanna (eds.), *Functioning and Management of European Beech Ecosystems*, Ecological Studies 208,
DOI: 10.1007/978-3-642-00340-0_20, © Springer-Verlag Berlin Heidelberg 2009

is given by Alef (1991). It was a major breakthrough in those days to be able to reliably quantify this total living microbial pool of a soil, rather than measuring separate component groups of organisms as had been the case with microscopic techniques. It opened the possibility for budget analysis of very diverse ecosystems (Hunt 1977; Jenkinson and Rayner 1977; Paul and Voroney 1980; McGill et al. 1981; Van Veen et al. 1985; Knapp et al. 1983). In general, agricultural soils have a microbial biomass-C (C_{mic}) content in the range of 100–1,500 μg C g^{-1} soil while grassland and forests show considerable higher values of between 1,000 and >3,000 μg C g^{-1} soil (Anderson and Domsch 1980, 1986a, 1989, 1993; Lynch and Panting 1980; Jenkinson and Ladd 1981; Insam et al. 1989).

At the beginning of 1990, data on the range of microbial biomass carbon (C_{mic}) were not available for German forest soils. The long-term programme of the Forest Ecosystem Research Centre at Göttingen allowed the thorough survey of C_{mic} contents of very diverse forests in Lower Saxony. The large amount of C_{mic} data recorded made it possible to empirically detect relationships between C_{mic} contents and soil or environmental parameters. This kind of acquired knowledge can help to better interpret effects of anthropogenic impacts.

Table 20.1 shows a selection of studied forest sites (beech and beech-oak stands) of Lower Saxony which are listed in a north-south direction together with their C_{mic} values. This survey included only large natural forests which are either Federal State owned or private property, but always under the management of forest departments. Since the goal of this intensive measuring campaign was to understand the underlying principle of microbial biomass development in forest soils, sampling was done with great care to avoid short-term effects on the micro-flora. This included, for instance, sampling only in early spring before the emergence of ground-cover to avoid extreme temperature or nutrient effects (root exudation of trees or plants) (Anderson and Domsch 1989, 1993). Further, measurements were concentrated on the mineral horizon (A_h), since it was assumed that the surface litter layers (L, O_h) are more prone to short-term impacts, particularly to the ongoing liming campaign, which would have made it difficult to separate such anthropogenic effects from the endemic interspecific soil effects on microbial biomass. A one-way ANOVA of these sites again revealed the strong influence of the soil pH on the level of microbial biomass as was reported earlier (Anderson and Domsch 1993). Soil pH is the main controlling factor of C_{mic}. Sites with a mean pH of ~3.2 ($n = 248$) had significantly ($p < 0.001$) lower C_{mic} values compared to neutral sites of mean pH of ~6.7 ($n = 150$). The mean value of microbial biomass for acidic sites amounted to 636 μg C_{mic} g^{-1} soil as compared to neutral sites with 1,294 μg C_{mic} g^{-1} soil which was almost twice as high. Independent of the location, the high microbial biomass levels can be expected in soils with a neutral pH together with high C_{org} contents. Despite the different soil types (i.e. Cambisol, Gleyic Cambisol, Podzol, Fluvisol, Vertic Cambisol, Rendzic Leptosol and Gleysols, according to FAO nomenclature) of the forest sites, soil type itself does not show any relation to C_{mic} values and soil genesis might indirectly affect C_{mic} by determining pH and C_{org} content in a particular climatic region. This has already been shown earlier for different soils under agricultural management

Table 20.1 Soil microbial biomass (C_{mic}), pH and C_{org} range of different broad-leaved forest sites of Lower Saxony (A_h horizon, depth 0–5 cm)

Site[a]	Geographical location[b]	C_{mic} range (μg C g^{-1} soil)	pH range (KCl)	C_{org} range (%)
Wingst (3), F.A. Wense	53.45°N, 9.05°E	377–533	2.8–3.4	4.3–8.1
Ummeln/Hepstedt (9), F.A. Bremervörde	53.17°N, 9.06°E	192–743	2.7–5.2	4.0–12.0
Meerdorfer Holz (163), F.A. Peine	52.22°N, 10.20°E	134–1,950	2.4–7.3	1.3–30.2
Fürstenauer Holz (37), F.A. Peine	52.18°N, 10.21°E	234–1,041	2.7–3.8	5.4–28.2
Lappwald/Dorm (14), F.A. Mariental	52.17°N, 11.00°E	382–945	3.1–5.4	4.3–11.7
Braunschweig (8), F.A. Baunschweig	52.15°N, 10.27°E	178–626	3.0–3.8	3.4–9.8
Elm (80), F.A. Königslutter	52.14°N, 10.46°E	354–2,082	3.2–7.3	2.7–9.1
Lecheln Holz (150), Rev. F. Wolfenbüttel	52.12°N, 10.32°E	180–1,631	2.9–7.0	1.6–9.7
Asse (25), Rev.F.Wittmar	52.08°N, 10.40°E	245–3,237	3.2–6.9	2.6–13.8
Oderwald (235), F.A. Liebenburg	52.06°N, 10.30°E	223–3,120	3.2–6.2	2.6–14.8
Siebenberge (5), F.A. Alfeld	52.03°N, 9.50°E	616–1,876	4.0–6.9	4.7–7.8
Hils (4), F.A. Grünenplan	51.56°N, 9.45°E	471–846	2.7–3.5	4.7–12.0
Harz (6), F.A. Bad Harzburg	51.54°N, 10.37°E	206–1,016	3.4–5.5	3.2–7.5
Vogler (6), F.A. Stadtoldendorf	51.54°N, 9.35°E	386–1,681	3.0–6.2	5.8–12.4
Solling (4), F.A. Neuhaus	51.46°N, 9.35°E	586–906	2.9–3.4	13.9–21.2
Göttinger Wald (20), F.A. Reinhausen	51.32°N, 10.03°E	1,027–2,853	3.6–6.8	5.5–12.9
Zierenberg (3), F.A. Wolfhagen	51.17°N, 9.15°E	723–1,330	4.1–6.4	4.3–7.1

C_{mic} was determined by the substrate-induced respiration technique (Anderson and Domsch 1978)

[a]Sites are listed by their historical names together with the names of the responsible district forestry departments (F.A. or Rev.F.) of locations where samples were taken. Figures in parentheses show numbers of locations sampled

[b]The co-ordinates of sites are approximations for large forest areas where many locations were visited and are ordered by decreasing latitude

Table 20.2 Comparison of C_{mic} contents of A_h and O_h horizons along with C_{org} and pH values of selected forest sites of Lower Saxony

Site[a]	Horizon	C_{mic} (μg C g^{-1} soil)	C_{org} (%)	pH (KCl)
Meerdorfer Holz (5)	A_h	361 ± 89	10.3 ± 4.8	3.1 ± 0.2
	O_h	2,625 ± 656	30.5 ± 5.9	4.8 ± 0.6
Fürstenauer Holz (10)	A_h	451 ± 97	11.9 ± 5.8	3.2 ± 0.2
	O_h	2,396 ± 477	30.0 ± 4.3	4.1 ± 0.8
Lecheln Holz (6)	A_h	267 ± 67	4.4 ± 0.9	3.3 ± 0.04
	O_h	1,340 ± 247	14.6 ± 2.3	3.2 ± 0.1
Oderwald (6)	A_h	604 ± 131	7.2 ± 0.9	3.9 ± 0.4
	O_h	2,522 ± 895	21.9 ± 3.3	4.7 ± 0.9
Ummeln/Hepstedt (9)	A_h	408 ± 185	7.6 ± 3.3	3.4 ± 0.7
	O_h	2,830 ± 485	27.6 ± 6.9	5.2 ± 0.8

[a]C_{mic}, C_{org} and pH values of sites are means (± SD) of the number of locations (in brackets) sampled per site. Five replicate samples per location were taken and bulked

(Anderson and Domsch 1989) and was also confirmed by Insam et al. (1989) and Kaiser et al. (1992).

The influence of the C_{org} content on microbial biomass development is not easy to detect because soil pH exerts a superimposing effect. In general, it was observed as stated above that high levels of C_{org} were related to high C_{mic} contents, provided that the soil pH is optimal (neutral) for microbial development. Table 20.2 shows this relationship by comparing the A_h and O_h horizons of sampling locations at five different forest sites. As expected, the C_{org} content of the A_h horizon is always considerably lower than the C_{org} content of the corresponding upper O_h horizon and the C_{mic} content in the O_h horizon is accordingly higher. O_h horizons can contain up to 7-fold more C_{mic} than A_h horizons. For example, Meerdorfer Holz and Ummeln/ Hepstedt, with a high pH together with a high C_{org} content in O_h horizons, show also the highest C_{mic} level. No C_{mic} values of O_h horizons of pH above 6.0 are available since such soils hardly develop any O_h horizon.

20.2.1 The Significance of the C_{mic}-to-C_{org} Relationship and the Specific Respiration Rate, qCO_2

Micro-organisms need carbon for survival and growth for which they depend on the soil organic matter. Micro-organisms form an integral component of the total soil organic matter itself. Early reports on this microbial biomass carbon (C_{mic}) to organic carbon (C_{org}) relationship showed ranges from 0.27 to 7.0% of the total C_{org} to be microbial carbon (literature search in Anderson and Domsch 1989), together with a linear relationship of these two parameters (Anderson and Domsch 1980). Further intensive studies on agricultural soils showed that a linear relationship between the level of microbial biomass carbon and soil carbon would be expected when the soils of similar soil pH and nutrient inputs are compared (Anderson and Domsch 1989). This was also observed with forest soils as reported by Domsch and Anderson (1993) and Anderson (2000). The sites listed in Table 20.2 have a percent C_{mic} in C_{org} content in a range between 0.35 and 1.15%. This wide range of microbial biomass build-up results from the different growth conditions which would exist on these sites. Microbial growth is not an erratic process but follows cell-physiological principles. For example, it is known from chemostat cultures that, under the same set of carbon, nutrients and specific environmental conditions, the same amount of cell mass is always produced, whereby the available carbon is always divided into two portions, a portion needed for maintenance and another used for growth (Pirt 1965, 1975). The maximum value for growth based on the amount of carbon used is called yield and is a constant under the particular set of environmental conditions. This constant gives an insight into the pattern of the total carbon utilised by the cells: the more is used for growth, the less is used for maintenance purposes, and the more efficient is a cell in its C use and vice versa (Pirt 1975). To what extent the soil pH will play a role with respect to the level of microbial build-up in relation to soil carbon can be seen in Fig. 20.1a. Under neutral soil conditions, up to four times more microbial biomass will develop than under acidic conditions for the same level (mg g^{-1}) of soil carbon. Multiple regression analyses show that the dependent variable microbial biomass-C can be predicted from a linear combination of C_{org} ($p < 0.001$) and pH ($p < 0.001$) of plots in the figure, with the exception of the plot beech-oak stand with neutral conditions where pH ($p < 0.136$) was no longer an influential factor. Particularly, under acidic soil conditions pH exerts a strong influence on a linear relationship between C_{mic} and C_{org}. Comparing acidic beech stands within a pH range of 2.6–3.9 (Fig. 20.1a) to beech stands of a very narrow pH range, such as pH 2.6–2.9 or pH 3.5–3.9, and considering only stands older than 75 years, the correlation coefficient will improve significantly (Fig. 20.1b).

If this cell-physiological principle as outlined above was applied to the higher aggregated level – the microbial community or the so-called synecological level, the C_{mic}-to-C_{org} relationship would be analogous to the growth yield. The parameter percent C_{mic}-to-C_{org} is therefore an internal indicator of how much of the organic carbon is used for growth by the indigenous microbial cell mass. Further, in order to assess the portion of carbon available for maintenance of microbial cells, Anderson and Domsch (1985a, b) suggested that basal respiration of the cell mass (unit CO_2–C released per unit microbial biomass, qCO_2) would provide a useful approximation of that amount. These so-called eco-physiological parameters have found a wide range of applications (e.g. Insam and Domsch 1988; Insam and Haselwandter 1989; Anderson and Domsch 1990; Höper and Kleefisch 2001; Dilly et al. 2001; Moscatelli et al. 2005) since their introduction to soil ecological studies (Anderson and Domsch 1986b). The following considerations will concentrate on these two parameters "percent C_{mic}-to-C_{org}" and the "specific respiration, qCO_2" which were tested on the ecological theory of Odum (1969) "the development of diversity in ecosystem concomitant with an efficient use of energy".

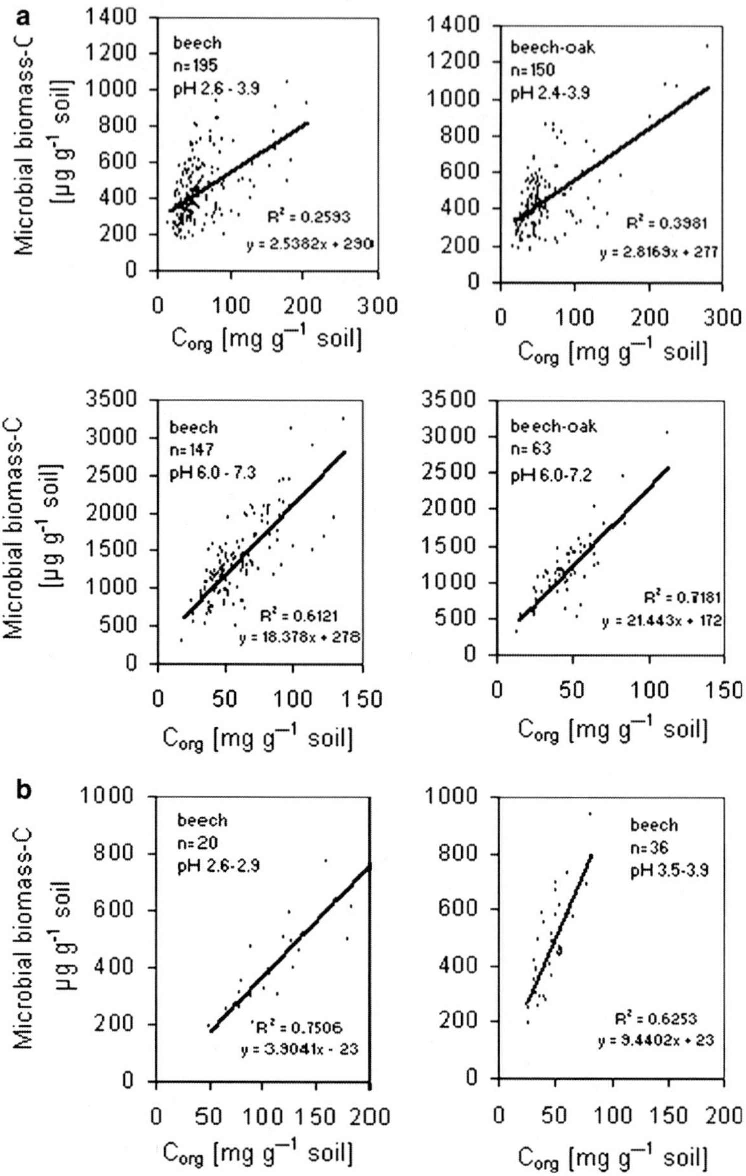

Fig. 20.1 Examples of the linear relationship between microbial biomass-C and soil organic carbon of (**a**) beech and beech-oak stands in acid and neutral soils (age 10–260 years) and (**b**) of beech stands below pH 3.0 and in the range of pH 3.5–3.9 (>75–260 years)

In Table 20.3, some criteria are extracted from Odum (1969) for ecosystem development which seem applicable and which could be extrapolated to the microbial biomass.

20.2.1.1 Beech Monocultures Versus Beech–Oak Stands

For soil microbial communities, the ecological theories put forth by Odum (1969) would imply that, together with an increase in diversity of substrate inputs from the primary producer, a more efficient utilisation by the established and adapted microflora could be expected. This would mean that, in more complex systems, percent C_{mic} in C_{org} should be higher and qCO_2 lower than in less complex systems (see Table 20.3). We tested this assumption by comparing forest sites of acidic and neutral stands of either beech or beech-oak stands over 75 years since, as was shown above (Fig. 20.1a), pH had a strong influence on microbial biomass development and, further, former investigations had indicated that the age of forest stands was also an influential factor (Anderson and Domsch 1993). Figure 20.2a demonstrates that percent C_{mic} in C_{org} of the more complex system of beech-oak stands was found to be higher than that of pure beech stands which suggested a more efficient use of the available carbon by the beech-oak micro-flora. Differences are small and more pronounced under neutral conditions but are statistically significant for each pH set when tested by one-way ANOVA ($p < 0.001 - p$ 0.004). The mean values of percent C_{mic} in C_{org} for old beech stands on acid soils ($n = 47$) was 0.61 compared to 0.75 for old beech-oak stands ($n = 97$); for old beech stands on neutral soils, the mean value was 2.31 ($n = 89$) compared to 2.7 for old beech-oak stands ($n = 66$). Also, as expected, unit respiration released per unit microbial biomass (qCO_2) is lower in beech-oak stands than in beech stands (Fig. 20.2b) ($p < 0.001 - p$ 0.024).

These results are further supported when pure old spruce stands on acid soils were analysed. In these stands, the type of organic matter input is even less diverse as in, for instance, beech stands. This is reflected in a still lower value of percent C_{mic} in C_{org} of only 0.44 concomitant to a higher qCO_2 of 2.5 μg CO_2–C mg^{-1} h^{-1} (not shown) than those of beech stands (see above). An earlier publication (Anderson and Domsch 1993) presented similar results with even fewer numbers

Table 20.3 Expected trends in ecosystem succession after Odum (1969)[a] as applicable to the soil microbial community

Ecosystem characteristics	Developmental stages	Mature stages
	Primary producer community	
Community respiration/total biomass	High	Low
Biomass supported/unit energy	Low	High
	Secondary producer community	
Microbial community respiration/unit C_{mic} (qCO_2)	High	Low
Microbial biomass supported/unit C (% C_{mic} in C_{org})	Low	High
	General	
Total organic matter	Low	High
Species diversity	Low	High
Biochemical diversity	Low	High
Nutrient exchange rate	High	Low

[a]Partially extracted and modified from Fig. 2.1 and Table. 2.1 (Odum 1969) and selected with relevance to the contents of this chapter

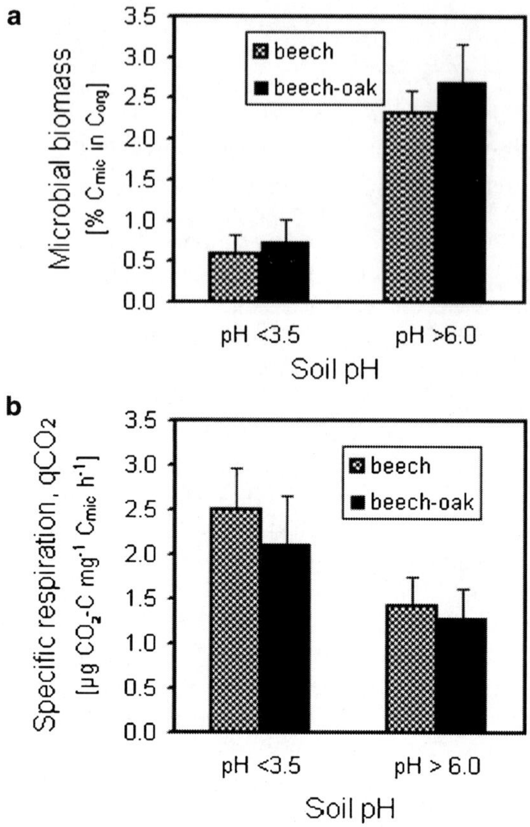

Fig. 20.2 Comparison between beech and beech-oak stands with (**a**) respect to build-up of microbial biomass expressed as percent C_{mic} in C_{org} and (**b**) specific respiration, qCO_2 in acid and neutral soils (A_h horizon)

of forest soils analysed. Parallel studies on Chinese forest soils support these results (Ding et al. 1992).

It was surprising that the data found for this C:C relationship of neutral forest soils were very close to agricultural soils where percent C_{mic} in C_{org} values of 2.3% for monocultures (=simple system) and 2.9 for crop rotation (=complex system) were found (Anderson and Domsch 1986a, 1989). These similar values suggest that microbial biomass build-up following decomposition of organic matter must be very similar in both terrestrial systems and may follow the same principle.

Taken all together, it comfirms the assumption of the manifestation of a more efficient microbial community in terrestrial systems with increasing diversity of organic matter input. Whether the observed differences of efficient carbon use is indeed related to a microbial community of greater species richness cannot be answered at the moment. However, this question of possible linkages between above-ground biodiversity as a controlling agent of below-ground species diversity has attracted a great deal of interest (Wall and Moore 1999; Hooper et al. 2000).

20.3 Microbial Communities as Affected by Soil pH

Considering that microbial cells have to keep up a species-specific ‘internal’ cell pH which lies on average at about pH 6.0 (Langworthy 1978), it is expected that more maintenance energy would be needed by the cells if the surrounding pH is lower than that of the internal cell pH. Chemostat-grown microbial cells stressed by lowering the pH had an increased maintenance carbon demand and a decreased growth rate (Brown et al. 1980). There is every reason to assume that the same cell-physiological principle would apply to micro-organisms when living under soil conditions.

20.3.1 pH Dependent Changes of the C_{mic}- to- C_{org} Relationship, qCO_2 and the Fungal–Bacterial Ratio of Broad-Leaved Forest Stands

In order to determine the underlying relationship between soil pH and microbial growth, percent C_{mic} in C_{org} and qCO_2 values of mature (>75 year) beech forest stands (A_h horizon) were related to soil pH as the independent variable. Figure 20.3 shows a very good linear correlation with a correlation coefficient of 0.88 ($p \leq 0.001$) for the relationship between percent C_{mic} in C_{org} and pH indicating that about 77% of the variances in the data were accounted for by this relationship, while other factors were responsible for the remaining 23%. Figure 20.3 also clearly demonstrates the interdependency between biomass build-up and respiration – with a decrease in the percent microbial C in total C_{org} (Fig. 20.3a), the qCO_2 rises in an equal manner (Fig. 20.3b). The coefficient of determination (r^2) for the regression on qCO_2 and pH indicated that 56% of the variances of the data were accounted for by this regression. This shows that production of CO_2 is linked to factors which do not effect C_{mic} to the same extent.

We assume that site-specific factors play an additional role since long-term measurements on particular sites repeatedly produced the so-called ‘outliers’, values which were always positioned far away either above or below the regression line. The regression line of Fig. 20.3a cuts the X-axis at about pH 2.0 providing an additional proof of the quality of the data, since below pH 2.0 no active soil microbial flora would be expected to exist. Bewley and Stotzky (1983) found no respiratory activity at this pH under controlled laboratory studies of soils. Similar good regressions (not shown) can be generated for beech-oak forest ($r^2 = 0.77$; $n = 272$) or pure spruce stands ($r^2 = 0.79$; $n = 152$) which verify the results given above where pH was identified as the major controlling factor for microbial growth.

This strong relationship between soil pH and the total microbial biomass was further investigated to find out if special groups of micro-organisms were particularly affected by low soil pH. For instance, it was investigated whether and to what

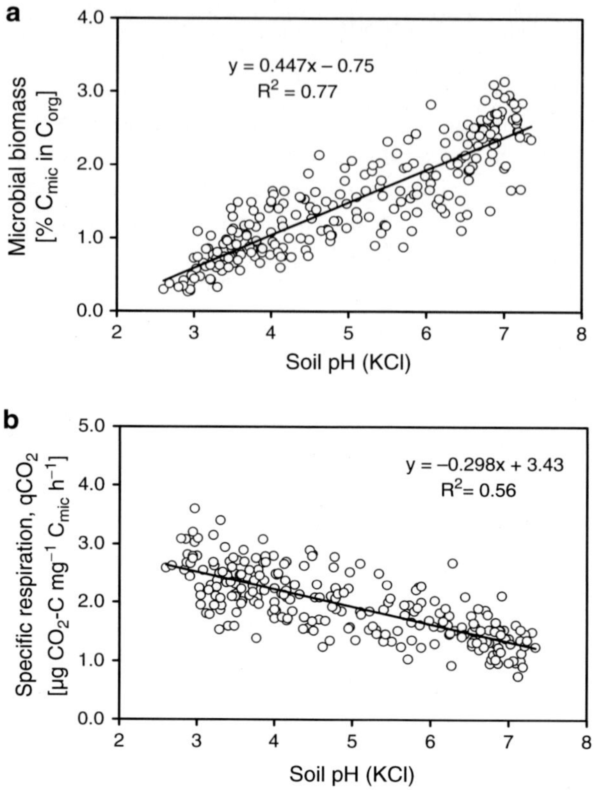

Fig. 20.3 Correlation of (**a**) microbial biomass (percent C_{mic} in C_{org}) and (**b**) qCO_2 with the pertaining soil pH of mature (>75 years) beech forest stands ($n \geq 260$)

extent the two major groups – bacteria and fungi – would react to pH changes. To study this, we used th*e selective inhibition method* (Anderson and Domsch 1973, 1975), where the respiratory activity of fungal and bacterial groups are separated by the application of specific inhibitors. Previous studies on the inhibitor concentrations for forest soils had shown that the optimal concentrations of both inhibitors depended on the level of the total prevailing biomass. We found a good selectivity of the inhibitor combinations of streptomycin (bacterial inhibitor) and cycloheximide (fungal inhibitor) with the following combinations: 1–3 mg streptomycin together with 6–10 mg cycloheximide g^{-1} soil for acid soils and 2–4 mg streptomycin together with 6–8 mg cycloheximide g^{-1} soil for neutral soils.

Figure 20.4 shows changes in the fungal-bacterial respiratory ratio where the bacterial respiratory activity is given as percentage of the total CO_2 evolution along a pH gradient. The generated correlation is best described (best fit) by an exponential function $y = cx^b$. The test is very time-consuming and only 61 broad-leaved

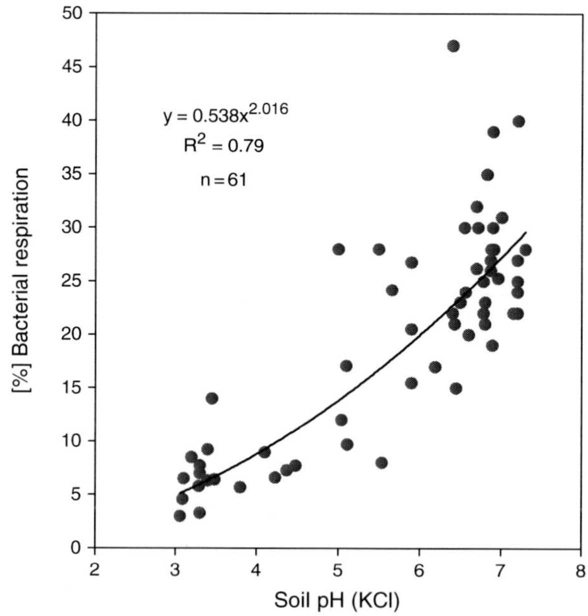

Fig. 20.4 Correlation between bacterial respiration (CO_2 as percent of total respiration and soil pH of broad-leaved forest stands (A_h horizon)

forest soils in total were investigated. Because of this limited number of observations no separation of the age of forest stands or type (beech or beech–oak) was done. Despite this, the coefficient of determination, $r^2 = 0.79$, was impressive. In earlier publications, soil fungi were found to be quite acid tolerant (Bååth et al. 1980; Domsch et al. 1980; Bewley and Parkinson 1985; Gams 1992). Surveys on acid tolerance of fungi or bacteria rested in many cases on observations from in vitro studies on single isolates. Comparative observations showing shifts in the fungal–bacterial respiratory ratio with changes in soil pH have so far not been investigated. Our data give clear evidence that bacteria are extremely sensitive to an acid environment, for instance at pH 3 only about 5% of the total CO_2 evolution would be attributable to the respiratory activity of bacteria (Fig. 20.4). This indicates a decrease in bacterial activity under acidic soil conditions. New studies on bacterial community structures by phospholipid fatty acids analysis point to a decrease of mainly Gram-negative bacteria under acidic conditions (Bååth and Anderson 2003).

Our data on the fungal–bacterial ratio of forest soils comply with published data for agricultural soils when neutral soils are compared (Anderson and Domsch 1975). Somewhat higher bacterial respiratory contributions were determined by Alphei et al. (1995) for A_h horizons of a beech stand with a pH range of 5.0–7.0 (H_2O) or by Parkinson et al. (1978) for the L, H layers of a spruce stand. For acidic sites, no information is available on fungal–bacterial ratio studies done by the *selective inhibition technique* other than those shown here.

20.4 Conclusion and Management Implications

Acidification of soils will affect the microbial community in two ways. Firstly, high H^+-ion input will affect the microbial cell, whereby the degree of the detrimental effect would depend on the specific cell-physiological make-up of the micro-organisms. Evidence hitherto can be seen from an experimental approach under controlled laboratory conditions (Blagodatskaya and Anderson 1999) where excess H^+-ions had immediately eradicated part of the (probably high proton sensitive) microbial biomass. This loss was irreversible as long as the acid environment prevailed. One major group of sensitive organisms was found to be bacteria (Sect. 20.3.1).

The immediate question which follows is will the changes in microbial biomass and composition change the capacities to mineralise soil organic matter? The preliminary study by Kreitz and Anderson (1997) has shown a decrease in mineralisation whereby bacteria from an acid environment utilised organic substrates which differed from those which bacteria used in a neutral environment. Particularly, the ability of bacteria from acid soils to mineralise carboxylic acids was lacking (Kreitz and Anderson 1997). Besides the loss in functional capacities of a soil community, the microbial diversity may be reduced with lowering of soil pH as shown by the fungal-bacterial respiration ratio of a microbial community (Sect. 20.3.1).

The second effect is of a possible change in the substrate quality on soil acidification. Many reports indicate that a loss of nutrients from the soil system (Ulrich 1980; Verstraten et al. 1990; Raubuch and Beese 1995) or a decrease in nutrient availability (Curtin et al. 1998; Miltner and Zech 1998) or Al^{+3} toxicity (Rosswall et al. 1986) may reduce the activity of the decomposer organisms. Although the amount of C_{org} was found to be very similar in forest soils of pH 3.5 and those of pH 6.0 of mature stands (see Table 20.1), their C availability may differ, as indicated by three times higher percent C_{mic} in C_{org} values under neutral conditions (Fig. 20.3). This difference cannot solely be explained by the optimal conditions for microbial growth which occurs under this pH regime, but points to a higher level of species diversity under neutral soil conditions.

These results provide evidence that the ecological theories can be applied to the soil microbial community in a promising way. This holistic approach is a subject of discussion (Wardle and Giller 1996; Ohtonen et al. 1997). Among many attempts to test Odum's postulates regarding soil biological properties, Skujins and Klubek (1982) were the first to examine successional trends relating to microbial diversity. Odum (1985) cited 18 effects which could be expected to appear in stressed ecosystems and, when adapted to field studies, stress effects were particularly prominent at the structural (species) level (Odum 1990). Our studies have shown that changes of both structural (shift in the fungal-bacterial ratio) and functional changes would occur under acid stress.

Acknowledgements We thank Kurt Steffens, Maria Bota and Kirsten Höpker for reliable technical assistance. We are very much indebted for the help of the many forest officers, who suggested forest stands and supplied maps.

References

Alef K (1991) Methodenhandbuch Bodenmikrobiologie. Ecomed, Landsberg

Alphei J, Bonkowski M, Scheu S (1995) Application of the selective inhibition method to determine bacterial: fungal ratios in three beechwood soils rich in carbon – optimization of inhibitor concentrations. Biol Fertil Soils 19:173–176

Anderson JPE, Domsch KH (1973) Quantification of bacterial and fungal contributions to soil respiration. Arch Microbiol 93:113–127

Anderson JPE, Domsch KH (1975) Measurement of bacterial and fungal contributions to respiration of selected agricultural and forest soils. Can J Microbiol 21:315–322

Anderson JPE, Domsch KH (1978) A physiological method for the quantitative measurement of microbial biomass in soils. Soil Biol Biochem 10:215–221

Anderson JPE, Domsch KH (1980) Quantities of plant nutrients in the microbial biomass of selected soils. Soil Sci 130:211–216

Anderson T-H (2000) Bewertung bodenmikrobiologischer Kenngrößen nach langjähriger Beobachtung von Waldstandorten – Vergleich zu Agrarböden. Mitteilungen der Deutschen Bodenkundlichen Gesellschaft 93:120–123

Anderson T-H, Domsch KH (1985a) Maintenance carbon requirements of actively-metabolizing microbial populations under *in situ* conditions. Soil Biol Biochem 17:197–203

Anderson T-H, Domsch KH (1985b) Determination of ecophysiological maintenance carbon requirements of soil microorganisms in a dormant state. Biol Fertil Soils 1:81–89

Anderson T-H, Domsch KH (1986a) Carbon link between microbial biomass and soil organic matter. In: Megusar F, Gantar M (eds) Proceedings of the 4th international symposium on microbial ecology, Slovene Society for Microbiology, Lubljana, pp 467–471

Anderson T-H, Domsch KH (1986b) Carbon assimilation and microbial activity in soil. Z Pflanzenernaehr Bodenk 149:457–468

Anderson T-H, Domsch KH (1989) Ratios of microbial biomass carbon to total organic-C in arable soil. Soil Biol Biochem 21:471–479

Anderson T-H, Domsch KH (1990) Application of eco-physiological quotients (qCO_2 and qD) on microbial biomasses from soils of different cropping histories. Soil Biol Biochem 22:251–255

Anderson T-H, Domsch KH (1993) The metabolic quotient for CO_2 (qCO_2) as a specific activity parameter to assess the effects of environmental conditions, such as pH, on the microbial biomass of forest soils. Soil Biol Biochem 25:393–395

Bååth E, Anderson T-H (2003) Comparison of soil fungal/bacterial ratios in a pH gradient using physiological and PLFA-based techniques. Soil Biol Biochem 35:955–963

Bååth E, Berg B, Lohm U, Lundgren B, Lundkvist H, Rosswall T, Söderström B, Wiren A (1980) Effects of experimental acidification and liming on soil organisms and decomposition in a Scots pine forest. Pedobiologia 20:85–100

Bewley RJF, Parkinson D (1985) Bacterial and fungal activity in sulphur dioxide polluted soils. Can J Microbiol 31:13–15
Bewley RJF, Stotzky G (1983) Simulated acid rain (H_2SO_4) and microbial activity in soil. Soil Biol Biochem 15:425–429
Blagodatskaya EV, Anderson T-H (1999) Adaptive responses of soil microbial communities under experimental acid stress in controlled laboratory studies. Appl Soil Ecol 11:207–216
Brown MH, Mayes T, Lelieveld HLM (1980) The growth of microbes at low pH values. In: Gould GW, Corry JEL (eds) Microbial growth and survival in extremes of environment. Academic, London, pp 71–98
Curtin D, Campbell CA, Jalil A (1998) Effects of acidity on mineralization: pH-dependence of organic matter mineralization in weakly acidic soils. Soil Biol Biochem 30:57–64
Domsch KH, Anderson T-H (1993) Stoffwechselkoeffizienten mikrobieller Sekundärproduzenten – Verhalten und Entwicklung der mikrobiellen Biomasse in Waldstandorten. In: Forschungszentrum Waldökosysteme der Universität Göttingen; Abschlussbericht 1989–1993 zum BMFT-Forschungsvorhaben Stabilitätsbedingungen von Waldökosystemen, Teil B, pp 146–160
Domsch KH, Gams W, Anderson T-H (1980) Compendium of soil fungi, vol 1. Academic, London
Dilly O, Winter K, Lang A, Munch J-C (2001) Energetic eco-physiology of the soil microbiota in two landscapes of southern and northern Germany. J Plant Nutr Sci 164:407–413
Ding Ming Mao, Yi Wie Min, Liao Lan Yu, Martens R, Insam H (1992) Effect of afforestation on microbial biomass and activity in soils of tropical China. Soil Biol Biochem 24:865–872
Gams W (1992) The analysis of communities of saprophytic microfungi with special reference to soil fungi. In: Winterhoff W (ed) Fungi in vegetation science. Kluwer, Netherlands, pp 183–223
Hooper DU, Bignell DE, Brown VK, Brussard L, Dangerfield JM, Wall DH, Wardle DA, Coleman DC, Giller KE, Lavelle P, van der Putten H, de Ruiter PC, Rusek J, Silver WL, Tiedje KM, Wolters V (2000) Interactions between aboveground and belowground biodiversity in terrestrial ecosystems: patterns, mechanisms, and feedbacks. BioScience 50:1049–1061
Höper H, Kleefisch B (2001) Untersuchung bodenbiologischer Parameter im Rahmen der Boden-Dauerbeobachtung in Niedersachsen. Bodenbiologische Referenzwerte und Zeitreihen. Arbeitshefte – Boden, Heft 4, E. Schweizerbart, Stuttgart, p 94
Hunt HW (1977) A simulation model for decomposition in grasslands. Ecology 58:469–484
Insam H, Haselwandter K (1989) Metabolic quotient of the soil microflora in relation to plant succession. Oecologia 79:171–178
Insam H, Parkinson D, Domsch KH (1989) Influence of macroclimate on soil microbial biomass. Soil Biol Biochem 21:211–221
Jenkinson DS, Ladd JN (1981) Microbial biomass in soil: measurement and turnover. In: Paul EA, Ladd JN (eds) Soil biochemistry, vol 5. Marcel Dekker, New York, pp 415–471
Jenkinson DS, Powlson DS (1976) The effect of biocidal treatments on metabolism in soil. V. A method for measuring soil biomass. Soil Biol Biochem 8:189–202
Jenkinson DS, Rayner JH (1977) The turnover of soil organic matter in some of the Rothamsted classical experiments. Soil Sci 123:298–305
Kaiser EA, Mueller T, Joergensen RG, Insam H, Heinemeyer O (1992) Evaluation of methods for soil microbial biomass estimations and their relation to soil texture and soil organic matter. Soil Biol Biochem 24:675–683
Knapp EB, Elliott LF, Campbell GS (1983) Carbon, nitrogen and microbial biomass interrelationships during the decomposition of wheat straw: a mechanistic simulation model. Soil Biol Biochem 15:455–461
Kreitz S, Anderson T-H (1997) Substrate utilization patterns of extractable and non-extractable bacterial fractions in neutral and acidic beech forest soils. In: Insam H, Rangger A (eds) Microbial communities. Functional versus structural approaches. Springer, Berlin, pp 149–160

Langworthy TA (1978) Microbial life in extreme pH values. In: Kushner DJ (ed) Microbial life in extreme environments. Academic, London, pp 279–315
Lynch JM, Panting LM (1980) Variations in the size of the soil biomass. Soil Biol Biochem 12:547–550
McGill WB, Hunt WH, Woodmansee RG, Reuss JO (1981) Phoenix – a model of the dynamics of carbon and nitrogen in grassland soils. In: Clark FE, Rosswall T (eds) Terrestrial nitrogen cycling, vol 33. Ecological Bulletin, Stockholm, pp 49–115
Miltner A, Zech W (1998) Carbohydrate decomposition in beech litter as influenced by aluminium, iron and manganese oxides. Soil Biol Biochem 30:1–7
Moscatelli MC, Lagomarsino A, Marinari S, De Angelis S, Grego S (2005) Soil microbial indices as bioindicators of environmental changes in a poplar plantation. Ecol Ind 5:171–179
Odum EP (1969) The strategy of ecosystem development. Science 164:262–270
Odum EP (1985) Trends expected in stressed ecosystems. BioScience 35:419–422
Odum EP (1990) Field experimental tests of ecosystem-level hypotheses. Tree 5:204–205
Ohtonen R, Aikio S, Väre H (1997) Ecological theories in soil biology. Soil Biol Biochem 29:1613–1619
Parkinson D, Domsch KH, Anderson JPE (1978) Die Entwicklung mikrobieller Biomassen im organischen Horizont eines Fichtenstandortes. Oecol Plant 13:355–366
Paul EA, Voroney RP (1980) Nutrient and energy flows through soil microbial biomass. In: Ellwood DC, Hedger JN, Latham MJ, Lynch JM, Slater JH (eds) Contemporary microbial ecology. Academic, London, pp 215–237
Pirt SJ (1965) The maintenance energy of bacteria in growing cultures. Proc R Soc Lond B 163:224–231
Pirt SJ (1975) Principles of microbe and cell cultivation. Blackwell, Oxford
Raubuch M, Beese F (1995) Pattern of microbial indicators in forest soils along an European transect. Biol Fertil Soils 19:362–368
Rosswall T, Schnürer J, Söderlund S (1986) Interaction of acidity, aluminium ions and microorganisms. In: Jensen V, Kjøller A, Sørensen LH (eds) Microbial communities in soil. Elsevier, London, pp 395–410
Skujins J, Klubek B (1982) Soil biological properties of a montane forest sere: corroboration of Odum's postulates. Soil Biol Biochem 14:505–513
Ulrich B (1980) Production and consumption of hydrogen ion in the ecosphere. In: Hutchinson TC, Havas M (eds) Effects of acid rain precipitation on terrestrial ecosystems. Plenum, New York, pp 255–282
Vance ED, Brookes PC, Jenkinson DS (1987) An extraction method for measuring soil microbial biomass C. Soil Biol Biochem 19:703–707
Van Veen JA, Ladd JN, Amato M (1985) Turnover of carbon and nitrogen through the microbial biomass in a sandy loam and a clay soil incubated with $^{14}C(U)$glucose and $^{15}N(NH_4)_2SO_4$ under different moisture regimes. Soil Biol Biochem 17:747–756
Verstraten JM, Dopheide JCR, Duysings JJHM, Tietema A, Bouten W (1990) The proton cycle of a deciduous forest ecosystem in the Netherlands and its implication for soil acidification. Plant Soil 127:61–69
Wall DH, Moore JC (1999) Interactions underground. Soil biodiversity, mutualism, and ecosystem processes. BioScience 49:109–118
Wardle DA, Giller KE (1996) The quest for a contemporary ecological dimension to soil biology. Soil Biol Biochem 28:1549–1554

Chapter 21
Soil Organic Carbon and Nitrogen in Forest Soils of Germany

R. Brumme, M. Egenolf, C. Aydin, J. Block, K.J. Meiwes, and K.v. Wilpert

21.1 Introduction

Soils contain the largest terrestrial organic carbon pool on the earth, about two to three times more organic carbon than is present in vegetation or atmosphere (IPCC 2001). The high fraction of organic carbon stored in soils is of concern because organic compounds could easily be oxidised to CO_2, N_2O and nitrate in a changing environment and would increase the greenhouse effect and nitrate losses. Several studies have indicated that the amount of organic matter in soils is highly variable due to possible losses or gains of soil organic carbon and nitrogen predominantly under changing environmental conditions (e.g. Schulze 2000; Puhe and Ulrich 2001; Matzner 2004). Testing the effect of environmental variables on soil organic carbon is a very complex task. The very long residence time of a large fraction of soil organic carbon requires long observation periods from the time when an environmental factor was changed experimentally. Other processes may alter under experimentally changed conditions and may increase or decrease the resultant effect and may require an extended period for a new steady-state condition to establish. Another approach which has frequently been used is to study soil C-stocks along environmental gradients (e.g. Schlesinger 1990; Schulze 2000; de Koning et al. 2003; Thuille and Schulze 2006). One advantage of such chronosequence studies is the assumption that the state of soil C present at an undisturbed site is the product of natural change and thus would include the adaptation to existing climate factors (steady-state conditions). All factors known to affect soil organic carbon have to be considered in such studies. Moreover, any order of predicting parameters may be questioned because it is commonly not possible to determine the valid cause and effect relationships in most cases as many potentially confounding factors may be involved in the changes in soil C observed along environmental gradients (Garten et al. 1999).

The goals of this study were: (1) to quantify the relationship among soil C, N, soil, and environmental variables for about 500 forest plots along a north-south gradient in Germany, and (2) to describe any changes in the quality and quantity of

R. Brumme and P.K. Khanna (eds.), *Functioning and Management of European Beech Ecosystems*, Ecological Studies 208,
DOI: 10.1007/978-3-642-00340-0_21, © Springer-Verlag Berlin Heidelberg 2009

surface organic matter layer in relation to environmental and anthropogenic changes.

Database. The dataset for this study included about 500 forest soil profiles which were sampled on a 8 × 8 km grid between 1987 and 1993 in Baden-Württemberg, Niedersachsen, Rheinland Pfalz, Schleswig-Holstein and Saarland, Germany. These data are part of a much larger dataset which were analysed in a joint initiative of the federal states of Germany represented by their forest research institutes and the Federal Ministry of Food, Agriculture, and Consumer Protection (BMELV) to evaluate the current forest soil conditions of Germany (Soil Conditions Survey, Bundeszustandserhebung im Wald, BZE) and were published as a report ("Deutscher Waldbodenbericht 1996"; Wolff and Riek 1997a, b).

Two of the federal states are located in the north of Germany (Niedersachsen, Schleswig-Holstein), primarily on the lowland, which was formed by glaciers and consists of Pleistocene sediments. Saarland and Rheinland Pfalz represent the middle and Baden-Württemberg the south of Germany and range from low to high mountain ranges. Of the four most dominant tree species, pine, oak, beech and spruce, pine dominate the lowland of Niedersachsen and Schleswig Holstein, and spruce dominate in Baden-Württemberg and higher altitudes in Saarland, Rheinland Pfalz, and Niedersachsen. Since mixed forests are common in Germany and the litter of conifer trees is known to be more resistant to decomposition than litter from deciduous trees, we use the fraction of conifer trees to consider this factor in the analysis. The clay content was not measured at the sites which are known to have a large effect on the C-sequestration in soils. Therefore, we calculated the particle surface area of the mineral soil using the "feel" method in the field which is not an absolute but a relative measure of the surface for binding soil organic carbon. The north–south gradient covers the whole range in altitude, climate, vegetation and soil conditions in Germany (Table 21.1).

Table 21.1 Percentiles of site characteristic values (0% = minimum, 50% = median, 100% = maximum (n = 534 for C in mineral soil)

	Unit	0%	10%	50%	90%	100%
Altitute	m	5	48	330	675	1,160
Mean annual air temperature	°C	4.5	6.5	8.0	8.6	9.7
Annual precipitation	mm per year	550	650	800	1,230	1,900
Fraction of conifer trees	%	0	0	85	100	100
Particle surface area[a]	$cm^2\ cm^{-3}$	1,080	1,200	5,050	9,900	22,200
Soil pH(H_2O) (0–30 cm)		2.9	3.7	4.2	5.2	8.3
Exchangeable Al (0–30 cm)	$mmol_c\ g^{-1}$	0.1	15	40	80	176
Exchangeable Ca (0–30 cm)	$mmol_c\ g^{-1}$	0	1	4	51	728
C-concentration (0–30 cm)	$mg\ kg^{-1}$	3.1	12.4	21.8	49.1	128
N-concentration (0–30 cm)	$mg\ kg^{-1}$	0.2	0.7	1.4	3.1	16.0
C/N (0–30 cm)		6.3	11	16	25	39
C stock (O horizon)	$Mg\ ha^{-1}$	0	3.3	17	50	128
N stock (O horizon)	$Mg\ ha^{-1}$	0	0.09	0.69	2.0	5.1
C/N (F horizon)		14.7	20	23	30	48

[a]Estimated from "feel" method in the field

21.2 Carbon and Nitrogen Stocks in Temperate Forest Soils

The mean amount of soil organic carbon in German forest soils stored in the surface organic horizon to 90 cm depth is 99 Mg C ha^{-1} (Wolff and Riek 1997a, Table 21.2). Of the total amount of soil, 60% of organic carbon is stored in the top 30 cm of the mineral soil, followed by 15% in 30–60 cm and 6% in 60–90 cm depth. The surface organic layers contribute about 19%. The total amount of nitrogen in forest soils of Germany is 6.8 Mg ha^{-1} as an overall median. Nitrogen followed the depthwise distribution of carbon in the soil profiles, but with a much larger proportion of N stored in deeper soil depths (12%, 53%, 22% and 13% in the surface organic layer, 0–30, 30–60 and 60–90 cm, respectively), probably due to a higher amount of fixed nitrogen in clay minerals and lower C/N ratios at greater depths.

Humus classes are commonly used to estimate the nutrient availability in forest ecosystems in Germany (Ponge and Chevalier 2006). The classification based on morphological studies of the surface organic layer which increased in thickness with decreasing pH and nutrient availability. Carbon and nitrogen stored in the surface organic layer increased from 6.2 Mg C ha^{-1} in mull to 31.7 Mg C ha^{-1} in mor humus (Table 21.3). Mull humus contribute with 6% C to the total stocks in L–90 cm which equals about three annual litterfalls. These soils are less acid (pH(KCl) 4, 0–10 cm depth) and had a high base content which promote the activity of earthworms and litter decomposition in the mineral soil. These soils are commonly interpreted as soils with a high nutrient turnover. In contrast, mor humus contributes 29% C to the total stocks which equals about 15 annual litterfalls and indicates a low nutrient turnover and low pH (pH(KCl) 3.2, 0–10 cm depth). The lowest total C-stock in L–90 cm was found in mull–moder soils and increased with decreasing pH up to 110 Mg C ha^{-1} in mor humus soils and with increasing pH to 97 Mg C ha^{-1} in mull humus soils. The low C-stock in mull–moder soils resulted from a low C-content in 0–30 cm depth and the surface organic layer.

The N-stock shows a similar pattern in the surface organic layer as observed for carbon and increased from 0.28 Mg N ha^{-1} in mull to 1.3 Mg N ha^{-1} in mor humus (Table 21.3). In contrast to carbon, the N-stocks decreased continuously from mull (8.3 Mg N ha^{-1}) to mor humus soils (5.4 Mg ha^{-1}) in L–90 cm depth. The decrease of nitrogen in the mineral soil with soil acidity is primary related to the 0–30 cm depth and is not balanced by a steady increase of nitrogen in the surface organic layer.

Table 21.2 Median values of carbon and nitrogen stocks in forest soils of Germany (10 and 90-percentile values are given in parentheses) (Wolff and Riek 1997a, b)

	Mg C ha^{-1}	Mg N ha^{-1}	*n*
O horizon	18.4 (4.4, 43)	0.77 (0.18, 1.84)	1,699
0–30 cm	58.5 (28.4, 105)	3.50 (1.33, 7.44)	1,631
30–60 cm	14.3 (5.0, 38.5)	1.43 (0.45, 3.25)	1,198
60–90 cm	5.8 (0.9, 19.3)	0.82 (0, 2.47)	1,111
Total	99.0 (53, 192)	6.8 (3.4, 13.4)	1,110

Table 21.3 Carbon and nitrogen storage (median) in soils of different humus classes in Germany (according to Forstliche Standortaufnahme (1996), mull-type humus combine L–mull and F–mull, mull–moder-type humus equals mullartiger moder, moder-type humus combine feinhumusarmer and feinhumusreicher moder, moder–mor-type humus equals rohhumusartiger moder, and mor-type humus equals Rohhumus. (Mg C ha^{-1}, Mg N ha^{-1}) (Wolff and Riek 1997a, b)

	Mull		Mull–Moder		Moder		Moder–Mor		Mor	
	C	N	C	N	C	N	C	N	C	N
O horizon[a]	6.2	0.28	10.2	0.46	20.5	0.88	27.2	1.11	31.7	1.30
0–30 cm	67.2	5.18	55.5	3.77	58.1	3.37	59.4	3.03	56.9	2.41
30–60 cm	15.2	1.88	15.4	1.54	14.0	1.42	14.9	1.63	16.3	1.23
60–90 cm	8.0	0.94	6.5	1.39	5.6	0.95	5.5	0.78	5.5	0.50
Total	96.6	8.28	87.6	7.16	98.2	6.62	106.9	6.55	110.4	5.44

[a] *n* refers to number of sites for C in O horizon: mull (n = 347); mull–moder (n = 177); moder (n = 577); moder–mor (n = 218); mor (n = 234)

For many of the sites included by Wolff and Riek (1997a), the density of the fine earth and the rock fraction had been estimated in the field, and were not measured by laboratory methods. This leads to considerable uncertainties of the figures in Table 21.2 and Table 21.3. Wirth et al. (2004) undertook a new determination of soil densities and rock fractions for the sites situated in the state of Thüringen using laboratory measurements. They found lower soil density values and about 40% less carbon than in the first study that had used site estimates. The rock fraction in soils was overestimated in the estimate of Wolff and Riek for the Thüringen sites and lowered the errors by 20%. Thus, the data presented in Tables 21.2 and 21.3 may lie in the lower range due to incorrect estimates of soil density in some federal states.

The amount of carbon and nitrogen stored in the soils of the world were estimated by Post et al. (1982, 1985). They aggregated the data from 3,100 soil profiles which were classified by natural vegetation as per Holdridge life-zone. For cool temperate moist forests, mean values of C- and N-stock were estimated to be 120 and 6.3 Mg ha^{-1} to a depth of 1 m. These values were similar to those estimated for Germany. Similarly, Batjes (1995, 1996) estimated C- and N-stocks in the soils of the world which were based on 4,353 soil descriptions (492 belonged to Europe) and were aggregated as per Holdridge life-zone classification. C-stock for temperate forest region was estimated to be 121 Mg C ha^{-1} and the N-stock 11.8 Mg N ha^{-1}. Most soils in this estimate were primarily under agricultural use which might explain the high N-stock of the soil. Except for the data presented for Germany, most data did not include the surface organic layer in the estimate. An estimate of the C-stock by Jobbágy and Jackson (2000) based on three global databases reported 174 Mg C ha^{-1} (n =60) in temperate deciduous and 145 Mg C ha^{-1} (n =123) in evergreen forests in the 0–100 cm soil depth. For the Swiss Alps, Perruchoud et al (2000) calculated for forest soil (8 $\times$ 8 km grid sampling) a mean mineral soil stock of 98 Mg C ha^{-1}.

21.3 Environmental Controls on C- and N-Sequestration

Complex relationships between SOM content and environmental variables were obtained as shown by correlation analysis. Carbon concentration in the mineral soil ranged from 3 to 128 mg g^{-1} and positively correlated with precipitation, pH, particle surface and altitude, and negatively with temperature (Table 21.4). One of the most important factors for C-sequestration is altitude although it is likely that altitude is not a direct cause of this relationship; rather, elevation is related to the variation in climate, fraction of conifer trees in the stands, C/N and particle surface area along the German altitudinal gradient (Table 21.4). Multiple regression analysis where predictors with the highest reduction in residual variance were added successively explained 37% of the variation (Table 21.5). Altitude was not included in the multiple regression analysis. Temperature covariates with precipitation and accounted for a similar percentage of the variation (Table 21.4). Soil pH reduced the residual variance by 9% if added to temperature as the first variable in multiple regression analysis and indicated the importance of soil chemical properties for C-sequestration in the mineral soil. Soil pH determines the C-sequestration in soils as controller of C-concentration in the multiple regression analysis because Al, Fe and Ca contents, which are related to soil pH, form complexes with organic compounds and may strongly bind humic substances to clay (Stevenson 1994) (see Sect. 21.4).

The altitudinal gradient in Fig. 21.1 revealed that precipitation might be responsible for increased C-sequestration with increasing altitude in the mineral soil. A review of dissolved organic matter in temperate forests indicated that precipitation had a positive effect on the leaching of dissolved organic carbon (DOC) and nitrogen (DON) from the surface organic layer to the mineral soil layers (Michalzik et al. 2001). Dissolved organic matter increases the acidity in the surface mineral soil and may well explain the observed increase of exchangeable Al (and Fe, not shown) in the surface mineral soil at higher altitudes (Fig. 21.1b). These conditions may encourage the sequestration of DOC and DON in the mineral soil by forming co-ordination complexes with Al and Fe (Stevenson 1994). Other parameters which affected the sequestration of DOC and DON were the particle surface area and the fraction of conifer trees which also increased with altitude (Fig. 21.1b). However, these factors did not change at an altitude higher than 800 m. Thus, climatic parameters and Al concentration are the most important factors controlling the C- and N-sequestration at higher altitudes.

Our results agree with other studies which were conducted along altitudinal gradients mostly showing increased C-sequestration in the mineral soil (Jenny 1980; Brady 1974; Garten et al. 1999; Perruchoud et al. 2000; Powers and Schlesinger 2002; de Koning et al. 2003; Thuille and Schulze 2006). The observations at the Fresno altitude transect (U.S.A.) indicated increased C-content from 0.8% at 300 m to 2.2% at 2,400 m (Jenny 1980). Although the vegetation composition varied, the main conclusion was that the accumulation of organic carbon in the soil was predominantly controlled by climate. Brady (1974) analysed SOM accumulation in grassland soils in a north–south transect in the Mississippi Valley region of

Table 21.4 Correlation matrices (r_{ij}) for C-concentration in the mineral soil (*Cm*) (0–30 cm), C-stock in the surface organic layer (*Co*) and predictors (×): annual precipitation, mean annual air temperature, fraction of conifer trees, pH(H_2O) (0–30 cm), C/N (0–30 cm), particle surface area (par. sur.; 0–30 cm), and altitude. (n = 509)

	Cm (mg kg^{-1})	Co (mg kg^{-1})	Prec (mm)	Temp (°C)	Conifer (%)	pH	C/N	Par. sur (cm^2 cm^{-3})	Altitude (m)
Cm	1.000	−0.092	0.415[a]	−0.424[a]	0.068	0.249[a]	0.021	0.248[a]	0.438[a]
Co	−0.092	1.000	0.003	−0.078	0.311[a]	−0.385[a]	0.408[a]	−0.337[a]	−0.150[a]
Prec	0.415	0.003	1.000	−0.594	0.257	−0.134	−0.124	0.047	0.704
Temp	−0.424	−0.078	−0.594	1.000	−0.228	0.085	0.085	−0.150	−0.706
Conifer	0.068	0.311	0.257	−0.228	1.000	−0.170	0.143	−0.055	0.267
pH	0.249	−0.385	−0.134	0.085	−0.170	1.000	−0.232	0.480	0.016
C/N	0.021	0.408	−0.124	0.085	0.143	−0.232	1.000	−0.364	−0.205
Par. sur.	0.248	−0.337	0.047	−0.150	−0.055	0.480	−0.364	1.000	0.325
Altitude	0.438	−0.150	0.704	−0.706	0.267	0.016	−0.205	0.325	1.000

[a] p value < 0.01 of testing the model $C = \alpha + \beta x$ against H_0: $\beta = 0$ by t-test

Table 21.5 Multiple regression for the prediction of soil C-concentration (mg kg^{-1}) in the top 30 cm of the mineral soil (*Cm*) ($r^2 = 0.37$) and of C-stock (Mg C ha^{-1}) of the surface organic layer (*Co*) ($r^2 = 0.31$) (arranged for decreasing values of reduction in residual variance, predictors with reduction of residual variance of $r^2 < 1\%$ were excluded) (particle surface area: par. sur.; fraction of conifer trees: conifer)

$Cm = 114 - 26\,temp + 0.57\,pH^2 + 0.0000085\,prec^2 + 0.51\,C/N^2 + 1.41\,temp^2 + 0.55\,par.sur$
$Co = 87 + 1.05\,C/N - 29.7\,pH + 0.09\,conifer + 2.12\,pH^2$

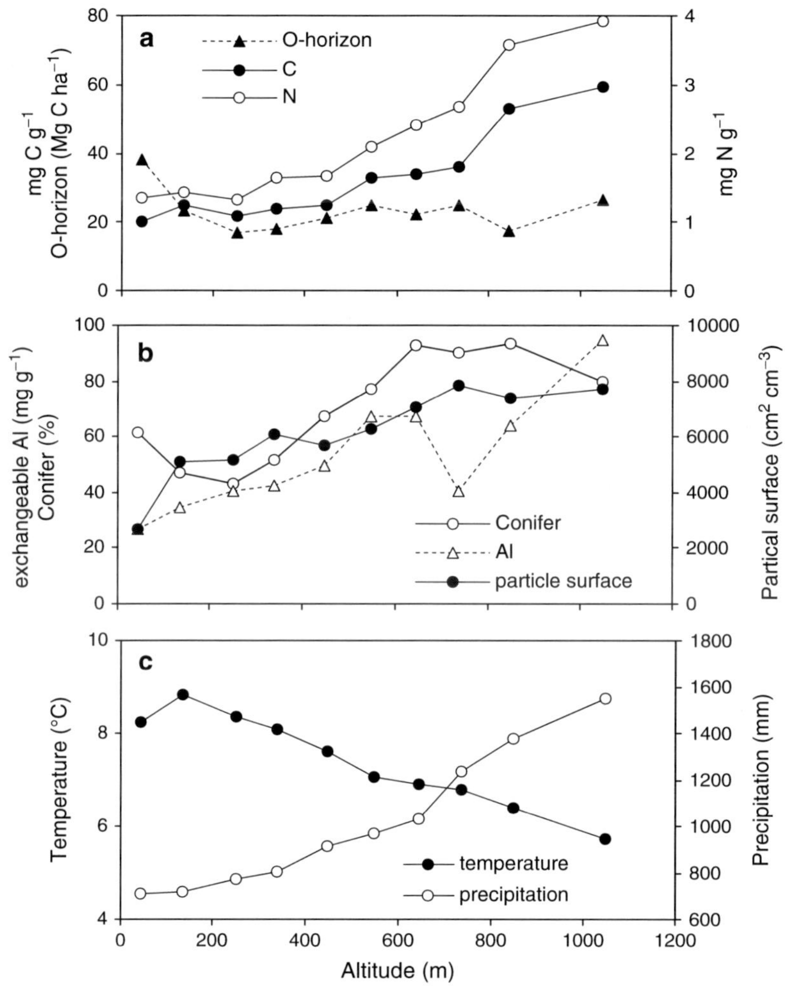

Fig. 21.1 C- and N-concentrations in the mineral soil (0–30 cm), C-stock in the organic O-horizon (**a**); content of exchangeable aluminium and conifers (0–30 cm), particle surface area (**b**); annual precipitation and mean annual air temperature (**c**); in relation to altitude above sea level

the USA and found a marked decrease with increasing temperature. Correspondingly, an increase in the annual rainfall from west to east was reflected by an increased accumulation of organic matter in soils. Along an elevation gradient between 235 and 1,670 m in the Appalachian Mountains in the USA, the C-concentration increased from 9 to 23 mg g^{-1} in 5–20 cm soil depth (Garten et al. 1999). Despite many interacting environmental factors that affect soil organic carbon along the elevation gradient, mean annual temperature was a reasonable predictor of the turnover time of unprotected soil organic carbon (O layer and labile C in the mineral soil). Powers and Schlesinger (2002) reported positive correlations between C-concentrations and elevation along an altitudinal gradient in Costa Rica. At the low elevation site, C was positively correlated to Al in organo-metal complexes and to the amount of non-crystalline clays in the high elevation soils, indicating the importance of changing soil properties along altitudinal gradients. Elevation changes in this study represented a weathering gradient where younger, allophanic soils occurred at higher elevations and older soils with gibbsite, goethite and kaolinite as dominant clay minerals in the lowlands (Powers and Schlesinger 2002).

At lower elevations (below 100 m), high values of surface organic layer stock of carbon may be explained by low temperature, higher fraction of conifers (pine) and the dominance of sandy soils with low pH, all the factors leading to reduced decomposition of organic matter (Fig. 21.1). High surface organic layer mass is commonly associated with stands located at higher elevations or latitudes (Wollum 1973; Vogt et al. 1986; Garten et al. 1999). For example, a study in forests of the Sacramento mountains in southern New Mexico indicated a strong increase in surface organic layer mass with altitude. The surface organic layer mass increased from 9.4 to 80.8 Mg ha^{-1} from the lower, dry and warm area (290 mm per year, 26°C in July) dominated by pine trees to the higher, cold and wet end (655 mm per year, 15°C in July) of the gradient dominated by white fir (Wollum 1973). A study with ^{14}C labelled standard plant material incubated along an altitude gradient in the tropical Andes showed a clear decrease in decomposition rate with increasing altitude (Coûteaux et al. 2002). The decrease in decomposition was primarily related to the recalcitrant fraction whereas the labile fraction decomposed very quickly independent of the altitude. However, in our study, the altitude gradient did not show any relationship with the mass of surface organic layer. The reasons may be related to relatively low temperature gradient of 5°C along the elevation gradient.

Spruce forests are present along the whole altitudinal gradient described here, but beech and pine forests are less important at elevations higher than 700 m, and oak forests commonly occurred below 500 m. Figure 21.2 shows the C- and N-stocks of the surface organic layer for a defined climate range (altitudinal range from 200 to 400 m) and similar soil chemical properties ($<$100 mg exchangeable Ca per kg) of deciduous ($<$20% conifer) and conifer ($>$80% conifer) forests. This stratification criterion included forests dominated by deciduous and conifer trees under the same climatic conditions and similar soil properties (except C/N ratio and exchangeable Fe content which were significantly lower under deciduous trees, not shown). Conifer forests showed significant lower element concentrations of base cations, P, Al, Fe but higher C-concentrations and C/N ratios in the F horizon (only

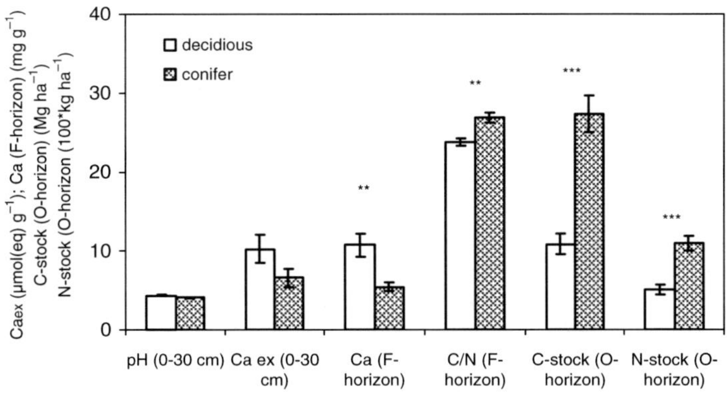

Fig. 21.2 Comparison between deciduous (<20% conifer content, $n = 75$) and conifer forests (>80% conifer content, $n = 62$) at 200–400 m altitude on soils with exchangeable Ca < 100 mg kg^{-1} for various soil properties in relation to C- and N-stocks. Properties of mineral soil in 0–30 cm depth (pH(H_2O), exchangeable calcium), surface organic F-layer (C/N ratio), and total surface organic O-layer (C- and N-stock) (mean values and standard error) (*$p < 0.05$, **$p < 0.01$, ***$p < 0.001$)

shown for Ca and C/N) and more than two times higher C- and N-stocks in the surface organic horizon than those of deciduous forests. C- and N-concentrations in the mineral soil were not different between deciduous and conifer sites although the C/N ratio was significantly higher ($p = 0.016$) under conifer trees. Thus tree species can have a significant effect on C- and N-stocks and a number of other related soil properties.

21.4 Soil Control on C- and N-Sequestration

Whereas much is known about the inhibitory effect of acid deposition and soil acidity on decomposition and microbial activity (Swift et al. 1979; Persson et al. 1989; Salonius 1990; Wolters 1991; Persson and Wirén 1993; Wolters and Schaefer 1994), little is known about the long-term effect of soil acidity on the dynamic of carbon and nitrogen sequestration and their depthwise distribution in the soil. Soil pH correlated positively with C in the mineral soil (Table 21.4) but indicated that there were two minima, at pH(H_2O) values of 4.8 and 3.0 (Fig. 21.3c). The increase in C-concentration in soils with pH of >4.8 correlated with an increase in exchangeable Ca^{2+} ($r^2 = 0.40$, $n = 80$, Fig. 21.4a). Less important parameters were altitude, precipitation, and temperature (r^2 of about 0.2). Below pH 4.8, the increase in C showed the best correlation with the increase in exchangeable Al^{3+} ($r^2 = 0.38$, $n = 454$) whereas precipitation and temperature showed lower values (r^2 of about 0.2). The correlation between C-concentration and exchangeable Al increased if sites with a fraction of conifer <95% and exchangeable Ca >15 mg kg^{-1} were excluded ($r^2 = 0.52$, $n = 235$, Fig. 21.4b).

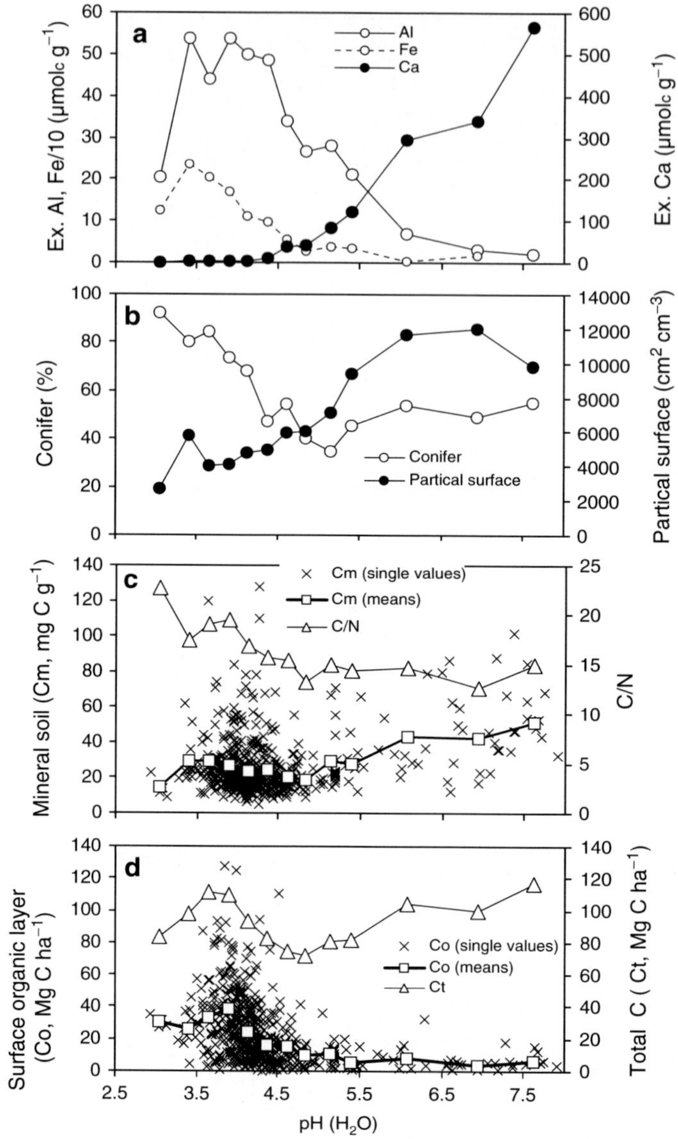

Fig. 21.3 Relationship between soil pH and mean values of exchangeable aluminium, iron and calcium in the mineral soil (**a**); mean values for the fraction of conifers in the stand and the particle surface area (**b**); single values and mean values of C-concentration (*Cm*) and C/N values in the mineral soil (**c**); single values and mean values of C-stock in the surface organic horizon (*Co*) and total Ct stock (*Ct*, sum of Co and Cm) (**d**). In all cases 0–30 cm soil depth was considered and mean values for 13 pH classes were included

It is well known that the composition of exchangeable cations on soils changed with soil acidity from prevailing base cations like magnesium and calcium in base-rich soils to aluminium, iron and protons in acid soils (Ulrich 1966; Prenzel and

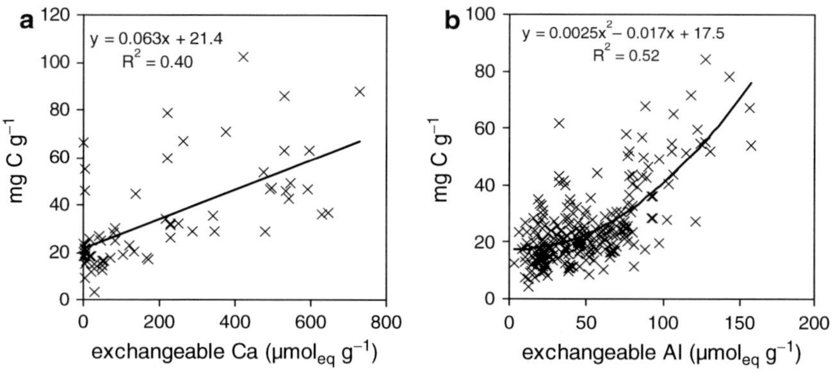

Fig. 21.4 Relationship between C-concentration in the mineral soil (0–30 cm) and exchangeable Ca in soils with pH values of >4.8 (**a**), and exchangeable Al in soils soil pH of <4.8, a content of exchangeable Ca < 15 mg kg^{-1}, and fraction of conifer >95% (**b**); all data for 0–30 cm soil depth

Schulte-Bisping 1994). The main polyvalent cations responsible for the binding of humic and fulvic acids to clay are Ca^{2+}, Fe^{3+} and Al^{3+} (Stevenson 1994; Sollins et al. 1996). Ca^{2+} does not form strong complexes with organic molecules and is only effective in making a bridge linkage; in contrast, Fe^{3+} and Al^{3+} form co-ordination complexes with organic compounds. Varadachari et al. (1991) observed a decreasing binding of humic acid on illite, montmorillonite, and kaolinite following the order: Al > Ca > Mg > K, Na. Adsorption isotherms of DOC on pure kaolinite, synthetic geothite and hematite showed that the maximum adsorption capacity of Fe oxides for DOC was about five times higher than the maximum adsorption capacity of kaolinite in tropical soils (Benke et al. 1999). Miltner and Zech (1998) reported from an incubation experiment with beech litter that decomposition of litter was lower in the presence of ferrihydrite (12%) than aluminium hydroxide (65%). Generally, toxic effects, adsorption processes or complexation processes with Fe^{3+} and Al^{3+} lowers the microbial degradation of organic matter and therefore increased their sequestration in acid soils.

The high C-stock in the base-rich mineral soils and the low C-stock in the surface organic horizon of such soils (Fig. 21.3) are maintained by the presence of deep burrowing anecic earthworms like *Lumbricus terrestris* (Chap. 7). Anecic earthworms incorporate leaf litter from the soil surface deep into the mineral soil and provide fresh organic resources for high activity of heterotrophic biomass. Anecic earthworms are highly sensitive to soil acidity and only a few individuals survive in soils with pH <4.3 (Edwards and Bohlen 1996). High soil pH and base saturation support their activity of steady litter incorporation into the mineral soil. In one of the first studies on the morphology of surface organic layer along an acid gradient in northern Germany, the base cation content in the mineral soil was the most important factor regulating the accumulation of humus in the surface organic layer (Wittich 1933). This was confirmed in a study based on 440 soil profiles in the state of Nordrhein Westfalen, Germany, by v. Zezschwitz (1968). He concluded that decomposition of organic matter increased with base saturation of the mineral soil through improvement in the microbial and earthworm activities. A detailed study on decomposition along soil acidity gradient is presented in Chap. 12. In our study, carbon sequestration is thus related to low Ca^{2+}, Fe^{3+} and Al^{3+} contents and low incorporation of leaf litter by anecic earthworms which have led to low C-content in forest mineral soils at pH 4.8. This pH is identical with the exchanger buffer range (pH 4.2–5) which is defined by the replacement of exchangeable base cations by aluminium (Ulrich 1981, 1987).

The amount of N in the mineral soil increased together with C at pH values of >pH 4.8, but was very small below pH 4.8, as indicated by increased C/N ratios in Fig. 21.3c, which could be explained by the different origin or quality of organic matter in the mineral soils of different pH values. Soils with a high pH had mull humus type where decomposition of organic matter occurred primarily in the mineral soil. Through intensive microbial decomposition, nitrogen is released and accumulated in the mineral soil whereas carbon is mostly lost as CO_2 leading to a decrease in the C/N ratio. Under acid soil conditions, leaf litter accumulates in the surface organic layer and the organic matter in the mineral soils originates from root detritus and from the dissolved organic matter that has eluviated from the surface organic layer to the mineral soil. Both of these organic matter fractions have high C/N ratios. Moreover, decomposition rate in acid mineral soils is low, releasing less carbon. Soils under conifer stands may differ from those under deciduous forests. For example, the amount of soluble organic carbon and nitrogen retained in the mineral soil was slightly higher at a coniferous site (260 kg C and 6.7 kg N ha^{-1} per year) than at a deciduous site (187 kg C and 6.2 kg N ha^{-1} per year) (Kalbitz et al. 2004).

The surface organic layer contained low amounts of C and N under high soil pH values of >4.8 (mean values were 7.9 mg C and 0.36 Mg N ha^{-1}), but this amount increased exponentially with soil pH of <4.8 (mean values were 26 Mg C and 1.1 Mg N ha^{-1}) (Fig. 21.3d). The C-stock in the surface organic layer showed relationships with soil properties in the mineral soil as C/N ratios, particle surface area, soil pH and the fraction of conifers (Table 21.4). Multiple regression analysis where C/N, pH and the fraction of conifer were added successively explained 31% of the variation (Table 21.5). Even though C/N and pH accounted for a similar percentage of the variation (Table 21.4), soil acidity was the key factor controlling C-sequestration in the surface organic layer since soil pH controls the bioturbation in soils.

The total C-stock (O horizon plus 0–30 cm soil depth) along the pH gradient had two maxima of similar amplitude that occurred due to higher C-stocks in the surface organic horizon in soils of pH <4.8 and due to a high amount of C-stocks in the mineral soils of pH >4.8 (Fig. 21.3d). The total C-stock in soils of pH 4.8 was lower by about 40 Mg C ha^{-1} suggesting that variation among soil properties may have a strong effect on the C-sequestration in forest soils which has so far not received sufficient attention. In this context, soils containing high content of silicates may be close to a balanced C budget, assuming other factors such as climate and management have not been changed over several decades. More susceptible to soil acidification are those soils which have pH values from 5 to 6.2 and occur in the silicate buffer range and may have lost considerable amounts of carbon from the mineral soil by humus degradation (Puhe and Ulrich 2001; Chap. 16). Most German forest soils have soil pH <5 (86%, Wolf and Riek 1997b) and the majority were acidified during recent decades by acid deposition (see next paragraph), thus humus degradation might have been an important process during that period. Thus C-sequestration in the surface organic horizon of acid forest soils may be the prevailing process for C-sequestration in the recent past.

21.5 Environmental and Anthropogenic Control on C- and N-sequestration in Surface Organic Horizons

Carbon and nitrogen stored in the surface organic layers increased from mull, moder to mor humus types (Table 21.6) and equalled to amounts of about 2, 6 and 14 times that of annual inputs of C through litterfall. This reduction in litter decomposition and nutrient turnover is related to soil pH in the mineral soil which decreased from 5.1 in mull to 4.0 in mor humus types.

The amount of carbon present in the surface organic layers of ten different humus types is presented in Fig. 21.5. The amount showed an exponential increase from mull to mor humus types. Carbon content in the L/F horizons followed a hyperbolic curve in contrast to the H horizon where an exponential increase in humus accumulation from mull to mor humus types was observed. In the L/F horizons, highest microbial activity and most of the mineralisation occurred (Chap. 6+13). The end products of the decomposition are fairly recalcitrant humic compounds which are then retained in the H horizons. The exponential increase in the H horizon indicated that the production of recalcitrant material produced in the F horizon exceeded the microbial decomposition in the H horizon, probably in response to soil acidification.

Acid depositions interfere in the decomposition and humification processes causing an increase of the humus sequestration in soils. Several studies have indicated that the addition of acid often reduced, whereas liming increased, the microbial activity in the short-term (Persson et al. 1989; Salonius 1990; Wolters 1991; Persson and Wirén 1993). In the long-term, an effect of soil acidity on microbial activity is evident from studies comparing soils of different pH (Chaps. 20 and 6). Soil acidification by atmospheric depositions may thus have contributed

Table 21.6 Mean stocks of C (Mg ha^{-1}), N (Mg ha^{-1}), P (kg ha^{-1}) of mull (*n* =132), moder (*n* =335) and mor (*n* =33) humus layers, and pH(H_2O) of the surface mineral soil (Sd standard deviation)

	pH		C		N		P	
	Mean	Sd	Mean	Sd	Mean	Sd	Mean	Sd
Mull	5.1	1.1	7.51	6.91	0.32	0.29	19.8	16.9
Moder	4.2	0.5	25.81	17.16	1.10	0.76	58.6	45.8
Mor	4.0	0.3	59.27	28.94	2.20	1.04	97.9	47.2

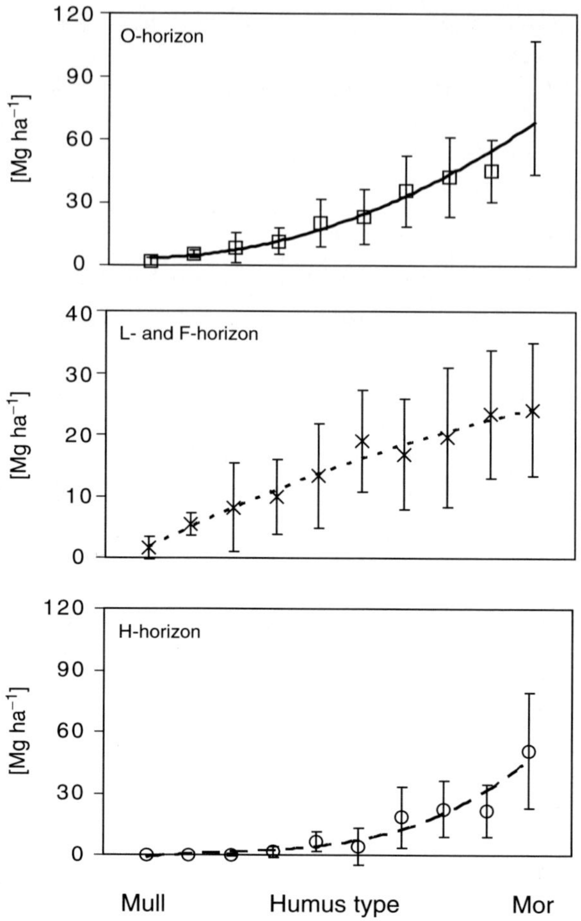

Fig. 21.5 Carbon stocks in the surface organic layer (O-horizon), the L + F-horizons and the H-horizon stratified into ten humus types from mull to mor humus in Germany

to the increase of humus in the surface organic layer of soils with low proton buffer capacity. Ulrich et al. (1989) found a decrease of soil pH(H_2O) by an average of 0.45 and a decrease of base saturation from 31 to 6% between 1954 and 1986 in Germany. Falkengren-Grerup et al. (1987) reported a reduction of base cation storage during a 15- to 35-year period in Sweden of on average 50%. In 90 soil profiles sampled for pH measurements in 1927 when resampled in 1986, indicated a general decrease in pH(H_2O) from 0.3 to 0.9 units in south-west Sweden (Hallbäcken and Tamm 1986). Veerhoff et al. (1996) summarised the literature on soil acidification and concluded that soils which had previously pH(KCl) of 3–4 decreased by 0.4–0.6 units, and those at pH >5 (except clayey and calcareous soils) decreased by 1–2 units within the last 20–30 years.

Several studies indicated that the higher deposition of nitrogen has decreased the C/N ratio in the surface organic horizons as the comparative values shown in Fig. 21.6 for the 1970s to those measured in the 1980s and 1990s. Comparison of

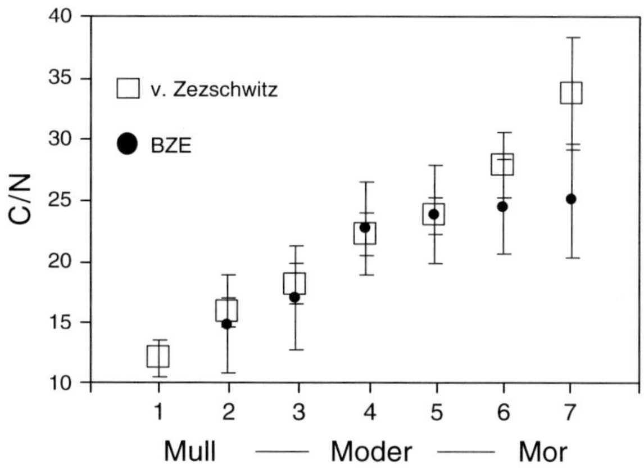

Fig. 21.6 C/N of different humus forms analysed in 1987–1993 in Germany (BZE, Wolff and Riek 1997) and 1976–1980 in western Germany (v. Zezschwitz 1980). *1–7* refer to different stages of surface organic layer from mull to mor; C/N of stages *1–3* (v. Zezschwitz 1980) and of *1–4* (BZE) refer to the upper mineral soil, other stages refer to those in the H horizons

data analysed in the 1970s by v. Zezschwitz (1980) in western Germany with those of the 1980s and 1990s by Wolff and Riek in Germany (1997a) showed that C/N ratios had decreased in the less active humus layers of moder and mor types (Fig. 21.6). A similar study where C/N ratios were analysed in 1961–1965 by Evers et al. (1968) and about 30 years later by Buberl et al. (1994) in Baden-Württemberg in southern Germany showed that the frequency distribution of C/N ratios had shifted towards lower values (Fig. 21.7). A decrease in C/N ratios with time was also observed in the surface organic layers in pine forests in northern Germany on Pleistocene substrate which were sampled first in 1959 and then again in 1997 (Nieder et al. 2000). In this study, the C/N ratios of seven pine forests which already had low values (C/N = 28) in 1959 did not change, but they decreased in 12 pine forests from 34.5 (n =3) to 29 and from 40.3 (n =9) to 28 after 38 years.

In addition to the changes in the quality of organic matter in the surface organic layer by atmospheric N-deposition, the long-term study at the Solling site showed an increase of C and N in the surface organic layer of 347 kg C and 21 kg N ha^{-1} per year during a 30-year period at the beech stand (Chap. 4; Meiwes et al. 2002). The accumulation was even higher at an adjacent spruce stand at the Solling site and amounted to 999 kg C and 42 kg N ha^{-1} per year. Both stands showed no significant change in C/N ratio between 1966 and 2001 indicating that N-sequestration was accompanied by an adequate amount of C since the beginning of the measurement. In a similar study, decreasing C/N ratios and increasing amount of C and N in the surface organic layer (mor humus type) was observed over a period of 22–30 years in two Scots pine stands (*Pinus sylvestris* L.) in southern Germany (Prietzel et al. 2006). At these sites, the soils were heavily degraded by historic litter-raking. Decomposition studies by B. Berg (Berg et al. 1995; Berg and McClaugherty 2003) indicated that a greater amount of litter with high N-concentration was lost through decomposition at the beginning but the loss was low in later stages of decomposition. The high amount of litter accumulation in surface organic horizons, as was observed at the Solling site (Chap. 4), may be a result of high N-concentrations

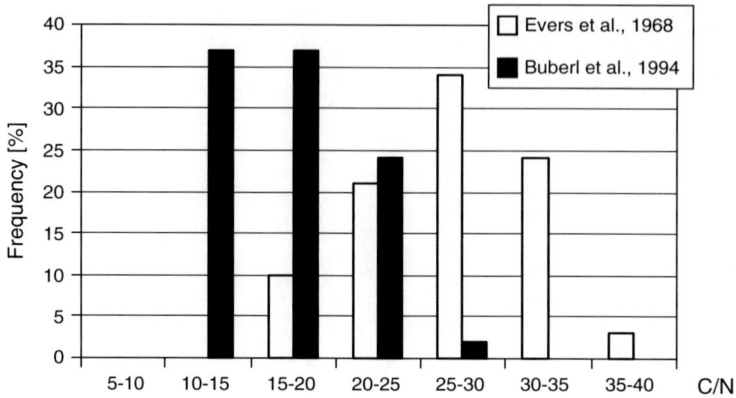

Fig. 21.7 Frequency distribution of C/N ratios in surface forest soils measured in 1961–1965 by Evers et al. (1968) (n = 96) and 1987–1993 by Buberl et al. (1994) (n =167) in southern Germany

in litter from N-depositions. Accumulation of the amount of organic matter in surface organic horizons may be a widespread process in polluted areas. A number of methods are used to study the sequestration rate of organic matter in soils. A study on the comparison of three different methods showed that similar sequestrations rates occurred in Swedish forest soils. The results obtained for the three methods were: mean value of 180 kg C ha^{-1} per year for the 'limited value approach', 96 kg C ha^{-1} per year for the 'N balance approach' and 180 kg C ha^{-1} per year for measurements of changes of humus depth over 40 years (Berg et al. 2007). An estimate for European forests based on the 'N balance approach' gave an average C-sequestration value of 143 kg C ha^{-1} per year in European forest soils for the period 1960–2000 (de Vries et al. 2006).

21.6 Conclusions

- The carbon and nitrogen stocks in Germany are 99 Mg C and 6.8 Mg N ha^{-1} between the L layer and 90 cm depth. The surface organic layer contributes up to 19% to the C– and 11% to the N-stocks. Classification of soils according to humus forms (mull, moder and mor type humus) revealed increased C-stocks from mull to mor humus type soils of about 13% down to 90 cm depth and a decrease in soil pH. Increased C-stocks are related to a decrease in the mineral soils which is more than balanced by increased sequestration of C in surface organic layer of acid soils. In contrast, nitrogen is 35% lower in mor (5.4 Mg N ha^{-1}) than in mull humus type soils (8.3 Mg N ha^{-1}). This decrease from mull to mor humus type soils is related to an enormous decrease of N in the surface mineral soil depth which is not balance by increased sequestration of N in the surface organic layer. A higher biological activity and bioturbation was made responsible for the high N-stocks in less acid soils. A forest management supporting mull humus soils would significantly increase the N-retention from atmospheric deposition in forest soils.

- Carbon concentrations in the mineral soil correlated positively with altitude, precipitation, soil pH and particle surface area and negatively with temperature. Of less importance is the fraction of conifers in the stand. At pH 4.8, C-concentration value was the lowest. A high concentration of C in soils with >pH 4.8 was related to the biological activity. Soils with <pH 4.8 also had high values which was explained by high DOC eluviation from the surface organic layer and high root turnover in acid mineral soils.
- The N-concentration increased together with C in mineral soils with >pH 4.8 but was less important in soils with <pH 4.8. This was related to the quality of organic matter (high C/N material from roots and DOC) and the low microbial activity in acid soils.
- The amount of surface organic layer is small in soils with >pH 4.8 but increases as the soil becomes more acid. Soil pH, the C/N ratio in the mineral soil and the fraction of conifers in the stand explained some of the variation of the C-stock in the surface organic layer. The surface organic layer of deciduous forests stored less than 50% of the carbon and nitrogen stored in this layer of conifer forest under similar climate and soil conditions.
- C/N ratios in moder and mor humus types of surface organic layers have decreased in Germany due to high atmospheric N-depositions. The amount of organic matter has not decreased but increased at the Solling site indicating that high amounts of N are probably retained by the surface organic layers in forest soils of Germany.

Acknowledgements Many thanks to Niels H. Batjes and Edzo Veldkamp for their valuable comments on an earlier version and to Jürgen Prenzel for statistical analysis.

References

Batjes NH (1995) World soil carbon stocks and global change. Working paper 95/11, ISRIC, Wageningen, Netherlands

Batjes NH (1996) Total carbon and nitrogen in the soils of the world. Eur J Soil Sci 47:151–163

Benke MB, Mermut AR, Shariatmadari H (1999) Retention of dissolved organic carbon from vinasse by a tropical soil, kaolinite, and Fe-oxides. Geoderma 91:47–63

Berg B, McClaugherty C, de Santo AV, Johansson MB, Ekbohm G (1995) Decomposition of litter and soil organic matter – Can we distinguish a mechanism for soil organic matter builtup? Scand J For Res 10:108–119

Berg B, McClaugherty C (2003) Plant litter. Springer, Berlin
Berg B, Gundersen P, Akselsson C, Johansson MB, Nilsson A, Vesterdal L (2007) Carbon sequestration rates in Swedish forest soils – a comparison of three approaches. Silva Fennica 41:541–558
Brady NC (1974) The nature and properties of soils, 8th edn. Macmillan, New York
Buberl HG, v Wilpert K, Trefz-Malcher G, Hildebrand EE, Wiebel M (1994) Der chemische Zustand von Waldböden in Baden-Württemberg. Ergebnisse der Bodenzustandserhebung im Wald 1989–92 (BZE). Mitteilungen der Forstlichen Versuchs- und Forschungsanstalt Baden-Württemberg 182, Freiburg im Breisgau, Germany
Coûteaux MM, Sarmiento L, Bottner P, Acevedo D, Thiéry JM (2002) Decomposition of standard plant material along an altitudinal transect (65–3968 m) in the tropical Andes. Soil Biol Biochem 34:69–78
de Koning GHJ, Veldkamp E, Lopez-Ulloa M (2003) Quantification of carbon sequestration in soils following pasture to forest conversion in northwestern Ecuador. Glob Biochem Cycles 17:1098, doi:10.1029/2003GB002099
de Vries WD, Reinds GJ, Gundersen P, Sterba H (2006) The impact of nitrogen deposition on carbon sequestration in European forests and forest sols. Glob Change Biol 12:1151–1173
Edwards CA, Bohlen PJ (1996) Biology and ecology of earthworms. Chapman and Hall London
Evers FH, Schöpfer W, Mikloss J (1968) Die Zusammenhänge zwischen Stickstoff, Phosphor– und Kaliummengen (in kg/ha) und den C/N-, C/P- und C/K-Verhältnissen der Oberböden von Waldstandorten. Mitteilung d. Vereins f. Standortskunde und Forstpflanzenzüchtung 18
Falkengren-Grerup U, Linnermark N, Tyler G (1987) Changes in acidity and cation pools of south Swedish soils between 1949 and 1985. Chemosphere 16:2239–2248
Garten CT, Post WM, Hanson PJ, Cooper LW (1999) Forest soil carbon inventories and dynamics along an elevation gradient in the southern Appalachian Mountains. Biogeochemistry 45: 115–145
Hallbäcken L, Tamm CO (1986) Changes in Soil Acidity from 1927 to 1982–1984 in a forest area of South-West Sweden. Scand J For Res 1:219–232
IPCC (Intergovernmental Panel on Climate Change) (2001) In: Houghton JT, Ding Y, Griggs DJ, Noguer M, van der Linden PJ, Xiaosu D (eds) Climate Change 2001 – The Scientific Basis. Cambridge University Press, UK, pp 944
Jenny H (1980) The soil resource. Ecol Stud 37:51–377
Jobbágy E, Jackson RB (2000). The vertical distribution of soil organic carbon and its relation to climate and vegetation. Ecol Appl 10:423–436
Kalbitz K, Zuber T, Park J-H, Matzner E (2004) Environmental controls on concentrations and fluxes of dissolved organic matter in the forest floor and in soil solution. In: Matzner E (ed) Biogeochemistry of forested catchments in a changing environment – a German case study. Ecological Studies 172, Springer, Berlin, pp 315–338
Matzner E (2004) Biogeochemistry of forested catchments in a changing environment. Ecological Studies 172, Springer, Berlin
Meiwes KJ, Meesenburg H, Bartens H, Rademacher P, Khanna PK (2002) Akkumulation von Auflagehumus im Solling, Mögliche Ursachen und Bedeutung für den Nährstoffkreislauf. Forst und Holz 57:428–433
Michalzik B, Kalbitz K, Park JH, Solinger S, Matzner E (2001) Fluxes and concentrations of dissolved organic carbon and nitrogen – a synthesis for temperate forests. Biogeochemistry 52:173–205
Miltner A, Zech W (1998) Carbohydrate decomposition in beech litter as influenced by aluminium, iron and manganese oxides. Soil Biol Biochem 30:1–7
Nieder R, Wachter H, Isermann K (2000) Erhöhte Stoffausträge bald auch aus Waldböden? AFZ/Der Wald 11:594–599
Perruchoud D, Walthert L, Zimmermann S, Luscher P (2000) Contemporary carbon stocks of mineral forest soils in the Swiss Alps. Biogeochemistry 50:111–136
Persson T, Wirén A (1993) Effects of experimental acidification on C and N mineralization in forest soils. Agric Ecosyst Environ 47:159–174

Persson T, Lundkvist H, Wirén A, Hyvönen R, Wessén B (1989) Effects of acidification and liming on carbon and nitrogen mineralization and soil organisms in mor humus. Water Air Soil Pollut 45:77–96
Ponge JF, Chevalier R (2006) Humus Index as an indicator of forest stand and soil properties. For Ecol Manage 233:165–175
Post WM, Emmanuel WR, Zinke PJ, Stangenberger AG (1982) Soil carbon pools and world life zones. Nature 298:156–159
Post WM, Pastor J, Zinke PJ, Stangenberger AG (1985) Global patterns of soil nitrogen storage. Nature 317:613–616
Powers JS, Schlesinger WH (2002) Relationships among soil carbon distributions and biophysical factors at nested spatial scales in rain forests of northeastern Costa Rica. Geoderma 109: 165–190
Prenzel J, Schulte-Bisping H (1994) Some chemical parameter relations in a population of German forest soils. Geoderma 64:309–326
Prietzel J, Stetter U, Klemmt HJ, Rehfuess KE (2006) Recent carbon and nitrogen accumulation and acidification in soils of two Scots pine ecosystems in Southern Germany. Plant Soil 289:153–170
Puhe J, Ulrich B (2001) Global climate change and human impacts on forest ecosystems. Ecological Studies 143. Springer, Berlin, Germany
Salonius PO (1990) Respiration rates in forest soil organic horizon materials treated with simulated acid rain. Can J For Res 20:910–913
Schlesinger WH (1990) Evidence from chronosequence studies for a low carbon-storage potential of soils. Nature 348:232–234
Schulze ED (2000) Carbon and nitrogen cycling in European forest ecosystems. Springer, Berlin
Sollins P, Homann P, Caldwell BA (1996) Stabilization and destabilization of soil organic matter: mechanisms and controls. Geoderma 74:65–105
Stevenson FJ (1994) Humus chemistry – genesis, composition, reactions. Wiley, New York
Swift MJ, Heal OW, Anderson JM (1979) Decomposition in terrestrial ecosystems. Studies in Ecology 5, Blackwell, Oxford
Thuille A, Schulze ED (2006) Carbon dynamics in successional and afforested spruce stands in Thuringia and the Alps. Glob Change Biol 12:325–342
Ulrich B (1966) Kationeneaustausch-Gleichgewichte im Boden. Z Pflanzenernähr Düng Bodenkd 113:141–159
Ulrich B (1981) Ökologische Gruppierung von Böden nach ihrem chemischen Bodenzustand. Z f Pflanzenernährung und Bodenkunde 144:289–305
Ulrich (1987) Stability, elasticity, and resilience of terrestrial ecosystems with respect to matter balance. In: Schulze ED, Zwölfer H (eds) Potential and limitations of ecosystem analysis. Ecological Studies 61. Springer, Berlin, pp 11–49
Ulrich B (1989) Effects of acidic precipitation on forest ecosystems in Europe. In: Adriano DC, Johnson AH (eds) Acidic precipitation, vol 2: Biological and ecological effects. Springer, New York Berlin Heidelberg
Ulrich B, Meyer H, Jänich K, Büttner G (1989) Basenverluste in den Böden von Hainsimsen-Buchenwäldern in Südniedersachsen zwischen 1954 und 1986. Sonderdruck aus "Forst und Holz", 44. Jahrgang, Nr 10:S251–S253
v. Zezschwitz E (1968) Beziehungen zwischen dem C/N-Verhältnis der Waldhumusformen und dem Basengehalt des Bodens. Fortsch Geol Rheinld u Westf 16:143–174
v. Zezschwitz E (1980) Analytische Kennwerte typischer Humusformen westfälischer Bergwälder. Z Pflanzenernaehr Bodenkd 143:692–700
Varadachari C, Mondal AH, Ghosh K (1991) Some aspects of clay-humus complexation – effect of exchangeable cations and lattice charge. Soil Sci 151:220–227
Veerhoff M, Roscher S, Brümmer GW (1996) Ausmaß und ökologische Gefahren der Versauerung von Böden unter Wald. Umweltbundesamt, Berichte 1/96. Erich Schmidt Verlag, Berlin
Vogt KA, Grier CC, Vogt DJ (1986) Production, turnover, and nutrient dynamics of above- and belowground detritus of world forests. In: MacFadyen A, Ford ED (eds) Advances in Ecological Research 15. Academic, London, pp 303–378

Wirth C, Schulze ED, Schwalbe G, Tomczyk S, Weber G, Weller E (2004) Dynamik der Kohlenstoffvorräte in Wäldern Thüringens. Mitteilungen der Thüringer Landesanstalt für Wald, Jagd und Fischerei 23, www.thueringenforst.de
Wittich W (1933) Untersuchungen in Nordwestdeutschland über den Einfluß der Holzart auf den biologischen Zustand des Bodens. Mitt aus Forstwirtschaft und Forstwissenschaft. Verlag Schaper, Hannover
Wolff B, Riek W (1997a) Deutscher Waldbodenbericht 1996, Band 1. Bundesministerium für Ernährung, Landwirtschaft und Forsten (BMELF), Bonn, Germany
Wolff B, Riek W (1997b) Deutscher Waldbodenbericht 1996, Band 2. Bundesministerium für Ernährung, Landwirtschaft und Forsten (BMELF), Bonn, Germany
Wollum AG (1973) Characterization of the forest floor in stands along a moisture gradient in Southern New Mexico. Soil Sci Soc Am Proc 37:637–640
Wolters V (1991) Effects of acid rain on leaf-litter decomposition in a beech forest on calcareous soil. Biol Fertil Soils 11:151–156
Wolters V, Schaefer M (1994) Effects of acid deposition on soil organisms and decomposition processes. Wiley, New York, pp 83–127

Chapter 22
Management Options for European Beech Forests in Relation to Changes in C- and N-Status as Described by the Three Study Sites

N. Bartsch and E. Röhrig

22.1 An Introduction to European Beech

22.1.1 Geographic Range and Current Distribution

The genus *Fagus* dominates large areas of the cool temperate forest regions in the northern hemisphere. It is divided into two main areas of distribution: *Fagus sylvatica* L. (and *Fagus orientalis* Lipsky, which is very similar) in Europe and in the western region of Asia Minor, and *Fagus grandifolia* Ehrh. in eastern North America. The occurrence of the genus in Japan (*Fagus japonica* Maxim. and *F. crenata* Bl.) and in China, represented by a number of species, is somewhat more limited (Schmucker 1942; Barnes 1991; Ching 1991; Jahn 1991; Peters 1997; Dierschke and Bohn 2004).

As one of the last broadleaved trees to survive the last ice age, European beech (*F. sylvatica*) spread out very slowly from the small refuges it occupied in southern France, southern Italy and the western Balkans into Central Europe (Magri et al. 2006). The expansion of beech into Central Europe started when the climate became increasingly colder and more humid (Tinner and Lotter 2001). Due to the limited distribution of its seeds and the slow growth rate of young trees, European beech required a period of about 7,000 years before it was able to compete with the light-demanding trees that were dominant at the time (Firbas 1949; Lang 1994; Gliemeroth 1995). European beech was very successful as it turned out (Bohn et al. 2004; Bolte et al. 2007). Leuschner et al. (2006) claims that beech is the "most successful Central European plant species".

The nature of its expansion probably explains why the genetic differences between stands across their entire distribution range are smaller than those for many other European tree species (Konnert et al. 2000; Sander et al. 2001), although essential genetic differences can occur within a single stand (Dähne et al. 1997; Konnert and Henkel 1997; Ziehe 2007).

R. Brumme and P.K. Khanna (eds.), *Functioning and Management of European Beech Ecosystems*, Ecological Studies 208,
DOI: 10.1007/978-3-642-00340-0_22, © Springer-Verlag Berlin Heidelberg 2009

F. sylvatica has been able to maintain its dominant role in most broadleaved stands due to the following biological characteristics:

- Broad site tolerance especially in relation to soil conditions (Ellenberg 1996; Leuschner 1998; Leuschner et al. 2006).
- High shade tolerance (Ellenberg 1996; Röhrig et al. 2006).
- Production of dense foliage to produce dark shade (Hagemeier 2002).
- Prolonged period of height growth increment and the capacity of dominant and predominant trees to expand their crown (Guericke 1999; Leuschner 1998).
- High plasticity and competitiveness in the rooting pattern (Hertel 1999; Horn 2002).
- Scant browsing attraction for ungulates compared to other broadleaved tree species (Ellenberg 1989; Ammer 1996; Schulze 1998; Partl et al. 2002).

In Germany, the proportion of beech of the total forest area has steadily decreased during the last two centuries. At the beginning of the Middle Ages, Germany was still densely forested and about 80% of the forest consisted of beech. As the population increased, many forest areas were cleared for agriculture and were increasingly used to fulfill other household needs. Thus, there was an expansion in the use of forest areas for grazing livestock, the collection of litter, obtaining wood for charcoal production and for extracting salt and other minerals. European beech was affected more than any other tree species. Beech was less suited than other tree species to coppice, and coppicing was the primary system of forest management until the mid-1900s. Thereafter, coinciding with industrial development, the demand for conifer wood increased leading to the reforestation of clear-felled areas and non-forest open areas with coniferous species, primarily with Norway spruce (*Picea abies* (L.) Karst.) and Scots pine (*Pinus sylvestris* L.). Reforestation with coniferous species proved to be simpler and cheaper to implement. Thus, by 1960, the percentage of Scots pine and European larch (*Larix decidua* Mill.) had increased to 35%, and of Norway spruce and other conifers to 37%, of the forest area in Germany. While beech forests on limestone soils were least affected by this conversion, beech forests on acidic soils were affected the most.

More recently, the proportion of broadleaved trees in the forested area in Germany, and of European beech in particular, is once again increasing. Between 1960 and 1990, this percentage increased from 21 to 25%, largely at the expense of pine and larch (which decreased from 35 to 31%) and to a lesser extent to spruce and other conifers (37% to 35%). In 1990, beech occupied 14% of the forest area in Germany (=1.465 million ha), of which pure stands comprised 46%, mixed forests with other broadleaves 28% and mixed forests with conifers 26% (Bick and Dahm 1993; BML 1994, 2003). The conversion of pure coniferous stands to mixed broadleaved–coniferous stands on sites unsuitable for growing broadleaved forests has continued in keeping with the stipulation that the mixed stands contain a significant proportion of spruce, Douglas fir (*Pseudotsuga menziesii* (Mirb.) Franco) and pine. A number of factors are taken into account when making such management decisions (Lüpke 2004; Knoke et al. 2008):

- In many areas of the temperate regions in Central Europe, broadleaved forests comprising a high percentage of beech are the best adapted forests in terms of their physiognomy, element cycling and ability to regenerate after disturbance.
- Species diversity and resistance to windthrow, forest fires and insect pests are usually higher in broadleaved and mixed stands than in pure coniferous stands.
- Broadleaved and broadleaved–conifer forests correspond to the characteristics of the natural landscape, particularly in the hilly to montane elevations, and as such fulfill most socio-cultural expectations and aspirations.
- Improved stand management and the application of site-specific regeneration methods have assisted in increasing the net income from appropriate beech forest management practices.

22.1.2 Environmental Amplitude

F. sylvatica has a wide distribution across the cool to moderately warm climatic zone with evenly distributed rainfall. In the northern reaches, there is no lower elevation limit to its distribution. From the Central Mountain Range to the south, *F. sylvatica* occurs predominantly in the sub-montane and montane elevations. It is excluded from areas where long severe winters and pronounced dry summer periods prevail. The minimal annual rainfall requirement for beech forests is assumed to be about 500 mm (Leuschner 1998). Beech is susceptible to the occurrence of late frost in the early growth stages, even where conditions are otherwise suitable.

In the temperate region, no other broadleaved species thrives on sites with such different pedological substrates as does beech, where it is found on soils derived from weathered limestone (calcium carbonate buffer zone) to sedimentary soils which are very poor in nutrients and are close to the iron buffer zone. The soil water regime tends to be the limiting factor to its distribution. Beech is less competitive than other species (primarily *Quercus robur* L., *Carpinus betulus* L., *Fraxinus excelsior* L.) on sites with temporary or permanent water-logging in the surface soils (Ellenberg 1996; Leuschner 1998; Fig. 22.1).

Innumerable attempts have been made to develop a systematic classification describing the ecological diversity found in beech forests. Some are based on the plant sociological characteristics (e.g. Oberdorfer 1992; Pott 1992) whereas others place more emphasis on the overall ecological parameters (Ellenberg 1996; Leuschner 1998). In this chapter, we have adopted the latter classification and have confined the discussion to three widely distributed forest types:

- *Beech forests on rendzina and pararendzina* soils occur primarily in limestone mountainous areas. The water regime differs considerably and the forest communities range from “moist beech forests on limestone” to “dry beech forests with a sedge under-storey on dry slopes”. Beech forests are found at almost all altitudes, from the hilly to the montane zones (with *Abies alba* Mill.) and in the

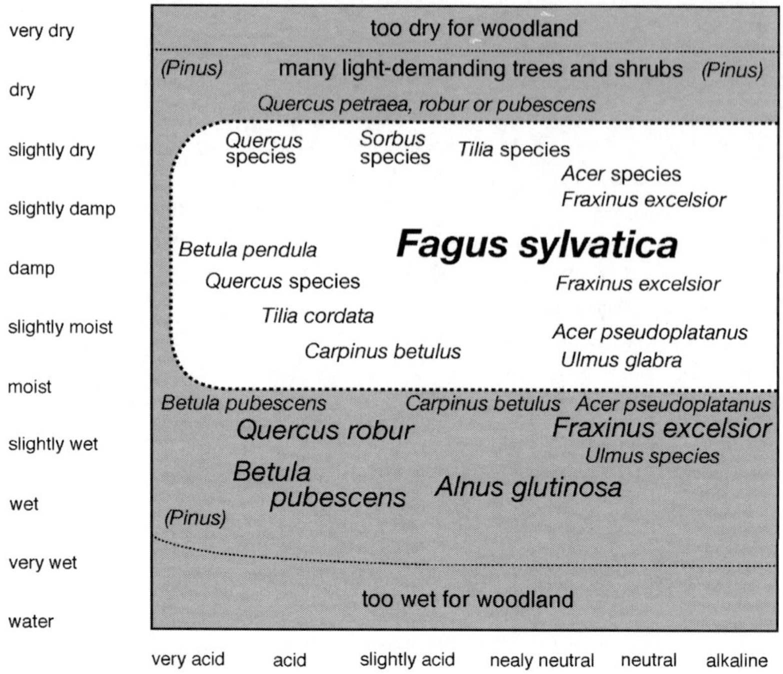

Fig. 22.1 Ecological characteristics of forest tree species in the sub-montane zone in Central Europe in relation to soil moisture content and acidity (from Ellenberg 1996; Leuschner 1998). The *dotted line* defines the approximate areas where *Fagus sylvatica* is dominant. The *size* of the print describes the approximate extent of its presence in the canopy layer

sub-alpine zone (forests comprising *Acer pseudoplatanus* L. and *F. sylvatica*). They are all characterised by a species-rich ground vegetation (number of species decreasing from south to north) and a mull type of litter layer (beech forests on mull). Example: Göttinger Wald research site.

- *Beech forests, and mixed beech forests on rich to moderate brown earths* occur primarily on (sandy) loams. Some soils are formed from glacial deposits but more often from the weathering of rocks containing mineral-rich silicates and are often covered by a thin loess layer. The water supply in these soils is variable, although most are primarily moist soils that sometimes have gley characteristics in the subsoil. Noble hardwood species (primarily *A. pseudoplatanus*, *F. excelsior*) may also be found on these sites, and even oak on flat and sub-montane sites. Plant sociologists have classified numerous ground vegetation communities in these forests that have developed in response to altitudinal differences and variability in water supply. The litter layer ranges from moder with mull characteristics to mull. Example: Zierenberg research site.
- *Beech forests, and mixed beech forests on very acidic soils* are derived from silicate rocks, sandstone or from partly sandy glacial sediments which are poor in nutrients. The soils often have a surface layer of decalcified loess. They occur

primarily in the Central Mountain Ranges (submontane to montane zone) and occasionally in the lowlands. The number of species in the ground vegetation is distinctly lower due to the low base cation content. Although differences in these communities are distinguishable and classified accordingly, they are often collectively referred to as *Luzulo-Fagetum* communities due to the frequent occurrence of *Luzula luzuloides*. In these communities, the litter layer is usually moder type, which can become mor like when soil conditions become unfavourable. A small percentage of oak is occasionally present, particularly where a moderately developed pseudogley occurs (*Q. robur*) or where dry summers limit the competitiveness of beech to oak (*Quercus petraea* Liebl.). Example: Solling research site.

22.1.3 Growth

European beech is a typical shade tolerant tree species. When the amount of light entering the stand is limited, generally shade leaves develop. These are conspicuously thinner than sun leaves and have a larger surface area. Young seedlings of the natural regeneration can survive when photosynthetically active radiation (PAR) is less than 10% of full sunlight, albeit with little if any growth increment. At radiation levels of up to 20%, shoot height growth is minimally whereas diameter increment, aboveground biomass and particularly root growth are significantly reduced. After about a decade, however, this growth pattern of beech places it at an advantage over other more light demanding species (with the exception of *A. alba* and *Taxus baccata* L.) in closed stands.

The formation of a large leafy, dense crown is a characteristic of dominant and predominant beech trees. Such a crown casts dark shade over the surface organic layer corresponding to the stand area it occupies. Ellenberg (1996) classified the beech alongside fir (*A. alba*), hornbeam (*C. betulus*) and yew (*T. baccata*) as one of the tree species which casts the darkest shade on the surface organic layer in Central Europe. In order to capture crown space and produce shade, beech consumes less nutrients and biomass than other tree species (particularly pioneer species) (Hagemeier 2002). Even as a mature tree it is capable of expanding into free canopy space by 10–30 cm annually when the opportunity arises (Guericke 1999).

Height and diameter growth of young beech trees is slow by comparison to more light-demanding tree species but the duration of its continual growth with age is comparatively longer (Fig. 22.2). Consequently, in pure stands, the volume of timber produced, although dependent on site conditions, is considerable. In classical forest mensuration, yield tables based on tree height are used to assess stand production. An example is provided in Table 22.1. It is now known that a significantly higher wood production can occur on good and moderately good sites (Spiecker et al. 1996; Pretzsch et al. 2000).

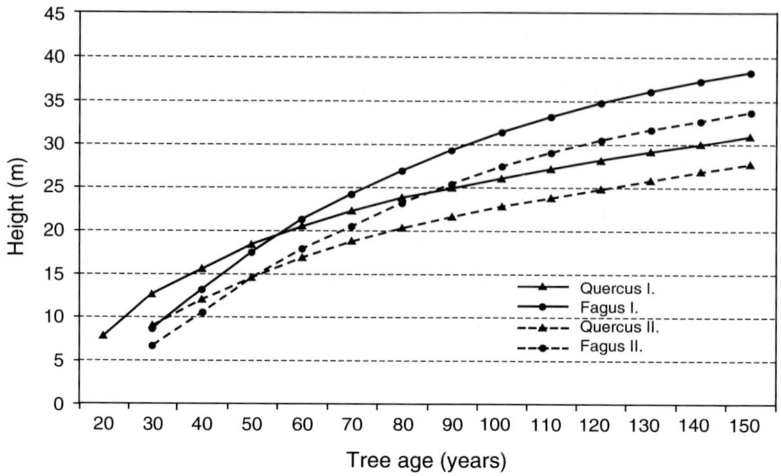

Fig. 22.2 Height growth of *Fagus sylvatica* and *Quercus* in yield classes I and II (moderate thinning) (data from yield table of Schober 1987)

Table 22.1 Yield table for *Fagus sylvatica* for moderate thinning treatment (DBH, diameter breast height) (from yield table of Schober 1987)

Yield class	Age (years)	Mean height (m)	Mean diameter (DBH) (cm)	Basal area (m^2)	Stand volume (m^3)	Mean annual increment (m^3)
I	100	32.4	33.6	31.0	472	7.8
	140	37.3	50.2	33.3	617	8.6
II	100	27.4	29.3	30.2	426	6.4
	140	32.7	43.7	32.5	533	7.2
III	100	23.3	25.2	29.3	338	5.0
	140	28.2	38.2	31.7	451	5.7
IV	100	19.2	21.1	28.4	270	3.7
	130	23.7	32.8	30.7	369	4.4

Beech develops a dense semicircular root system (heart root system) with an irregular pattern of coarse and fine root growth. In contrast to *P. abies* and *P. sylvestris*, the horizontal roots do not extend far beyond the breadth of the crown. However, on flat limestone sites, beech root systems do extend further with deep roots in joints of the bedrock. The intensity of fine roots in the humus-rich surface soil is normally high. If oxygen levels in soils are low, or the soil acidity is high, or where a mechanical impediment occurs in the soil, then beech develops a flat root system.

The beech root-ball is small compared to that of other tree species (including *P. abies*). The large number of highly ramified fine roots indicates that beech has a large root surface area and a heavy root-ball that provides a stable anchor for trees on most sites (Polomski and Kuhn 1998, 2001; Kuhr 2000; Fehrmann et al. 2003). This root morphology renders beech very stable in storms during the winter months when leaves are absent and most storms occur (Otto 2000).

22.1.4 Crown Damages and Tree Mortality

After it had become apparent towards the end of the 1970s that the forests were in a poor condition, environmental monitoring was introduced in Germany in 1984 (and later in other European countries). As part of the process, an estimation of the crown condition for the most common tree species was undertaken in all German states. These data on the annual state of the forest are provided in reports published regularly by the German Federal government since 1984. Beech stands have revealed variable crown damage, with 30% of trees on average having extensive visible crown damage. Only 20–25% of the trees were found to have no visible damage in the last 20 years. Initially, it was assumed that this damage was caused by the impact of air pollution on the trees and on the soil. Since then, however, crown damage has been attributed to other causes that may be related indirectly to pollution, for example weather conditions, fructification, insects and diseases. Damage caused by insects and diseases is now termed the "complex beech disease". It begins with the infestation of the stem by the beech woolly aphid (*Cryptococcus fagisuga*), which then often attracts the fungus *Nectria coccinea*. As a result, necroses develop on the stem with very different outcomes, varying from complete tree recovery to death of the tree following infestation with white rot and xylobiotic beetles (Lunderstädt 1992, 2002; Eisenbarth et al. 2001). This disease occurs in beech stands in epidemic proportions every few decades. Such stands need careful monitoring to identify any latent infestation, with particular attention to stands on moist sites and to dominant trees. About 40 years ago in southern Lower Saxony, major damage was recorded in beech stands (Wagenhoff and Wagenhoff 1975). Stands occurring on sites with dolomite and red clay soils rich in base cation soils, such as those found at the research site in Göttinger Wald, were most affected. Dead trees were rarely found in stands on sandstone soils poor in base cations.

22.1.5 Wood Quality

Wood from beech is not particularly valuable because it does not possess any outstanding technical or decorative characteristics. Moreover, wood defects often occur in older stems, reducing the wood value (spiral grain, large branch knots, wood discoloration; see Knoke and Schulz-Wenderoth 2001). Although small stems and stems with major defects are difficult to sell, good stem cuts are in high demand for plywood, and good sawn timber can be readily marketed.

The percentage of usable stemwood during a single forest generation is generally lower for beech than for some other species (for example, spruce and Douglas fir).

22.2 Common Forest Management Practices for European Beech

22.2.1 Regeneration

Natural regeneration is of paramount importance for beech, more so than for any other tree species in Central Europe. As a typical shade tolerant species in closed forests, beech regeneration can occur under relatively limited light conditions. It is, however, sensitive to late frost as a young plant if it becomes too exposed and is susceptible to competition from ground vegetation. The natural regeneration of beech is usually more economical than its planting, which is time consuming and is associated with high risk of failure.

Large quantities of seed (heavy crops have more than 250 beech nuts m^{-2}) occur in beech stands every 5–15 years (Körner 2005). In places, one can find more than 1,000 nuts m^{-2} (Röhrig et al. 2006). The frequency of the fructification for beech (and also for other forest trees) appears to have increased in the last 30 years compared to earlier years. Reasons for this change in the frequency of fructification may be an increase in nitrogen inputs, intensification of thinning practices and climate change (Schulze and Eichhorn 2000; Eichhorn et al. 2003; Paar et al 2004; Röhrig et al. 2006, Schmidt 2006). The quantity of seed in different cropping years at Solling is provided in Table 22.2. In the stand at Zierenberg, more than half the mature beech trees had developed fruit in the period 1989–2000 and in the years 1992, 1995 and 1998 (Schulze and Eichhorn ; Paar et al. 2004). In 2000, heavy crop was produced on all the three sites (Chap.11).

Only a modest proportion of the large quantity of seed reaching the surface organic layer in the crop years germinates as observed from seedling numbers in the following spring. This occurs primarily due to the seed being removed by birds and

Table 22.2 Quantity and quality of beech seed in a beech stand in Solling in different years

Year	Number of seeds (n m^{-2})		Empty or parasitised seeds
	Mean	Maximum	(% of mean)
1987	356	722	15
1989	70	120	25
1990	451	668	15
1992	652	1,276	13
1994	28	132	87
1995	996	2012	21
1998	128	348	22

rodents, the infestation of seedlings by fungus and desiccation. Soil cultivation may restrict seed losses (see below).

Natural regeneration is regarded as successful if at least 20 seedlings m^{-2} fairly evenly distributed across the regeneration area are found in the following autumn, and if five seedlings m^{-2} are found in the 10th year after regeneration (Huss 1972; Dohrenbusch 1990).

The high shade tolerance of the young growth, the rapid increase in diameter growth increment of beech when the stand is opened up and the relatively low danger of windthrow in a storm suggest that regeneration under the canopy of a mature stand is particularly suited to beech. Shelterwood harvesting and group selection harvesting methods have proven to be appropriate regeneration methods. Both methods, depending on intensity and frequency of the seed crop, on the micro-site conditions and on harvesting progress in the mature stand, result in more or less uneven-aged regeneration areas. As this growth dynamic resembles what would occur in virgin forests dominated by beech (see Sect. 22.4), these regenerated forests are regarded as close-to-nature forests. However, in contrast to unmanaged natural forests, forest management practices prevent managed beech forests from attaining the mature and senescent growth phases of stand development, as the trees are removed once they have reached certain optimal sizes for which the dimensions are specified (Thomasius and Schmidt 2003).

The aim of the shelterwood harvesting method is to obtain regeneration over a larger area ($\geq$1 ha) within several years or a few decades. By opening up the crown fairly evenly in the mature stand, uniform ecological conditions are created to encourage the development of even-aged pure stands (Fig. 22.3). Open regeneration areas ease further harvesting operations (reduction of damage to old trees and young saplings) and fencing, which is often necessary for the protection of seedlings from grazing. It also assists additional silvicultural practices in the young stand (replanting, removal of poor quality plants).

The group selection harvesting method of regeneration differs from the shelterwood system in that the crown is deliberately opened up in an irregular manner to encourage regeneration. In some areas, canopy gaps are created either by cutting groups of trees (0.03–0.1 ha) or clumps of trees (0.1–0.5 ha). In other parts of the stand, smaller openings are created, or a closed crown is retained. As the harvesting progresses, the size of stand area that has been opened up increases until the remaining mature stand is eventually cleared (Fig. 22.4). In this way, from the beginning of the regeneration felling to the clearing of the last old trees, ecologically diverse areas are created within the stand. At some stage in the regeneration period, most tree species will encounter the specific ecological requirements that favour its growth. Group selection logging is thus first and foremost a method applied for the natural regeneration of mixed stands. It is largely used for mixed stands of fir, spruce and beech, but also for other species mixtures, e.g. noble hardwoods with beech or for the artificial introduction of admixed tree species.

In the last 10 years, efforts have been made by the State Forest Departments in Germany to replace the traditional regeneration methods of shelterwood harvesting and group selection harvesting with *target diameter harvesting*. This method is

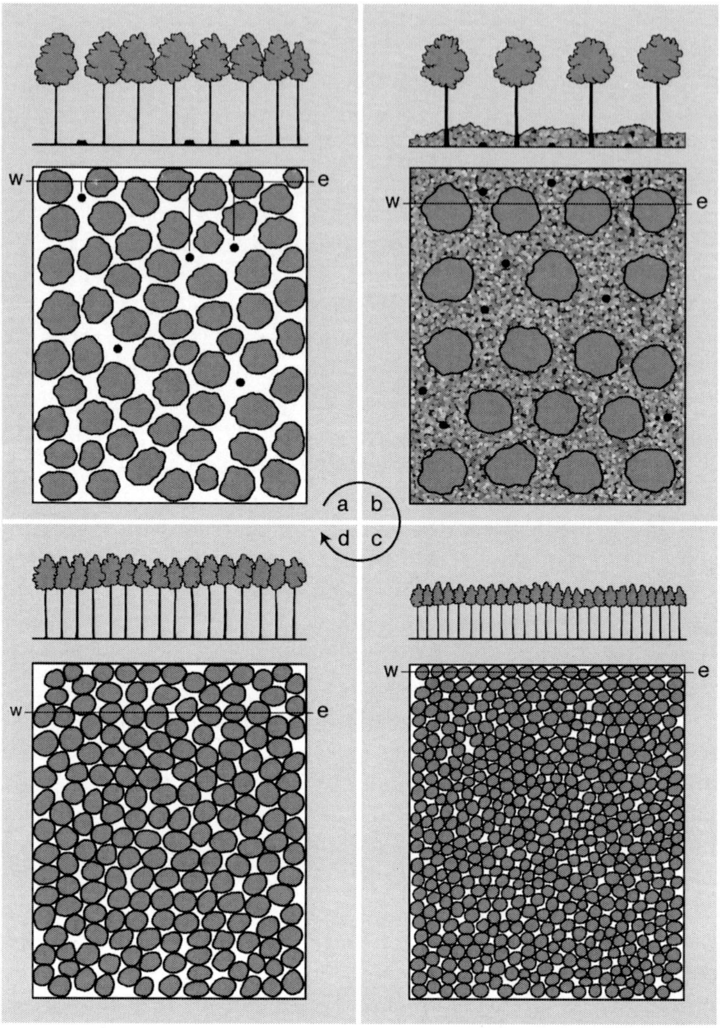

Fig. 22.3 Shelterwood cutting diagram (from Röhrig et al. 2006). (**a**) In a crop year after the seeds have fallen, a shelterwood harvesting operation is carried out fairly evenly across the entire area (seed harvesting). The intensity of the operation is dependent on the actual crown density, the anticipated development of the ground vegetation and the light requirements of the regeneration in the first few years. Skidding by harvesting vehicles often causes some soil disturbance which may favour regeneration. If a thick surface organic layer is present or dense ground vegetation is likely, additional soil amelioration is undertaken. (*w* west, *e* east). (**b**) Following the earliest harvesting operations, the regeneration commences in the second year after the seed harvesting (light secondary harvesting). The intensity and the frequency of the secondary harvesting depend upon the light requirements of the regeneration and its sensitivity to frost. (**c**) In a final harvest the last remaining trees in the mature stand above the regeneration are removed. (**d**) Result of the shelterwood cutting: predominantly even-aged stands with evenly spaced trees

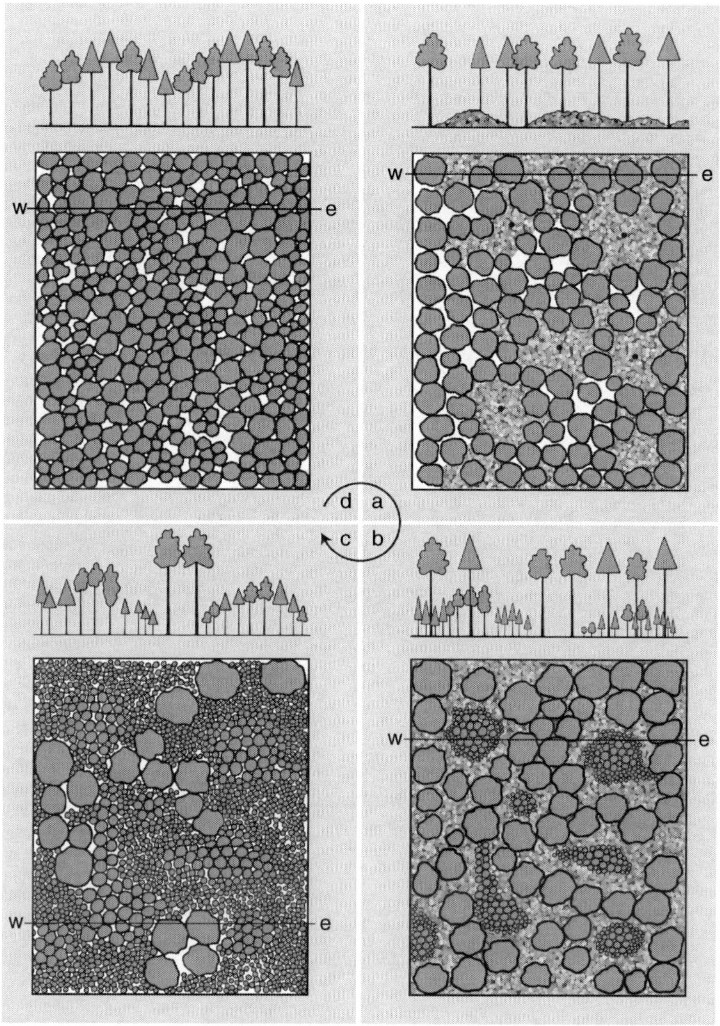

Fig. 22.4 Group selection cutting diagram (from Röhrig et al. 2006). (**a**) The tree regeneration is initiated with an irregular harvest to open up the canopy. The first cutting (preparatory harvesting) is conducted mainly to free the regeneration from slow growing shade tolerant trees in the sub-canopy (fir, beech). Groups of stunted trees exist in most mature stands. If these trees have the potential to grow and develop they are freed from competition. Otherwise, the stand is opened up in mast years (seed harvesting) in areas that appear to be suitable for regeneration depending on the condition of the mature stand (species composition, age, crown cover) and the soil (humus type, soil cover). This may be achieved through the creation of small gaps in the stand (gap cutting) or small-scale shelterwood harvesting. (**b**) and (**c**) The regeneration nuclei (selected groups) are expanded by a secondary harvest in the mast years. This involves harvesting of trees from small circular patches primarily in the shady boundaries (on the south to the west side) of the regeneration core, and a somewhat wider peripheral felling of trees to open up the canopy. In addition to the expansion of the existing selected group, new groups are laid down during the regeneration life span. (**d**) Result of the group selection harvesting: uneven-aged stands with irregularly spaced trees

based on the principle of utilising each tree individually once the diameter has attained a specific economically optimal size, that is its target diameter (for beech: 60–70 cm diameter at breast height, 1.3 m above the ground, commonly at 100–110 years of age). The trees that are still too small to harvest are left until they have reached the target diameter. Beech is capable of regenerating adequately in the small gaps that are created through the application of this method. However, the light conditions at the surface organic layer are inadequate for the regeneration of tree species that are less shade tolerant (see below). In addition to opportunities to increase the production yield from the harvest, the aim of using the target diameter harvesting is to create a structurally diverse forest (Otto 1992). However, target diameter harvesting initially reduces structural diversity due to the absence of the larger diameter trees that had been removed because they reached the target diameter. The average size of the trees harvested is lower and more uniform as a result. The aim to increase the structural diversity can be achieved in the next generation, however, as the next generation grows on small areas under canopy cover for long periods.

When tree species with a lower shade tolerance than beech are present in the re-growth, it is necessary to achieve a balance between the shelterwood and target diameter harvesting methods so that the crown density is appropriate for the regenerating species. To correspond with the higher light demands of the admixed species, the density of the canopy needs to be reduced by creating larger stand gaps (Table 22.3). By harvesting small areas so that the light intensity reaching the surface organic layer is 7–15% of full sunlight, favourable conditions are created for shade-tolerant tree species (order of increasing light demand: *A. alba, F. sylvatica, C. betulus, Tilia cordata* Mill. and *T. platyphyllos* Scop., *Pseudotsuga menziesii*). If moderately-sized gaps are cut with an associated light intensity of 15–30% of full sunlight, species requiring more light such as *P. abies, Q. robur*, and *Q. petraea, A. pseudoplatanus, Acer platanoides* L., and *A. campestre* L., *F. excelsior, Sorbus aucuparia* L., and *Sorbus torminalis* (L.) Crantz can regenerate and survive for long periods, though beech on such sites remains more competitive. For larger cuttings with 30–70% of full sunlight reaching the surface organic layer, the middle group of

Table 22.3 Light intensity on the surface organic layer (PAR in % of full sunlight) at the centre of gaps for different harvesting methods in mature beech stands (from Lüpke 2004)

Cutting method	Gap size		Light intensity (% full sunlight)	Height of mature stand (m)	Reference
	Diameter (m)	Area (m^2)			
Single tree selection	11	97	7	32	Tabaku (1999)
Shelterwood	n/a	n/a	11	34	Lüpke (1982)
Target diameter harvesting	13	140	14	34	Tabaku (1999)
Group selection	17	250	12–15	34	Lüpke (1998)
Group selection	30	710	20–30	32	Schmidt et al. (1996)
Group selection	50	2,000	45–60	34	Lüpke (1982)

trees mentioned above can grow well and can hold their own against beech. The growth conditions are also sufficient for the light-demanding tree species *Betula pendula* Roth. and *B. pubescens* Ehrh., *S. aucuparia*, *P. sylvestris*, *Larix decidua*, *Populus tremula* L., *Salix alba* L. and *Alnus glutinosa* (L.) Gaertn. (Lüpke 2004). The ground vegetation often becomes increasingly dense as the stand develops and forms a major hindrance for regeneration, especially in stands comprising light demanding species. This lack of regeneration is not always the consequence of errors in stand management (lack of shade-tolerant admixed tree species in the under-storey, opening the stand up too quickly). It often corresponds to natural successional processes with or without prior disturbance of the stand structure.

22.2.2 Soil Scarification

Soil scarification, where ground vegetation and humus layer are mechanically disturbed and mixed into the upper mineral soil, has been found to be most effective for the success of natural regeneration especially during the years when a heavy seed crop is produced. In some cases, soil scarification is a necessity, for example on sites with a deep moder humus layer (Solling research site) or where ground vegetation inhibits regeneration (Zierenberg research site). In shelterwood harvesting, the soil is treated in strips. The benefits of soil treatment increase when conducted after seedfall as the seeds are then worked into the upper mineral soil (Burschel et al. 1964). The depth of soil treatment is usually between 10 and 20 cm and the percentage area treated lies between 30 and 70% (Dohrenbusch 1990). However, soil scarification may damage fine root systems of standing trees and induce an over-stimulation of mineralisation (see Sect. 22.5).

22.2.3 Liming

Humus types on sites with acidic soils (Solling research site) that do not favour germination and initial growth can be improved by liming. This is commonly carried out in conjunction with soil treatment (Rehfuess 1990; Ammer and Makeschin 1994; Eberl 2001; Nilsson et al. 2001; Schüler 2002). Generally 3–5 t ha^{-1} of lime, mostly dolomitic limestone, is spread over acid soils which are poor in base cations to prevent, or at least impede, acidification due to atmospheric pollutants and to improve the availability of the base cations, Ca and Mg. Larger amounts of lime may accelerate mineralisation, causing nutrient losses. Liming may reduce soil acidity, improve litter decomposition and the availability of nutrients to plants, improve soil humification processes, increase activity of micro-organisms and soil flora and improve bioturbation and the physical soil structure (Rehfuess 1990; Marschner 1993; Wolters et al. 1995; Blume et al. 2002,

Ammer and Huber 2007). Generally, liming becomes effective in improving the soil conditions after a number of years depending on the ecological conditions and may remain effective for a few decades (Immer et al. 1993; Nilsson et al. 2001; Ammer and Huber 2007). Where Mg is more readily exchanged in the soil, the response to liming is quicker (Hüttl 1990; Munk and Rex 1992; Meiwes 1994; Illmann et al. 1997; Blume et al. 2002) and can affect soils to deeper depths.

22.2.4 Thinning

For thinning operations, single trees are selected and removed to increase the high valued growth increment in the remaining stand. However, in the above-mentioned target diameter harvesting method, the best valued timber is removed. Thinning operations in young beech stands (diameters <15 cm at 1.3 m height) are exclusively implemented to promote the growth of admixed species and to remove particularly vigorous trees of poor quality. Only minimum silvicultural measures are carried out for economic reasons since the timber obtained is non-merchantable. In beech stands, a major reduction in stand density is avoided until the trees, by virtue of growing up in the shade of the crown, have developed a clean bole of about 6–8 m length. This is reached when the trees in the upper canopy have attained a height of about 15 m (tree age 40–60 years). Afterwards, the good quality trees (80–160 trees ha^{-1}) are favoured by the removal of the competition trees every 4–6 years to promote diameter growth increment. The crowns of the less shade-tolerant tree species that are potentially more valuable than beech, present on sites with soils rich in base cations such as in Göttinger Wald and at Zierenberg, are freed up from the more competitive beech. With increasing stand age, silvicultural operations provide merchantable timber. These operations are continued until all the trees are finally removed in a target diameter cutting, or until shelterwood or group selection harvesting methods are initiated to promote natural regeneration.

22.3 Specific Silvicultural Characteristics and Treatments of the Three Beech Ecosystems

The three beech forest ecosystems presented in this volume differ essentially in relation to soil conditions (Chap. 3), in species mix in the canopy and herbaceous layers (Chap. 5), as well as in management practices within the forests to date (Chap. 5). These factors determine the specific forest management practices for the stands.

22.3.1 Solling

Stand establishment and silvicultural practices (Chap. 5) carried out in the approximately 165-year-old beech stand at the Solling site are typical for forests growing on soils poor in base cations in that area. These conditions are considered optimal for beech as it outgrows all other indigenous tree species. The shelterwood harvesting system in conjunction with soil treatment to remove the moder humus layer that inhibits regeneration results in a pure single-storey stand in the optimal growth phase.

Only a fraction of *beechnuts* germinate and survive for more than one growing season. An investigation of the 1983 seed crop at the Solling research site showed that 18% of the seeds (177 beech nuts m^{-2}) germinated in May 1984 under 95% canopy cover. Of these, 11.4% survived until November. On a limed area (60 tons ha^{-1} dolomite), the germination rate of 1.7% was much lower due to greater losses during the winter. Yet by the end of the growing season, the rate of survival of 50% was considerably higher. Liming had increased the biomass of seedlings by 50%. This was entirely due to an increase in the aboveground biomass (Jansen 1987).

Light conditions are of minor importance for the development of beech regeneration. However, the amount of light at the surface organic layer is critical for seedling development. In a beech stand bordering directly onto the research site at Solling, with similarities to the Solling site in terms of its age, the nature of its formation, and past silvicultural practices, an investigation of the beech regeneration was carried out in gaps 30 m in diameter (Bartsch et al. 2002). Ten years after the stand was opened up, about eight beech seedlings m^{-2} were found both at the centre of the gap and under the crowns of the boundary trees at the north edge. Under the crowns of the boundary trees at the south edge of the gap, there were 13.5 seedlings m^{-2}. The seedlings originated from several seed crops and had reached a maximum height of 50 cm. In contrast, in the closed stand, only isolated seedlings of up to 2-years-old were found with a maximum height of 10 cm. Sparse soil vegetation that developed in the gaps presented no impediment to natural regeneration. This differed in the gaps where lime was applied after harvesting (3 tons ha^{-1} dolomite was spread during cultivation of surface soil layer). On these sites, beech regeneration was more prolific after the first three seed-crops than on the unlimed gaps. However, in the following years, regeneration was almost entirely absent on the limed gap where a dense layer of *Rubus fruticosus* and *R. idaeus* vegetation had developed and provided major competition. The light conditions on the surface organic layer under the ground vegetation cover emulated those in a closed beech stand. The plant densities under the boundary trees of the limed gap at the north (20.4 seedlings m^{-2}) and south edge (13.8 seedlings m^{-2}) were higher than for the corresponding areas in the unlimed gap (Bartsch 1997; Bartsch et al. 2002).

The aims of the current silvicultural strategies adopted by the state forest departments are to promote mixed stands which have structural diversity (NMELF 1996a, b). In the Solling area, *P. abies* is considered the most important admix species in beech stands due to its relatively high growth rates, although the species is not native to the area. On sites with better base cation availability, *A. pseudoplatanus* is also a suitable admix species (Otto 1991). In order to introduce admixed tree species with higher light demands, it is proposed to open up of the existing beech stands with higher intensity and more irregularity than has so far been achieved by the shelterwood harvesting system (Pampe 2000; Kühne and Bartsch 2003; Lüpke 2004). The absence of seed trees for the admixed species would need the planting of desired species.

22.3.2 Göttinger Wald

On the limestone sites in southern Lower Saxony, a small-scale mosaic of soils with shallow to moderate rooting depth occurs. These soils are biologically very active with low water storage capacity but a high nutrient supply, particularly of base cations. There is little inhibition to regeneration. Thus, measures to promote regeneration such as soil treatment and liming are unnecessary.

The Göttinger Wald is comprised of a species-rich *mixed forest of broadleaved tree species* with different light demands and different growth patterns. The shade-tolerant beech is dominant in the stands and covers up to 90% of the surface area (Chap. 5). The light-demanding tree *F. excelsior* is the most common admixed species in the stand. Additional noble hardwoods with different levels of shade tolerance are present in isolation and include *A. pseudoplatanus*, *A. platanoides*, *Ulmus glabra* Huds., *Tilia platyphyllos*, *T. cordata*, *Prunus avium* L., *S. torminalis*, *Acer campestre* and *C. betulus*. The oak species (*Quercus petraea*, *Q. robur*), which were favoured in past coppice of standard forest systems, are present on isolated areas. *T. baccata*, the only coniferous species present, was of common occurrence in the past.

The overriding dominance of beech in Göttinger Wald (Chap. 5) may be attributed to past silvicultural practices. Until 1855 the stand, presently 120–135 years old, was managed as a coppice where beech was suppressed and oak and ash species fostered. Well-stocked areas within the stand that promised a good supply of timber were kept whereas other areas were subjected to a large-scale shelterwood harvesting to promote natural regeneration. The under-storey trees were thinned lightly to keep the canopy closed. This regeneration method and the thinning regime were more favourable for beech than for the noble hardwoods, which created forests of low structural diversity. The cessation of harvesting more than 40 years ago has further reduced the species mix (Wagenhoff 1975, Chap. 5).

Beech–noble hardwood forests on limestone sites produce ample regeneration under the appropriate light conditions. The composition of tree species in the *natural regeneration* is usually extremely different from that of the mature stand. At the research site in Göttinger Wald, with its closed canopy, regeneration consists primarily of *F. excelsior* and beech. *A. pseudoplatanus*, *A. platanoides*, *U. glabra* and *S. aucuparia* are present in decreasing proportions. In 2001, about one-third of the area was covered with regeneration. Some regeneration formed part of the shrub layer (Chap. 5).

If the canopy is opened up further, such as may occur in group selection harvesting, the density, the height increment and the proportion of noble hardwoods would remain higher. In two gaps with diameters of about 30 m which were established in a neighbouring mixed beech stand in 1989 the amount of regeneration cover was closely linked to light availability (Schmidt 1997; Lamberts and Schmidt 1999). On the micro-sites within the gap with no canopy cover, the noble hardwoods dominated beech for tree height and density. Ash, in particular, proved to be very competitive. It was able to survive in the shade for many years and, once gaps were created, grew rapidly to form the upper tree canopy. Height growth of young beech seedlings was comparatively slow but outgrew the admixed species in later years. Height and density of regeneration trees indicated that *F. excelsior* and other noble hardwood species were suppressed at the gap edges due to their higher demand for light than of beech. When *A. pseudoplatanus, A. platanoides* and *U. glabra* were present in the gap regeneration they were able to outgrow ash (Wagner 1996).

The competition among tree species may be altered by ungulate game. The regenerating beech with noble hardwood species is less affected by selective browsing of ungulates, mainly roe deer (*Capreolus capreolus*) (Brötje and Schmidt 2005; Huss and Butler-Manning 2006). *Ulmus* and *Fraxinus* are often more browsed than *Acer* (Ellenberg 1989; König 1997).

22.3.3 Zierenberg

The 157-year-old stand on basalt at Zierenberg was converted to high forest from a coppice with standards forest system, partly by planting beech (Chap. 5). In addition to beech, other species such as *F. excelsior* and *A. pseudoplatanus* are also present in the mature stand. Oak trees from the previous coppice are no longer present. In contrast to the stands of Solling and Göttinger Wald which when opened by harvesting did not result in the development of ground vegetation that restricted regeneration, the stand at Zierenberg developed a dense under-storey of *Urtica dioica* affecting the success of regeneration practices.

Despite many years of heavy beech fructification and silvicultural practices in the Zierenberg stand to encourage natural regeneration over the last three decades, the quantity of regeneration remained inadequate. In 1970, practices to open up the stand were commenced. In the following years, three additional harvesting operations were carried out. In 1972, stinging nettles were sprayed with herbicide. In 1980, the soil was cultivated in some areas. The natural regeneration of beech and other tree species (*F. excelsior*, *A. pseudoplatanus*, *A. platanoides*) is especially poor in the areas dominated by *U. dioica* (Fig. 22.5). Beech regeneration was inhibited where the degree of *U. dioica* cover was more than 30%. Beech cannot regenerate on areas where *U. dioica* cover exceeds 90% (Fenner et al. 2000). Such areas are characterised by a high fraction of magnesium on exchange sites in the

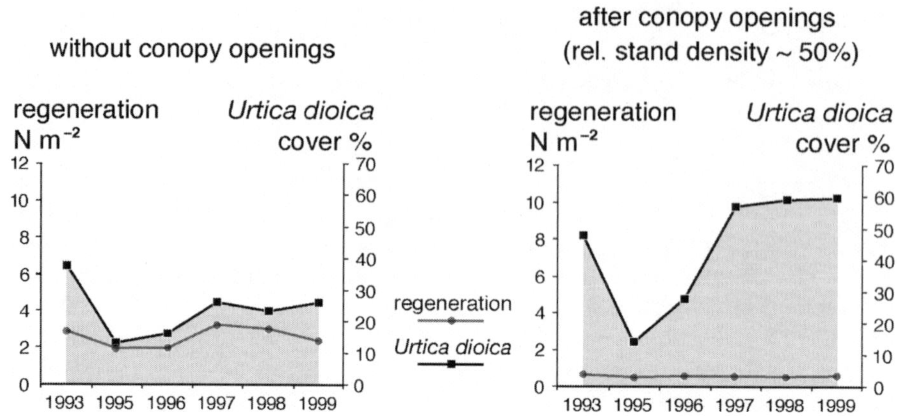

Fig. 22.5 Density of beech regeneration and degree of cover of *Urtica dioica* in areas within the stand with different degrees of light on the surface organic layer of beech stands in Zierenberg (from Eichhorn et al. 2003)

soil and by a higher light intensity (Mrotzek 1998; Schulze and Eichhorn 2000; Eichhorn et al. 2003). Light conditions of 2–4% of full sunlight are necessary for *U. dioica* growth and, where they exceed 8–10% of full sunlight in the stand, this species can become dominant in the under-storey (Mrotzek 1998; Fenner 1999).

Thus, control over the intensity of light in the stand is required for tree regeneration to be successful. Values of under canopy cover below 60% and big gaps should not occur. This may be achieved by maintaining a sub-canopy layer and under-storey. Once the number and quality of the established regeneration is adequate, the stand can be opened up further with due care.

Areas with dense *U. dioica* growth can only be regenerated by planting. Tall seedlings (*Fraxinus* and *Acer* as well as *Fagus)* need to be introduced under the canopy of the mature stand. On the fenced areas with rendzina soils in Zierenberg, the natural regeneration of *Fraxinus* and *Acer* is as successful as on the limestone sites in Göttinger Wald.

Soil treatment did not eradicate *Urtica*. It did, however, cause excessive nitrification leading to nitrate leaching and the loss of base cations. Application of lime is unnecessary on this soil as this basalt soil is rich in base cations.

22.4 Naturalness of the Forest Management Practices

Due to the absence of virgin forests or forests that have remained unmanaged for long periods (natural forests) in Central Europe, the knowledge of structure and processes in such close-to-nature beech forest ecosystems is primarily based on the virgin forests still found in Southern and Eastern Europe (Albania, Slovenia, Slovakia, Poland, and Ukraine). These forests are located at higher elevations in mountainous regions and represent forest types that are comparable to forest communities in Central Europe. From research undertaken in these forests, the following *characteristics of close-to-nature beech forests* can be derived (Falinski 1986; Korpel 1995; Tabaku and Meyer 1999; Saniga and Schütz 2001; Meyer et al. 2003; Standovár and Kenderes 2003; Christensen et al. 2005):

- A clear dominance of beech. Depending on successional processes and site conditions admixed species may be present periodically or continuously in small numbers.
- Major differences in tree diameter and stand structure in small areas. Beech forest regeneration occurs in small canopy gaps resulting in a mosaic of small areas at different stages of development. The development of single-layered stands tends to occur in the young to early mature development stages.
- Presence of mature trees as single trees or in groups.
- Presence of standing or fallen deadwood of large dimensions in isolation or in clumps.

At the three research sites, the ecological conditions are optimal for the growth of beech (Sect. 22.1). Without forest management, beech (with the exception of micro-sites at Göttinger Wald with shallow rendzinas and a dry summer period) would dominate the forests in the long-term, and the admixed species that had previously been favoured by forest management practices would regress.

In order to evaluate how natural an ecosystem is, the degree of *hemerobity* is determined. Hemerobity is a measure of human cultural influence on the development of an ecosystem that hinders an ecosystem reaching the final stage of succession. Criteria for the determination of the degree of hemerobity include species composition, indicator values, the continuity of landuse and structural parameters (including deadwood) (Kowarik 1999). The influence of liming has proved difficult to assess since, as yet, there is no known way of reversing the effects of liming on species composition (Schmidt 2002). Research in the Solling area has shown that in limed stands only a few ruderal species that are generally absent in forests have increased in number. The *Luzulo-Fagetum* species composition is maintained with or without liming treatments.

Due to intensity and timing of the harvesting operations in the three above-mentioned beech forests and the application of regeneration methods adapted to the natural forest dynamics, these forests demonstrate the lowest degree of hemerobity. As such, they constitute forests with the highest degree of naturalness that may be achieved in managed forests (Dähne et al. 1997; Kowarik 1988, 1999; Weckesser 2002; Thomasius and Schmidt 2003).

Canopy gaps in close-to-nature beech forests originate primarily from the death of individual trees as a result of old age or windthrow. The expansion of these small gaps due to sun scald or windthrow is uncommon. Thus, the expanse of gaps is, on average, smaller than that of the projected crown area of mature beech trees (Tabaku and Meyer 1999; Standovár and Kenderes 2003). Large-scale disturbances are rare in the broadleaved forests of Central Europe (Bengtsson et al. 2000), but they may result from storms of extraordinary severity. In 1990 a severe storm

Table 22.4 Gap size (arithmetic mean and maximum) and gap area (percentage area per 10 ha research plot) in Albanian virgin forests, unmanaged forest reserves and managed stands in Central Germany (from Tabaku and Meyer 1999)

Stand	Gap size (m^2)		Gap area
	Mean	Maximum	(% of stand area)
Virgin forest			
Miridita	74.2	273.3	6.6
Puka	60.6	132.9	3.4
Rajca	68.7	208.6	3.3
Forest reserve			
Heilige Hallen	183.0	1094.4	13.3
Limker Strang	102.4	808.0	4.9
Managed stands			
Bleicherode	97.4	348.5	8.4
Dassel	142.3	530.3	17.9
Stauffenburg	175.4	682.6	28.7

caused windthrow in large areas across southern Lower Saxony and northern Hessen, but the research sites were not affected (Winterhoff et al. 1995).

Gaps created in beech stands by shelterwood harvesting, group selection harvesting or target diameter harvesting are similar in size to those occurring in virgin forests and close-to-nature forests (Table 22.4, compare Table 22.3). However, in contrast to gap creation in a natural forest, which takes place primarily in the overmature and senescent phases (Thomasius and Schmidt 2003), gaps in managed forests are created in the optimal phase before wood quality decreases (stand age about 120 years).

European climate-warming projections predict a large increase in temperature anomalies in central Europe, which will result in more frequent and intensive drought periods (IPCC 2007). An increase in temperature should not be an ecophysiological problem for beech in Central Europe, but drought stress could be critical for growth (Manthey et al. 2007). There are some indications that beech, similarly to oak, maple and ash, are likely to be less affected by climate change than the widely cultivated conifers *P. abies* and *P. sylvestris* (Bolte et al. 2007; Kölling and Zimmermann 2007; Kölling et al. 2007; Manthey et al. 2007). However, growing conditions could become critical for beech in areas where the precipitation is less than 500 mm per year or about 250 mm during the months of May to September (Bolte et al. 2007).

22.5 Effects of Forest Management Practices on C- and N-Dynamics

Silvicultural practices in beech forests are confined to light thinnings to create small openings in the canopy for regeneration and harvesting. The large-scale clear-felling for regeneration of beech is not practiced in Central Europe

Table 22.5 Gap cutting and liming in beech stands in the Solling (Bartsch et al. 2002), Göttinger Wald (Schmidt 2002) and Zierenberg sites (Godt 2002)

	Solling	Göttinger Wald	Zierenberg
Date of gap cutting	October 1988	March 1988	September 1990
Gap size (diameter)	30 m	25 m	30 m
Number of gaps	4	6	1
Liming	2 gaps with 3 t ha^{-1} dolomite and soil scarification	–	–
Monitoring period	1990–1998	1988–1998	1990–1996

(Sect. 22.2). The establishment of regeneration on sites with acidic soils may be assisted by liming and soil treatments. In order to investigate the effects of these management practices on nutrient cycling, gaps in the canopy were created in the three beech stands. At the Solling site, liming in conjunction with soil treatment was carried out (Table 22.5). Total *C- and N- pools* in soil, measured from the litter layer to 40 cm depth, did not change at the Solling site. However, significant changes in the distribution of soil C and N occurred within the profile.

Although the gaps were found to have substantially reduced litter inputs, there was no significant decline in *surface organic layer* mass over a 10-year period in the unlimed gaps (Bauhus et al. 2004). The apparent absence of accelerated decomposition in gaps was supported by field measurements of CO_2 emissions, which showed that the differences in CO_2 values between gap and stand were largely attributable to root respiration (Brumme 1995, Chap. 17). Further micro-climate changes did not accelerate litter decomposition on this site.

As in the Solling stand, the creation of gaps in Göttinger Wald did not lead to a definite long-term increase in *nitrogen mineralisation* (Fig. 22.6). While there was only a small difference in mineralisation rates measured in the stand, at edge and gap positions in the Göttinger Wald, they were significantly lower in the gaps on the acid soils at the Solling site than in the stand. This may be explained by the very small soil temperature differences between the gaps and the surrounding stand and by consistently high soil moisture content in the gaps (Bauhus and Bartsch 1996). This, in turn, may have resulted in seasonally anaerobic conditions in the surface organic layer indicated by two to five times higher N_2O losses in the gap compared to the stand (Chap. 17). Losses of N_2 are much lower than losses of N_2O in the centre of the gap at Solling (Wolff and Brumme 2003) while no measurements have been conducted on NO. Additional causes of reduced N-availability may be immobilisation of nitrogen and a reduced mycorrhiza activity (Bauhus 1994; Vor 1999).

A reduction in surface organic layer mass was only achieved in the Solling stand when the creation of gaps was accompanied by the *amelioration* of the acid soil conditions which increased earthworm population and activity, produced lush under-storey vegetation and provided high quality litter (Chap. 12, 13). While the turnover rate of 3.8 years for beech litter in limed gaps is still larger than would

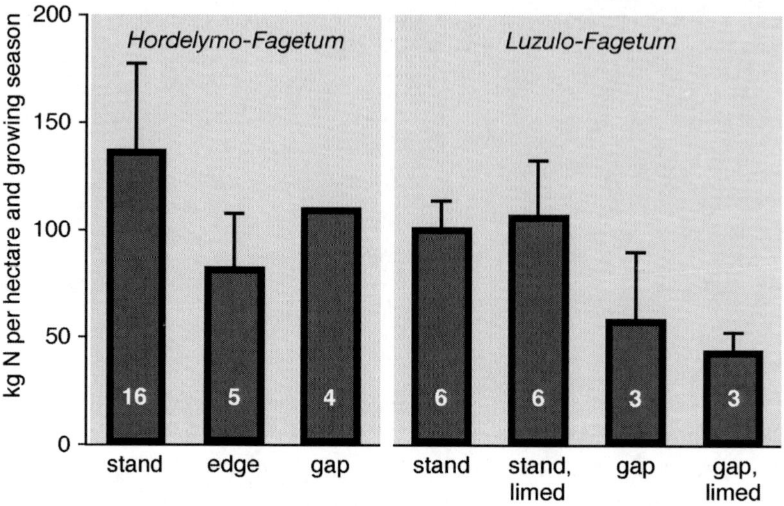

Fig. 22.6 Nitrogen net mineralisation (mean and standard deviation) of beech stands in Göttinger Wald (*Hordelymo-fagetum*) and the Solling (*Luzulo-Fagetum*) (data from Runge 1974; Gadow 1975; Grimme 1977; Eickmeyer 1978; Reichhardt 1981, 1982; Ibrom and Runge 1989; Schneider 1992; Bauhus 1994; Vor 1999; Schmidt 2002; Weckesser 2002). The *numbers* below the *columns* represent the number of research plots (growing season of mostly 30, partly 38 or 52 weeks). In situ-incubation experiments using polyethylene bags or steel cylinders in the biologically active top soil, mostly to a depth of 20–40 cm including the top humus layer (O horizon)

occur in mull humus, a change in humus type from a typical moder to a mull-moder during 8 years was evident. The decrease in surface organic layer mass in limed gaps corresponded with an increase in soil C- and N-concentrations in the surface (1–10 cm) mineral soil (Bauhus et al. 2004).

The unlimed gaps in the Solling stand showed increased N losses in the form of *nitrate leaching* (Fig. 22.7) and N_2O emissions when compared with the limed gaps and the intact stand (Bauhus and Bartsch 1995; Brumme 1995; Bartsch 2000; Bartsch et al. 2002, Chap. 18), Low N-retention in unlimed gaps may be related to low microbial biomass (Bauhus and Barthel 1995; Vor 1999) and fine-root density (Bauhus and Bartsch 1996). At Zierenberg, the nitrate concentration in the seepage water below the root zone increased in the first year only after the gaps were created (Godt 2002), while N_2O emissions did not change (Chap. 18). In the closed mature stand at Zierenberg, nitrate concentration of 500 $\mu mol_c\ l^{-1}$ (middle of the year 1990) were considerably higher than those at Göttinger Wald (69 $\mu mol_c\ l^{-1}$) and Solling (9 $\mu mol_c\ l^{-1}$) (Chap. 16). At the Solling site, losses of N in the seepage water in the unlimed gap ranged between 118 kg ha^{-1} per year in 1991 and 64 kg ha^{-1} per year in 1996 (Bartsch et al. 2002), and this compared to about 1 kg ha^{-1} per year in the closed stand. At Zierenberg, the N-outputs doubled in the four years after the gap was cut when compared to the undisturbed plot (Godt 2002).

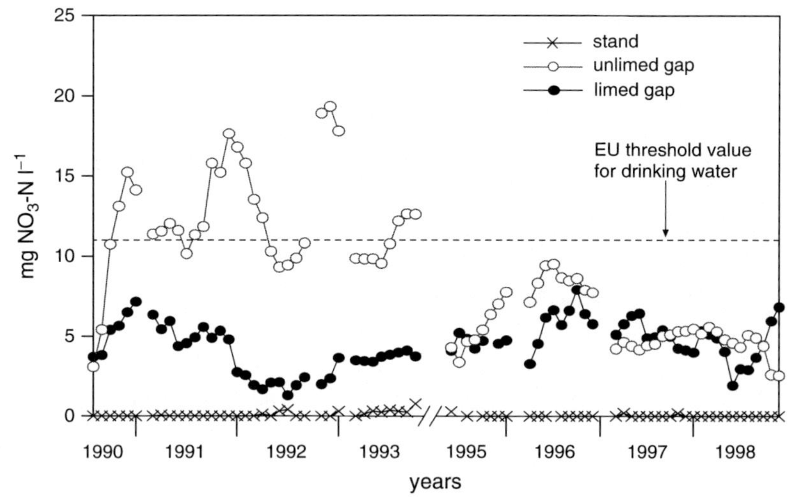

Fig. 22.7 NO_3–N concentrations in soil solution below the rhizosphere (80 cm soil depth) of the stand at the centre of the unlimed gap and the limed gap in the Solling beech forest for the years 1990 to 1993 and 1995 to 1998. The European Union standard for maximum tolerable NO_3–N concentrations is indicated by the *dotted line* (from Bartsch et al. 2002). The gap was established in October 1988

N-output data for the gaps at Göttinger Wald were not available, but Schmidt (2002) estimated that the maximum value of N-output in the gaps 10 years after they were created was similar in magnitude to that of N-inputs.

At Solling, the disruption of the nitrogen cycle as a result of gap cutting lasted longer and was greater than on the Göttingen and Zierenberg sites. The high N losses may have resulted from increased nitrification and insufficient plant N-uptake (Bartsch et al. 2002; Bauhus et al. 2004).

At the Göttinger Wald and Zierenberg sites, there was dense *ground vegetation* and under-storey in the gaps which may have taken up N, taken otherwise by the forest stand (Godt 2002; Schmidt 2002). There was a substantial delay in the colonisation of unlimed gaps by plants at the Solling site (Bartsch et al. 2002). In the fourth year after gap creation, the aboveground and belowground biomass of the ground vegetation at the centre of the gap amounted to about 600 kg ha^{-1}. In the seventh year, it was 1,420 kg ha^{-1}. This corresponded to the N-retention of 9.5 kg ha^{-1} in the fourth and 22.2 kg ha^{-1} in the seventh year after gap creation (Bartsch et al. 2002). The concentration of N in seepage water over time reflected the vegetation development in the gaps (Fig. 22.7). In the fourth year after creating the gap, N stored by ground vegetation and under-storey at the centre of the gap was 70 times higher at Zierenberg than at Solling (Godt 2002).

At Solling, *liming* had a major influence on biomass development, the amount of nitrogen stored by the ground vegetation, and the N-outputs in the seepage water (Fig. 22.7) and into the atmosphere (Chap. 18). In the fourth year after gap creation and lime application, the above ground and below ground biomass of the ground vegetation (herbaceous vegetation and tree regeneration) at the centre of the limed gap was about 4,016 kg ha^{-1} and in the seventh year 6,460 kg ha^{-1}. At the centre of the unlimed gap the biomass was 597 kg ha^{-1} in the fourth year and 1,427 kg ha^{-1} in the seventh year. This corresponded to a N-storage in the ground vegetation of 63 kg N ha^{-1} (fourth year) and 118 kg N ha^{-1} (seventh year) in the limed gap and 9.5 kg N ha^{-1} (fourth year) and 22.2 kg N ha^{-1} (seventh year) in the unlimed gap (Bartsch et al. 2002).

The ground vegetation on the limed and unlimed gaps was fundamentally different at the Solling site. At the centre of the limed gaps, the biomass of the herbaceous vegetation, comprised predominantly of nitrophilic species, was 10 times (3,701 kg ha^{-1} in the fourth year) to 7 times (6,448 kg ha^{-1} in the seventh year) higher than on the unlimed gaps. In contrast, the tree regeneration biomass was higher in the unlimed gaps. The lush herbaceous vegetation in the limed gaps out-competed the beech regeneration. At the centre of the limed gaps, the beech seedlings from the 1989 and 1990 seed crops disappeared almost entirely (Sect. 22.3). In the unlimed gaps, in contrast, the biomass of beech regeneration was highest at the gap centre.

The gap experiments demonstrated that even small-scale disturbance in mature beech stands can cause substantial changes in element fluxes and may lead to high element losses. In this context, the resilience of sites on limestone and basalt with regard to nutrient retention was higher than on the base cation poor site in the Solling. Because the gaps represented only a small part of a watershed and the changes occurred in a limited space of time only, the overall effect on the quality of ground and spring water may not be significant. However, the regeneration of mature beech stands takes place over a long period of time (more than 40 years) during which small areas of the stand are sequentially cleared, and this operation may affect the ground water quality.

The high losses of nitrate and N_2O on the unlimed Solling gaps corresponded to the N losses in large-scale clear-cuts and cleared sites of northern hardwoods in the Hubbard Brook study (New Hampshire, USA) (Bormann and Likens 1979). In both cases, the removal of nutrient uptake by plants played a major role in changing the soil and seepage water chemistry, indicating the importance of effective and early coupling of decomposition and nutrient uptake by new vegetation after tree removal. If soil acidity or gap size inhibit rapid re-vegetation of the gap or an expansion of the roots of surrounding trees into gaps (Bauhus and Bartsch 1996), liming may be regarded as an important vegetation management practice. However, in such cases, the strong competition by ground vegetation may hinder beech regeneration.

The *carbon* stored in *beech forests under different management regimes* were investigated in 16 different stands of different ages on soils developed from limestone and overlayed by loess of varying thickness in Thüringen (about 100–150 km east of the research sites in southern Lower Saxony and northern Hessen) (Mund 2004; Mund and Schulze 2006). The shelterwood and selection management regimes showed no difference in the amount of C stored in the soil. However, the absence of forest management has led to an increase in the carbon pool (Fig. 22.8). With the exception of the possible influence of tree species

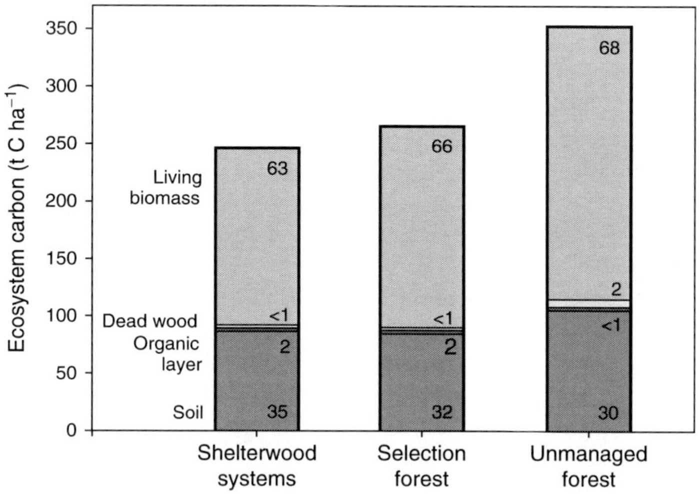

Fig. 22.8 Total amount of carbon stored in differently managed beech forest ecosystems at the Hainich-Dün region, Central Germany. The *numbers* given at the *right corner of the bars* represent the proportion of carbon (%) stored in the different compartments. Total soil organic carbon includes the entire mineral soil down to the transition zone to the bedrock (C horizon), and is not corrected for site-specific variations in soil properties (e.g. clay content, C:N ratio, soil depth) (from Mund and Schulze 2006)

composition on carbon storage in the mineral soil, no direct relationship was found between changes in the characteristics of the tree stand (density, area, woody biomass) through forestry practices and the carbon content in mineral soil. The organic carbon content in the mineral soil in shelterwood systems ranged from 70 to 98 tons C ha^{-1}. In selection systems, it ranged from 65 to 107 tons C ha^{-1}. In the natural forest, which has not been managed for 35 years, the carbon content in the mineral soil was about 20% higher than in the managed forests.

22.6 Conclusions

Of the around 50 tree species indigenous to Central Europe, *F. sylvatica* stands out because of its ecological characteristics and its broad environmental amplitude. As a tree species with a high shade tolerance, beech is particularly suited for regeneration under the canopy of a mature stand in shelterwood and group selection harvesting methods. Target diameter harvesting in the absence of additional measures to assist regeneration may promote the development of pure beech stands as the crown gaps created are inadequate for light-demanding species.

The silvicultural goal for the three sites is to develop a continuous stocking of broadleaved trees (Sect. 22.2) in order to maintain high levels of nutrient cycling and minimum nutrient losses from forest ecosystems. These aims are largely met through the application of shelterwood and group selection harvesting practices presently used in these stands. The relatively new silvicultural concept of target diameter cutting will also prove useful, as it mimies gap dynamics in virgin forests. A reduction of losses of nitrate and N_2O under high N-input systems during regeneration require active root uptake possibly by trees and to some extent by the under-storey vegetation. Some of the essential silvicultural practices may include appropriate thinning operations for the maintenance of an under-storey and sub-canopy layer, and assistance to develop advanced regeneration under the canopy of the mature stand, which may require planting of tall seedlings where competition from ground vegetation is high and where natural regeneration is absent (Zierenberg), or through application of liming materials on sites poor in base cations (Solling). Large gaps that do not allow canopy closure within a few years should only be cut if light-demanding mixed forest species are to be encouraged.

The regeneration and timber utilisation methods employed in beech stands do not show major differences in the C retention by soils. As the carbon retention by mineral soil represents the balance of carbon inputs and outputs over many decades and centuries, its relationship to the actual stand characteristics is difficult to assess.

References

Ammer C (1996) Impact of ungulates on structure and dynamics of natural regeneration of mixed mountain forests in the Bavarian Alps. For Ecol Manage 88:43–53

Ammer S, Huber C (2007) Die Regenwurmlebensgemeinschaft im Höglwaldexperiment 21 Jahre nach Kalkung. Allg Forst u Jagdztg 178:213–220

Ammer S, Makeschin F (1994) Auswirkungen experimenteller saurer Beregnung und Kalkung auf die Regenwurmfauna und die Humusform in einem Fichtenaltbestand (Höglwaldexperiment). Forstwiss Cbl 113:70–85

Barnes BV (1991) Temperate deciduous forests of North America. In: Röhrig E, Ulrich B (eds) Temperate deciduous forests. Ecosystems of the world 7. Elsevier, Amsterdam, pp 219–344

Bartsch N (1997) Auswirkungen femelartiger Auflichtung von Buchenaltbeständen auf Stoffhaushalt, Zersetzertätigkeit, Krautschicht und Buchenverjüngung. Gutachten im Auftrag des Nieders Min Ernährung, Landwirtschaft u Forsten, Hannover (unpublished)

Bartsch N (2000) Element release in beech (*Fagus sylvatica* L.) forest gaps. Water Air Soil Pollut 122:3–16

Bartsch N, Bauhus J, Vor T (2002) Effects of group selection and liming on nutrient cycling in an European beech forest on acidic soil. In: Dohrenbusch A, Bartsch N (eds) Forest development – succession, environmental stress and forest management. Springer, Berlin, pp 109–142

Bauhus J (1994) Stoffumsätze in Lochhieben. Ber Forschungszentrum Waldökosysteme Univ Göttingen, A 113

Bauhus J, Barthel R (1995) Mechanisms for carbon and nutrient release and retention in beech forest gaps. II. The role of soil microbial biomass. Plant Soil 168–169:585–592

Bauhus J, Bartsch N (1995) Mechanisms for carbon and nutrient release and retention in beech forest gaps. I. Microclimate, water balance and seepage water chemistry. Plant Soil 168–169:579–584

Bauhus J, Bartsch N (1996) Fine root growth in beech (*Fagus sylvatica*) forest gaps. Can J For Res 26:2153–2159

Bauhus J, Vor T, Bartsch N, Cowling A (2004) The effects of gaps and liming on forest floor decomposition and soil C and N dynamics in a *Fagus sylvatica* forest. Can J For Res 34: 509–518

Bengtsson J, Nilsson SG, Franc A, Menozzi P (2000) Biodiversity, disturbance, ecosystem function and management of European forests. For Ecol Manage 132:29–50

Bick U, Dahm S (1993) Bundeswaldinventur 1986–1990. BML, Bonn

Blume H-P, Brümmer GW, Schwertmann U, Horn R, Kögel-Knabner I, Stahr K, Auerswald K, Beyer L, Hartmann A, Litz N, Scheinost A, Stanjek H, Welp G, Wilke B-M (2002) Scheffer/Schachtschabel – Lehrbuch der Bodenkunde. 15. Aufl Spektrum Akademischer, Heidelberg

BML (Bundesmin Ernährung, Landwirtschaft u Forsten) (1994) Der Wald in den neuen Bundesländern. Eberswalde

BML (Bundesmin Ernährung, Landwirtschaft u. Forsten (2003) Gesamtwaldbericht der Bundesregierung. Berlin

Bohn U, Gollub G, Hettwer C, Neuhäuslová Z, Raus T, Schlüter H (2004) Karte der natürlichen Vegetation Europas. Bonn-Bad Godesberg

Bolte A, Czajkowski T, Kompa T (2007) The north-eastern distribution range of European beech – a review. Forestry 80:413–429

Bormann FH, Likens GE (1979) Pattern and process in a forested ecosystem. Springer, New York

Brötje J-H, Schmidt W (2005) Spitzahorn (*Acer platanoides*) im Kalkbuchenwald – Struktur, Streufall, Samenproduktion und Fraßschäden. Allg Forstzeitschr 60:497–502

Brumme R (1995) Mechanisms for carbon and nutrient release and retention in beech forest gaps. III. Environmental regulation of soil respiration and nitrous oxide emissions along a microclimatic gradient. Plant Soil 168–169:593–600

Burschel P, Huss J, Kalbhenn R (1964) Die natürliche Verjüngung der Buche. Schr Forstl Fakultät Univ Göttingen u Nieders Forstl Versuchsanst 34. Sauerländer's, Frankfurt a M

Ching KK (1991) Temperate deciduous forests in East Asia. In: Röhrig E, Ulrich B (eds) Temperate deciduous forests. Ecosystems of the world 7. Elsevier, Amsterdam: pp 539–555

Christensen M, Hahn K, Mountford EP, Ódor P, Standovár T, Rozenberger D, Diaci J, Wijdeven S, Meyer P, Winter S, Vrska T (2005) Dead wood in European beech (*Fagus sylvatica*) forest reserves. For Ecol Manage 210:267–282

Dähne J, Eichhorn J, Gossenauer-Marohn H, Janßen A, Paar U, Rau H-M, Weisgerber H (1997) Beiträge zur Stabilität der Buche (*Fagus sylvatica* L.) in Hessen. Forst u Holz 52: 414–420

Dierschke H, Bohn U (2004) Eutraphente Rotbuchenwälder in Europa. Tuexenia 24:19–56

Dohrenbusch A (1990) Die Verjüngungsentwicklung der Buche (*Fagus sylvatica* L.): Schr Forstl Fakultät Univ Göttingen u Nieders Forstl Versuchsanst 97. Saueränder's, Frankfurt a M

Eberl C (2001) Wirkungen der praxisüblichen Kompensationskalkung auf Indikatoren des Humus-, Mineralboden- und Kronenzustandes auf Dauerbeobachtungsflächen in Niedersachsen und Schleswig-Holstein. Freiburger Forstl Forsch 33:167–181

Eichhorn J, Paar U, Schulze, I-M (2003) Exkursionsführer Level II Fläche Zierenberg. Hessen-Forst FIV, Hann-Münden(unpublished)

Eickmeyer B (1978) Zum Nährstoffhaushalt von *Allium ursinum*-Beständen in einem Kalkbuchenwald. Master thesis Univ. Göttingen (unpublished)

Eisenbarth E, Wilhelm GJ, Berens A (2001) Buchen-Komplexkrankheit in der Eifel und den angrenzenden Regionen. Allg Forstzeitschr 56:1212–1217

Ellenberg H jun (1989) Eutrophierungsveränderungen der Waldvegetation: Folgen für und Rückwirkungen durch Rehwildverbiß. Verh Ges Ökologie 17:425–435

Ellenberg H (1996) Vegetation Mitteleuropas mit den Alpen in ökologischer, dynamischer und historischer Sicht. 5 Ulmer, Stuttgart

Falinski JB (1986) Vegetation dynamics in temperate lowland primeval forest. Ecological studies in Bialowieza forest. Geobotany 8:1–537

Fehrmann L, Kuhr M, Gadow K von (2003) Zur Analyse der Grobwurzelsysteme großer Waldbäume an Fichte (*Picea abies* (L.) Karst.] und Buche (*Fagus sylvatica* L.). Forstarchiv 74:96–102

Fenner S (1999) Die Konkurrenz der Brennessel (*Urtica dioica* L.) um Licht und Wasser gegenüber der Buchennaturverjüngung auf Basalböden im hessischen Bergland. Master thesis Fachhochschule Hildesheim/Holzminden Göttingen

Fenner S, Hetsch W, Eichhorn J, Schulze I-M (2000) Die Konkurrenz der Großen Brennnessel (*Urtica dioica* L.) um Licht und Wasser gegenüber der Buchennaturverjüngung auf Basaltböden im hessischen Bergland. Forst u Holz 55:441–445

Firbas F (1949) Spät- und nacheiszeitliche Waldgeschichte Mitteleuropas nördlich der Alpen. Fischer, Jena

Gadow A von (1975) Ökologische Untersuchungen in Ahorn-Eschenwäldern. Dissertation Univ Göttingen

Gliemeroth AK (1995) Paläologische Untersuchungen über die letzten 20000 Jahre in Europa. Fischer, Stuttgart

Godt J (2002) Canopy disintegration and effects on element budgets in a nitrogen-saturated beech stand. In: Dohrenbusch A, Bartsch N (eds) Forest development – succession, environmental stress and forest management. Springer, Berlin, pp 143–165

Grimme K (1977) Wasser- und Nährstoffversorgung von Hangbuchenwäldern. Scripta Geobotanica 12

Guericke M (1999) Untersuchungen zur Kronenplastizität der Buche. Tagungsber Sektion Waldbau im DVFF 8.9.-10.9.1999 Trippstadt, pp 66–75

Hagemeier M (2002) Funktionale Kronenarchitektur mitteleuropäischer Baumarten am Beispiel von Hängebirke, Waldkiefer, Traubeneiche, Hainbuche, Winterlinde und Rotbuche. Diss Botanicae 361. J Cramer, Berlin

Hertel D (1999) Das Feinwurzelwerk von Rein- und Mischbeständen der Rotbuche. Diss Botanicae 317. J Cramer, Berlin

Horn A (2002) Konkurrenz zwischen natürlich verjüngten Eschen und Buchen in Bestandeslücken: Wachstum, Feinwurzelverteilung und ökophysiologische Reaktion auf Austrocknung. Ber Forschungszentrum Waldöksysteme Univ Göttingen, A 177

Huss J (1972) Die Entwicklung von Buchenjungwüchsen auf einer Naturverjüngungsfläche. Forst u Holz 27:56–58

Huss J, Butler-Manning D (2006) Entwicklungsdynamik eines buchendominierten "Naturwald"-Dauerbeobachtungsbestandes auf Kalk im Nationalpark Hainich-Thüringen. Waldökologie online 3:59–73

Hüttl RF (1990) Vergleichende Analyse von Düngungsversuchen in der Bundesrepublik Deutschland und in den USA. Forst u Holz 45:580–587

Ibrom A, Runge M (1989) Die Stickstoff-Mineralisation im Boden eines Sauerhumus-Buchenwaldes unter dem Einfluss von Kalkung oder Stickstoffdüngung. Ber Forschungszentrum Waldökosysteme Univ Göttingen, A 49

Illmann J, Geissen V, Brümmer GW (1997) Auswirkungen von Kalkungs- und Düngungsmaßnahmen auf die bodenbewohnenden Collembolenfaunen in Laub- und Nadelwald. Verh Ges Ökol 27:427–434

Immer A, Schmidt W, Meiwes KJ, Beese F (1993) Langzeitwirkungen von Kalkung und Düngung auf den chemischen Zustand im Oberboden, die Humusformen und die Bodenvegetation in einem Fichtenforst. Forstwiss Cbl 112:334–346

IPCC (2007) Climate change 2007: The physical science basis. Summary for policy makers (http://www.ipcc.ch)

Jahn G (1991) Temperate deciduous forests of Europe. In: Röhrig E, Ulrich B (eds) Temperate deciduous forests. Ecosystems of the world 7. Elsevier, Amsterdam, pp 377–502

Jansen M (1987) Der Einfluß von Kalkungen auf die Entwicklung von Keimlingen in Buchenbeständen. Master thesis Univ Göttingen (unpublished)

Knoke T, Schulz-Wenderoth S (2001) Ein Ansatz zur Beschreibung von Wahrscheinlichkeit und Ausmaß der Farbkernbildung bei Buche (*Fagus sylvatica* L.). Forstwiss Cbl 120:154–172

Knoke T, Ammer C, Stimm B, Mosandl R (2008) Admixing broadleaved to coniferous tree species: a review on yield, ecological stabilitiy and economics. Eur J For Res 127:89–101

Kölling C, Zimmermann L (2007) Die Anfälligkeit der Wälder Deutschlands gegenüber dem Klimawandel. Gefahrenstoffe/Reinhaltung der Luft 67:259–268

Kölling C, Zimmermann L, Walentowski H (2007) Klimawandel: Was geschieht mit Buche und Fichte? AFZ/Der Wald 62:584–588

König E (1997) Einfluß des Rehwildverbisses auf die natürliche Verjüngung. Allg Forstzeitschr 52:320–323

Konnert M, Henkel W (1997) Untersuchungen zur genetischen Variation der Buche in Thüringen. Allg Forst- u Jagdztg 168:182–190

Konnert M, Ziehe M, Tröber U, Maurer W, Janßen A, Sander T Hussendörfer E, Hertel H (2000) Genetische Variationen der Buche (*Fagus sylvatica* L.) in Deutschland: Gemeinsame Auswertung genetischer Inventuren über verschiedene Bundesländer. Forst u Holz 55:403–408

Körner C (2005) An introduction to the functional diversity of temporate forest trees. In: Scherer-Lorenzen M, Körner C, Schulze E-D (eds) Forest diversity and function – temperate and boreal systems. Ecol. Studies 176, Springer, Berlin, pp 13–37

Korpel S (1995) Die Urwälder der Westkarparten. Fischer, Stuttgart

Kowarik I (1988) Zum Einfluß des Menschen auf Flora und Vegetation: Theoretische Konzepte und ein Quantifizierungsansatz am Beispiel von Berlin (West). Landschaftsentw u Umweltfor 56

Kowarik I (1999) Natürlichkeit, Naturnähe und Hemerobie als Bewertungskriterien. In: Konold W, Böcker, R, Hampicke U (eds) Handbuch Naturschutz und Landschaftspflege V-2. 1:1–18. Ecomed, Landsberg

Kühne C, Bartsch N (2003) Zur Naturverjüngung von Fichten-Buchen-Mischbeständen im Solling. Forst u Holz 58:3–7

Kuhr M (2000) Grobwurzelarchitektur in Abhängigkeit von Baumart, Alter, Standort und sozialer Stellung. Dissertation Univ Göttingen

Lambertz B, Schmidt W (1999) Auflichtungen in Buchen- und Buchenmischbeständen auf Muschelkalk und Buntsandstein: Untersuchungen zur Verjüngungs- und Vegetationsstruktur. Verh Ges Ökol 29:81–88

Lang G (1994) Quartäre Vegetationsgeschichte Europas: Methoden und Ergebnisse. Fischer, Jena

Leuschner C (1998) Mechanismen der Konkurrenzüberlegenheit der Rotbuche. Ber Reinh-Tüxen-Ges 10:5–18

Leuschner C, Meier IC, Hertel D (2006) On the niche breadth of *Fagus sylvatica*: soil nutrient status in 50 Central European beech stands on a broad range of bedrock types. Ann For Sci 63:335–368

Lunderstädt J (1992) Stand der Ursachenforschung zum Buchensterben. Forstarchiv 63:21–24

Lunderstädt J (2002) Langzeituntersuchung zur Befallsdynamik der Buchenwollschildlaus (*Cryptococcus fagisuga* L.) und der nachfolgenden Nekrosebildung in einem Buchen-Edellaubholz-Mischbestand. Allg Forst- u Jagdztg 173:193–200

Lüpke B von (1982) Versuche zur Einbringung von Lärche und Eiche in Buchenbestände. Schr Forstl Fak Univ Göttingen u Nieders Forstl Versuchsanst 74. Sauerländer's, Frankfurt a M

Lüpke B von (1999) Silvicultural methods of oak regeneration with special respect to shade tolerant mixed species. For Ecol Manage 106:19–26

Lüpke B von (2004) Riskominderung durch Mischwälder und naturnaher Waldbau – ein Spannungsfeld. Forstarchiv 75:43–50

Magri D, Vendramin GG, Comps, B, Dupanloup I, Geburek T, Gömöry D, Latalowa M., Litt T, Paule L, Roure JM, Tantau I, van der Knaap WO, Petit RJ, de Beaulieu J-L (2006) A new scenario fort he Quaternary history of European beech populations: palaeobotanical evidence and genetic consequences. New Phytol 171:199–221

Manthey M, Leuchner C, Härdtle W (2007) Buchenwälder und Klimawandel. Natur u Landschaft 82:441–445

Marschner B (1993) Die Rolle der mikrobiellen Aktivität bei der Sulfatfreisetzung nach Kalkung eines sauren Waldbodens. Mitt Dtsch Bodenkdl Ges. 72:605–608

Meyer P, Tabaku V, Lüpke BV (2003) Die Struktur albanischer Rotbuchen-Urwälder: Ableitungen für eine naturnahe Buchenwirtschaft. Forstwiss Cbl 122:47–58

Meiwes KJ (1994) Kalkungen. In: Matschullat J, Heinrichs H, Schneider J, Ulrich B (eds) Gefahr für Ökosysteme und Wasserqualität. Springer, Berlin, pp 415–431

Mrotzek R (1998) Wuchsdynamik und Mineralstoffhaushalt der Krautschicht in einem Buchenwald auf Basalt. Ber Forschungszentrum Waldökosysteme Univ Göttingen, A 152

Mund M (2004) Carbon pools of European beech forests (*Fagus sylvatica*) under different silvicultural management. Dissertation Univ Göttingen

Mund M, Schulze E-D (2006) Impacts of forest management on the carbon budget of european beech (*Fagus sylvatica*) forests. Allg Forst u Jagdztg 177:47–63

Munk H, Rex M (1992) Zur Moblität des Magnesiums in Böden und Düngemitteln. Allg Forstzeitschr 47:796–799

Nilsson SI, Andersson S, Valeur I, Persson T, Bergholm J, Wiren A (2001) Influence of dolomite lime on leaching and storage of C, N, and S in a Spodosol under Norway spruce. For Ecol Manage 146:55–73

NMELF (Nieders Min Ernährung, Landwirtschaft u Forsten) (1996a) Waldentwicklung Solling. Schriftenr Waldentwicklung in Niedersachsen 5

NMELF (Nieders Min Ernährung, Landwirtschaft u Forsten) (1996b) Waldentwicklungstypen für Planung und Vollzug des LÖWE-Programms. Runderlass des ML vom 01.10.1996

Oberdorfer E (1992) Süddeutsche Pflanzengesellschaften: Wälder und Gebüsche 2 Aufl Fischer, Jena

Otto H-J (1991) Langfristige, ökologische Waldbauplanung für die Niedersächsischen Landesforsten. Aus dem Walde 43

Otto H-J (1992) Rahmenbedingungen und Möglichkeiten zur Verwirklichung der ökologischen Waldentwicklung in den niedersächsischen Landesforsten. Forst u Holz 47:75–78

Otto H-J (2000) Waldbauliche Erfahrungen nach Sturmkatastrophen: eine Rückschau in Niedersachsen. Forst u Holz 55:371–376

Paar U, Kuhr M, Rockel A, Westphal J, Eichhorn J (2004) Die Fruktifikation der Buche. In: Schmidt M (ed) Ergebnisse und Perspektiven angewandter Buchenforschung in Hessen. Forschungsber Hessen-Forst FIV 31:29–37

Pampe A (2000) Zur Konkurrenz von Buche und Fichte in der montanen Stufe des Harzes. Tagungsber Sektion Waldbau im DVFF 13.9.-15.9.2000 Dessau/Klieken: pp 95–109

Partl E, Szinovatz, V, Reimoser F, Schweiger-Adler J (2002) Forest restoration and browsing impact by roe deer. For Ecol Manage 159:87–100

Peters R (1997) Beech forests. Kluwer, Dordrecht

Polomski J, Kuhn N (1998) Wurzelsysteme. Haupt, Bern

Polomski J, Kuhn N (2001) Wurzelhabitus und Standfestigkeit der Waldbäume. Forstwiss Cbl 120:303–317

Pott R (1992) Die Pflanzengesellschaften Deutschlands. Ulmer, Stuttgart

Pretzsch H, Dursky J, Pommerening A, Fabrika M (2000) Waldwachstum unter dem Einfluss großregionaler Standortveränderungen. Forst u Holz 55:307–314

Rehfuess KE (1990) Waldböden: Entwicklung, Eigenschaften und Nutzung. 2 Aufl. Parey, Hamburg und Berlin

Reichardt C (1981) Mikroklima, Lichtverteilung und Stickstoff-Mineralisation unter Buche, Eiche und Fichte auf vergleichbarem Standort im Solling. Master thesis Univ Göttingen (unpublished)

Reichardt C (1982) Stickstoff-Mineralisation im Boden eines Kalkbuchenwaldes. Mitt Dtsch Bodenk Ges 34:33–38

Röhrig E, Bartsch N, Lüpke B von (2006) Waldbau auf ökologischer Grundlage. Ulmer, Stuttgart

Runge M (1974) Die Stickstoffmineralisation im Boden eines Sauerhumus-Buchenwaldes. Oecol Plant 9:201–218

Sander T, Rothe GM, Weisgerber H, Janssen A (2001) Allelic and genotypic variation of 13 European beech populations in Hessen/Germany. For Gen 8:13–24
Saniga M, Schütz J-P (2001) Dynamics of changes in dead wood share in selected beech virgin forests in Slovakia within their development cycle. J For Sci 47:557–565
Schmidt W (1997) Zur Vegetationsdynamik von Lochhieben in einem Kalkbuchenwald. Forstwiss Cbl 116:207–217
Schmidt W (2002) Stickstoffkreislauf in Schlaglücken eines Kalkbuchenwaldes. Allg Forst- u Jagdztg 173:67–76
Schmidt W (2006) Zeitliche Veränderung der Fruktifikation bei der Rotbuche (*Fagus sylvatica* L.) in einem Kalkbuchenwald (1981–2004). Allg Forst- u Jagdz 177:9–19
Schmidt W, Weitemeyer M, Holzapfel C (1996) Vegetation dynamics in canopy gaps of a beech forest on limestone – the influence of light gradient on species richness. Verh Ges Ökol 25: 253–260
Schmucker T (1942) Die Baumarten der nördlich-gemäßigten Zone und ihre Verbreitung. Silvae Orbis 4
Schneider B (1992) Vergleichende bodenökologische Untersuchung zweier Auflichtungen im Kalkbuchenwald. Master thesis Univ Göttingen (unpublished)
Schober R (1987) Ertragstafeln wichtiger Baumarten. 3. Aufl. Sauerländers, Frankfurt a M
Schüler G (2002) Schutz versauerter Böden in nachhaltig bewirtschafteten Wäldern: Ergebnisse aus 10jähriger interdisziplinärer Forschung. Allg Forst- u Jagdztg 173:1–7
Schulze K (1998) Wechselwirkungen zwischen Waldbauform, Bejagungsstrategie und der Dynamik von Rehwildbeständen. Ber Forschungszentrum Waldökosysteme Univ Göttingen, A 150
Schulze I-M, Eichhorn J (2000) Veränderungen im Stickstoffhaushalt von Buchenwäldern auf Basalt: Die Ausbreitung der Großen Brennnessel (*Urtica dioica* L.) und ihr Einfluss auf die natürliche Verjüngung der Buche. Forst u Holz 55:435–441
Spiecker H, Mielikäinen K, Köhl M, Skovsgaard JP (eds) (1996) Growth trends in European forests. Springer, Berlin Heidelberg New York
Standovár T, Kenderes K (2003) A review on natural stand dynamics in beechwoods of East Central Europe. Appl Ecol Environ Res 1:19–46
Tabaku V (1999) Struktur von Buchen-Urwäldern in Albanien im Vergleich mit deutschen Buchen-Naturwaldreservaten und -Wirtschaftswäldern. Diss Univ Göttingen
Tabaku V, Meyer P (1999) Lückenmuster albanischer und mitteleuropäischer Buchenwälder unterschiedlicher Nutzungsintensität. Forstarchiv 70:87–97
Thomasius H, Schmidt PA (2003) Waldbau und Naturschutz. In: Konold W, Böcker R, Hampicke U (eds) Handbuch Naturschutz und Landschaftspflege VII-3:1–44. Ecomed, Landsberg
Tinner W, Lotter AF (2001) Central European vegetation response to abrupt climate change at 8.2 ka. Geology 29:551–554
Vor T (1999) Stickstoffkreislauf eines Buchenaltbestandes nach Auflichtung und Kalkung. Ber Forschungszentrum Waldökosysteme Univ Göttingen, A 163
Wagenhoff A (1975) Die Wirtschaft in Edellaubholz/Buchen-Mischbeständen auf optimalen Standorten im Forstamt Bovenden. Aus dem Walde 24:5–60
Wagenhoff A, Wagenhoff E (1975) Verlauf und Auswirkungen des Buchenrindensterbens im Forstamt Bovenden in den Jahren 1959 bis 1965. Aus dem Walde 24:111–168
Wagner S (1996) Height increment of single plants in mixed beech and ash regeneration beneath old growth. In: Skovsgaard JP, Johannsen VK (eds) Modelling regeneration success and early growth of forest stands. Proc IUFRO Conf 10–13 June 1996 Copenhagen, pp 518–529
Weckesser M (2002) Die Bodenvegetation von Buchen-Fichten-Mischbeständen im Solling: Struktur, Diversität und Stoffhaushalt. Cuvillier, Göttingen
Winterhoff B, Schönfelder E, Heiligmann-Brauer G (1995) Sturmschäden des Frühjahrs 1990 in Hessen. Forschungsbericht Hessische Landesanstalt für Forsteinrichtung, Waldforschung und Waldökologie 20:1–176
Wolf I, Brumme R (2003) Dinitrogen and nitrous oxide formation in beech forest floor and mineral soils. Soil Sci Soc Am J 67:1862–1868

Wolters V, Ekschmitt K, Scholle G (1995) Wirkung auf Bodenorganismen und biologische Umsetzungsprozesse. Allg Forstzeitschr 50:936–941

Ziehe M (2007) Genomische Assoziationen durch Selbst- und Fremdbefruchtung und ihre Bedeutung für die Interpretation genetischer Strukturen am Beispiel der Buche (*Fagus sylvatica* L.) Schr Forstl Fak Univ Göttingen u Nieders Forstl Versuchsanst 143. Sauerländer's, Frankfurt a M

Part D
Synthesis

Chapter 23
Stand, Soil and Nutrient Factors Determining the Functioning and Management of Beech Forest Ecosystems: A Synopsis

R. Brumme and P.K. Khanna

23.1 Background and Issues

The three European beech ecosystems described in this volume differ in their soil chemical properties which are related to the differences in their soil parent materials. The ecosystems are located in an undulating landscape primarily formed by different geological formations of Triassic limestone (Göttinger Wald) and sandstone (Solling) with locally interspersed tertiary volcanic materials (Zierenberg) and loess. The difference in elevation between the three sites is less than 100 m and the distance is about 70 km, which provided comparable climatic conditions at the three sites. *Fagus sylvatica* L. is the dominating tree species at all three sites. *F. sylvatica* possesses wide physiological amplitude and can occupy a variety of sites, though its growth performance varies depending upon a number of factors such as genetics, climate and soil. The chemical state of a forest soil affects decomposition and bioturbation processes through its effects on the litter quality and faunal activity, and thus is the main driving variable for nutrient availability, acidity related stress, the productivity and stability of beech ecosystems.

The natural development of soils has, however, undergone significant changes in recent times due to atmospheric depositions which have accelerated soil acidification processes, have changed the activity of soil biota, have affected decomposition and nutrient turnover processes (short-term changes) and may have led to long-term changes in the soil organic matter content. High inputs of nitrogen by atmospheric deposition have increased forest growth. These inputs together with the increased application of dolomitic lime to improve soil acidity may have further affected the forest productivity. Emission control measures have reduced the deposition of H, S, N and base cations since the late 1980s causing additional impact on the future development of forest ecosystems (Meesenburg et al., Chap. 15). Considering these recent changes and their interactions, it still remains very difficult to predict the future development with respect to the productivity and stability of these forests.

R. Brumme and P.K. Khanna (eds.), *Functioning and Management of European Beech Ecosystems*, Ecological Studies 208,
DOI: 10.1007/978-3-642-00340-0_23,

We envisioned that the wide differentiation in various ecosystem characteristics among the three beech sites described in this volume might provide ample background material to describe the functioning of these and other similar beech ecosystems. Therefore, data on N status from these three beech sites were compared with those of other beech forests in Germany, providing useful information on future trends in ecosystem development and relationships for developing site and stand management strategies. This synthesis section deals with some of these relationships where site factors and nutrient turnover processes are used to describe productivities (health and vitality) of these three ecosystems under the following headings:

(1) Productivity, C stocks, C balances, and their relations to soil chemical state of the three beech ecosystems
(2) N stocks and N balances at the three beech ecosystems
(3) Nutrient status and productivity of European beech ecosystems in a changing environment
(4) N-status and its significance for European forest ecosystems
(5) Dynamics of C and N sequestration in European forests
(6) Bioturbation as a central process of C and N dynamics: role of soil biota
(7) Forest management strategies – future perspectives

23.2 Productivity, C Stocks, C Balances, and Their Relations to Soil Chemical State of Three Beech Ecosystems

The three *F. sylvatica* stands are of similar age and have accumulated 520–690 m^3 ha^{-1} of merchantable wood in the above ground biomass (>7 cm diameter; Rademacher et al. Chap. 8). Annual increments in the long-lasting tree biomass compounds (branches, bark, wood and coarse roots) at the three sites were 1.15 Mg C ha^{-1} (Solling), 1.7 Mg C ha^{-1} (Zierenberg) and 3.3 Mg C ha^{-1} (Göttinger Wald) (Table 23.1). The annual net primary production (NPP) which included the annually recycled biomass compounds was lower at the Solling and Zierenberg sites (5.1 and 5.0 Mg C ha^{-1} per year) compared to the Göttinger Wald site (6.5 Mg C ha^{-1}). These NPP values are in the range reported for four beech forests (>100 years old) along a transect from Denmark to Italy (4.4–6.5 Mg ha^{-1} per year, Scarascia-Mugnozza et al. 2000). Differences in the carbon allocated to the leaves did not explain long-lasting biomass increment differences among the three beech sites, as their contribution to the NPP of the trees was similar (2.6–2.7 Mg C ha^{-1} per year) (Table 23.1). However, the production of fine roots on the three sites was different, it was about 1.3 Mg C ha^{-1} per year at the acid Solling site, and that compared to 0.6 Mg C ha^{-1} per year at the calcareous Göttinger Wald site. Positive effects of base cations on the growth and functioning of roots (Murach et al., Chap. 9) is expected to increase the uptake of nutrients and water, and to decrease the C-allocation to

Table 23.1 Net primary production (NPP) (Mg C ha^{-1} per year), annual C-balances (tree litterfall, Khanna et al., Chap. 11; tree fine roots, Murach et al., Chap. 9; herb litter, Schulze et al., Chap. 10), and changes in C-pools per ha at Solling, Göttinger Wald, and Zierenberg (DOC losses, Brumme et al., Chap. 16; above ground increment, Rademacher et al., Chap. 8; C-changes in soil, Meiwes et al., Chap. 4). (sd or range in parenthesis)

		Solling	Zierenberg[a]	Göttinger Wald
NPP of vegetation		5.1	5.4	6.7
NPP of trees		5.1	5.0	6.5
		Soil C-Balances		
Litter production	Tree litterfall	2.6 (0.7)	2.7	2.6 (0.8)
	Fine-roots from trees	1.3	ND	>0.6[b]
	Herb[e]	0.01	0.42 (0.36–0.82)	0.22 (0.04–0.43)
	Total	3.9	≈3.7	>3.4
Decomposition[d]		2.6/1.6	−/2.1	−/2.4
DOC losses		0.01	0.02	0.02
Total C-production – decomposition		+1.3/+2.3	−/+1.6	−/+1.0
		C-pool changes		
Tree increment[c]		+1.15	+1.7[f]	+3.3
O-horizon increment (1966–2001)		+0.35	ND	ND
Mineral soil increment (1966–2001)		0	ND	ND

[a]Assuming a fine-root litter production similar to that at Göttinger Wald
[b]This value could be higher since it considers only 0–20 cm soil depth (Solling 0–50 cm) due to high stone content
[c]Including branches, bark, wood and coarse roots
[d]Decomposition refers to field measurements of soil respiration at rooted and nonrooted locations (Brumme et al., Chap. 17) and to C-mineralisation studies (values given behind the slash; Brumme et al., Chap. 13)
[e]Weighted mean values and range of means of different facies calculated by the use of C-concentrations of 45% for Göttinger Wald and Solling and 42% for Zierenberg (Bolte 2006)
[f]Calculated by using an average of the functions between biomass of various tree components and tree basal area for Göttinger Wald and Solling (Chap. 8, personal communication, Rademacher)

the fine root biomass by increasing the longevity of fine roots. The higher amount of NPP required for the production and the maintenance of fine roots at the Solling site may partly explain the relatively low carbon gain for aboveground growth at this site. The high soil acidity at this site may create additional soil chemical stress for the aboveground growth by limiting the availability of base cations. Low soil availability of base cations has resulted in very low content of Ca, Mg and K in the leaves, whereas high values of leaf N content indicated its luxurious nutrition (Khanna et al., Chap. 11). Relationships between N uptake for growth increment, an index of NPP, soil pH and the content of Mg in leaves of 17 beech forests in Germany indicated that growth rates were low on soils with low base saturation (Fig. 23.3). However, the low NPP value at the base-rich Zierenberg site may be the result of low tree density in comparison with the Göttinger Wald site (Rademacher et al., Chap. 8) which in turn favoured the growth of ground vegetation by providing high light intensity below the tree stands (Schmidt, Chap. 5). The luxuriant growth of ground vegetation has increased the NPP of the Zierenberg site by about 0.4 Mg C ha^{-1} per year and of the Göttinger Wald site by 0.2 Mg C ha^{-1} per year.

Data on balancing of C fluxes and on periodic inventories of soil C (Table 23.1) indicated that the soil at the Solling site was accumulating C. The total carbon input to the soil at the Solling site was 3.9 Mg C ha^{-1} per year and consisted primarily of litterfall (67%) and fine-root litter (33%). The litter input by ground vegetation is low. The total carbon loss by decomposition processes ranged from 1.6 to 2.6 Mg C ha^{-1} per year and the remaining 1.3–2.3 Mg ha^{-1} would have accumulated annually. Soil inventories conducted periodically at the Solling site partly confirmed these results. The organic carbon stock in the surface organic layer increased by about 0.35 Mg ha^{-1} per year over a period of 35 years (Meiwes et al. 2002, Chap. 4). No change in the carbon stocks of the mineral soil layer at the Solling site was observed due to large spatial variation primarily in the 0–10 cm of the mineral soil. Amount of leaffall, however, appears to be constant at the Solling site as shown by the data since 1990 despite the temporal annual variation in these values (Khanna et al., Chap. 11). However, fructification has more than doubled the total litterfall and frequency of fructification has increased during the last few decades. Thus, litter production and its decomposition may not be in equilibrium. Additional factors reducing decomposition are related to atmospheric depositions. Many studies on N fertilisation has shown that fertilisation leads to retarded decomposition (e.g. Fog 1988; Berg and Matzner 1997; Olsson et al. 2005). Therefore, high N depositions are expected to have a similar effect in retarding litter decomposition. However, Sjöberg et al. (2004) suggested that humus accumulation under high acid depositions might result from more than one factor.

The carbon dynamics at the Göttinger Wald and Zierenberg sites differed from that at the Solling site due to higher productivity of the vegetation on the former sites. At the Göttinger Wald and Zierenberg sites, the surface organic layer was small and was mostly covered by ground vegetation. Ground vegetation produced about 0.2 and 0.4 Mg C ha^{-1} per year litter at these sites (Table 23.1). Production of high amounts of ground vegetation litter and low amount of fine root litter may commonly occur on less acid soils (Murach et al., Chap. 9 and Schulze et al., Chap. 10). Amounts of C retained when assessed from soil-C balance indicated values of +1.0 Mg at Göttinger Wald and +1.6 Mg C ha^{-1} per year at Zierenberg site.

Amounts of C retained annually by the soils of the three beech sites was similar to the value estimated by Schulze (2000) (1.44 Mg C ha^{-1} per year) for a European transect. The values obtained by using models indicated that 100–600 kg C ha^{-1} per year were retained in soils which were primarily related to differences in litterfall and thus tree growth (Nabuurs and Schelhaas 2002; Liski et al. 2002). Our estimates of C retention are not confirmed by the nitrogen balance measured on these three beech sites (see Sect. 23.3) suggesting that C mineralisation rates may be underestimated which would overestimate the C sequestration in the soils. Our C mineralisation estimates are based on differences between rooted and root-free areas at the Solling site which would depend on a number of factors, e.g. the dynamic of roots and associated root exudates after clearing (Brumme et al., Chap. 17). Incubation studies may under- or overestimate C mineralisation, e.g. by excluding root related inputs of C or by causing higher decomposition through physical soil disturbance (Brumme et al., Chap. 13). However, N balances seem to be a useful alternative to estimate C sequestration in forest ecosystems (Holland et al. 1997; Nadelhoffer et al. 1999; deVries et al. 2006). Based on calculated N retention rates in soils (N deposition minus net N uptake minus N leaching) and their C/N ratios, deVries et al. (2006) estimated an annual net C sequestration of 143 kg ha^{-1} for forest soils in Europe. Berg et al. (2007) reported similar values of between 96 and 180 kg C ha^{-1} per year for Sweden which they estimated with three different methods. Since an annual sequestration rate of 120 kg C ha^{-1} equals only 3% of litter production which is lower than the error involved in such measurements, for example, by soil respiration measurements, it illustrates the performance of using soil N changes as a useful method.

23.3 N Stocks and N Balances at Three Beech ecosystems

Atmospheric depositions of N at the three beech sites ranged from 21 to 25 kg N ha^{-1} per year in the period 1990–2002 and were not significantly different among the sites (Table 23.2). However, N-balance on these sites indicated a completely different fate of the atmospheric deposited nitrogen. Losses of N by seepage water at the three sites were 1.7 kg ha^{-1} per year (Solling), 21 kg ha^{-1} per year (Zierenberg) and 4.5 kg N ha^{-1} per year (Göttinger Wald). The gaseous N-losses as nitrous oxide were of the same magnitude as the leaching losses at the Solling site but were negligible at theGöttinger Wald and Zierenberg sites. Therefore, about 84 and 76% of the atmospheric deposited nitrogen were retained at Solling and Göttinger Wald sites, respectively, with much lower values (13%) at the Zierenberg site.

When the N-balance method is employed to calculate the amount of N retained at the Solling site, the value was quite high, being 21 kg N ha^{-1} per year for the period 1990–2002 and 36 kg N ha^{-1} per year for 1981–1989 (Table 23.2). Similar values were obtained by periodic soil N-inventories conducted on the site since 1966. The amount of N-retention in the surface organic layer was large (about 21 kg N ha^{-1} per year) and was three to four times higher than that observed in the long-lasting tree components (branches, bark, wood and coarse roots) at the Solling site (6 kg N ha^{-1} per year). During the period 1969–2002, the total N deposition decreased from 40 (1969–1989) to 25 kg N ha^{-1} per year (1990–2002) (mean for the whole period was 35 kg N ha^{-1} per year). The mean value was close to the amount that was calculated (27 kg N ha^{-1} per year) to accumulate in the tree growth and the surface organic layer increments during the same period. The decrease in atmospheric N depositions has resulted in an equivalent reduction in the amount retained by the soil while the leaching losses did not change during this period (Fig. 23.1). The soil N-balance conducted by the amount of annual nitrogen input by litter minus the net N-mineralisation confirmed the observation from the ecosystem N-balance that nitrogen is sequestered in this soil at the Solling site.

Table 23.2 N-balances on the three beech sites for the period 1990 – 2002 (period 1981–1989 in parenthesis for Solling and Göttinger Wald). Values were taken for input (Meesenburg et al., Chap. 15); output (Brumme et al., Chap. 16); N_2O (Brumme and Borken, Chap. 18); tree litterfall (Khanna et al., Chap. 11); tree fine roots (Murach et al., Chap. 9); herb litter (Schulze et al., Chap. 10); uptake by trees (Rademacher et al., Chap. 8, for Zierenberg Brumme et al., Chap. 16); and C-changes in soil (Meiwes et al., Chap. 4). All values are given in kg N ha^{-1} per year

		Solling	Zierenberg	Göttinger Wald
	Ecosystem N-Balances			
Uptake by trees		110–113	98–110[a]	99–107
Input	Wet and dry deposition	25 (40)	24	21 (26)
Output	Leaching losses	1.7 (2.0)	21	4.5 (9.5[b])
	N_2O losses	1.9	0.4	0.2
Input – Output		+21 (+36)	+3	+16 (+16)
	Soil N-Balances			
Litter-N input	Tree litterfall	68	75	63
	Fine roots from trees[d]	36–39	ND	18–26
	Herb	<0.4	36	21–31
	Total	104–107	129–137[a]	102–120
NMM[e]		74/90	80/128–167	99/ND
	N-Pool Changes			
Tree increment[c]		6.0	5–9	18
O horizon increment		21	ND	ND
Mineral soil increment		0	ND	ND

[a]Assuming a fine-root production similar to that on the Göttinger Wald site
[b]First year of measurements with untypical high N losses (see Chap. 16) was excluded
[c]Including branches, bark, wood and coarse roots
[d]Range refer to dead and living roots containing different N concentrations
[e]NMM (net N mineralisation, is given for "long-term soil column field experiment" and the "in situ sequential coring method,"; Brumme et al., Chap. 13)

The sequestration of N in the surface organic layer provided an estimate of C sequestration at the Solling site. Assuming a C/N ratio of the sequestrated organic matter to be 19, the same as in the surface organic layer, the annual amount of C sequestration was calculated to be 580 kg ha^{-1} for the period 1981–1989 and 313 kg ha^{-1} for 1990–2002. The mean C sequestration value for the whole observational period is thus 422 $\pm$ 170 kg ha^{-1} which comes close to the amount estimated by soil inventories (347 kg C ha^{-1}; Meiwes et al., Chap. 4). It also suggests that the sequestered organic matter may have a lower C/N ratio. A C/N ratio of 15.5 in the sequestrated organic matter would meet the amount of carbon estimated by inventories.

The amount of N retained annually (input-output) at the Göttinger Wald site for the period 1990–2002 was 16 kg N ha^{-1} which was 2 kg N ha^{-1} lower than the annual amount of N retained by the aboveground tree increment (18 kg N ha^{-1} Table 23.2). For the period 1981–1989, total N-deposition was about 5 N ha^{-1} per year higher. The higher depositions during 1981–1989 led to equivalently higher leaching losses indicating that the capacity to immobilise additional N in the soil and through plant uptake at this site is independent of the amount of N present in atmospheric depositions (Fig. 23.1). This is quite a different result from that observed at the highly acid Solling site where the amount of N input by depositions was nearly completely retained in the soil.

At the Zierenberg site, a small fraction (13%) of the total N deposited as atmospheric inputs was retained (3 kg N ha^{-1} per year) because most of it was lost by leaching (Table 23.2). Considering the amount of N required for plant growth at the Zierenberg site, stored soil N remains the main source of nitrogen supply (2–6 kg N ha^{-1} per year). Nitrate leaching increased with an increase in precipitation suggesting that the general trend to higher precipitation during the non-growing season (Panferow et al., Chap. 2) is expected to increase leaching losses at this site. Assessing N-retention from litter production and mineralisation studies does not always provide reliable values. By using the leaching method, low

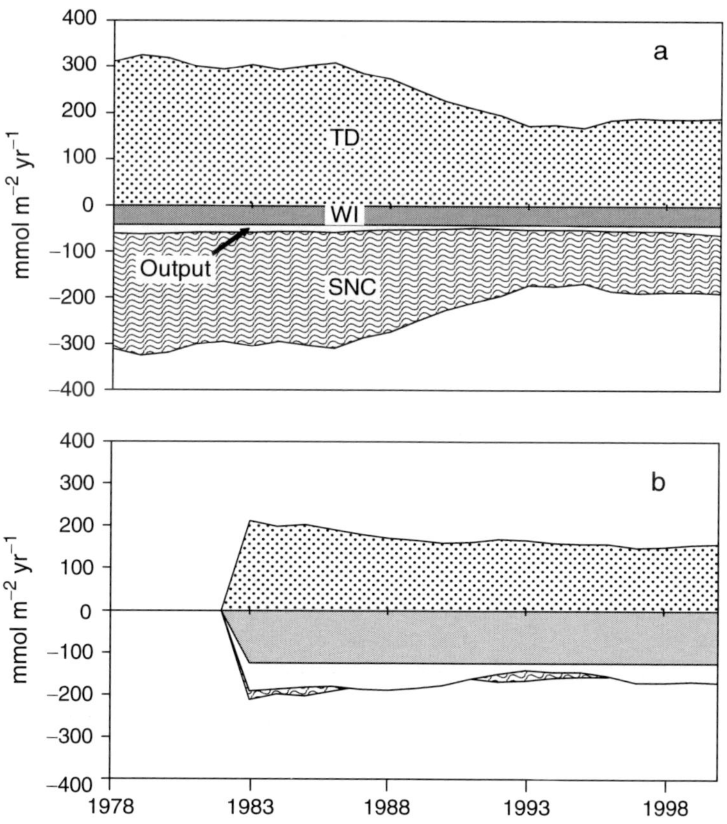

Fig. 23.1 Time series of budgets (5-year moving average) of *TD* (total deposition, dry and wet deposition), soil-N-change (*SNC*), output with seepage (*output*) and wood increment (*WI*) at (**a**) Solling and (**b**) Göttinger Wald (mmol$_c$ m^{-2}) (adopted from Meesenburg et al. Chap. 15 and Brumme et al. Chap. 16)

mineralisation rates (74 ± 14 kg N ha^{-1} per year) suggest an accumulation of N at Zierenberg site. The high mineralisation rates of 128–167 kg N ha^{-1} per year measured with the in situ sequential coring method would cover the amount of N-input by litterfall. Ulrich (1987) suggested that high N mineralisation rates might result from humus degradation in this soil. The degradation would occur when soil is acidified and is shifted from the silicate type of proton buffer range to the exchanger type of proton buffer range (see Sect. 23.4 for more information).

23.4 Nutrient Status and Productivity of European Beech Ecosystems in a Changing Environment

Atmospheric depositions of acidity and nitrogen have changed the soil chemical and nutritional status of forests, particularly of those having low buffering to soil acidification changes. The trend of increasing atmospheric depositions peaked in 1980–1985 across Europe (Ferrier et al. 2001). By that time, it has reduced soil pH by up to 2 units and base saturation by up to 50% and has increased the N content in the litter layer (Brumme et al., Chap. 21). Long-term monitoring at the Solling and Göttinger Wald sites indicated that atmospheric depositions have declined when emission control measures were introduced in Europe and structural changes in the industry had occurred. Total depositions decreased for sulphur (-60% at the Solling and Göttinger Wald sites) and H (−72% at the Solling and Göttinger Wald sites) between the 1981–1989 and 1990–2002 periods (Meesenburg et al., Chap. 15). These changes in the atmospheric inputs caused an increase in the dissolution of aluminum sulphate stored in soils and led to the release of stored acids in the drainage water. However, this decrease in acid inputs also simultaneously decreased the inputs of base cations (−60% at Solling and Göttinger Wald). Long-term annual budgets indicated that the Solling soil which was previously a sink for Ca and sulphate became a net source of these elements during the last three decades (Fig. 23.2), and thus delayed the recovery of this soil from acidification.

The nutritional status of 17 beech forests in Germany (Haußmann and Lux 1997), which belonged to the intensive forest monitoring programme (Level II) of the international cooperative program on assessment and monitoring of air pollution effects on forests, ICP Forests, indicated that Ca was insufficient at seven beech sites, Mg at four sites, P at three sites and K at two sites, whereas no forest site had insufficient nutrition of N (Table 23.3). Among these sites, eight had moder type humus, of which five sites had Ca deficiency. Similar results were observed, with an exception for P, for the German forest inventory sites (Wolff and Riek 1997a, b) which included a higher number of beech stands (n = 75) in Niedersachsen, Brandenburg, and Bayern States. Among these stands, about 54% of the beech forests were described as P-deficient. Despite a reduction of N deposition caused by emission control in Europe (−38% at Solling and −20% at Göttinger Wald), N deposition was in excess of the demand for plant increment at

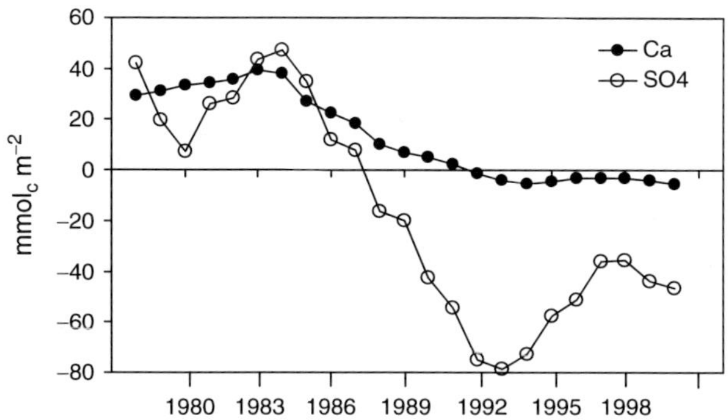

Fig. 23.2 Time series of Ca budgets (total deposition + weathering – plant increment – seepage losses) and sulfate budgets (total deposition – seepage losses) (5-year moving averages, mmol$_c$ m^{-2}) at the Solling site (adopted from Brumme et al. Chap. 16)

Table 23.3 Number of beech forests (17 German forest sites (Level II) including Solling (*SO*), Göttinger Wald (*GW*), Zierenberg (*ZB*), Haußmann and Lux 1997) with element content (1996–2003) in leaves in the ranges of insufficient, sufficient, or luxurious plant nutrition for N (<20.3, 20.3–23.8, > 23.8 mg g^{-1}), Ca (<5.8, 5.8–8.6, 8.6 mg g^{-1}), Mg (<0.99, 0.99–1.43, > 1.43 mg g^{-1}), P (<1.14, 1.14–1.52, > 1.52 mg g^{-1}), K (<5.4, 5.4–7.3, > 7.3 mg g^{-1}), according to Krauß and Heinsdorf (2005)

	N	Ca	Mg	P	K
Insufficient	0	7, SO	4, SO	3, GW	2
Sufficient	10, ZB	6	6, GW	13, SO, ZB	9, SO, ZB
Luxurious	7, SO, GW	4, GW, ZB	7, ZB	1	6, GW

all stands of the Level II sites by up to 27 kg N ha^{-1} per year. The net N retention by beech trees ranged from 6 to 20 kg N ha^{-1} per year (mean value of 13 kg N ha^{-1} per year, Becker et al. 2000). The Solling site thus has N retention value at the lower end (6 kg N ha^{-1} per year), whereas the Göttinger Wald site is at the upper end (18 kg N ha^{-1} per year), and the Zierenberg site is between the two (5–9 kg N ha^{-1} per year), of the range of values given for annual amounts of N retained (Table 23.2).

The retention of N at the German Level II beech forests is correlated significantly with Mg (r^2 = 0.45) and Ca contents ($r^2 = 0.33$) in leaves and with the soil acidity given as pH in soil solution ($r^2 = 0.41$), pH in humus layer ($r^2 = 0.48$) and the base saturation in the surface mineral soil layer ($r^2 = 0.36$) (Fig. 23.3). Since plant N increment is not correlated with indices of N availability (total N depositions, N content in the foliage and soils, C/N of surface organic layers, N content in soil solutions), base cations may be considered as an important factor determining plant growth in these high N ecosystems. This would suggest the need for an additional supply of Ca and Mg by liming the forest sites. Liming has significantly increased

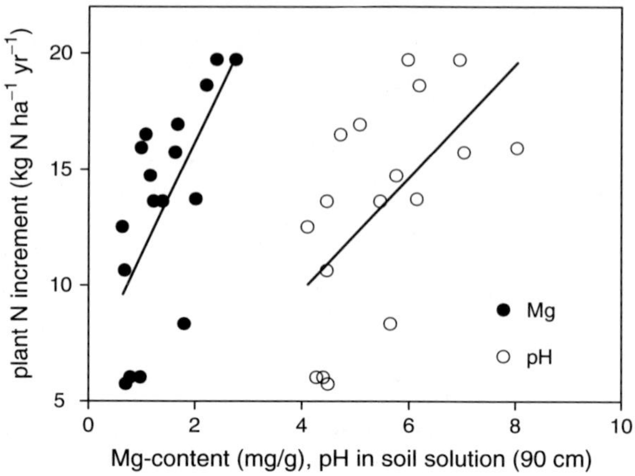

Fig. 23.3 N increment (kg N ha^{-1} per year) in plant growth components (Becker et al. 2000) in relation to Mg content in leaves and soil pH($CaCl_2$) in 20–40 cm depth for 17 beech forests of Level II sites in Germany

the Ca and Mg content in foliage at several sites in the Solling area which received dolomite (Rademacher et al., Chap. 8) and has contributed to an increase in growth (Fig. 23.3). The mean foliage level of the 17 beech sites indicates sufficient N concentrations (23.9 mg g^{-1}) indicating clearly that atmospheric N depositions have removed the N limitation in these forests.

In recent decades, aboveground growth of trees has increased (Spiecker 1999; Kauppi et al. 1992), with values of 25–30% increase in Europe during 1971–1990 (Kauppi et al. 1992). For most cases, tree growth increments since the 1950s have much higher values than could be predicted by common yield tables (Pretzsch 1996). Forest inventories have indicated an increase of 10–20% in the standing volume during 1971–1987 in southern Germany which exceeded the yield table values by 12–43% (12% for broadleaved, 27% for oak, 31% for spruce and fir, 43% for pine and larch). A weighted mean growth increment considering the proportions of tree species showed an annual increase of 29% in Germany. By assuming an annual aboveground wood increment of 2.6 Mg C ha^{-1} per year between 1987 and 1993 (Dieter and Elsasser 2002), the additional C-sequestration can be calculated to be about 0.58 Mg C ha^{-1} per year in Germany. This increase in the growth of wood may be due to many reasons. Firstly, N status of European forests has improved due to high atmospheric inputs. Availability of some other nutrients in soils may have been reduced by acid depositions, but depositions of base cations prior to emission control measures were implemented might have compensated for any losses. Secondly, dolomitic lime has been extensively applied to many forests in Germany to improve the calcium and magnesium availability in soils. Thirdly, an increase in the concentration of atmospheric CO_2 and nitrogen depositions may have increased the photosynthetic activity to compensate for losses of foliage and fine roots turnover due to acid depositions.

23.5 N-Status and Its Significance for European Forest Ecosystems

N enrichment of forest ecosystems has a number of consequences relating to their productivity, vitality, soil acidification, N use, N losses via different means and effects on neighboring systems. Input–output analysis has been shown to be a useful tool to quantify small changes of elements which occur in large stocks of elements in soils (Ulrich 1992; Matzner 2004) that could otherwise remain undetected by repeated sampling commonly done in ecological studies. Brumme and Khanna (2008) used 8-year records (1996–2003) of input–output analysis to evaluate changes in the N status of 53 German forest sites (Haußmann and Lux 1997; Block et al. 2000) belonging to the intensive forest monitoring program (Level II) of the international cooperative program on assessment and monitoring of air pollution effects on forests (ICP Forests) operating under the UNECE Convention on Long-range transboundary air pollution. Of the 53 forest sites, 17 sites are occupied by beech, 15 by spruce, 12 by pine and nine by oak. Trees are 60 to more than 130 years old. Some important site and stand characteristics are given in Table 23.4. Of the 17 beech forests, nine had mull type humus and eight had moder type humus with C/N ratios of 14–21 in the upper mineral soil (mull type humus) and 19–39 in the surface organic layer (moder type humus).

These forest sites were distinguished into three groups based on the following criteria where total deposition (TD) referred to the annual amount of total N deposition including wet and dry depositions, WI to the amount retained by wood increment and O to the amount lost by nitrate leaching.

$TD - WI = O$	(quasi-) Steady-state type	(I)
$TD - WI > O$	Accumulation type	(II)
$TD - WI < O$	Degradation type	(III)

Amounts of gaseous N-losses were not included here because estimates of fluxes of all gases (N_2O, NO, N_2) do not exist on the regional scale. High fluxes of gaseous N losses are estimated to occur at only a few sites (Barton et al. 1999; Brumme et al. 2005; Kesik et al. 2005) and are assumed to be equal to atmospheric N fixation at most of the sites. Considering this uncertainty and other errors in the estimates of water fluxes, total N deposition, wood increment and leaching losses, an annual uncertainty of ±5 kg ha^{-1} above and below the 1:1 line in Fig. 23.4 was used to distinguish the patterns of N changes among different sites.

Forest sites between the two parallel lines in Fig. 23.4 represent those where the excess of total deposition (TD_{ex}) is balanced by nitrate leaching assuming an uncertainty of ±5 kg N ha^{-1} per year. Such forest sites are considered to be of the (quasi-) Steady-State Type with respect to N (n = 23 of which eight were occupied by beech). Sites located on the outside to the left of the lines lose nitrogen in excess of total N-deposition larger than 5 kg N ha^{-1} per year suggesting that soil

Table 23.4 Stand characteristics of forests at different N states (means, min and max in parenthesis) (three sites of the degradation type are not shown)

	N 0–5 cm	Total deposition	Plant N-increment	N leaching	Soil N change	pH 20–40 cm	base sat.	N 0–5 cm	C/N[a]	Foliage N	Precipitation	Mean annual temperature
		kg N ha^{-1} per year				(CaCl$_2$)	%	%		%	mm per year	°C
						Beech forests						
All beech sites (quasi-) steady state	16	21 (18–31)	13 (6–20)	5 (0–25)	3 (−24–11)	4.4 (3.7–7.3)	39 (4–99)	0.43 (0.09–0.7)[b]		2.4 (2.2–2.7)	850 (700–1,200)	8.0 (6.4–9.9)
Mull	4	20 (19–22)	17 (14–20)	1 (0–5)	2 (−2–5)	4.9 (4–7.3)	65 (14–99)	0.41 (0.2–0.7)	17 (14–21)	2.3 (2.2–2.6)	720 (700–780)	8.2 (7.6–8.9)
Moder	4	21 (18–24)	14 (8–17)	8 (2–16)	−1 (−5–4)	4.1 (4–4.2)	13 (4–21)	0.25 (0.2–0.3)[b]	22 (20–23)	2.5 (2.4–2.6)	920 (820–1,040)	7.8 (6.4–9.9)
Accumulation type												
Mull	4	21 (19–27)	10 (6–16)	3 (1–6)	8 (6–10)	4.3 (3.7–4.9)	59 (11, 84)	0.50 (0.3–0.7)	17 (16–18)	2.5 (2.3–2.7)	850 (760–980)	7.9 (7.1–8.6)
Moder	4	22 (15–31)	11 (6–15)	2 (0–7)	9 (6–11)	4.1 (3.9–4.2)	12 (4, 24)	0.27 (0.1–0.5)	27 (19–36)	2.3 (2.2–2.4)	930 (740–1,200)	8.2 (6.5–9.1)
						50 forest sites						
(quasi-) steady state	23											
Mull	8	18 (12–23)	15 (9–20)	2.3 (0.1–7.6)	1.1 (−3–5)	5.2 (4–7.6)	70 (8–100)	0.43 (0.2–0.7)	16 (13–21)	2.3 (1.9–2.7)	780 (700–1,080)	8.7 (7.6–11)
Moder-mor	15	17 (7–33)	10 (5–17)	5.8 (0.1–19)	0.7 (−5–4.5)	4.2 (4–4.4)	13 (4–29)	0.35 (0.1–0.9)[b]	25 (20–29)	spruce: 1.5 (1.3–1.6) deciduous: 2.5 (2.4–2.6)	780 (490–1,200)	7.9 (6.0–9.9)
Accumulation type	27											
Mull	6	21 (19–27)	10 (6–16)	2.5 (0.9–5.8)	9.1 (7–14)	4.1 (3.7–4.9)	53 (8–84)	0.49 (0.3–0.7)	16 (14–18)	deciduous: 2.4 (2.1–2.7)	850 (620–1,080)	8.1 (7.1–9.1)
Moder-mor	21	23 (10–37)	8 (4–15)	3.8 (0–13)	11 (6–21)	4.1 (3.2–4.6)	6.4 (2–16)	0.22 (0.02–0.6)	26 (19–36)	spruce: 1.4 (1.2–1.5) deciduous: 2.4 (2.2–2.6)	935 (560–1,320)	7.5 (5.5–9.7)

[a]C/N is given for the top mineral soil (mull) and the litter layer (moder)
[b]Excluded is one unusually high value of 1.4%

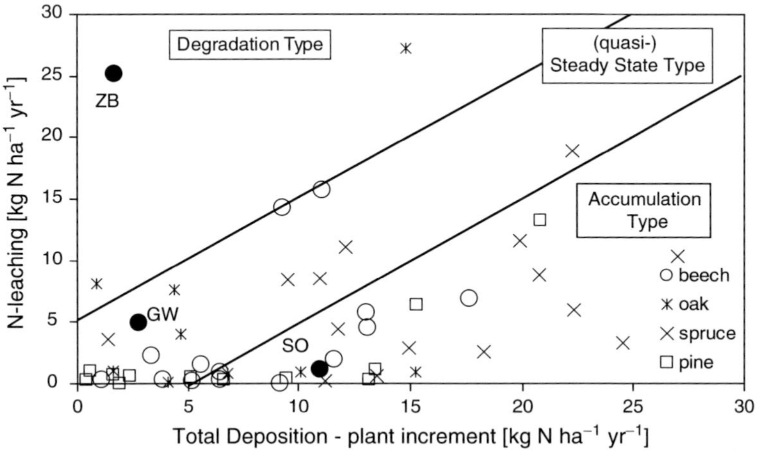

Fig. 23.4 Nitrate leaching (Block et al. 2000) versus total nitrogen deposition (Gehrmann et al. 2001) in excess of N-uptake by aboveground wood increment (Becker et al. 2000) of 53 Level II forests in Germany occupied by beech, oak, spruce and beech trees. The *lines* divide forests into three N-states, forests with decreasing amount of soil organic matter and N losses from soil organic matter (degradation type), forests with additional N-sinks in the soil (accumulation type), and forests showing no change [(quasi-) steady-state type] (*lines* provide estimates of uncertainties of ± 5 kg ha^{-1} above and below the 1:1 *line*). Beech sites at Solling (*SO*), Göttinger Wald (*GW*), and Zierenberg (*ZB*) are highlighted by *solid circles*. (Figure adopted from Brumme and Khanna 2008, modified)

N loss is associated with a decrease of the amount of soil organic matter (Degradation Type, $n = 3$, one site with beech) whereas those on the right of the lines accumulate more than 5 kg N ha^{-1} per year (accumulation type, $n = 27$, eight sites with beech).

23.5.1 Forests of the Accumulation Type

Of the 53 forest sites, 27 sites accumulated nitrogen in the soil and annual accumulation values ranged from 6 to 21 kg N ha^{-1} (Table 23.4). Plants accumulated 4–16 kg N ha^{-1} for annual plant increment which is lower than the amount required for the soil. Out of 21 sites with moder type humus, 11 had C/N ratio values <25 (two were occupied by beech) and showed the same mean annual N retention in the soil (11 kg N ha^{-1}) as those with C/N ratios >25 (two of these sites were occupied by beech) indicating that factors other than C/N ratio are responsible for N retention in these soils. The C/N ratio has often been shown to be a potentially useful predictor of the level of nitrate leaching from forest ecosystems (Matzner and Grosholz 1997; Dise et al. 1998a, b; Gundersen et al. 1998; MacDonald et al. 2002; Borken and Matzner 2004). A new analysis by van der Salm et al. (2007) revealed that N deposition by

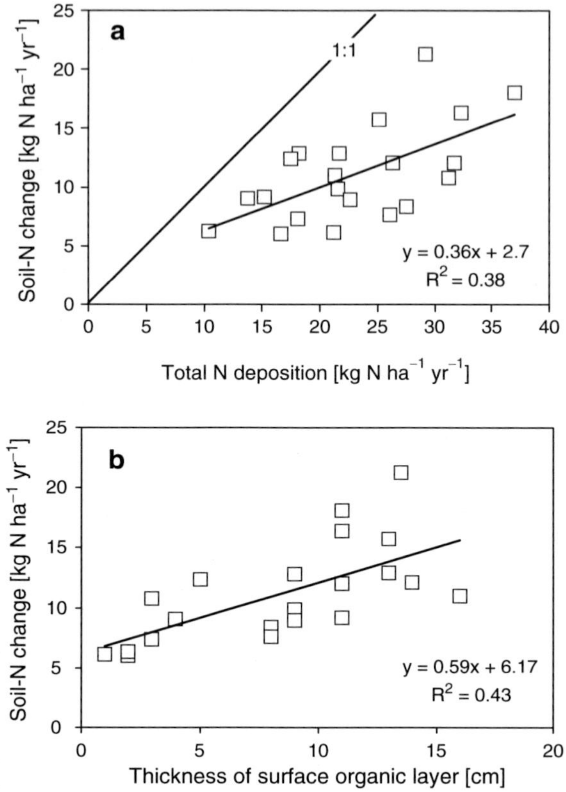

Fig. 23.5 Soil-N change (kg N ha^{-1} per year) versus (**a**) total depositions (kg N ha^{-1} per year) and (**b**) thickness of surface organic layer (cm) for 21 Level II sites retaining annually more than 5 kg N ha^{-1} deposited N in moder–mor type humus soils (accumulation type)

throughfall was more important than C/N ratio of the organic layer. However, the sinks in plant and soil have not been studied exclusively on these sties.

Accumulation type forests of the moder–mor type humus showed a significant correlation between N retention and total N deposition ($r^2 = 0.38$; Fig. 23.5a). Factors affecting organic matter accumulation in mature stands are not well known but may probably relate to the high N depositions (Fog 1988; Berg and Matzner 1998). A decrease in decomposition activity following the addition of ammonium fertiliser to the litter layer was reported in the literature review (Berg and Matzner 1997) suggesting the stabilisation of the humus in the surface organic layer in the presence of high N content. Decomposition studies by Berg et al. (1995) showed that high N content of the litter accelerated its decomposition in the beginning, but reduced the decomposition in the long-term leading to the accumulation of litter. It therefore implies that high N depositions may probably stabilise the organic matter of moder and mor humus after its enrichment in leaves and needles or by microbial immobilisation during decomposition. Whether N stabilises organic matter or changes the microbial community or the litter quality cannot be deduced from the literature (Fog 1988; Berg and McClaughterty 2003; Olsson et al. 2005). However, positive correlations between N retention and the thickness of the surface organic layer ($r^2 = 0.43$), and the atmospheric depositions of nitrogen ($r^2 = 0.38$) (Fig. 23.5) of the 21 forests of the accumulation type suggest that factors other than N depositions might also be involved.

One such factor may be soil acidity. Soil pH has decreased during the last decades by up to 2 units in Europe, especially in soils of low proton buffer capacities. This decrease in soil pH may have caused an imbalance in the processes of decomposition and sequestration of soil organic matter since the thickness of the surface organic layer generally increased with soil acidity (Brumme et al., Chap. 21). Soils of the accumulation type showed positive correlations between surface organic layer thickness and pH in soil solution ($r^2 = 0.43$). Additional N sequestration may occur in the H-horizon when its amount in L and F horizons remains unchanged when moder type of litter layer is being formed (Fig. 23.12). Moreover, acid deposition might directly be involved in N sequestration as is suggested by the correlation between N retention and sulphate deposition ($r^2 = 0.39$) and has been shown to reduce C-mineralisation in experimental acidification studies when pH was reduced with diluted sulphuric acid (Persson and Wirén 1993).

Solling beech site presents itself as a typical accumulation type. It accumulated in the soil annually by 36 kg N ha^{-1} from 1981 to 1989 and by 21 kg N ha^{-1} from 1990 and 2002 (Table 23.2) following a decrease in total depositions of similar amounts. However, during this period, leaching losses of N remained low and unchanged at about 2 kg N ha^{-1} per year and point to the potential for further sequestration of atmospheric deposited nitrogen. N retention in the Solling soil significantly increased with N deposition with a rate of 0.96 kg per kg deposited N (Fig. 23.6). Since the beginning of Solling project in 1968, the C/N ratio of the litter layer has remained unchanged but surface organic layer mass has increased as indicated by the repeated inventories during the last 35 years of the observation period (Meiwes et al., Chap. 4). The same process may have caused N retention in other moder type humus soils as 11 sites of the accumulation type have C/N ratio <25. At sites of the accumulation type with high C/N ratio >25 ($n = 10$), additional

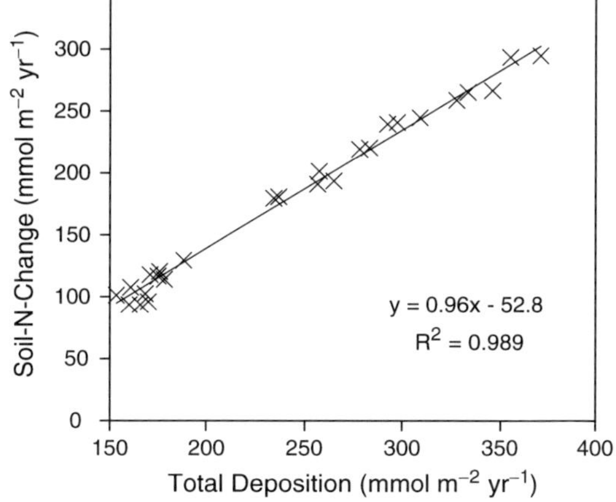

Fig. 23.6 Annual soil-N change versus annual total N deposition for the Solling site for 1976–2002

N sequestration may occur through an increase in N concentration in the organic layer. Repeated inventories conducted between 1974 and 2004 indicated that C/N ratios had decreased and C stocks increased in two Scots pine ecosystems in Southern Germany (Prietzel et al. 2006). On the two pine sites, carbon had increased by about 210 and 400 kg ha^{-1} per year and nitrogen by 13 and 18 kg ha^{-1} per year, respectively, while C/N ratio declined from about 36 in 1974 to 23 and 29 in 2004. These changes were attributed to a recovery of the degraded ecosystems through former litter-raking practices.

Of the 27 forest sites of the accumulation type, six had mull type humus, and four of them were occupied by beech which retained 6–10 kg N ha^{-1} per year (Table 23.4). These beech forests had high soil N content and high base saturation in the mineral soil. However, soil pH values were as low as were observed on soils with moder type humus. Such conditions are considered typical for soils where humus degradation may be occurring. It is not clear if this process is active in these soils as the amount of humus accumulation in the surface organic layer may exceed that of humus degradation in the mineral soil.

N balance has been shown to be a highly sensitive parameter for calculating changes in C content in soils (Holland et al. 1997; deVries et al. 2006). Annual N retention in soils of the accumulation type with moder–mor type humus (n = 21) ranged from 6 to 21 kg N ha^{-1} (mean 11 $\pm$ 5 kg N) which would mean C sequestration of 150–590 kg C ha^{-1} per year (mean 290 $\pm$ 113 kg C) assuming no change in N to C ratios. For the whole of Germany, forest soils with moder–mor type humus will sequester about 1 Tg C per year when 66% of the forested area of 10.15×10^6 ha in Germany are considered to have such soils and of these 51% of sites accumulate N. The total C sequestration rate in German forest soils may diverge from the calculated value as other soils such as those with mull type humus may also accumulate N or accumulate N without C sequestration (constant C/N ratio). However, Hyvönen et al. (2008) reported a mean increase in soil C-stock of 11 $\pm$ 1.7 kg C per 1 kg added N in long-term fertilisation experiments in northern European forests. This number is close to the value of 13 $\pm$ 5 kg C (sequestered) kg N^{-1} (added) calculated for the accumulation type with moder–mor type humus by C sequestration divided by total N deposition.

23.5.2 Forests of the Degradation Type

Of the 53 forests presented in Fig. 23.4, three sites belonged to the Degradation type where the amount of N loss exceeded the inputs in excess of plant increment. Two of these sites are occupied by oak and have untypical thick surface organic layers of a moder-humus type. Commonly, the oak forests have thin mull humus type litter layers. Low C/N ratios <20 in the surface organic layer material suggest degradation of humus in this layer. The third site which is losing N from the soil through degradation in the mineral soil is the beech Zierenberg site (see Sects. 23.2 and 23.3).

23.5.3 *Forests of the (quasi-) Steady State Type*

Among this collective of forest sites, 23 sites are of the (quasi-) Steady-State Type (Table 23.4). Eight forests of this type with mull humus type had higher N content in mineral soil than those of the moder humus type ($n = 15$). They also had a higher pH and a much higher base saturation in the mineral soil. Mg and Ca contents in the leaves were also higher by about 30 and 60%, respectively, than in the leaves of trees on moder humus type soils. Uptake of N by forest trees which have developed on mull humus type soils is higher than on moder humus type soils at the (quasi-) Steady State Type sites and higher than by forest trees at the accumulation type sites. The negative relationship between soil N change and plant N increment of pine, spruce, oak (Brumme and Khanna 2008) and beech trees (Fig. 23.7) in the (quasi-) Steady State and the accumulation types is explained by soil acidity which positively influenced the retention of N in the soil with increase in surface organic layer thickness (Fig. 23.5b) and negatively influenced the plant growth with decreasing base cations (Fig. 23.3).

The Göttinger Wald site is a typical case of the (quasi-) Steady-State Type (see Sects. 23.2 and 23.3). N dynamics on this site followed the amount of N deposited annually. A decrease in total N depositions after emission control measures were introduced decreased the leaching losses which were in equivalent amounts to the decrease in depositions (Table 23.2). This indicates that the soil at the Göttinger Wald site has a limited capacity to retain any more nitrogen. It is not clear why those forests with mull humus type very low base saturation did not accumulate organic matter in the surface organic layer.

Low mineral N content values in forest soils of moder humus type (Table 23.4, $n = 15$) indicate that these forests had gone through the phases of humus degradation and N accumulation in the surface organic layer and have approached the (quasi-) Steady State phase through a decrease in C/N ratio of about 25 in the surface organic layer (Fig. 23.10). The beech forests in this group currently lose deposited N in excess of plant increment of up to 16 kg N ha^{-1} per year indicating that N accumulation in such forests is of a transient nature.

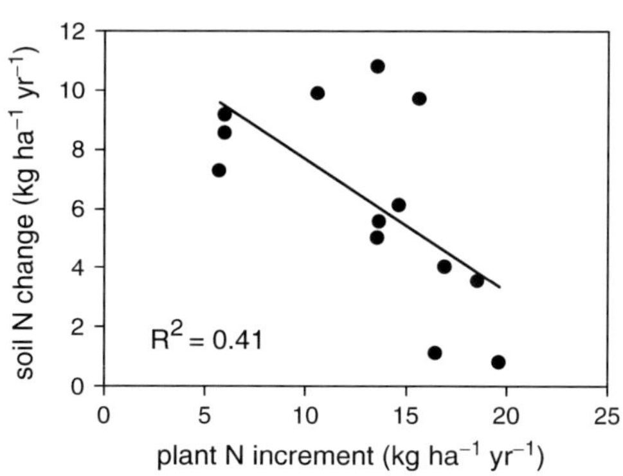

Fig. 23.7 Soil N change in relation to plant N increment (kg N ha^{-1} per year) in beech forests of the (quasi-) steady state and accumulation type

23.6 Dynamics of C and N Sequestration in European Forests

Chemical status of soils may affect the soil C- and N-stocks in a forest. Temporal changes of C- and N-stocks of soils are difficult to follow because the mechanisms involved in the sequestration of C and N and humus degradation processes are not well understood.

The Göttinger Wald site has a thin surface organic layer and contains 115 Mg C ha^{-1} in the surface soil (L layer to 30 cm depth, pH(H_2O) 6.7) which compares well with the mean C-stock of German forest soils with pH(H_2O) >5.5 (110 Mg C ha^{-1}). The N-stock in the surface soil of Göttinger Wald site is about 30% more than the mean for German sites of similar pH (7 Mg N ha^{-1} in German forest soils at pH(H_2O) >5.5). Soils in the carbonate buffer range had the highest carbon and nitrogen stocks in the mineral soil layer (Figs. 23.8, 23.9). This is related to high earthworm and decomposer activity and the stabilisation processes through clay-humus complex formation in the mineral soil. Although the carbonate content in the surface mineral soil of the Göttinger Wald site occurs as small gravels, and the surface soil layer at Göttinger Wald lies between the carbonate and silicate buffer ranges, further acidification of this soil is expected to be a slow process. This slow process may be related to root uptake of bases from deeper soil depths and to the bioturbation by earthworms, and to the relatively high clay content of the surface soil. The input–output balance of the Göttinger Wald site, a typical representative

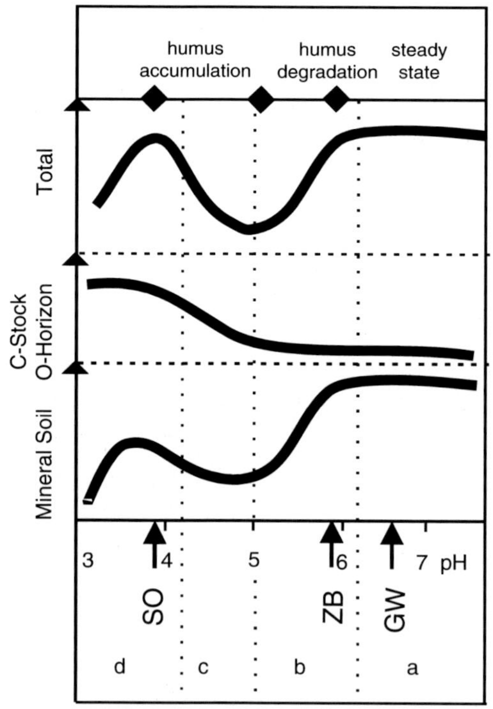

Fig. 23.8 Schematic diagram showing C-stock in the mineral soil, the surface organic layer (O horizon), and the total C-stock of German forest soils along the soil pH (H_2O) gradient (adopted from Brumme et al., Chap. 21, by smoothing the curves). The beech forests at Göttinger Wald, Zierenberg and Solling are arranged according to their soil pH. The buffer ranges are indicated by *dotted lines:* **a** carbonate buffer, **b** silicate buffer, **c** exchanger buffer, **d** Al/Fe buffer. The ranges of humus states "humus accumulation, humus degradation," and "steady-state" are indicated at the *top* of the figure

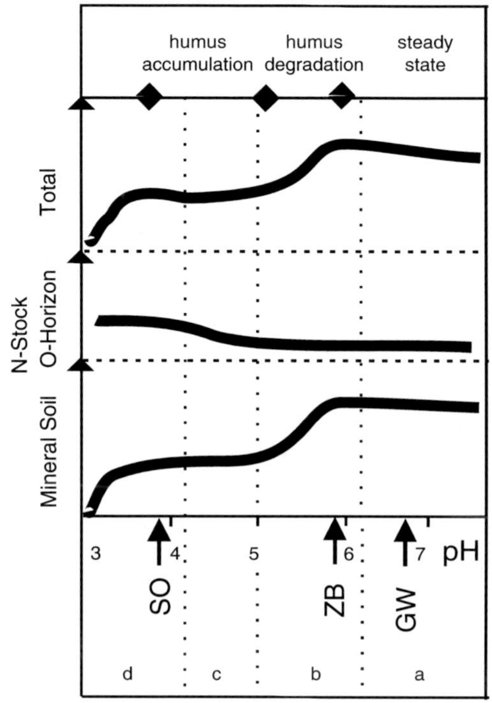

Fig. 23.9 N-stock in the mineral soil, the surface organic layer (O horizon), and the total in German forest soils with changes in soil pH(H_2O) (adopted from Brumme et al., Chap. 21, by smoothing the curves). The beech forests at Göttinger Wald, Zierenberg, and Solling are shown on pH scale. The buffer ranges are indicated by *dotted lines*: **a** carbonate buffer, **b** silicate buffer, **c** exchanger buffer, **d** Al/Fe buffer. Stages of "humus accumulation, humus degradation," and "steady-state" are indicated at the *top* of the figure

of the (quasi-) Steady-State Type, indicated that nitrogen was not retained in the soil (Table 23.2). Thus, soils in the carbonate buffer range may show the (quasi-) steady-state conditions where decomposition of organic matter equals litter production and the amount of soil organic matter does not change with time.

In an earlier study by Wittich (1952), similar values of N content as reported here were observed in the surface mineral soil which had developed on calcareous or basaltic bedrock. Thus, atmospheric N depositions did not change N levels in such soils. Wittich (1952) reported that, in two Rendzic Leptosol soils, total N values of 4% and 4.8% were observed which were in the similar range (3.9%) as observed at the Göttinger Wald site. Similarly, for a Eutric Cambisol developed on basaltic bedrock (Meißner site), Wittich (1952) reported a value of 4.03% which was also close to that for the Zierenberg soil (4.3%).

Soil at the Zierenberg (0–30 cm depth, pH(H_2O) of 5.5) site occurs in the transitional range of high and low C- and N-stocks among the soil collective of different pH values (Figs. 23.8, 23.9). Below soil pH value of 5.7, the C- and N-stocks in soils decreased with a minimum value approaching in soils of pH 4.4 to 5.2. Zierenberg soil occurs in the silicate buffer range which, due to moderate buffer rate, has undergone acidification through high atmospheric acid loads (Eichhorn and Hüttermann 1994, 1999). Acidification would reduce exchangeable base cations and increase the content of aluminum. Ca-humates are not stable as Al inhibits polymerisation of low to high molecules through formation of soluble complexes of organic matter and Al. Consequently, the humus in the mineral soil starts to degrade (humus degradation or humus disintegration) as indicated by high nitrate leaching (Table 23.2) causing further acidification (Ulrich 1984; Eichhorn and Hüttermann 1994). The C-stock at Zierenberg (76 Mg C ha^{-1} for L-30 cm depth) equals the minimum value of C-stocks observed in Germany for soils at pH 4.4–5.2; but the amount of N-stock (6.6 Mg N ha^{-1} for L–30 cm depth) is about 35% higher than the mean for forest soils in Germany.

C-stock at the Solling site is slightly higher than the overall mean observed for German forest soils at pH(H_2O) 3.9 (126 Mg C ha^{-1} at Solling compared to mean of 110 Mg C ha^{-1} for German sites), and the N-stock is 35% higher (7.2 Mg N ha^{-1} at Solling compared to 5.3 Mg N ha^{-1} in German sites for L–30 cm depth) (Figs. 23.8, 23.9). At this site, high retention of nitrogen by the surface organic layer causes very low nitrate leaching despite high N depositions (Table 23.2). The high N-input of 40 kg ha^{-1} per year during the period 1981–1989 resulted in only a slight increase in N-leaching of 0.3 kg N ha^{-1} per year when the values are compared with the period 1990–2002. Organic matter accumulation on such sites which have moder type of litter layer may therefore retain high amounts of N caused through an increase in the thickness of the surface organic layer.

Of the forested area in Germany, 86% of sites have soils with pH(H_2O) < 5 (Wolff and Riek 1997b). They may have undergone soil acidification, humus degradation or organic matter accumulation in the surface organic layer, and some have reached a new (quasi-) Steady-State of moder humus type (Fig. 23.10). Ulrich in his forest ecosystem theory (1984, 1987, 1992, 1994) suggested that most forest soils (except some sandy soils) in Germany, due to the influence of glaciation processes, were not acidified during the Sub Atlantic period (0–2,500 B.P.). At that time, soils had developed mull type humus of litter layers (Puhe and Ulrich 2001) and contained high C and N contents in the mineral soil. Further acidification which moved the soils from the exchanger to the Al/Fe buffer ranges is primarily related to acid precipitation of recent times (Puhe and Ulrich 2001). Soil acidification is accompanied by losses of base cations, nitrogen and carbon in the mineral soil (Humus Degradation Type) (Fig. 23.10). After the degradation phase, there is a small accumulation of C in the mineral soil and a large accumulation in the surface organic layer (accumulation type) as indicated in Figs. 23.8 and 23.11. Nitrogen is accumulated in the surface organic layer together with C while its accumulation in the mineral soil is negligible (Fig. 23.9). This may relate to low nitrogen sources of organic matter in acid mineral soils, such as dissolved organic carbon and roots (Brumme et al., Chap. 21). Acid soils thus lose their retention function for nitrogen in the mineral soil.

Thus, the soil chemical state, which is closely linked to the stabilisation processes of soil organic matter, has changed on the majority of the sites during decades by atmospheric deposition, and has caused C and N dynamics of transient nature (accumulation and degradation phase) in 9 of 17 beech sites (29 of the total of 53 sites) (Table 23.4). There are only four beech sites (in total eight forest sites) which have resisted these constraints [(quasi-) steady state phase with mull humus type]. These forests are saturated with nitrogen and are losing atmospheric-deposited

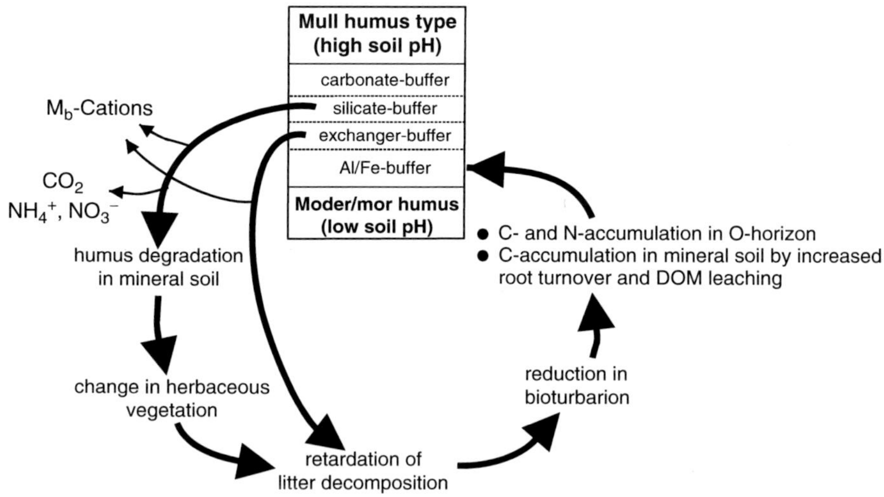

Fig. 23.10 Effects of acid load on ecosystem processes affecting the organic matter and N dynamics in forest soils in relation to the chemical state of the soils (M_b cations: Ca^{2+}, Mg^{2+}, K^+, Na^+) (Ulrich 1992, modified)

nitrogen in excess of that taken up by plants. In addition to these sites, there are four beech forests (in total 15 forest sites) which have a moder or mor type humus in the (quasi-) steady state phase and which are also losing any additional nitrogen deposited in excess of that taken up by plants. However, they differ in the dynamics of C and N sequestration. The sites with mull humus type may not have changed much by atmospheric deposition. They have not gone through the phase of N accumulation as proposed by Aber et al. (1998) and have been enriched with nitrogen since the Sub Atlantic times. The (quasi-) steady state phase of the moder humus type sites results from humus degradation followed by C and N sequestration in the surface organic layer.

23.7 Bioturbation as a Central Process of C and N Dynamics: Role of Soil Biota

The soil chemical state has been suggested here as one of the main driving variables which determines forest productivity and the carbon and nitrogen sequestration in the soil. It controls the cycling and distribution of nutrients in forest ecosystems, the availability of plant nutrients, the activity and growth of fine roots, distribution of soil biota species and activity of soil biota (Fig. 23.11). However, the key process relating to element cycling in the temperate biomes appears to be the bioturbation by earthworms. Earthworms together with termites and ants are sometimes described as soil engineers because of their role in soil-forming processes through soil bioturbation. Soil pH is one of the important factors determining the presence

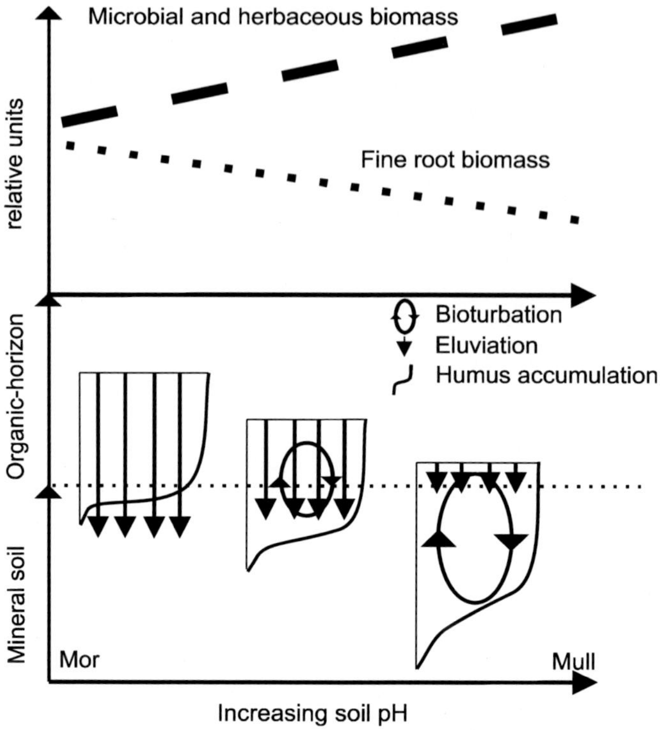

Fig. 23.11 Three states of forest soils at less acid (mull humus), intermediate and acid conditions (mor humus) as affected by faunal bioturbation, eluviation of dissolved organic carbon from top organic horizon to mineral soil, by litter layer accumulation, and by production of microbial, herbaceous and fine root biomass. (adopted from Beese, personal communication)

or the absence of earthworms in soils. Acid-intolerant species of earthworms have usually been absent from acid soils (Edwards and Bohlen 1996). In addition to low soil pH, reduced availability of energy resources for earthworms due to low quality of tree litter and the absence of herbaceous species may form another factor affecting the type and density of earthworm populations. The faunal studies at the three experimental sites showed that the three sites have varying amount and activity of soil biota.

The soils with high pH contained 205 individuals m^{-2} at Göttinger Wald and 114 individuals m^{-2} at Zierenberg of *Lumbricidae* whereas the acid soil at the Solling site had only 19 individuals m^{-2} (Schaefer and Schauermann, Chap. 7). The absence of *Lumbricidae* at the Solling site reduced the litter incorporation into the mineral soil causing the litter to accumulate on the surface of the mineral soil forming a moder humus type (Fig. 23.12).

The soil chemical state at the Göttinger Wald and Zierenberg sites acts positively on the microbial and herbaceous biomass, and thereby increases the activity of deep-burrowing anecic and surface-burrowing endogeic earthworms. The soils at the two sites are completely covered by herbaceous species which provide diverse

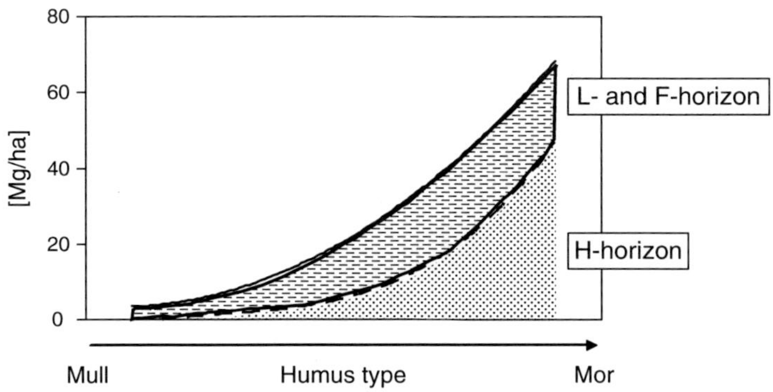

Fig. 23.12 Theoretical build-up of total surface organic layer, the L/F horizon and in the H horizon stratified from mull to mor humus in Germany (from Chap. 21)

and energy-rich resources of 0.2 Mg at Göttinger Wald and 0.4 Mg C ha^{-1} per year at Zierenberg (Schulze et al., Chap. 10) (Table 23.1). Through the activity of soil biota, the total aboveground litter fall of 2.6 Mg at Göttinger Wald and 2.7 Mg C ha^{-1} per year at Zierenberg (Khanna et al., Chap. 11) is consumed during a year from the soil surface (Schaefer et al., Chap. 12). At both sites, a thick A_h-horizon and a thin litter layer of mull-humus type were formed through earthworm activity. Incorporation of plant litter by earthworms produces high amount of microbial biomass in the mineral soil (Brumme et al., Chap. 6). Thus, soil micro-organisms have access to diverse and readily available energy source at these sites. Mixing plant litter with mineral soil while passing through the guts of earthworms may accelerates the stabilisation process of soil organic matter (Edward and Bohlen 1996) and the formation of organo-mineral complexes (Zech and Guggenberger 1996). Organo-mineral complexes consist of high proportions of less humified organic carbon in soils with mull humus layer and provide a much better nutrient source for microorganisms than C present in the mineral layers of soils of the mor humus type. Available C sources in acid mineral layers of soils are dead roots and some amounts of dissolved organic carbon through eluviation of the surface organic layer, as is the case at the Solling site. Lack of easily available C sources in the mineral soil at the Solling site and the presence of high acidity have reduced the microbial biomass to half the amount present at the Göttinger Wald and Zierenberg sites.

Moreover, soils with mull humus have a higher cation exchange capacity, a higher proportion of essential plant available base cations, a higher N-stock and more plant-available water than those with the mor humus. Deep burrowing anecic and endogeic earthworms form stable soil aggregates in the mineral soil and establish permanent burrows with high C and N contents accumulating in the burrow linings (Don et al. 2008). Vertical burrows increase the gas exchange between deeper soil layers and atmosphere. At the Göttinger Wald site, high gas diffusivity in the mineral soil was observed which promoted the methane and oxygen diffusion from the atmosphere into the mineral soil causing high methane oxidation (Borken and Brumme, Chap. 19) and low N_2O emissions (Brumme and

Borken, Chap. 18). In contrast, the low diffusivity in the mineral soil of the Solling site reduced the methane diffusion and increased the N_2O emissions. The soil pH is a good indicator for the trace gas exchange between the soil and atmosphere because there is a positive relationship between soil pH and soil diffusivity.

Diversity of the soil biota is higher at the forest sites with high degree of bioturbation and mull humus type of litter layer. For example, shell-bearing *Gastropoda* are lacking at the Solling site in contrast to 487 individuals m^{-2} at the Zierenberg and 120 individuals m^{-2} at the Göttinger Wald site (Schaefer and Schauermann, Chap. 7). *Gastropoda glomeris* feeds on intact litter material without much digestion and assimilation and provides a highly attractive food substrate for endogeic earthworms and microbes. However, the litter material at the Solling site may be of low quality (low Ca content) for the shell-bearing *Gastropoda*. Under highly acid soil conditions, populations and activity of earthworms are low and fungi dominate the microbial population (Anderson, Chap. 20). This specifically affects the production and loss of nitrate from ecosystems. For example, the highly efficient autotrophic nitrifiers may be replaced by less efficient heterotrophic nitrifiers under highly acid conditions reducing the nitrification in soils. Under such conditions, mineral nitrogen may remain in the ammonium form and the nitrate losses through leaching and denitrification are reduced.

The central role of faunal bioturbation in beech soils is evident from the many nutrition and soil-related processes. Appropriate management measures are thus required to develop soil conditions which are conducive to the establishment of active faunal populations.

23.8 Forest Management Strategies: Future Perspectives

European beech (*F. sylvatica* L.) has regained its importance in the last few decades mainly due to its wide ecological amplitude enabling it to grow under a range of site conditions, which under possible future changes of the climate and other environmental and social factors would provide it with an advantage over other tree species for forests of Germany. In future, beech may be replacing some of the mono-species stands of pine and spruce. The aim of the forest management in Germany is to increase the proportion of beech, especially as mixed stands, to achieve higher diversity and stability of forest stands in the long-term. This change in the stand structure through regeneration or planting is one of the most difficult tasks that forest management is facing today. In managed beech forests of Central Europe, natural regeneration is the dominant form of stand development. The success of any silvicultural treatment is measured in terms of a quick and problem-free conversion of mature stands into younger stands. From an ecological stand point, any conversion of stands should have minimum effects on the neighbouring ecosystems such as through leaching of nitrates or emission of greenhouse gases. It is, however, difficult to predict changes in the site and environmental factors during the course of a single rotation of a forest stand, such as, for example, changes in soil acidity, increase in N supply, amelioration practices and climate change phenomenon. There is also another unknown factor which is the lack of experience with regard to management decisions taken on a stand. For given climatic conditions, the regeneration of beech stands will primarily depend upon the soil conditions and other vegetation components as well as on the protection measures taken against deer browsing. Each of the three beech stands represents a different soil and site condition and can thus be viewed as a unique set for describing and developing silvicultural options (Bartsch and Röhrig, Chap. 22).

Usually, the amount of element losses during the conversion phase of a stand depends upon the degree of disturbance on the site and the period required for reversing the disturbance. Harvesting of a mature stand leads to a decrease in the uptake of water and nutrients. This may cause additional leaching losses of nitrate which is commonly turned over as leaf litter and fine roots annually prior to harvesting. Moreover, additional amounts of organic matter may be mineralised due to environmental changes in the surface soil. Additional losses of nitrate may also occur through denitrification as soils become wetter and contain excessive amounts of nitrates. Any amount of N_2O produced during denitrification will be of special concern due to its greenhouse gas character (Brumme and Borken, Chap. 18). There may be additional changes in the soil which act as a natural sink of methane gas due to changes in moisture status reducing the oxidation processes associated with the methane uptake (Borken and Brumme, Chap. 19). Also of major concern is the compaction of soils with harvesting machines which will create high emission areas of greenhouse gases, the so-called "hot spots."

For the regeneration of a stand, a rational approach would be to provide additional exposure of organic layer to sunlight for the growth of the saplings, but this rational management practice goes against the demand for minimum site disturbance on many regeneration sites. However, this is not a major requirement for the light-tolerant beech saplings. Recent experiments have shown that beech could be regenerated successfully by creating small gaps in the mature stand. This has led to replacing the commonly practiced shelterwood harvesting and group selection methods of regeneration with the single tree selection and target diameter harvesting of beech stands (Bartsch and Röhrig, Chap. 22). One single important factor for obtaining successful regeneration of beech is the control of browsing by deer. Moreover, it has been observed on the Göttinger Wald site that the presence of calcareous bedrock causing high biological activity in the soil would not need any soil tillage (disturbance) or liming for the regrowth. The high density of herb under-storey does not provide any specific hindrance to the germination and growth of beech saplings. Shelterwood harvesting assists the shade-tolerant beech saplings to develop and form pure species stands whereas group selection would also help the growth of other light-loving noble hardwood trees and can be employed to increase the proportion of these trees in the mixed species stands. Rapid growth of young noble hardwood trees (e.g., *Fraxinus excelsior, Acer pseudoplatanus, Ulmus glabra, Acer platanoides*) leads to strong competition for light and rooting volume of the soil to reduce ground vegetation and shrubs. The establishment of these species is highly restricted due to deer browsing, and insufficient regulation of deer populations.

On the nutrient-rich sites, as at Zierenberg, regeneration of beech has caused problems because of the dense and rapid growth of stinging nettles (*Urtica dioica*) or other under-storey plants. At Zierenberg, dense growth of stinging nettles cannot be kept in check due to the lack of noble hardwood tree species. Moreover, other contributory factors such as high amounts of available nitrogen, phosphorus and magnesium may be important to support thickets of stinging nettles. On such sites, small gaps in the mature stand may be required for successful regeneration. Large gaps should not be created. Small gaps may be achieved by maintaining a sub-canopy layer and understorey. Gas exchange on this site is not restricted due to high proportion of air pore spaces in soils, so the production of N_2O is low and the sink for methane will not be limited.

Regeneration of beech on acid sites, as is the case for the Solling site, requires special silvicultural efforts. The regeneration remains restricted on such acid sites due to the thick litter layer and highly acid mineral soils. Creating large-sized gaps in the canopy of the mature stand would lead to leaching losses of nitrate, emissions of N_2O and reductions in methane uptake. A surface application of lime would be required to improve the conditions for germination and initial growth of beech sampling but may also increase the growth of strongly shading shrub layer (e.g. *Rubus idaeus, R. fruticosus*). Through the competition by the dense shrub layer, beech seedlings may be completely lost in due course. However, the losses of N would be reduced and the uptake of methane enhanced. For such a site, a slowly advancing shelterwood harvesting and a lime application on the soil surface without soil disturbance should provide the most useful technique for regeneration that would serve both the ecological and silvicultural requirements.

Frequency of fructification has increased in recent decades due to high N inputs and the general vitality of the stands (Khanna et al., Chap. 11). However, the seeds are mostly lost to wild animals and soil fauna. Frequent and heavy fructification has also led to high requirements for other elements especially P, whose concentrations are higher in fruit components than in leaves. Moreover, high N inputs may have led to a decrease in litter decomposition and accumulation of organic matter in the litter layer which would also retain some amount of P in the litter layer, by reducing its turnover rate. The consequences of the high P requirements due to fructification, and high retention due to increase in productivity, and low turnover rates of P may further affect the N to P balance in plants, especially on those sites which are already low in P supply. Further experimental studies are required in future to assess the effect of such changes on the nutrition of beech stands.

23.9 Conclusions

- The three beech forests at Solling, Göttinger Wald and Zierenberg receive similar atmospheric inputs and are growing under similar climatic conditions as the long-term observations indicated. The three beech sites, however, differ in soil chemical conditions. The soils are rich in base contents at the Zierenberg site which is derived from weathered Tertiary basaltic debris and at the Göttinger Wald site which has Triassic limestone as parent material. These two sites are characterised by high soil pH values and a high diversity of plants and decomposers. High levels of decomposer activity developed a mull humus type, a high nutrient turnover and high tree productivity at the two sites. In contrast, the Solling site has an acid moder humus type soil which has developed from loess material overlaying Triassic sandstone. This soil has a very low pH, has low nutrient turnover and low tree productivity.

- The low productivity at the Solling site is caused by low availability of base cations (K, Ca, Mg) and high soil acidity which has increased the belowground allocation of biomass to maintain a high level of fine roots in the acid soil. Production of herbaceous plants is low at the Solling site, whereas it is high at the base-rich Göttinger Wald and Zierenberg sites which therefore act as an additional sink for nutrients and a luxuriant substrate for decomposers.
- Emissions of SO_2 and NO_x have decreased significantly since the end 1980s in Europe through political decisions to control emissions which have led to a noticeable decline in atmospheric depositions of sulphuric acid but to a lesser extent of nitrogen. The long-term observation at Solling and Göttinger Wald indicated that total depositions had decreased for sulphur (−60% at Solling and Göttinger Wald), H (−72% at Solling and Göttinger Wald), and nitrogen (−38% at Solling, −20% at Göttinger Wald) and base cations (−60% at Solling and Göttinger Wald) between the two periods 1981–1989 and 1990–2002.
- Effects of atmospheric depositions differed on each of the three sites. On the Göttinger Wald site with calcareous bedrock material, acid depositions affected slightly the surface soil layer. There was a small effect of acid inputs on the productivity of this stand. It had high soil biological activity and almost closed C and N cycles. However, on the base-rich Zierenberg site, acid depositions and soil acidification caused significant soil changes. This site was experiencing humus degradation where a net loss of N from the mineral soil occurred. At the Solling site, further addition of acidity through acid depositions to the acid soil decreased the availability of base cations. Prior to the 1980s, there were some inputs of base cations through atmospheric depositions which helped to reduce nutrient imbalances. Emission control measures at the end of the 1980s have changed this scenario by decreasing the supply of base cations with atmospheric depositions. Moreover, a decrease in acid inputs initiated a change in soil chemistry where the net release of previously stored sulphate has occurred and caused a delay in any recovery from further soil acidification.
- High atmospheric N inputs have removed any growth limitation due to N on beech forests in Germany (including the Göttinger Wald, Solling and Zierenberg sites). However, foliage levels showed an insufficient supply of Ca, Mg, P and K on some of these sites. Due to the significance of base cations for the optimum growth of N-enriched forests on acid soils, liming will show a positive effect on Ca and Mg contents of leaves and on tree productivity. Under certain conditions, other elements like P and K may become deficient.

- The amount of nitrogen retained in forests is controlled by plant uptake and sequestration in soils. Plant growth has increased during recent decades and will thus act as a sink for deposited N. Moreover, the frequency and the amount of fructification have increased in recent years affecting the plant uptake of deposited N. This is related to very high amounts of nutrients present in the fruit components and to a doubling of mast production compared to earlier studies. Elements transferred through litterfall to the organic layer have more than doubled (nitrogen) during the mast years.
- Forests with less acid soils showed high N retention by plants and low retention by soils whereas those on acid soils had low N retention by plants but high retention by soils, as indicated by 8-year records of input and output analysis of 53 German forests. Base-rich soils promote decomposer activity and improve nutrient cycles, and develop mull type humus with intensive faunal burrowing activity. In such soils, litter production and litter decomposition are in equilibrium and N content in the soil is at a maximum. Thus, any N input in excess of plant increment is leached from forests as is indicated by the long-term observations made at the Göttinger Wald site. Such forests are described as the (quasi-) Steady State Type with mull type humus (8 of 53 forest sites evaluated).
- Another type are forests where a part of atmospherically-deposited nitrogen is accumulated in the soil (accumulation type). The annual retention rates in soils (half of 53 forests sites evaluated) ranged from 6 to 21 kg N ha^{-1} depending upon total N depositions, acid–sulphate depositions, and the thickness and C/N ratio of the surface organic layer (moder-mor type humus). The thickness of the surface organic layer is related to soil acidity which has changed during the last decades and is supposed to be far from a steady state. At the long-term monitoring beech site at Solling (accumulation type), an increase in the mass of the surface organic layer with a constant C/N ratio of 19 over 35 years of observation was noted. This reduced the leaching losses of N, even when total N depositions of more than 40 kg N ha^{-1} per year occurred prior to the introduction of emission control measures.
- Another group of forests is where soils are currently losing C and N from the mineral soil layers by humus degradation (Degradation Type). However, this group has only a few sites due to its transient nature but includes the Zierenberg site, one of the three main sites of this study. However, most soils have crossed this state and are now accumulating N and C in the soil. There are some sites which have recovered to a new (quasi-) Steady State Type with moder–mor type humus (15 of 53 forests sites evaluated).
- The use of N balance has turned out to be a sensitive method of estimating soil C sequestration despite the high amount of C already present in forest soils. Mean annual C sequestration was calculated to be 290 ± 113 kg C ha^{-1} at sites of the accumulation type where moder–mor type humus is developing. Total sequestration rate for such sites may amount to 1 Tg C per year in German forest soils.
- Bioturbation of soil organic matter by earthworms and soil fauna appears to be a key process relating to the element cycling in the temperate biomes. Bioturbation increased C and in particular the N content in the upper mineral soil and

develop a mull humus type. A soil survey of forests in Germany indicated high C and N contents in the mineral soil of mull humus type soils. The lower C content in the mineral soil of moder and mor humus type soils that have occurred through humus degradation is compensated by a higher C content in the surface organic layer. The total N content did not recover from N loss in the mineral soil and is in total lower (2.8 Mg ha^{-1}) in mor than in mull humus type soils. Restricted bioturbation in highly acid soils may be improved by liming which may shift organic matter and N retention from the surface organic layer to the mineral soil in the long-term. Liming may also reduce the load of greenhouse gases (CO_2, N_2O, CH_4) in the atmosphere.

- In future, beech will be replacing some of the mono-species stands of pine and spruce in Germany, especially as mixed stands in order to achieve higher plant diversity and stability of forest stands in the long-term. Any conversion of stands should have minimum effects on the neighbouring ecosystems such as through leaching of nitrates or greenhouse gases exchange with the atmosphere by maintaining closed element cycling. This study of three diverse beech sites with similar climatic conditions indicated that the regeneration of beech stands will primarily depend upon the soil conditions and other vegetation components as well as on the protection measures taken against deer browsing.

Acknowledgment For developing this chapter we have extensively used the data and information provided by various authors of the preceding chapters of this book, to whom we wish to express our sincere thanks. We are grateful to the Johann Heinrich von Thünen Institute (Eberswalde, formerly Federal Research Centre for Forestry and Forest Products, BFH) and the various Forest Research Institutes of different federal States of Germany for providing the data in Chapter 23.4 to 23.6. We acknowledge with gratitude the contributions by Prof. Dr F. Beese who provided many useful ideas on various topics and his general encouragement and support.

References

Aber J, McDowell W, Nadelhoffer K, Magill A, Berntson G, Kamakea M, McNulty S, Currie W, Rustad L, Fernandez I (1998) Nitrogen saturation in temperate forest ecosystems. BioScience 48:921–934

Barton L, McLay CDA, Schipper LA, Smith CT (1999) Annual denitrification rates in agricultural and forest soils: a review. Aust J Soil Res 37:1073–1093

Becker R, Block J, Schimming CG, Spranger T, Wellbrock N (2000) Critical Loads für Waldökosysteme – Methoden und Ergebnisse für Standorte des Level II-Programms, Arbeitskreis A der Bund-Länder Arbeitsgruppe Level II. Bundesministerium für Ernährung, Landwirtschaft und Forsten, Bonn, Germany

Berg B, McClaugherty C, de Santo AV, Johansson MB, Ekbohm G (1995) Decomposition of litter and soil organic matter – Can we distinguish a mechanism for soil organic matter builtup? Scand J For Res 10:108–119

Berg B, Matzner E (1997) Effect of N deposition on decomposition of plant litter and soil organic matter in forest systems. Environ Rev 5:1–25

Berg B, McClaugherty C (2003) Plant litter. Springer, Berlin
Berg B, Gundersen P, Akselsson C, Johansson MB, Nilsson A, Vesterdal L (2007) Carbon sequestration rates in Swedish forest soils: a comparison of three approaches. Silva Fennica 41:541–558
Bolte A (2006) Biomasse- und Elementvorräte der Bodenvegetation auf Flächen des forstlichen Umweltmonitorings in Rheinland-Pfalz. Ber Fosrschungszentrum Waldökosysteme B 72, Universität Göttingen
Block J, Eichhorn J, Gehrmann J, Kölling C, Matzner E, Meiwes KJ, v. Wilpert K, Wolff B (2000) Kennwerte zur Charakterisierung des ökochemischen Bodenzustandes und des Gefährdungspotentials durch Bodenversauerung und Stickstoffsättigung an Level II-Waldökosystem-Dauerbeobachtungsflächen. Bundesministerium für Ernährung, Landwirtschaft und Forsten, Arbeitskreis C der Bund-Länder Arbeitsgruppe Level II, Bonn, Germany
Borken W, Matzner E (2004) Nitrate leaching in forest soils: an analysis of long-term monitoring sites in Germany. J Plant Nutr Soil Sci 167:277–283
Brumme R, Verchot LV, Martikainen PJ, Potter CS (2005) Contribution of trace gases nitrous oxide (N_2O) and methane (CH_4) to the atmospheric warming balance of forest biomes. In: Griffiths H, Jarvis PG (eds) The carbon balance of forest biomes. Taylor&Francis, Oxon, New York, pp 293–318
Brumme R, Khanna PK (2008) Ecological and site historical aspects of N dynamics and current N status in temperate forests. Glob Chang Biol 14:125–141
deVries WD, Reinds GJ, Gundersen P, Sterba H (2006) The impact of nitrogen deposition on carbon sequestration in European forests and forest soils. Glob Chang Biol 12: 1151–1173
Dieter M, Elsasser P (2002) Carbon stocks and carbon stock changes in the tree biomass of Germany's forests. Forstw Cbl 121:195–210
Dise NB, Matzner E, Forsius M (1998a) Evaluation of organic horizon C:N ratio as an indicator of nitrate leaching in conifer forests across Europe. Environ Pollut 102:453–456
Dise NB, Matzner E, Gundersen P (1998b) Synthesis of nitrogen pools and fluxes from European forest ecosystems. Water Air Soil Pollut 105:143–154
Don A, Steinberg B, Schoening I, Pritsch K, Joschko M, Gleixner G, Schulze ED (2008) Organic carbon sequestration in earthworm burrows. Soil Biol Biochem 40:1803–1812
Edwards CA, Bohlen PJ (1996) Biology and ecology of earthworms. Chapman & Hall London
Eichhorn J, Hüttermann A (1994) Humus disintegration and nitrogen mineralization. In Godbold DL, Hüttermann A (eds) Effects of acid rain on forest processes. Wiley, New York, pp 129–162
Eichhorn J, Hüttermann A (1999) Mechanisms of humus dynamics and nitrogen mineralisation. In: Rastin N, Bauhus J (eds) Going underground – ecological studies in forest soils. Research Signpost, Trivandrum, India, pp 239–277
Ferrier RC, Jenkins A, Wright RF, Schöpp W, Barth H (2001) Assessment of recovery of European surface waters from acidification 1970–2000: an introduction to the Special Issue. Hydrol Earth Syst Sci 5:274–282
Fog K (1988) The effect of added nitrogen on the rate of decomposition of organic matter. Biol Rev 63:433–462
Gehrmann J, Andreae H, Fischer U, Lux W, Spranger T (2001) Luftqualität und atmosphärische Stoffeinträge an Level II-Dauerbeobachtungsflächen in Deutschland. Bundesministerium für Verbraucherschutz, Ernährung und Landwirtschaft, Arbeitskreis B der Bund-Länder Arbeitsgruppe Level II, Bonn, Germany
Gundersen P, Callesen I, de Vries W (1998) Nitrate leaching in forest ecosystems is related to forest floor C:N rations. Environ Pollut 102:403–407
Haußmann T, Lux W (1997) Dauerbeobachtungsflächen zur Umweltkontrolle im Wald, Level II, Erste Ergebnisse. Bundesministerium für Ernährung, Landwirtschaft und Forsten, Bonn, Germany
Holland EA, Braswell BH, Lamarque JF et al (1997) Variations in the predicted spatial distribution of atmospheric nitrogen deposition and their impact on carbon uptake by terrestrial ecosystems. J Geophys Res 102:15849–15866
Hyvönen R, Persson T, Andersson S, Olsson B, Agren GI, Linder S (2008) Impact of long-term nitrogen addition on carbon stocks in trees and soils in northern Europe. Biogeochemistry 89:121–137

Kauppi PE, Mielikäinen K, Kuusela K (1992) Biomass and carbon budget of European forests, 1971 to 1990. Science 256:70–74

Kesik M, Ambus P, Baritz R, Bruggemann NB, Butterbach-Bahl K, Damm M, Duyzer J, Horvath L, Kiese R, Kitzler B, Leip A, Li C, Pihlatie M, Pilegaard K, Seufert G, Simpson D, Skiba U, Smiatek G, Vesala T, Zechmeister-Boltenstern S (2005) Inventories of N_2O and NO emissions from European forest soils. Biogeosciences 2:353–375

Krauß HH, Heinsdorf D (2005) Ernährungsstufen für wichtige Wirtschaftsbaumarten. Beitr Forstwirtsch u Landsch ökol 39:172–179

Liski J, Perruchoud D, Karjalainen T (2002) Increasing carbon stocks in the forest soils of western Europe. For Ecol Manage 169:159–175

MacDonald JA, Dise NB, Matzner E, Armbruster M, Gundersen P, Forsius M (2002) Nitrogen input together with ecosystem nitrogen enrichment predict nitrate leaching from European forests. Glob Chang Biol 8:1028–1033

Matzner E, Grosholz (1997) Beziehung zwischen NO3--Austrägen, C/N-Verhältnissen der Auflagen und N-Einträge in Fichtenwald (*Picea abies* Karst.)-Ökosystemen Mitteleuropas. Forstwiss Centralbl 116:39–44

Matzner E (2004) Biogeochemistry of forested chatchments in a changing environment. Ecological Studies 172, Springer, Berlin

Meiwes KJ, Meesenburg H, Bartens H, Rademacher P, Khanna PK (2002) Akkumulation von Auflagehumus im Solling, Mögliche Ursachen und Bedeutung für den Nährstoffkreislauf. Forst und Holz 57: 428–433

Nabuurs GJ, Schelhaas MJ (2002) Carbon profiles of typical forest types across Europe assessed with CO2FIX. Ecol Indic 1:213–223

Nadelhoffer KJ, Emmett BA, Gundersen P et al. (1999) Nitrogen deposition makes a minor contribution to carbon sequestration in temperate forests. Nature 398:145–148

Olsson P, Linder S, Giesler R, Högberg P (2005) Fertilization of boreal forest reduces both autotrophic and heterotrophic soil respiration. Glob Chang Biol 11:1745–1753

Persson T, Wirén A (1993) Effects of experimental acidification on C and N mineralization in forest soils. Agric Ecosyst Environ 47:159–174

Pretzsch H (1996) Growth trends of forests in Southern Germany. In: Spiecker H, Mielikäinen K, Köhl M, Skovsgaard J (eds) Growth trends in European forests. European Forest Institute Research Report No 5, Springer, New York

Prietzel J, Stetter U, Klemmt HJ, Rehfuess KE (2006) Recent carbon and nitrogen accumulation and acidification in soils of two Scots pine ecosystems in Southern Germany. Plant Soil 289:153–170

Puhe J, Ulrich B (2001) Global climate change and human impacts on forest ecosystems. Ecological Studies 143, Springer, Berlin, Germany

Scarascia-Mugnozza G, Bauer GA, Persson H, Matteucci G, Masci A (2000) Tree biomass, growth and nutrient pools. In: Schulze ED (ed) Carbon and nitrogen cycle in european forest ecosystems. Ecological Studies 142, Springer, Berlin, pp 49–62

Schulze ED (2000) Carbon and nitrogen cycle in european forest ecosystems. Ecological Studies 142, Springer, Berlin

Sjöberg G, Knicker H, Nilsson SI, Berggren D (2004) Inpact of long-term N fertilization on the structural composition of spruce litter and mor humus. Soil Biol Biochem 36:609–618

Spiecker H (1999) Overview of recent growth trends in European forests. Water Air Soil Pollut 116:33–46

Ulrich B (1984) Stability and destabilization of central European forest ecosystems – A theoretical, data based approach. In: Cooley JH, Golley FB (eds) Trends in ecological research for the 1980s. Plenum, New York, pp 217–237

Ulrich B (1987) Stability, elasticity and resilience of terrestrial ecosystems with respect to matter balance. In: Schulze ED, Zwölfer H (eds) Ecological Studies 61. Springer, Berlin, pp 11–49

Ulrich B (1992) Forest ecosystem theory based on material balance. Ecol Model 63:163–183

Ulrich B (1994) Process hierarchy in forest ecosystems: An integrative ecosystem theory. In: Godbold DL, Hüttermann A (eds) Effects of acid rain on forest processes. Wiley, New York, USA, pp 353–398
van der Salm C, de Vries W, Reinds GJ, Dise NB (2007) N leaching across European forests: Derivation and validation of empirical relationships using data from intensive monitoring plots. For Ecol Manage 238:81–91
Wittich W (1952) Der heutige Stand unseres Wissens vom Humus und neue Wege zur Lösung des Rohhumusproblems im Walde. In: Schober R (Hrsg) Schriftenreihe der Forstlichen Fakultät der Universität Göttingen, Band 4, JD Sauerländer's Verlag, Frankfurt am Main, Germany
Wolff B, Riek W (1997a) Deutscher Waldbodenbericht 1996. BMELF, Bd 1
Wolff B, Riek W (1997b) Deutscher Waldbodenbericht 1996. BMELF, Bd 2
Zech W, Guggenberger G (1996) Organic matter dynamics in forest soils of temperate and tropical ecosystems. In: Piccolo A (ed) Humic substances in terrestrial ecosystems. Elsevier, Amsterdam, pp 101–170

Index